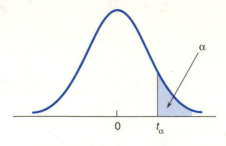

Table B The *t*-distribution

df	$t_{0.050}$	$t_{0.025}$	$t_{0.010}$	$t_{0.005}$	df
1	6.314	12.706	31.821	63.657	1
2	2.920	4.303	6.965	9.925	2
3	2.353	3.182	4.541	5.841	3
4	2.132	2.776	3.747	4.604	4
5	2.015	2.571	3.365	4.032	5
6	1.943	2.447	3.143	3.707	6
7	1.895	2.365	2.998	3.499	7
8	1.860	2.306	2.896	3.355	8
9	1.833	2.262	2.821	3.250	9
10	1.812	2.228	2.764	3.169	10
11	1.796	2.201	2.718	3.106	11
12	1.782	2.179	2.681	3.055	12
13	1.771	2.160	2.650	3.012	13
14	1.761	2.145	2.624	2.977	14
15	1.753	2.131	2.602	2.947	15
16	1.746	2.120	2.583	2.921	16
17	1.740	2.110	2.567	2.898	17
18	1.734	2.101	2.552	2.878	18
19	1.729	2.093	2.539	2.861	19
20	1.725	2.086	2.528	2.845	20
21	1.721	2.080	2.518	2.831	21
22	1.717	2.074	2.508	2.819	22
23	1.714	2.069	2.500	2.807	23
24	1.711	2.064	2.492	2.797	24
25	1.708	2.060	2.485	2.787	25
26	1.706	2.056	2.479	2.779	26
27	1.703	2.052	2.473	2.771	27
28	1.701	2.048	2.467	2.763	28
29	1.699	2.045	2.462	2.756	29
inf.	1.645	1.960	2.326	2.576	inf.

This table is abridged from Table **A** of R. A. Fisher and F. Yates: *Statistical Tables for Biological, Agricultural, and Medical Research*, published by Longman Group, Ltd., London (previously published by Oliver & Boyd, Edinburgh), by permission of the authors and publishers.

Statistics and Probability in Modern Life

FIFTH EDITION

Joseph Newmark
The College of Staten Island of the City University of New York

Saunders College Publishing
Harcourt Brace College Publishers
Fort Worth   Philadelphia   San Diego
New York   Orlando   Austin   San Antonio
Toronto   Montreal   London   Sydney   Tokyo

Text Typeface: Caledonia
Compositor: General Graphic Services
Acquisitions Editor: Bob Stern
Developmental Editors: Ellen Newman, Janet Nuciforo
Managing Editor: Carol Field
Project Editor: Nancy Lubars
Copy Editor: Elaine Honig
Manager of Art and Design: Carol Bleistine
Art Director: Christine Schueler
Art and Design Coordinator: Caroline McGowan
Text Designer: Butler Udell Design
Cover Designer: Lawrence R. Didona
Text Artwork: Tech Graphics
Director of EDP: Tim Frelick
Production Manager: Charlene Squibb

Cover Credit: © 1991 Bishop/Phototake, Inc. NYC

Printed in the United States of America

STATISTICS AND PROBABILITY IN MODERN LIFE

ISBN   0-03-072867-3

Library of Congress Catalog Card Number: 91-053162

 4 016 987654

TO
Trudy
and our
children

Preface

Today, more than ever, statistics plays a very important role in our lives. This book has been designed to provide the reader with a comprehensive treatment of introductory statistics and probability in such areas as sociology, business, ecology, economics, education, medicine, psychology, and mathematics. Students in these fields must frequently demonstrate a knowledge of the language and methods of statistics. Methodology and applications have been integrated throughout the text.

A special effort has been made in this edition to make the concepts of statistics available to students who are not prepared for elaborate symbolisms or complex arithmetic. Although the mathematical content is complete and correct the language is elementary and easy to understand. Expressions that students have difficulty comprehending have been avoided, and the introduction of new terminology has been held to a minimum. In frequent **COMMENTS,** various points that students often misunderstand or miss completely are carefully discussed. Yet mathematical rigor has not been sacrificed. Introductory high school algebra is a sufficient prerequisite to use this text satisfactorily.

Each chapter starts off with a chapter objectives section and a newspaper clipping to set the stage for the material to be covered, and concludes with a summary and study guide to reinforce the ideas covered in the chapter. Additionally, each idea introduced is first explained in terms of an example chosen from everyday real-life situations with which the student can identify.

CHANGES IN THE FIFTH EDITION

This revised edition of the text reflects the many suggestions and recommendations of users of the earlier editions; the changes are outlined below.

1. A brief introductory paragraph describing the chapter contents has been added to the beginning of each chapter.
2. Chapter objectives are now keyed to the appropriate section of the text to make it easier for the student to locate.
3. Key words appear in color in the margin where they are introduced.

4. The MINITAB programs have been expanded and are based on the latest version, Release 7.1 (1990). Questions involving MINITAB now appear in the exercises at the end of the chapters.

5. In response to reviewer suggestions, several exercises are included in each chapter that are verbal rather than computational; such questions often ask the student to explain or give an example to support the answer.

6. The exercises have been revised and increased by more than 50 percent. There are now well over 1400 exercises (of all types) throughout the book. All exercises have been checked for accuracy.

7. The newspaper articles and data in the Statistics in the News sections have been updated.

8. Substantial material has been added to certain chapters, while other chapters have been reorganized and streamlined: **Chapter 1** includes a further discussion and elaboration on discrete or continuous numerical data. In **Chapter 2**, the following have been added: a discussion of the newest graphical technique of dotplots; a summary of rules for rounding decimals (as used in statistics); and additional descriptions of ogives, cumulative frequencies, and the difference between line graphs and frequency polygons. **Chapter 3** includes greater depth and variety of exercises involving summation notation. The discussion of Chebyshev's theorem has been expanded, and discussions of the trimmed mean, box plots, and/or box-and-whisker diagrams have been included, as well as MINITAB programs involving each of these new ideas. In **Chapter 6** a new discussion on how MINITAB can be used to generate and compute binomial probabilities has been added, thus eliminating the need for the binomial distribution formula and binomial probability charts. In **Chapter 7**, Section 7.5, which discusses how the normal distribution can be used with correction factors for calculating binomial distribution probabilities, has been considerably expanded. **Chapter 8** includes new discussions involving the ideas of systematic sampling and cluster sampling in Section 8.2, as well as stratified sampling. New MINITAB programs for generating random samples and for generating repeated samples to be used in conjunction with the Central Limit Theorem are presented. In response to reviewer suggestions, the population proportion symbol and discussion in **Chapter 9** has been changed; it is now denoted by p . In **Chapter 10**, the concept of P values has been introduced. In **Chapters 10** through **14**, procedures to be used when performing hypothesis tests have been summarized and are now presented in a box format. The discussion involving contingency tables in **Chapter 12** has been rewritten and expanded. Limitations on when the chi-square test can be used are given.

SUPPLEMENTS

The following supplements are available for use with this book:

1. A MINITAB supplement specifically keyed to this text has been prepared by Dr. Ernest Blaisdell of Elizabethtown College to teach the reader how to use the popular statistical package MINITAB. No prior computer knowledge is assumed.

2. *20/20 Statistics* is a visual interactive statistics courseware package that can be used both as a classroom teaching aid and as supplemental material for the students to work on independently. The package contains a tutorial workbook with software instructions and a module for each topic, a program disk that runs on Apple or IBM computers, and an Instructor's Manual that provides answers to exercises in the tutorial workbook and suggestions for use of the courseware by both students and instructors. The *20/20 Statistics* software and Instructor's Manual are written by Professor George Bergeman of Northern Virginia Community College, and the tutorial workbook is written by Professor Bergeman and Professor James P. Scott of Creighton University. No prior computer knowledge is required for use of this package.

3. An *Instructor's Resource Manual*, written by the author, gives detailed line by line solutions to *all* of the exercises, as well as to the questions in the Testing Your Understanding and Thinking Critically sections. Four additional tests for each chapter are included to provide flexibility in testing.

4. A *Student Solutions Manual*, written by Robert K. Smidt of California State Polytechnic University, San Luis Obispo, includes worked-out solutions to every other odd-numbered problem in the exercises and end-of-chapter tests.

ACKNOWLEDGMENTS

I would like to thank the many instructors and students who received the earlier editions of this book so warmly. I am grateful to those who took the time to send comments, suggestions, and corrections. I am also grateful to the following people who reviewed this edition of the text and who made valuable suggestions for its improvement:

Marlene J. Kovaly, Florida Community College at Jacksonville
Gary Kulis, Mohawk Valley Community College
Donald Sisson, Utah Agricultural Experiment Station, Utah University
Robert K. Smidt, California State Polytechnic University, San Luis Obispo
Linn Stranak, Union University
Marie M. Vanisko, Carroll College

Special thanks to the following people for their accuracy reviews of all examples and exercises:

Donald Sisson, Utah Agricultural Experiment Station, Utah University
Robert K. Smidt, California State Polytechnic University, San Luis Obispo

I would also like to express my gratitude to the following people for their excellent work on the various ancillary items that accompany this text:

George W. Bergeman, Northern Virginia Community College (20/20 Statistics)
Ernest Blaisdell, Elizabethtown College (MINITAB Supplement)

James P. Scott, Creighton University (20/20 Statistics)
Robert K. Smidt, California State Polytechnic University, San Luis Obispo

Many thanks also go to the authors and publishers who granted permission to use the statistical tables so necessary for this work. Among others, I am indebted to the Literary Executor of the late Sir Ronald A. Fisher, F.R.S., and to Oliver and Boyd, Ltd., Edinburgh for their permission to reprint tables from *Statistical Methods for Biological, Agricultural, and Medical Research*.

Finally, and most important, I wish to thank my wife Trudy and our children for their understanding and patience as they endured the enormous strain associated with completing this project. Without their encouragement it could not have been undertaken or completed.

Joseph Newmark

December, 1991

To The Instructor

In order to write a statistics and probability book that is easily understood by readers with varying backgrounds and help the student become actively involved in the learning process, the text contains the following features:

1. **Statistics in the News**   Each chapter begins with an article taken from the daily press that presents the ideas discussed in the chapter in applied context. This is intended to motivate the student by showing how the ideas mentioned in the chapter are applied in real-life situations. Moreover, there are numerous clippings and illustrations throughout the book to further clarify the material.

2. **Chapter Objectives**   Keyed to the appropriate section of the text, chapter objectives highlight the key ideas to be discussed.

3. **Introduction**   Each chapter's introductory section sets the stage for the ideas contained in the chapter.

4. **Comments**   Throughout the text "Comments" are included to emphasize ideas that students often miss or find confusing.

5. **Historical Notes**   Most chapters contain brief historical notes on mathematicians who contributed to the ideas discussed within the chapter. These should help humanize mathematics.

6. **Use of Boldface and Italic Type**   Certain words or definitions are set in italic type for emphasis; marginal notes are set in boldface type in text as well as color in the margin. Formulas and rules are highlighted for the student by a box.

7. **Examples and Exercises**   All of the applied examples and exercises have been carefully written to capture student interest. The exercises (graded from easy to more challenging) are grouped into two categories: (1) routine exercises that help build skills and understanding of concepts and (2) problem-solving exercises. As suggested by the NCTM, the use of calculators or computers is highly recommended. Throughout the book, some of the examples or exercises are specifically designed to be worked out using a hand-held calculator or computer. Such exercises are designated by the symbol .

8. **End-of-Chapter Material** Each chapter concludes with a Study Guide and Formulas to Remember. The Study Guide highlights and summarizes the main points of the chapter and includes page numbers for easy reference. Students should find these extremely useful when studying and preparing for exams.

9. **Testing Your Understanding of This Chapter's Concepts** Each chapter contains a section which presents questions that are often asked in class and that are designed to see if the student truly understands the concepts.

10. **Thinking Critically** Each chapter contains a section designed to make the reader think very carefully. These questions are somewhat more challenging.

11. **Chapter Review Exercises** Each chapter concludes with two comprehensive sets of exercises that can be used both as a review and as a student self-test.

12. **Suggested ·Further Reading** Each chapter has an updated suggested further reading list for students interested in pursuing any particular topic.

13. **Answers** Answers are provided at the end of the book for selected exercises as well as for all of the questions in the Testing Your Understanding and Thinking Critically sections and all of the Chapter Review Exercises.

14. **Ancillary Package** A complete set of ancillaries is available to supplement this text, including an Instructor's Resource Manual, Student Solutions Manual, MINITAB Supplement, and interactive statistics courseware for Apple and IBM computers.

Suggested Course Outline An instructor should have no difficulty in selecting material for a one- or two-semester course. The following outlines indicate how this text can be used:

One-Semester Course (meets 40 times per semester, 40 min. per session)

Text material	Approximate amount of time	Prerequisite needed for each chapter
Chapter 1	1 lesson	none
Chapter 2	6 lessons	none
Chapter 3	6 lessons	Chapter 2, Section 2.2 or the equivalent
Chapter 4	5 lessons	none
Chapter 5 (skip 5.5)	4 lessons	Chapter 4
Chapter 6 (skip 6.7 & 6.8)	5 lessons	Chapter 4
Chapter 7	5 lessons	Chapter 2
Chapter 8	4 lessons	Chapter 7
	36*	

*The remaining meetings can be devoted to exams and review.

Two-Semester Course—Semester 1

Text material	Approximate amount of time	Prerequisite needed for each chapter
Chapter 1	1 lesson	none
Chapter 2	6 lessons	none
Chapter 3	6 lessons	Chapter 2, Section 2.2 or the equivalent
Chapter 4	5 lessons	none
Chapter 5	5 lessons	Chapter 4
Chapter 6	7 lessons	Chapter 4
Chapter 7	5 lessons	Chapter 2
Chapter 8	<u>4</u> lessons 39*	Chapter 7

*The remaining meetings can be devoted to exams and review.

Two-Semester Course—Semester 2

Text material	Approximate amount of time	Prerequisite needed for each chapter
Chapter 9	5 lessons	Chapter 7
Chapter 10	7 lessons	Chapter 7
Chapter 11	8 lessons	Chapter 10
Chapter 12	4 lessons	Chapter 10
Chapter 13	7 lessons	Chapter 10
Chapter 14	<u>7</u> lessons 38*	Chapters 8, 10, 11, and 12

*The remaining meetings can be devoted to exams and review.

Since comments and suggestions from instructors using previous editions of this text proved invaluable in completing this revision, I would be grateful to continue receiving them. Please address all correspondence to me directly at The College of Staten Island, 715 Ocean Terrace, Staten Island, N.Y. 10301.

To The Student

This is a nonrigorous mathematical text on elementary probability and statistics. If you are afraid that your mathematical background is rusty, don't worry. The only math background needed to use this book is introductory high school algebra or the equivalent; however, those of you who are fairly knowledgeable in math will see how statistics can be used in interesting and challenging problems.

The examples are plentiful and are chosen from everyday situations. Also, the ideas of statistics are applied to a variety of subject areas. This variety indicates the general applicability of statistical methods. In frequent COMMENTS, appropriate explanations of statistical theory are given and the "why's" of statistical methods answered.

Occasionally a section or exercise is starred. This means that it is slightly more difficult and may require some time and thought.

Some of the exercises are designated with the symbol 🖩. These exercises have been specifically designed to be worked out using a hand-held calculator. However, they can also be easily worked out without such a calculator.

Each chapter concludes with a summary and a review of the formulas and major terms and concepts introduced in the chapter; page numbers are included for easy reference. In addition, there are Mastery Tests which will help in preparing for exams. Answers to selected exercises and all Mastery Test questions are given in the back of the book.

I hope that you will find reading and using this book an enjoyable and rewarding experience, since a basic knowledge of statistics is essential. Good luck.

Contents

XII   ANALYZING COUNT DATA: THE CHI-SQUARE DISTRIBUTION

XIII   ANALYSIS OF VARIANCE

1

The Nature of Statistics— What is Statistics?

By tagging birds, statisticians are able to predict the population of endangered species of birds. See discussion in Section 1.3. (© Degginger, Fran Heyl Associates)

When you hear the word "statistics" what do you think of? The word statistics as defined in many dictionaries and encyclopedias has several meanings.

In this introductory chapter we describe the general nature of statistics and discuss some of its uses (and abuses). We present some basic terminology that will be used throughout the book.

CHAPTER OBJECTIVES

After studying the material in this chapter, you should be able:

- **To discuss** the nature of statistics and numerous examples of how they are used. (Section 1.1)

- **To identify** the two major areas of statistics—descriptive statistics and inferential statistics. Descriptive statistics involves collecting data and tabulating it in a meaningful way. Inferential statistics involves making predictions based on the sample data. (Section 1.2)

- **To distinguish** between part of a group, called a sample, and the whole group, called the population. (Section 1.2)

- **To present** a brief discussion of the historical development of statistics and probability. (Section 1.4)

A DECADE OF CLEANING THE AIR

The first Great American Smokeout took place on Nov. 20, 1976 . At that time, 37 percent of the U.S. adult population were smokers. A decade later, 30 percent of the population still smokes. A total of 37 million Americans are ex-smokers. As the American Cancer Society reports, however, "The battle is far from over. Those who still smoke seem to be smoking more heavily, and smoking among women and young people is still a particular cause for concern. The harmful effects of involuntary, or passive smoking — the inhaling of cigarette smoke by non-smokers — have come to light. Cigarette companies are mounting an aggressive campaign against the anti-smoking movement."

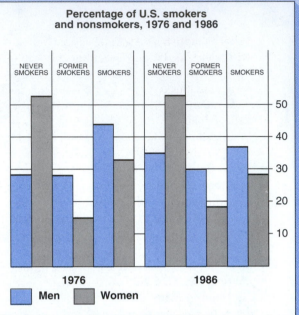

Percentage of U.S. smokers and nonsmokers, 1976 and 1986

Source : National Health Interview Survey (U.S.DHHS NCHS 1984). Reported in "Facts and Figures On Smoking, 1976-1986" American Cancer Society.

HIGH SCHOOL DEGREES

Percentage of seventeen-year-olds graduating from high school in the United States:

1870	2%
1900	6%
1920	17%
1930	29%
1940	51%
1950	59%
1960	65%
1970	76%
1980	72%
1989	70%

Source: U.S. Department of Education, National Center for Education Statistics.

The article above indicates that new statistics have been gathered regarding the effects of even passive cigarette smoking. Obviously, the cigarette companies interpret such statistics in a different manner. It is apparent that such statistical data must be analyzed very carefully before arriving at any conclusions.

Now consider the article to the left. Since peaking in 1970 the percentage of seventeen-year-olds graduating from high school in the United States has been declining. Such statistics have important implications for our educational system.

1.1 INTRODUCTION

Statistics

Data

Qualitative Data

Quantitative Data

Discrete Numerical Data

Continuous Numerical Data

What is **statistics?** Most of us tend to think of statistics as having something to do with charts or tables of numbers. While this idea is not wrong, mathematicians and statisticians use the word "statistics" in a more general sense. Roughly speaking, the term statistics, as used by the statistician, involves collecting any information called **data**, analyzing it, and making meaningful decisions based on the data. Collected data, which represent observations or measurements of something of interest, can be classified into two general types: qualitative and quantitative. **Qualitative data** refer to observations that are attributes. For example, a person's sex, eye color, or blood type, or the brand of whiskey preferred by an individual, and so on. Each person can be placed in one and only one category depending on the situation and how the categories are defined.

On the other hand, **quantitative data** represent observations or measurements that are numerical, for example, the weight of a student, the percentage of homes that are contaminated with radon gas in New Jersey, or the number of college students receiving financial aid at your college. In each of these cases an individual observation measures some quantity. This gives us quantitative data.

Numerical or quantitative data can be further subdivided into two categories, **discrete numerical data** and **continuous numerical data**. When data result from a *count*, we have discrete numerical data. The number of fans attending a football game in the New Jersey Meadowlands Sports Arena on a particular day is 0, 1, 2, 3, . . . , but it cannot be 2.4 or 6.95. On the other hand, data resulting from a *measure* of a quantity will usually be continuous. Thus, if a bag of sugar in a supermarket weighs 5 pounds (lb) (to the nearest pound), all we can be sure of is that the weight of the sugar is some value between 4.5 and 5.5 lb. Actually the sugar can weigh 5.000 lb, 4.998 lb, or any value in the interval 4.5 to 5.5.

Often data may appear to be discrete but in reality they are continuous. For example, Jennifer may claim that she is married for 12 years. Although she was married for 12 years on her last wedding anniversary day, today she is married 12 years plus some part of a year. We will have more to say about discrete and continuous data in later chapters.

The role played by statistics in our daily activities is constantly increasing. As a matter of fact, the nineteenth-century prophet H. G. Wells predicted that "statistical thinking will one day be as necessary for efficient citizenship as the ability to read and write."

The following examples on the uses of statistics indicate that a knowledge of statistics today is quickly becoming an important tool, even for the layperson.

1. Department of Commerce statistics show that we experienced a double-digit rate of inflation throughout 1980.
2. Statistics show that there may be an oversupply of doctors in the 1990s in certain areas of the United States.

3. Statistics collected by the Social Security Administration show that people, on the average, are living longer today than did their parents. This has important implications for predicting the future fiscal soundness of our social security system. (See the following article.)

AMERICA GROWING OLDER

Baltimore: According to the survey conducted by Balcy and Smith for the Social Security Administration, the number of Americans over 65 years of age increased by 6.3% over last year at this time. The survey further indicated that if current trends continue then by the end of this century, approximately 25% of the American population will be over 65 years of age.

These findings have far reaching implications for the financial solvency of the Social Security System.

Monday, January 7, 1991

4. Statistics show that the number of hours in the American workweek is constantly shrinking and that if the present trend continues, the average workweek will decline to 34 hours. (See Figure 1.1.)

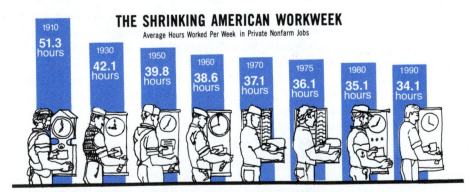

THE SHRINKING AMERICAN WORKWEEK
Average Hours Worked Per Week in Private Nonfarm Jobs

1910 — **51.3** hours
1930 — **42.1** hours
1950 — **39.8** hours
1960 — **38.6** hours
1970 — **37.1** hours
1975 — **36.1** hours
1980 — **35.1** hours
1990 — **34.1** hours

Figure 1.1
Note: Figures for 1910 and 1930 are for workers in manufacturing.
Source: U.S. Department of Labor.

5. Statistics show that if the air we breathe is excessively polluted, then undoubtedly some people will become ill.
6. Statistics show that the world's population is growing at a faster rate than the availability of food. (See Figure 1.2.)
7. Statistics show that the United States is facing a serious energy shortage and that alternative sources of energy must be found to meet the ever-increasing demand.
8. Statistics show that any student, no matter what his or her high school background, will succeed in college if properly motivated.
9. Statistics are used by both the Gallup and Harris polls in determining public opinion on controversial issues.
10. Statistics given in Table 1.1 show the victimization rates for persons age 12 and over by race, annual family income, and type of crime.
11. Statistics show that it is often impossible to obtain accurate records on the prevalence of crime because many victims simply do not report the crime to police. This makes apprehension of the offenders impossible. (See Table 1.2.)
12. Federal Reserve Bank statistics indicate that the number of bank failures has been increasing over the past few years, with more than 100 such failures pre-

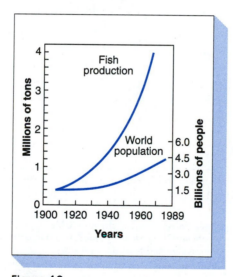

Figure 1.2
Over the years both the world population and the production of fish have increased considerably. Will the world's fish production continue to increase indefinitely, or will we reach a maximum point beyond which production will decrease?

Source: Fisheries of North America.

Table 1.1 Personal Crimes: Victimization Rates for Persons Age 12 and Over, by Race, Annual Family Income of Victims, Type of Crime, 1988*

Race and Income	Crimes of Violence	Rape	Robbery			Assault			Crimes of Theft	Personal Larceny	
			Total	With Injury	Without Injury	Total	Aggravated	Simple		With Contact	Without Contact
White											
Less than $3,000 (10,907,000)	54.4	3.1	10.7	4.4	6.3	40.5	15.9	24.6	83.9	4.3	79.6
$3,000–$7,499 (30,186,000)	35.3	1.0	7.2	2.8	4.4	27.1	11.3	15.8	80.3	3.6	76.7
$7,500–$9,999 (16,700,000)	33.8	[1]0.6	6.6	1.9	4.7	26.6	11.6	15.0	95.3	3.0	92.3
$10,000–$14,999 (38,650,000)	27.2	0.5	4.3	1.2	3.1	22.4	9.2	13.1	93.8	1.7	92.1
$15,000–$24,999 (29,168,000)	27.3	0.4	5.2	1.6	3.6	21.7	8.6	13.2	115.6	2.5	113.1
$25,000 or more (9,888,000)	25.6	[1]0.6	5.7	1.5	4.3	19.3	5.8	13.5	128.3	2.7	125.6
Black											
Less than $3,000 (3,377,000)	48.0	4.1	15.9	7.8	8.1	27.9	15.9	12.0	65.5	7.6	58.0
$3,000–$7,499 (6,469,000)	39.9	1.9	14.8	3.8	11.0	23.1	13.7	9.4	73.0	5.5	67.5
$7,500–$9,999 (1,926,000)	46.5	[1]1.3	17.2	8.8	8.4	28.0	15.4	12.6	91.2	8.2	83.0
$10,000–$14,999 (2,914,000)	32.9	[1]0.5	10.6	[1]2.5	8.2	21.8	11.7	10.1	93.3	4.5	88.4
$15,000–$24,999 (1,533,000)	42.0	[1]2.5	15.0	[1]5.1	9.9	24.5	10.8	13.6	129.5	[1]6.1	123.5
$25,000 or more (242,000)	10.7	[1]0.0	[1]10.7	[1]5.1	[1]5.5	[1]0.0	[1]0.0	[1]0.0	110.7	[1]4.5	106.1

Note: Detail may not add to total shown because of rounding. Numbers in parentheses refer to population in the group; excludes data on persons whose income level was not ascertained.

* Rate per 1000 population age 12 and over

[1] Estimate, based on zero or on about ten or fewer sample cases, is statistically unreliable.

Source: Criminal Victimization in the United States—A National Crime Survey Report. U.S. Department of Justice, Law Enforcement Assistance Administration, National Criminal Justice Information and Statistics Service.

dicted in each of the next few years. Such statistics have important implications for the integrity of our banking and monetary system.

Very little formal mathematical knowledge is needed to collect and tabulate data. However, the interpretation of the data in a meaningful way requires careful analysis. If this is not done by a statistician or mathematician who has been trained to interpret data, statistics can be misused. The following example indicates how statistics can be misused.

Table 1.2 Personal Crimes of Violence: Percent Distribution of Reasons for Not Reporting Victimizations to the Police, by Victim-Offender Relationship and Type of Crime, 1988*

Victim-Offender Relationship and Type of Crime	Total	Nothing Could Be Done; Lack of Proof	Not Important Enough	Police Would Not Want to Be Bothered	Too Inconvenient or Time Consuming	Private or Personal Matter	Fear of Reprisal	Reported to Someone Else	Other and Not Given
Involving strangers									
Crimes of violence	100.0	24.2	23.6	8.1	3.7	9.7	3.9	8.2	18.6
Rape	100.0	34.0	[1]1.8	[1]13.3	[1]0.0	[1]9.9	[1]6.9	[1]10.2	24.0
Robbery	100.0	31.0	17.1	10.3	4.6	7.5	5.5	7.6	16.2
Assault	100.0	21.4	26.7	7.1	3.6	10.4	3.2	8.3	19.2
Involving nonstrangers									
Crimes of violence	100.0	8.7	17.6	6.0	1.4	29.3	6.2	14.5	16.3
Rape	100.0	[1]12.1	[1]2.3	[1]9.3	[1]2.8	22.8	[1]10.7	[1]5.3	34.9
Robbery	100.0	12.5	17.8	[1]2.9	[1]0.0	21.6	10.4	17.3	17.5
Assault	100.0	8.2	18.2	6.2	1.5	30.4	5.5	14.6	15.4

Note: Detail may not add to total shown because of rounding.

[1] Estimate, based on zero or on about ten or fewer sample cases, is statistically unreliable.

Source: Criminal Victimization in the United States—A National Crime Survey Report. U.S. Department of Justice, Law Enforcement Assistance Administration, National Criminal Justice Information and Statistics Service.

In the city of Bushtown the occurrence of polio, thought to be nonexistent, increased by 100% from the year 1989 to 1990. Such a statistic would horrify any parent. However, upon careful analysis it was found that in 1989 there were two reported cases of polio out of a population of five million and in 1990 there were four cases.

1.2 CHOICE OF ACTIONS SUGGESTED BY STATISTICAL STUDIES

Since the word statistics is often used by many people in different ways, statisticians have divided the field of statistics into two major areas called *descriptive statistics* and *inferential statistics.*

Descriptive Statistics

Descriptive statistics involves collecting data and tabulating results using, for example, tables, charts, or graphs to make the data more manageable and meaningful. Very often, certain numerical computations are made that enable us to analyze data more intelligently. Most of us are concerned with this branch of statistics. For example, even if we know the income of every family in California, we are still unable to analyze the figures because there are too many to consider. These figures somehow must be condensed so that meaningful statements can be made.

Inferential statistics or **statistical inference,** on the other hand, is much more involved. To understand why, imagine that we are interested in the average height of students at the University of California. Since there are so many students attending the university, it would require an enormous amount of work to interview each student and gather all the data. Furthermore, the procedure would undoubtedly be costly and could take too much time. Possibly we could obtain the necessary information from a sample of sufficient size that would be accurate for our needs. We could then use the data based on this sample to make predictions about the entire student body, called the **population.** This is exactly what statistical inference involves. We have the following definitions.

Population

DEFINITION 1.1
Sample

A **sample** is any small group of individuals or objects selected to represent the entire group called the **population,** where the population is the set of all measurements of interest to the sample collector.

DEFINITION 1.2
Inferential Statistics

Inferential statistics is the study of procedures by which we draw conclusions and make decisions or predictions about a population on the basis of a sample.

Of course, we would like to make the best possible decisions about the population. To do this successfully, we will need some ideas from the theory of probability. Statisticians must therefore be familiar with both statistics and probability. Exactly how such decisions are made will be discussed in later chapters.

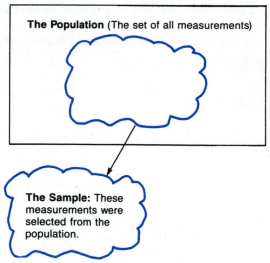

Figure 1.3
The relationship between a population and its sample.

1.3 STATISTICS IN MODERN LIFE

Statistics is so important to our way of living that many of us often use statistical analysis in making decisions without even realizing it. Today statistics are used by the nonmathematician as well as by the mathematician in such areas as psychology, ecology, sports, insurance, education, biology, business, agriculture, music, and sociology, to name but a few. The fields of study to which statistics and probability are being applied is constantly increasing. Their usefulness in the fields of biology, economics, and psychology are so enormous that the subjects of biometrics, econometrics, and psychometrics have come into being.

As statistics developed, probability began to assume more importance because of its wide range of applications. Today the application of probability in gambling is but one of its minor uses. In recent years statistics has even been applied to determine the total population of various species of living things. In particular, by using very simple procedures, statisticians have been able to predict the total population of such endangered species as the whooping crane and various fish. In each case a number of birds or fish are caught, tagged with a label or some other form of identification, and then released for breeding. When they are recaptured or sighted at a later date, the proportion of tagged fish or birds out of the total catch is calculated and used to predict the total population of the species.

Statistics and probability have recently been used to answer such questions as "Did Shakespeare author a newly discovered poem?" (*Science*, January 24, 1986, p. 335) or "What is the probability that a randomly selected person who has tested positive in a drug screening actually uses drugs?" (*Newsweek*, November 29, 1986, p. 18.)

In the future many new and interesting applications of statistics and probability are likely to be found.

1.4 HISTORICAL NOTES

Although statistics is one of the oldest branches of mathematics, its use did not become widespread until the twentieth century. Originally, it involved summarizing data by means of charts and tables. Historically, the use of statistics can be traced back to the ancient Egyptians and Chinese who used statistics for keeping state records. The Chinese under the Chou Dynasty, 2000 B.C., maintained extensive lists of revenue collection and government expenditures. They also maintained records on the availability of warriors.

The study of statistics was really begun by the Englishman John Graunt (1620–1674). In 1662 he published his book, *Natural and Political Observations upon the Bills of Mortality*. Graunt studied the causes of death in different cities and noticed that the percentage of deaths from different causes was about the same and did not change considerably from year to year. For example, deaths from suicide, accidents, and

JOHN GRAUNT (1620–1674)

Figure 1.4
Illustration "Bills of Mortality" redrawn from *Devils, Drugs, and Doctors* by Howard W. Haggard, M.D.

Copyright 1929 by Harper and Row, Publishers, Inc.; renewed 1957 by Howard W. Haggard. Reprinted by permission of the publisher.

certain diseases occurred not only with surprising regularity but with approximately the same percentage from year to year. Furthermore, Graunt's statistical analysis led him to discover that there were more male than female births. But since men were more subject to death from occupational hazards, diseases, and war, it turned out that at marriageable age the number of men and women was about equal. Graunt believed that this was nature's way of assuring monogamy.

After Graunt published his *Bills of Mortality*, many other mathematicians became interested in statistics and made important contributions. Pierre-Simon Laplace (1749–1827), Abraham De Moivre (1667–1754), and Carl Friedrich Gauss (1777–1855) studied and applied the **normal distribution** (see page 355). Karl Pearson (1857–1936) and Sir Francis Galton (1822–1911) studied the **correlation coefficient** (see page 535). These are but a few of the many mathematicians who made valuable contributions to statistical theory. In later chapters we will further discuss their works.

Although a great deal of modern statistical theory was known before 1930, it was not commonly used, simply because the accumulation and analysis of statistical data involved time-consuming, complicated computations. However, all this changed with the invention of the computer and its ability to perform long and difficult calculations in a relatively short period of time. Statistics soon began to be used for **inference,** that is, in making generalizations on the basis of samples. Also, probability theory was soon applied to the statistical analysis of data. The use of statistics for inference resulted in the discovery of new techniques for treating data.

Interestingly enough, the principles of the theory of probability were developed in a series of correspondences between Blaise Pascal (1623–1662) and Pierre Fermat (1602–1665). Pascal was asked by the Chevalier de Méré, a French mathematician and professional gambler, to solve the following problem: In what proportion should two players of equal skill divide the stakes remaining on the gambling table if they are forced to stop before finishing the game? Although Pascal and Fermat agreed on the answer, they both gave different proofs. It is in these correspondences during the year 1654 that they established the modern theory of probability.

A century earlier the Italian mathematician and gambler Girolomo Cardan (1501–1576) wrote *The Book on Games of Chance*. This is really a complete textbook for gamblers since it contains many tips on how to cheat successfully. The origins of the study of probability are to be found in this book. Cardan was also an astrologer. According to legend, he predicted his own death astrologically and to guarantee its accuracy he committed suicide on that day. (Of course, that is the most convincing way to be right!) He also had a temper and is said to have cut off his son's ears in a fit of rage.

1.5 WHAT LIES AHEAD

In the preceding sections we mentioned several uses of statistics to convince you that the development of statistics and probability is not static. It is constantly changing. Who knows what a beginning course in statistics will be like by the year 2000? Undoubtedly, different things will be stressed and new applications for statistics and probability will be found. Yet certain basic ideas of probability and its uses in statistical studies will not be changed. Such ideas are too fundamental. It is with these ideas that we will concern ourselves in this text.

As mentioned earlier, with the advent and widespread use of both the micro-computer and the large mainframe computer in the 1980s, many time-consuming and tedious calculations can now be performed quickly and efficiently by the computer. Very little background in computer programming is actually required since the programs are usually incorporated as part of routine statistical tasks. There are many statistical computer packages currently on the market such as COMP-STAT, MINITAB, and SPSS (Statistical Package for the Social Sciences) Batch System. One of the most widely used packages is MINITAB. Almost all the statistical computer programs given in this book are based on Release 7.1 of MINITAB.*

Along with the widespread acceptance and use of statistical concepts has come the unfortunate use of statistics for deception. Statistics can often be manipulated so that they show what a person wants them to show. In the next chapter we will indicate how statistical graphs can distort information presented by data. Statistical mistakes can be honest also. In its January 2, 1967 issue, *Newsweek* reported on page 10 that Mao-Tse Tung cut the salaries of certain Chinese government officials by 300%. Is it possible for a salary to be cut by 300%?

Also, consider the advertisement by the Morgan Nursing Home chain, which claims that the medical care rendered by its medical staff is so superior that the nursing home has not had any stroke victim deaths in its 25 years of operation. What the advertisement fails to mention is that all seriously ill patients are transferred to a nearby hospital. The patient deaths are then recorded on the hospital records where the deaths actually occurred, rather than on the nursing home records.

These are only two examples of the misuses (intentionally or not) of statistics. Read the book *How to Lie with Statistics* by Darrell Huff for a number of interesting examples on the misuses of statistics. (See the list of suggested further reading at the end of this chapter.)

In the following chapters we will develop the techniques used in all applications of statistics and discuss the role played by probability in these applications. If you follow the well-defined statistical rules that are given, you will avoid falling into some of the aforementioned traps.

EXERCISES FOR SECTION 1.5

1. The word "statistics" is often used by newscasters in radio and television newscasts. Classify these as descriptive or inferential statistics. Explain your answer.

---

*MINITAB is a registered trademark of Minitab, Inc., 3081 Enterprise Drive, State College, PA 16801. Telephone: (814) 238-3280 Telex: 881612

2. After carefully analyzing 10,000 pregnancies, a medical research team concludes that cigarette smoking or drug use by *any* pregnant woman is harmful to the unborn child. This conclusion involves
   a. quantitative data.
   b. qualitative data.
   c. descriptive statistics.
   d. inferential statistics.

3. A counselor at Dorby College is interested in determining the composition of the student body at the college. Which variables of interest measure characteristics that result in
   a. attribute data.
   b. discrete numerical data.
   c. continuous numerical data.

4. Several of the questions asked by the counselor mentioned in Exercise 3 are as follows:
   a. How do you get to school—walk, drive, use public transportation, or ride a bicycle?
   b. How many days per week do you go to school?
   c. How long does it take you to get from your home to school?
   Classify the responses to parts (a), (b), and (c) as attribute data, discrete numerical data, or continuous numerical data.

5. Consider the newspaper article given on page 4.
   a. What part of the survey is an example of descriptive statistics?
   b. What part of the survey is an example of inferential statistics?

6. A student who is registered for a math course claims that she will not do well in the course since math has never been an easy subject for her. This conclusion involves
   a. descriptive statistics only.
   b. inferential statistics only.
   c. both descriptive and inferential statistics.
   d. no statistics at all.
   e. none of these.

Exercises 7–10 indicate how statistics can be misused. After carefully analyzing them, try to find any possible errors in the reasoning.

7. A real estate broker made the following claim: Between 1970 and 1990 the salary of the average American rose by the same rate as did the cost for a new home. The broker concluded that the more money an American worker earns, the higher the price will be for a new home. Do the facts support this claim? Explain your answer.

8. A television commercial claims that "our razor blades are manufactured to such high standards that they will give you a shave that is 50% closer." If you are considering purchasing razor blades, should you buy this brand of blades because of its superior "closeness" property?

9. A national health food magazine claims that "95% of its subscribers who follow the magazine's recommendation and take megadoses of vitamin C are healthy and vigorous." Should you begin taking the vitamin C supplement?
10. During 1990 there were 234 accidents involving drunken drivers and 15,897 accidents involving drunken pedestrians reported in Danville. Can we conclude that it is safer in Danville to be a drunken driver than a drunken pedestrian? Explain your answer.
11. Can you find examples of how you use statistics in your daily experiences?
12. Imagine that you are interested in determining the number of people in San Francisco who write exclusively with their left hand. How could you go about doing this; that is, how would you select a sample?
13. Read the article "Statistics-Watching: A Guide for the Perplexed" in the *New York Times* (November 23, 1975, page 45). This article discusses how economic statistics can often be confusing.
14. Read the book *How to Lie with Statistics* by Darell Huff (New York: Norton, 1954) for a number of interesting examples on the misuses of statistics.

1.6 USING COMPUTER PACKAGES

We live in a computer age. The computer is a very important work machine. At home, as well as at the office or in school, personal computers are appearing in greater numbers and being used for a variety of purposes, such as word-processing, record-keeping, accounting, and other workaday tasks. The use of the computer in the field of statistics is no exception. The widespread availability of computers today (both mainframe and personal) and the supporting software allow the user to perform many detailed statistical calculations.

As with learning to drive a car, operating a computer (mainframe or personal) is an easy task once you learn how to navigate the keyboard, how to negotiate the disk operating system (DOS), what each applications program can do, and which key gets it to do these things. Fortunately, it is rarely necessary for you to write your own computer program to perform statistical analysis. An ample supply of computer software is currently on the market to perform most statistical tasks. These programs are ideally suited to perform the tedious and time-consuming statistical calculations. Many mainframe computers already have such programs stored in memory and can be called upon for use with a few simple commands.

Although there are many readily available statistical computer programs, the most popular of these are MINITAB, SAS (Statistical Analysis System), SPSS (Statistical Package for the Social Sciences), and BMDP.* Check with your local computer center to see which programs are available. (These centers have consultants who can

---

*For a detailed discussion of some of these packages, see J. Lefkowitz, *Introduction to Statistical Computer Packages* (Boston: Duxbury Press, 1985).

show you how to use these programs if you need help.) Throughout this book you will see examples of computer printouts illustrating particular ideas.

At this point we merely mention the existence of statistical packages. We will present the details of the packages at the end of most chapters of this text. In this way you can see how the packages can be used to do the calculations discussed within the chapter.

1.7 SUMMARY

In this chapter we discussed the nature of statistics and how they are used. We distinguished between descriptive statistics and statistical inference. A brief discussion on the origins of statistics and probability was given. Finally, we pointed out how statistics and probability are constantly gaining importance because of their ever-increasing wide range of applications.

Study Guide

The following is a chapter outline in capsule form. You should now be able to demonstrate your knowledge of the ideas mentioned by giving definitions, descriptions, or specific examples. Page references are given in parentheses.

Statistics involve collecting numerical information called **data,** analyzing it, and making meaningful decisions based on the data. (page 3)

Qualitative data refer to observations that are attributes. (page 3)

Numerical or **quantitative data** represent observations or measurements that are numerical. When data result from a *count*, we have **discrete numerical data.** On the other hand, data resulting from a *measure* of a quantity will usually be **continuous numerical data.** (page 3)

Descriptive statistics involve collecting data and tabulating results using, for example, charts, tables, or graphs to make the data more manageable and meaningful. (page 8)

A **sample** is any small group of individuals or objects selected to represent the entire group called the **population,** where the population is the set of all measurements of interest to the sample collector. (page 8)

Inferential statistics or **statistical inference** is the study of procedures by which we draw conclusions and make decisions or predictions on the basis of a sample. (page 8)

Testing Your Understanding of This Chapter's Concepts

1. In order to avoid any layoffs, Governor Rogers asked all state workers earning over $50,000 yearly to agree to a 10% cut in salary. Reluctantly, the workers

agreed. A year later each worker received a 10% increase in yearly salary. Are the workers now earning as much money as they were before? Explain your answer.

2. When a student applies for admission to college, he or she must also submit a high school transcript that gives the high school average. Explain how the high school average is both descriptive and inferential.

3. A recent survey by Haber and Sullivan (1990) concluded that 80% of all juvenile delinquents arrested in Parksville for various crimes during the years 1985–1989 were from broken homes. Can one conclude that a juvenile from a broken home in Parksville is likely to be arrested for committing various crimes? Explain your answer.

4. In a study* of the incidence of heart attacks in men a researcher noted that men who had only one job, on the average, had somewhat fewer heart attacks than men who had two jobs. The researcher concluded that having more than one job increases a man's chance of having a heart attack. Do you think that the facts support the conclusion? Explain your answer.

5. Suppose you are interested in determining the percentage of married men who get along well with their mother-in-law. Can you obtain this percentage by selecting all your relatives and your friends at work as a valid sample? Explain your answer.

THINKING CRITICALLY

1. The Stapelton Medical Association claims that 8 out of every 11 of its doctors carry medical malpractice insurance. If you select 11 Stapelton doctors at random, would you expect 8 of them to carry this type of insurance? Explain your answer.

2. Consider the newspaper article given at the top of page 17. Do you agree that the new net helps to protect dolphins or is it possible that the number of dolphins in the sea has already been reduced from earlier catches?

3. Suppose we are interested in determining the percentage of women who have at least as many children as their mothers did. Can we obtain this percentage by selecting all of our friends and relatives as a random sample? Explain your answer.

4. *Is Smoking Dangerous to Your Health?* The American Cancer Society claims that statistics overwhelmingly support this claim. Yet the tobacco industry claims that the statistics are not conclusive. Explain how both sides in the controversy interpret the results. (See newspaper article on page 2)

5. Suppose you are interested in determining who are safer drivers in Florida, men or women. Can you go to the Motor Vehicle Bureau and simply analyze the number of accidents involving men or women drivers? Explain.

*Rogers and Smith, 1989.

NEW TUNA FISH NET PROVES SUCCESSFUL

Los Angeles (Sept 3): Responding to claims that tuna fish nets have been inadvertently capturing and killing dolphins also, the tuna fish industry has developed a new kind of net for catching tuna fish. The new net has been in use for one year. Statistics show that the percentage of dolphins out of the total tonnage of tuna fish caught has dropped radically when compared to the average percentage over the preceding twenty-five years.

Said a spokesperson for the industry "We think that we have the problem under control."

Tuesday—September 3, 1988

6. After analyzing the police records of Sun Valley, a young and inexperienced statistician comes to the following conclusion. "Since two-thirds of all rape and murder victims on the town's college campus were relatives or former friends of their assailants, a female college student is much safer going out at night with strangers than remaining in her room." Do you agree with this conclusion? Explain your answer.

Review Exercises for Chapter 1

Some of the following questions involve statistical ideas to be discussed in greater detail in later chapters. Nevertheless, they are presented here to alert you to the type of thinking that is necessary in the statistical analysis of a problem. Based on your own experiences, you should be able to discuss and answer most, if not all, of them.

1. After carefully analyzing the available food supply and the world population, several environmentalists have concluded that the world will experience a world-wide food shortage by the year 2010. This conclusion involves
   a. quantitative data.
   b. qualitative data.
   c. descriptive statistics.
   d. statistical inference.
   e. all of these.
   f. none of these.

> ## BLOOD SUPPLY AT DANGEROUSLY LOW LEVELS
>
> *Nov. 22:* According to the greater Metropolitan Blood Council, the blood levels at various hospitals throughout the city stood at 20% of their normal level. Only 837 pints of blood were donated last week. This represents an all-time low record. Administrator Halsey Davis claimed that the drop in blood donations is due to an erroneous fear of the AIDS virus.
>
> These findings have far-reaching implications for those planning to have major surgery in the future and for the general health of our citizens.
>
> Wednesday—November 22, 1989

2. Consider the newspaper article shown here. What part of the article is an example of descriptive statistics?

3. In the previous exercise, what part of the article is an example of inferential statistics?

4. A researcher decided to investigate the reason for the drop in blood donations. He interviewed 5000 college students and determined that the reason was indeed the fear of contracting the AIDS virus. He concluded that this reason seems to be true for the entire population. This conclusion involves
   a. quantitative data only.
   b. qualitative data only.
   c. descriptive statistics.
   d. statistical inference.
   e. none of these.

5. After analyzing accident statistics from an insurance company, an analyst concludes that "since there are more automobile accidents when the roads are dry than when the roads are covered with ice, it is safer to drive when there is ice on the roads." Do you agree with this conclusion?

6. A researcher claims that "Now that women have entered the job market and assumed stress-ridden managerial jobs, they also will be subject to the same type of illnesses which until now primarily afflicted men only." Do you agree?

During the month of July the three families who live on Main Street accumulated 470, 220, and 330 pounds of newspapers for recycling, whereas the three families who live on South Street accumulated 360, 340, and 350 pounds of newspapers for recycling. Using this information, determine which of the statements in Exercises 7–9 are true or false.

7. The conclusion "The average amount of newspapers collected on Main Street is less than the average amount collected on South Street." involves descriptive statistics only.

8. The conclusion "For the month of August the families on South Street will contribute more pounds of newspapers for recycling than the families on Main Street." involves statistical inference.

9. The conclusion "The amount of newspapers collected from the families on Main Street is generally more consistent than the amount collected from the families on South Street." involves statistical inference.

10. The Cuban National Commission for the Propaganda and Defense of Havana Tobacco once noted that the human life span has more than doubled since the tobacco plant was discovered. Is it true that increased use of tobacco will result in a longer life? Explain your answer.

Chapter Test

1. In its annual report, the Aviation Council reported that there were 3 separate incidents in which a plane was struck by lightning during a blinding rainstorm. Furthermore, on clear sunny days there were 25 incidents in which engine failure resulted from birds being sucked into the plane's engines. In view of these findings the Council concluded that it is less dangerous to fly an airplane during a blinding rainstorm than on a clear sunny day when birds are present. Do you agree with this conclusion? Explain your answer.

2. To decrease the number of accidents, the Highway Patrol randomly stops cars driving on the Clearpark Expressway and subjects the drivers to sobriety tests to determine whether the driver is under the influence of alcohol. In 1988 the Highway Patrol arrested 782 motorists for drunken driving. During 1989 only 645 motorists were arrested for drunken driving. Can one conclude that more people are obeying the law and not driving while intoxicated? Explain your answer.

3. Which of the following statements most likely involves statistical inference and which most likely involves descriptive statistics?
   a. The warning label which appears on packages of cigarettes. "Warning: The Surgeon General has determined that cigarette smoking is dangerous to your health."
   b. The average annual cost of malpractice insurance for a neurosurgeon is $239 more in New York City than in Chicago.
   c. The world consumption of oil is increasing so rapidly that new sources of energy will have to be found or we will not have an adequate supply of energy in the near future.

For test questions 4–5 use the following information:

A consumer's group collected data on the prices charged by two large supermarket chains for a dozen of large eggs over a four-week period as indicated in the accompanying chart:

Week	Price Charged by George's Supermarket (cents)	Price Charged by Bandy's Supermarket (cents)
January 1–7	79	88
January 8–14	84	85
January 15–21	73	72
January 22–28	81	80

4. The conclusion "The average price charged by George's supermarket for a dozen of large eggs is less than the average price charged by Bandy's supermarket" involves descriptive statistics. True or false.
5. The conclusion "George's supermarkets will generally give you more for your money than Bandy's supermarkets" involves statistical inference. True or false.
6. In 1980 there were 23 licensed day-care centers in Dover. In 1990 there were 49 licensed day-care centers. Can one conclude that the reason for the increase in the number of licensed day-care centers in Dover is because a greater percentage of the population is now sending their children to these centers? Explain your answer.
7. What part of the following article involves descriptive statistics?

RADON GAS A SILENT KILLER

Dec. 2: Agents from the state's Environmental Protection Agency (EPA) inspected 527 houses yesterday and found 14 of them to be contaminated with the deadly and naturally occurring gas, radon. The gas is entering the houses from underground sources.

Bill Class estimates that about 15% of the homes in the region are contaminated with the radon gases, and called upon the governor to investigate the matter further.

Sunday—December 2, 1990

---

8. What part of the article involves inferential statistics?
9. A cardiologist claimed: "Statistics show that the average blood serum cholesterol level of Americans is dropping because the foods we eat contain less fatty items."

Based on your own research, would you say that the facts support this claim? Explain your answer.

10. Over the years numerous claims have been made that fluoridation of our water supply is associated with a higher incidence of cancer than occurs with use of nonfluoridated water. After a careful and detailed analysis of the water supply systems of 20 American cities, P. D. Oldham* concluded that there is no significant difference in the cancer rates after adjustments for the nonhomogeneity of the populations are made. Does the accompanying newspaper article indicate that Oldham's analysis of the water supply systems and his conclusions are wrong? Explain your answer.

CANCER AND FLUORIDATION OF OUR WATER SUPPLY

June 1: Medical researchers reported yesterday that preliminary studies of the fluoridation of our water system showed an apparent increase in the death rate from cancer per 100,000 of population. Further studies will be undertaken before any definite conclusions can be drawn.

June 1, 1990

11. Arlene takes her daughter Heather to the dentist. Heather observes that the first 100 children treated by the dentist before her begin to cry as the dentist drills their teeth. She concludes that *all* children cry as the dentist drills their teeth. This conclusion involves
    a. foolish statistics
    b. no statistics at all
    c. descriptive statistics
    d. inferential statistics

12. Statistical data that are non-numerical and that represent attributes are referred to as
    a. quantitative data
    b. qualitative data
    c. continuous data
    d. approximate data
    e. none of these

*Oldham, P. D., "Fluoridation of Water Supplies and Cancer: A Possible Association?" *Applied Statistics* 26, (1977), pp. 125–135.

13. The collection of data and the tabulation of the results involves
    a. inferential statistics
    b. quantitative statistics
    c. descriptive statistics
    d. hypothesis testing
    e. none of these

14. The study of procedures by which we draw conclusions and make decisions or predictions about a population on the basis of a sample is known as
    a. descriptive statistics
    b. inferential statistics
    c. quantitative statistics
    d. qualitative statistics
    e. none of these

15. After analyzing the available data, a college loan officer states that 12% of the students at this college default on their government-backed education loans. The officer then claims that 12% of the students at *all* colleges nationwide default on their government backed education loans. This conclusion involves
    a. foolish statistics
    b. no statistics at all
    c. descriptive statistics
    d. inferential statistics

Suggested Further Reading

1. Beniger, J. R., and D. L. Robyn. "Quantitative Graphs in Statistics: A Brief History," *The American Statistician* (February 1978), pp. 1–11.

2. Campbell, Stephen K. *Flaws and Fallacies in Statistical Thinking*. Englewood Cliffs, NJ: Prentice-Hall, 1974.

3. Dale, Edwin L. Jr. "Statistics-Watching: A Guide for the Perplexed," *The New York Times*, November 21, 1975, p. 45.

4. Galton, Francis. "Classification of Men According to Their Natural Gifts" in *The World of Mathematics* edited by James R. Newman. New York: Simon & Schuster, 1956. Vol. 2, part VI, chap. 2.

5. Graunt, John. "Foundations of Vital Statistics" in *The World of Mathematics* edited by James R. Newman. New York: Simon & Schuster, 1956. Vol. 3, part VIII, chap. 1.

6. Huff, Darrell. *How to Lie with Statistics*. New York: Norton, 1954.

7. Martin, Thomas L. Jr. *Malice in Blunderland*. New York: McGraw-Hill, 1973.

8. Newmark, Joseph, and Frances Lake. *Mathematics as a Second Language*, 4th ed. Reading, MA: Addison-Wesley, 1987.

9. Tanur, Judith M., F. Mosteller, W. H. Kruskal, R. F. Link, R. S. Pieters, and G. R. Rising. *Statistics: A Guide to the Unknown*. (E. L. Lehmann, Special Editor.) San Francisco, CA: Holden-Day, 1978.

2 The Description of Sample Data

The results of the analysis of blood samples collected from large groups of individuals can be analyzed using stem-and-leaf diagrams. See discussion on p. 69. (© *DiMaggio/Kalish, Fran Heyl Associates*)

As indicated in Chapter 1, descriptive statistics involves collecting data and tabulating them in a meaningful way. In this chapter we indicate how to organize the data and then present the results graphically.

CHAPTER OBJECTIVES

After studying the material in this chapter, you should be able:

- **To discuss** what is meant by a frequency distribution. This is a convenient way of grouping data so that meaningful patterns can be found. (Section 2.2)

- **To draw** a histogram, which is nothing more than a graphical representation of the data in a frequency distribution. (Section 2.2)

- **To analyze** circle graphs or pie charts, which are often used when discussing distributions of money. (Section 2.3)

- **To apply** frequency polygons and bar graphs, which are used when we wish to emphasize changes in frequency, for example in business and economic situations. (Section 2.3)

- **To introduce** the graph of a frequency distribution, which resembles what we call a normal curve. (Section 2.3)

- **To analyze** how the area under a curve is related to relative frequency and hence to probability. (Section 2.3)

- **To present** an alternate way of analyzing data graphically by means of stem-and-leaf diagrams, which are particularly well suited for computer sorting techniques. (Section 2.4)

- **To show how** graphs can be misused. (Section 2.5)

- **To explain** the use of index numbers, which are used to show changes over a period of time. (Section 2.6)

- **To show how** we use such words as random sample, frequency, class mark, relative frequency, area under a curve, and pictograph. These words are used frequently when studying statistics. (Throughout chapter)

Statistics in the News

DOLLAR'S BUYING POWER DECLINED MORE THAN 60% SINCE 1970

Washington (*June 18*): Despite slowing inflation, the purchasing power of the dollar has eroded nearly 60% since 1970, the Conference Board said. The business information firm said an average family of four that earned $10,000 in 1970 now needs $26,450 to maintain the same after-tax buying power. While inflation was cited as the major cause of the change, higher federal income and Social Security taxes were said to account for about one-third of the drop.

June 19, 1989

Many unions negotiate an escalator clause in their contract with management. When the Consumer Price Index (CPI) or some other index rises, then so will an employee's salary. Thus, it is hoped that the employee's purchasing power can keep up with inflation. How are such index numbers computed? In this chapter we indicate how index numbers are calculated.

Now consider the second article shown below concerning American cars and fuel economy. Is the picture presented by this graph fair? What should the height of the 1987 bar be as compared to the height of the 1979 bar? From these data, can we conclude that American car makers have doubled their average fuel economy?

In this chapter we indicate how to draw and interpret graphs properly.

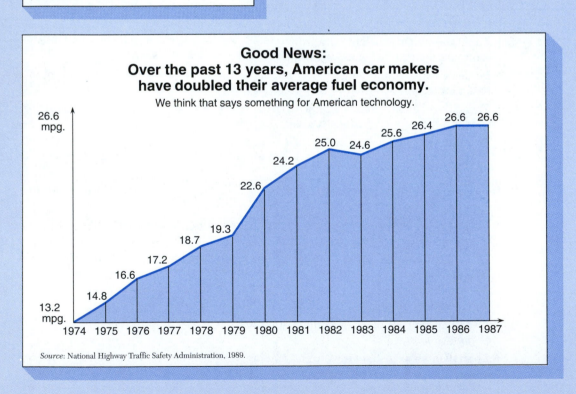

Good News:
Over the past 13 years, American car makers have doubled their average fuel economy.

We think that says something for American technology.

Source: National Highway Traffic Safety Administration, 1989.

2.1 INTRODUCTION

Suppose that John, a student in a sociology class, is interested in determining the average age at which women in New York City marry for the first time (see the accompanying newspaper article). He could go to the Marriage License Bureau and obtain the necessary data. (We will assume that all marriages are reported to the Bureau.) Since there are literally thousands of numbers to analyze, he would want to organize and condense the data so that meaningful interpretations can be drawn from them. How should he proceed?

> ### NEWLYWEDS GET OLDER
>
> Washington (*July 19*): According to statistics released yesterday by the U.S. Census Bureau, the average age at which Americans marry is on the rise. Fifteen years ago, the median age was 23.2 for men and 20.8 for women.
>
> But in 1984, the age for first marriage rose to 25.4 for men and 23 for women. The highest ever median age at marriage for men was 25.9 years in 1900. For women the previous high was 22 in 1890.
>
> Wednesday—July 20, 1988

One of the first things to do when given a mass of data is to group it in some meaningful way and then construct a frequency distribution for the data. After doing this, we can use various forms or graphs of the distribution so that different distributions can be discussed and compared.

In this chapter we consider some of the common methods of describing data graphically. In the next chapter we will discuss the numerical methods for analyzing data. In each case we will use examples from everyday situations so that you can see how and where statistics is applied.

2.2 FREQUENCY DISTRIBUTIONS

Rather than analyze thousands of numbers, John decides to take a sample. He will examine the records of the Marriage Bureau on a given day and then use the information obtained from this sample to make inferences about the ages of *all* women who marry for the first time in New York City.

The most important requirement in taking a sample of this sort is that it be a

TABLE 2.1 Age at Which 150 Women in New York City Married for the First Time

20	17	24	18	16	23	21	17	21	27	23	17	16	19	14
23	26	38	33	20	33	26	22	26	33	26	35	28	35	24
21	18	21	23	22	21	19	23	19	18	19	15	23	21	18
27	40	27	26	22	25	27	25	25	25	30	19	26	32	22
22	19	22	19	24	15	20	22	23	26	23	18	21	20	21
34	41	35	20	29	20	29	27	29	32	29	29	32	28	31
19	22	23	25	23	23	18	19	24	24	21	20	24	22	20
30	31	39	43	38	37	30	37	33	30	36	34	36	32	26
25	23	17	24	18	24	24	21	16	20	18	22	25	24	17
28	29	34	31	25	34	25	36	28	27	31	27	28	30	28

Random Sample   **random sample.** This means that each individual of the population, a woman who applies for a marriage license in our case, must have an equally likely chance of being selected. If this requirement is not satisfied, one cannot make meaningful inferences about the population based on the sample. Since we will say more about random samples in Chapter 8, we will not analyze here whether or not John's sampling procedure is random.

John has obtained the ages of 150 women for a given day from the Marriage Bureau. Table 2.1 lists the ages in the order in which they were obtained from the Bureau records.

The only thing we can say for sure is that most women were in their 20s and that one was as young as 14 while another woman was as old as 43 years when they first married. However, since the ages are not arranged in any particular order, it is somewhat difficult to conclude anything else. Clearly, the data must be reorganized. We will use a frequency distribution to do so.

DEFINITION 2.1
Frequency
Distribution

Frequency

A **frequency distribution** is a convenient way of grouping data so that meaningful patterns can be found. The word **frequency** will mean how often some number occurs.

A frequency distribution is easy to construct. We first make a list of numbers starting at 14, the youngest age, and going up to 43, the oldest age, to indicate the age of each woman. This list is the first column. In the second column we indicate tally marks for each age; that is, we go through the list of original numbers and put a mark in the appropriate space for each age. Finally, in the third column we enter the total number of tally marks for each particular age. The sum for each age gives us the frequency column. When applied to our example, we get the distribution shown in Table 2.2.

A number of interesting facts can be seen at a glance from Table 2.2. Since 23 occurred most often, this is the age at which most women married. Also, the frequency began to decrease as age increased beyond the age of 23.

TABLE 2.2 Construction of a Frequency Distribution for Age at Which 150 Women in New York City Married for the First Time

Column 1	Column 2	Column 3
Age	*Tally*	*Frequency*
14	|	1
15	||	2
16	|||	3
17	ⅣⅡ	5
18	ⅣⅡ |||	8
19	ⅣⅡ ||||	9
20	ⅣⅡ ||||	9
21	ⅣⅡ ⅣⅡ	10
22	ⅣⅡ ⅣⅡ	10
23	ⅣⅡ ⅣⅡ ||	12
24	ⅣⅡ ⅣⅡ	10
25	ⅣⅡ ||||	9
26	ⅣⅡ |||	8
27	ⅣⅡ ||	7
28	ⅣⅡ |	6
29	ⅣⅡ |	6
30	ⅣⅡ	5
31	||||	4
32	||||	4
33	||||	4
34	||||	4
35	|||	3
36	|||	3
37	||	2
38	||	2
39	|	1
40	|	1
41	|	1
42		0
43	|	1

Classes or Intervals

The data in Table 2.2 can also be arranged into a more compact form as shown in Table 2.3. In this table we have arranged the data by age groups, also called **classes** or **intervals.** We select a group size of 3 years since this will result in 10 different groups. Although any number of intervals can be used, for this example we have chosen to work with 10. With the possible exception of the first and last interval, classes are usually of equal size. Generally speaking, if we subtract the

TABLE 2.3 Age at Which 150 Women in New York City Married for the First Time

Class Number	Ages	Class Mark	Tally	Class Frequency	Relative Frequency
1	14–16	15	ℍℍ |	6	6/150
2	17–19	18	ℍℍ ℍℍ ℍℍ ℍℍ ||	22	22/150
3	20–22	21	ℍℍ ℍℍ ℍℍ ℍℍ ℍℍ ||||	29	29/150
4	23–25	24	ℍℍ ℍℍ ℍℍ ℍℍ ℍℍ ℍℍ |	31	31/150
5	26–28	27	ℍℍ ℍℍ ℍℍ ℍℍ |	21	21/150
6	29–31	30	ℍℍ ℍℍ ℍℍ	15	15/150
7	32–34	33	ℍℍ ℍℍ ||	12	12/150
8	35–37	36	ℍℍ |||	8	8/150
9	38–40	39	||||	4	4/150
10	41–43	42	||	2	2/150
			Total frequency =	150	

smallest age from the largest age and divide the results by 10, we will get a number, rounded off if necessary, that can be used as the size of each group. In our case we have

$$\frac{43 - 14}{10} = \frac{29}{10} = 2.9 \quad \text{or 3 when rounded}$$

COMMENT Some people prefer to have each class begin with a multiple of 5 so as to make information easy to read.

Class Number

We have labeled the first column of Table 2.3 **class number** because we will need to refer to the various age groups in our later discussions. Thus, we will refer to age group 26 to 28 as class 5. Notice also that each class has an *upper boundary* or *upper limit*, the oldest age, and a *lower boundary* or *lower limit*, the youngest age. A **class mark** represents the point that is midway between the boundaries of a class; that is, it is the midpoint of a class. Thus, 18 years is the class mark of class 2. We have indicated the class mark for each class in the third column of Table 2.3. This table, containing a series of intervals and the corresponding frequency for each interval, is an example of **grouped data.**

Class Mark

Grouped Data

COMMENT To obtain the class mark, we have

$$\text{class mark} = \frac{\text{lower limit} + \text{upper limit}}{2}$$

Thus, the class mark does not necessarily need to be a whole number.

COMMENT Although we mentioned that equal class intervals are commonly used, this does not necessarily include the first and last classes.

Class Frequency

In the fifth column of Table 2.3 we have the **class frequency,** that is, the total tally for each class. Finally, the last column gives the relative frequency of each class. Formally stated, we have the following definition.

DEFINITION 2.2
Relative Frequency

The **relative frequency** of a class is defined as the frequency of that class divided by the total number of measurements (the total frequency). Symbolically, if we let f_i denote the frequency of class i where i represents any of the classes, and we let n represent the total number of measurements, then

$$\text{Relative frequency} = \frac{f_i}{n} \qquad \text{for class } i$$

Since in our example the total frequency is 150, the relative frequency of class 6, for example, is 15 divided by 150, or $\frac{15}{150}$. Here $i = 6$ and $f_i = 15$. Similarly, the relative frequency of class 7 where $i = 7$ and $f_i = 12$ is $\frac{12}{150}$. The relative frequency of class 10 is $\frac{2}{150}$.

Histogram

Once we have a frequency distribution, we can present the information it contains in the form of a graph called a **histogram.** To do this, we first draw two lines, one horizontal (across) and one vertical (up-down). We mark the class boundaries along the horizontal line and indicate frequencies along the vertical line. We draw rectangles over each interval, with the height of each rectangle equal to the frequency of that class. All of our rectangles have equal widths. Generally, the areas of the rectangles should be proportional to the frequencies.

The histogram for the data of Table 2.3 is shown in Figure 2.1. The area under the histogram for any particular rectangle or combination of rectangles is proportional to the relative frequency. Thus, the rectangle for class 4 will contain $\frac{31}{150}$ of the total area under the histogram. The rectangle for class 7 will contain $\frac{12}{150}$ of the total area under the histogram. The rectangles for classes 9 and 10 together will contain $\frac{6}{150}$ of the total area under the histogram since

$$\frac{4}{150} + \frac{2}{150} = \frac{4 + 2}{150} = \frac{6}{150}$$

The area under a histogram is important in statistical inference and we will discuss it in detail in later chapters.

COMMENT By changing the number of intervals used, we can change the ap-

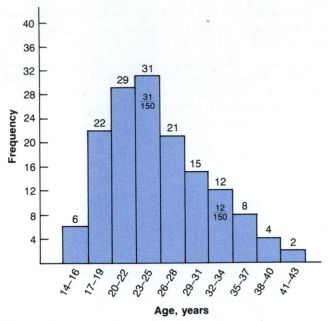

Figure 2.1

pearance of a histogram and hence the information it gives. (See Exercise 15 at the end of this section.)

We can summarize the procedure to be used in the following rules.

RULES FOR GROUPING DATA AND FOR DRAWING HISTOGRAMS

When unorganized data are grouped into intervals and histograms are drawn, the following guidelines should be observed.

1. The intervals should be of equal size (with the possible exception of the end intervals).
2. The number of intervals should be between 5 and 15. (Using too many intervals or too few intervals will result in much of the data not effectively presented.)
3. The intervals can never overlap each other. If an interval ends with a counting number, then the following interval begins with the next counting number.
4. Any score to be tallied can fall into one and only one interval.
5. The histogram is then drawn by using rectangles placed next to each other. The rectangles are placed together to indicate that the next interval begins as soon as one interval ends. The height of each rectangle represents its frequency (assuming the widths of the rectangles are equal). There are no gaps between the rectangles drawn in a histogram (with the exception of an interval having a frequency of zero).

To illustrate further the concept of a frequency distribution and its histogram, let us consider the following example.

EXAMPLE 1   Stock Transactions

Manya, a stock broker, keeps a daily list of the number of transactions where her customers buy the stock on margin. The following number of stocks was bought on margin by her customers during the first ten weeks of 1991.

30	28	16	23	22	10	8	15	16	23
29	30	22	15	24	9	5	4	15	24
21	24	16	14	21	13	6	2	23	22
24	21	27	13	20	8	4	3	24	20
22	18	23	23	17	7	6	1	16	17

Graphical analysis often helps a stockbroker analyze transactions. See discussion on p. 32. (*Courtesy of UNISYS*)

Construct a frequency distribution for the preceding data and then draw its histogram.

Solution

We will use 10 classes. The largest number is 30 and the smallest number is 1. To determine the class size, we subtract 1 from 30 and divide the result by 10, getting

$$\frac{30 - 1}{10} = \frac{29}{10} = 2.9 \quad \text{or 3 when rounded}$$

Thus, our group size will be 3. We now construct the frequency distribution. It will contain six columns as indicated.

Class Number	Number of Stocks	Class Mark	Tally	Class Frequency	Relative Frequency
1	1–3	2	|||	3	3/50
2	4–6	5	⩭	5	5/50
3	7–9	8	||||	4	4/50
4	10–12	11			1	1/50
5	13–15	14	⩭		6	6/50
6	16–18	17	⩭ ||	7	7/50
7	19–21	20	⩭	5	5/50
8	22–24	23	⩭ ⩭ ||||	14	14/50
9	25–27	26			1	1/50
10	28–30	29	||||	4	4/50

Total frequency = 50

The relative frequency column is obtained by dividing each class frequency by the total frequency, which is 50. We draw the following histogram, which has the frequency on the vertical line and the number of stocks on the horizontal line.

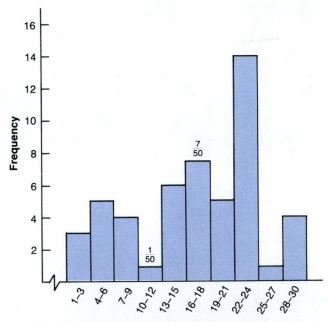

The rectangle for class 6 will contain $\frac{7}{50}$ of the total area under the histogram. Similarly, the rectangle for class 4 contains $\frac{1}{50}$ of that area. Finally, the rectangles for classes 1 and 2 contain

$$\frac{3}{50} + \frac{5}{50} = \frac{3+5}{50} = \frac{8}{50}$$

of the total area under the histogram. What part of the area is contained in *all* the rectangles under the histogram?

Relative Frequency Histogram

Often we may be interested in picturing data that have been arranged in relative frequency form. We can then draw a **relative frequency histogram** by constructing a histogram in which the rectangle heights are relative frequencies of each class. The procedure is illustrated in the following example.

EXAMPLE 2   Young Business Owners

According to Rick Hendricks, regional manager of First National Bank, the age at which business owners first open their businesses has been decreasing. In a random survey of 1000 business owners in the Hampton District the following results were obtained.

Age at Which Owner Started Business	Frequency	Relative Frequency (in decimal form)
16–20	20	20/1000 = 0.02 or 2%
21–25	120	120/1000 = 0.12 or 12%
26–30	230	230/1000 = 0.23 or 23%
31–35	200	200/1000 = 0.20 or 20%
36–40	160	160/1000 = 0.16 or 16%
41–45	80	80/1000 = 0.08 or 8%
46–50	70	70/1000 = 0.07 or 7%
51–55	60	60/1000 = 0.06 or 6%
56–60	50	50/1000 = 0.05 or 5%
61–65	10	10/1000 = 0.01 or 1%
Total Frequency = 1000		*Total* = 100%

The frequency histogram as well as the relative frequency histogram for the preceding data are shown on the following pages.

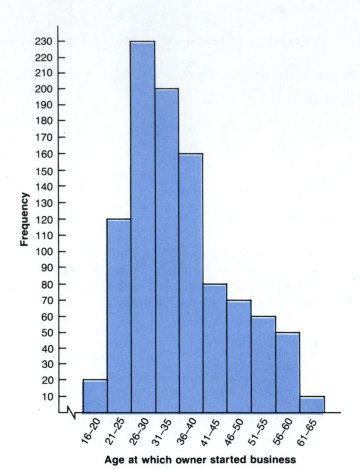

Frequency histogram for age at which owner started business.

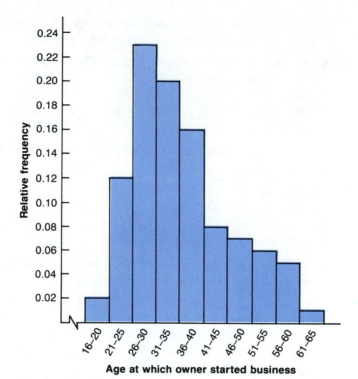

Relative frequency histogram for age at which owner started business.

EXERCISES FOR SECTION 2.2

For each of Exercises 1–6, construct a frequency distribution table and then draw its histogram. (Use ten classes in each.)

1. According to the Cadburry Police Department, the number of parking tickets issued daily during the first 100 days of 1990 was as follows:

34	23	61	28	79	37	38	47	28	58
36	30	78	43	66	36	61	35	36	26
51	33	73	49	55	47	76	36	40	34
34	24	36	61	45	28	44	44	43	42
39	25	32	28	36	36	33	47	46	74
41	37	51	24	31	76	75	50	32	37
28	41	40	23	34	74	67	28	67	74
34	34	46	38	58	67	70	32	76	33
31	48	62	34	47	56	56	24	28	26
35	56	74	72	36	59	55	23	47	27

2. Gwendolyn Jackson placed an ad in the college newspaper offering to sell her old car. The following bids for the car were received:

$3200	$2375	$3440	$3500	$3230	$3200	$3475	$3200	$3550	$3600
4000	3550	3300	3600	3480	3600	3525	3300	3750	3800
3800	3400	3950	3430	3840	3950	3800	3500	3600	3925
3750	3625	3575	3250	3980	3700	3200	3950	3350	3200
3900	3200	3200	3700	3550	3375	3900	3700	3975	3200

3. The ages of 50 applicants for a high-paying job were as follows:

36	32	46	19	26	71	70	45	27	32
23	36	35	18	29	69	62	23	62	69
29	29	41	33	53	62	65	27	71	28
26	43	57	29	42	51	51	19	23	21
30	51	69	67	31	54	50	18	42	22

4. In 1990 the All-Boro Construction Co. renovated 42 homes. The costs to customers for these renovations were as follows:

$20,000	$75,000	$29,000	$70,000	$54,000	$70,000	$72,000
80,000	60,000	80,000	70,000	89,000	98,000	85,000
100,000	80,000	65,000	68,000	94,000	115,000	151,000
150,000	29,000	148,000	120,000	142,000	20,000	108,000
75,000	81,000	25,000	38,000	65,000	58,000	98,000
125,000	100,000	76,000	45,000	115,000	43,000	37,000

5. All passengers on overseas flights must have their luggage weighed. The weights (in kilograms) of the luggage of 60 randomly selected passengers were as follows:

79	108	91	118	101	83	108	109	114	117
91	107	89	117	83	77	91	129	130	129
121	111	103	111	79	94	76	139	137	131
111	114	79	114	74	88	84	141	144	115
101	97	83	107	83	91	94	97	101	103
114	110	108	121	91	92	107	102	107	111

6. A survey conducted by the U.S. Energy Information Administration of 50 randomly selected households in the Denver area found that the annual energy consumption per household [in millions of BTU's (British Thermal Units)] was as follows:

257	362	409	314	217	329	230	191	320	400
262	340	143	246	369	114	265	388	179	245
378	121	264	273	395	209	263	186	346	228
285	237	299	354	253	346	291	295	350	269
296	261	306	279	261	379	264	229	306	279

7. The number of pounds of recyclable plastic collected from 60 soft drink crushing machines located in various places over a two-week period is as follows:

120	200	62	140	43	146	169	114	65	88
79	145	78	21	64	73	195	109	63	86
46	28	85	37	99	154	153	152	91	95
150	69	96	61	106	79	61	179	64	29
95	48	108	81	121	62	12	68	88	77
121	111	103	111	79	94	76	139	137	131

a. Construct a frequency distribution and its histogram using 10 classes.
b. Construct a frequency distribution and its histogram using only 5 classes.
c. Construct a frequency distribution and its histogram using 15 classes.
d. Compare the histograms in parts (a), (b), and (c). What information, if any, is lost by using fewer classes? by using too many classes?

8. A dress company employs 40 workers, each of whom completes similar products in the company's two factories. In each factory there are 20 employees. The number of products completed by each employee during a given time period in each factory is as follows:

Factory A						Factory B				
17	24	16	28	11		7	26	11	9	24
25	5	7	10	30		17	14	32	5	28
9	4	31	21	13		19	25	28	31	12
35	26	21	19	28		31	27	25	37	19

a. Construct frequency distributions and histograms for each factory. (Use 8 intervals)
b. Combine the data for both factories and construct a frequency distribution and its histogram. (Use 8 intervals)
c. Compare the histograms in parts (a) and (b). Comment.

9. Refer to Exercise 8. The personnel manager decides to change the working conditions in Factory A to determine what effect this will have on production. She installs new lighting facilities, new air conditioning, carpeting, piped music, and a new coffee machine. She now notices that the production of the employees in Factory A is:

11	21	37	15	17	18	25	19	28	28
14	21	29	17	21	16	35	16	28	21

a. Construct the frequency distribution and its histogram for the new data. (Use 8 intervals)
b. Compare the new histogram with the original histogram for Factory A and the histogram for the combined data. Comment.

10. Official records indicate that the number of summonses issued for littering on

the highways in the cities of Lanceville and Burgess over the last 30 days is as follows:

Lanceville						Burgess					
13	8	6	11	6	9	11	6	11	14	14	8
11	14	5	8	5	12	9	5	13	12	13	5
9	6	8	7	14	11	7	15	7	15	12	15
7	7	12	12	10	10	8	10	12	10	6	12
10	9	9	14	12	13	12	13	8	9	11	11

a. Construct a frequency distribution and histogram for each city.

b. By looking at histograms, can we determine which city enforces its antilittering laws more consistently?

11. Why is it important that classes be of equal width?

12. Jim Harris and Peter McKenzie are forest rangers. During the month of June each was assigned to prune trees in a particular tract of land. The number of trees pruned daily by each worker is as follows:

Jim Harris						Peter McKenzie				
47	61	27	38	27		37	44	58	39	58
39	47	38	41	55		29	38	60	53	53
51	38	43	57	53		43	36	53	47	56
29	63	52	61	41		41	27	55	41	59
38	42	27	33	27		53	38	39	38	53
52	71	32	43	63		33	45	37	35	45

a. Construct a histogram for each forest ranger. (Use 10 intervals)

b. Who is a better forest ranger (in terms of tree pruning)? Explain.

13. The ages of 100 runners in a marathon were recorded. The results are given here:

Age (in years)	Frequency
Over 78	1
70–77 inclusive	4
62–69 inclusive	9
54–61 inclusive	11
46–53 inclusive	12
38–45 inclusive	14
30–37 inclusive	17
23–29 inclusive	24
15–22 inclusive	8

Draw the histogram for the data and answer the following:

a. What part of the area is below, that is, to the left of the 46 to 53 years category?

b. What part of the area is in the 38 or above category?

c. What part of the area is in the 30 to 53 years category?
d. What part of the area is in the 70 to 77 years category?
e. What part of the area is *not* in the 70 to 77 years category?

14. Fifty parents of the students attending Washington Elementary School were randomly selected and asked to indicate the number of times that they called the school (for any reason) during the school year. Their responses were as follows:

11	21	29	15	10	7	6	23	17	11
10	16	12	14	14	2	12	7	16	17
9	11	18	2	7	3	10	9	8	16
14	10	3	8	3	5	8	0	12	22
17	15	7	11	15	1	14	4	10	10

Draw the histogram for the data (using ten intervals, each of length 3) and then answer the following:
a. What part of the area is below, that is, to the left of 12?
b. What part of the area is 18 or above?
c. What part of the area is between 3 and 26 inclusive?
d. What part of the area is either below 3 or above 26?
e. What part of the area is *not* between 12 and 20?
f. What part of the area is above 29?

15. a. Refer back to the data of Example 1 on page 31. Construct a frequency distribution and histogram using only (i) 5 classes (ii) 15 classes.
   b. Compare the histograms obtained in part (a) with the histogram given on page 32. Comment.

16. Many airlines require that passengers check in at least 45 minutes before the scheduled departure time. A recent survey by an FAA (Federal Aviation Administration) official at Los Angeles International Airport of 50 passengers traveling overseas indicated that these passengers had arrived the following number of minutes before their scheduled departure times:

57	83	41	55	59	7	77	75	79	84
46	75	68	38	70	82	53	45	60	10
54	23	47	56	79	38	61	23	45	54
21	68	78	74	42	38	97	65	28	67
83	32	54	86	46	45	82	38	42	31

a. Construct a relative frequency distribution for this data.
b. Draw a relative frequency histogram for this data.

2.3 OTHER GRAPHICAL TECHNIQUES

In the preceding section we saw how frequency distributions and histograms are often used to picture information graphically. In this section we discuss some other forms of graphs that are often of great help in picturing information contained in data.

Bar Graphs and Pictographs

Bar Graph

Pictograph

The **bar graph** and a simplified version, the **pictograph,** are commonly used to describe data graphically. In such graphs *vertical bars* are usually (but not necessarily) used. The height of each bar represents the number of members, that is, the frequency, of that class. The bars are often also drawn horizontally. We illustrate the use of such graphs with the following examples.

EXAMPLE 1

The number of people calling the police emergency number in New York City for assistance during a 24-hour period on a particular day was as follows:

Time		Number of Calls Received
Starting at	Ending at (up to but not Including)	
12 midnight–	2 A.M.	138
2 A.M.–	4 A.M.	127
4 A.M.–	6 A.M.	119
6 A.M.–	8 A.M.	120
8 A.M.–	10 A.M.	122
10 A.M.–	12 noon	124
12 noon–	2 P.M.	125
2 P.M.–	4 P.M.	128
4 P.M.–	6 P.M.	131
6 P.M.–	8 P.M.	139
8 P.M.–	10 P.M.	141
10 P.M.–	12 midnight	140

The bar graph for these data is shown in Figure 2.2.

From the bar graph we see that the least number of calls was received during the hours of 4 A.M. to 6 A.M. The number of calls received after that period steadily increased until a maximum occurred during the 8 P.M. to 10 P.M. period.

When such statistical information is presented in a bar graph rather than in a table, it can be readily used by police officials to determine the number of police officers needed for each time period.

COMMENT Although the time periods in Example 1 are consecutive, the bars in the bar graph are drawn apart from each other for emphasis and ease in interpretation.

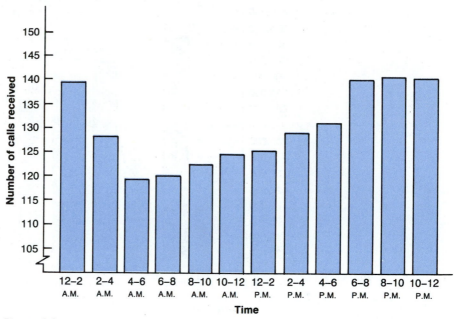

Figure 2.2

COMMENT   Note that the beginning of the vertical line in Figure 2.2 is drawn with a jagged edge. We should always start a vertical scale with 0. However, when all the frequencies are large, we insert this jagged edge in the beginning to indicate a break in the scale.

EXAMPLE 2

The number of Asians residing in California during 1970 and 1980 is shown in Table 2.4. The bar graph for these data is given in Figure 2.3.

	1980	1970	Percent Increase
Filipino	357,514	138,859	157
Chinese	322,340	170,131	89
Japanese	261,817	213,280	23
Korean	103,891	15,756	559

Source: U.S. Census Bureau, 1985.

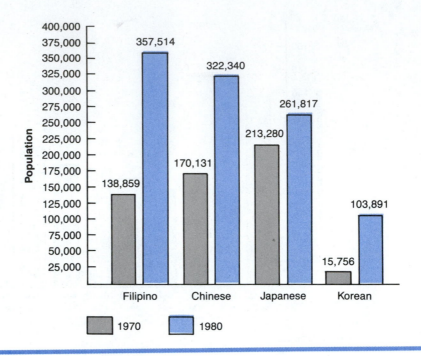

Figure 2.3

1970   1980

Who has high blood pressure

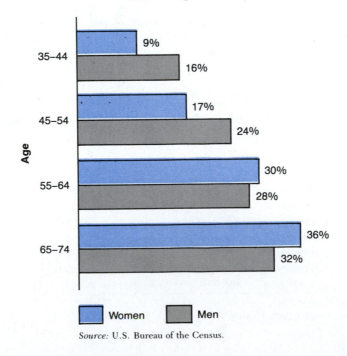

Women   Men

Figure 2.4

Source: U.S. Bureau of the Census.

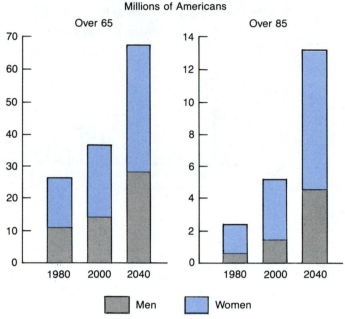

**Population Explosion
of Old People**

The rate of increase for those over 85
will be almost triple the rate of increase
for those over 65 by the year 2040.

Figure 2.5

Source: Social Security Administration.

When both bar graphs are drawn on the same scale, the U.S. Census Bureau can determine easily which segment of the Asian population increased the most between 1970 and 1980. Often the bars of the bar graph are placed side by side or superimposed one upon the other for easy comparison. This is shown in the bar graphs in Figures 2.4 through 2.6. Also, note that the bars in Figure 2.4 have been drawn horizontally.

Several variations of the bar graph are commonly used. In such modifications columns of coins, pictures, or symbols are used in place of bars. When symbols and pictures are used, the bars are sometimes drawn horizontally. We call the resulting graph a **pictograph.** Pictographs do not necessarily have to be drawn in the form of a bar graph. Several pictographs are shown in Figures 2.7 through 2.9.

Pictograph

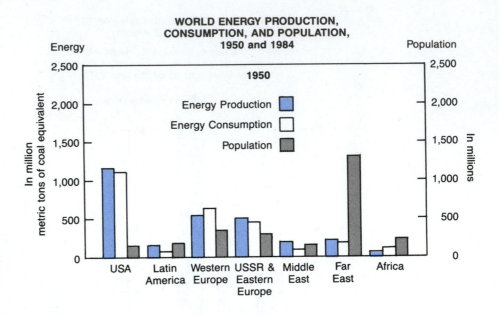

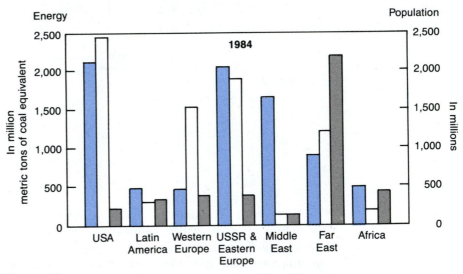

Figure 2.6

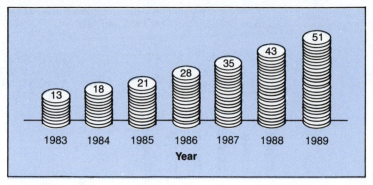

Figure 2.7
Expenditure for social services by a large northeastern city government.
Each symbol represents 6 million dollars.

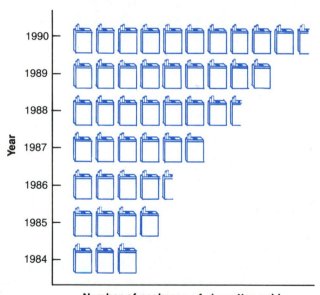

Number of packages of cigarettes sold

Figure 2.8
The number of packs of cigarettes sold weekly in the Magway
Bus Terminal. Each symbol represents 2000 packages.

Public and Nonpublic School Enrollment, K–12
New York State, 1970, 1980 and 1990

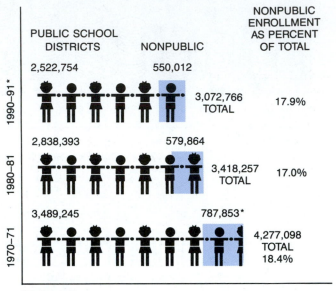

*Estimated

Figure 2.9

Enrollment in elementary and secondary schools increased annually during the 1960s. However, beginning in 1971 enrollment began to decline. This declining trend, which continued through the 1980s, was due both to reductions in the birth rate in recent years and outmigration from the state. Nonpublic school enrollment was expected to decline at a much slower rate than public school enrollment. Between 1983 to 1984 and 1990 to 1991 nonpublic school enrollment declined by 2.6%, whereas public school enrollment decreased by 4.4%.

Source: "Education Statistics New York State," January 1990, State Education Department.

SUMMARIZING

Sometimes we use a bar graph to picture numerical facts, whereas other times we use a histogram for the same purpose. In a bar graph we use a series of bars in the same direction, either all vertically or all horizontally. The length of a bar that represents some numerical fact depends on the size of the number that it represents. We must start the scale at zero and use equally spaced intervals so that the approximate size of a number can be read. A histogram is simply a bar graph in which the bars are placed next to each other in order to show that when one interval ends the other begins.

COMMENT   Often, to emphasize changes in a particular item such as the price of a stock, or temperature changes, we use a **line graph** (sometimes called a **broken line graph**) where a point is used to represent each numerical fact. The consecutive

Line Graphs
Broken Line Graph

points on the graph are joined together with line segments. If the line segment rises, this indicates that the item being analyzed is increasing. The use of such line graphs will be illustrated in the exercises. It should be noted that line graphs are not used to compare different items.

 ## Frequency Polygons

Another alternate graphical representation of the data of a frequency distribution is a *frequency polygon.* Here again, the vertical line represents the frequency and the horizontal line represents the class boundaries.

DEFINITION 2.3 Frequency Polygon	If the midpoints (class marks) of the tops of the bars in a bar graph or histogram are joined together by straight lines, then the resulting figure, without the bars, is a *frequency polygon.*

EXAMPLE 3

Draw the frequency polygon for the data of Example 1 on page 31.

Solution

Since the histogram has already been drawn (page 32), we place dots on the midpoints of the top of each rectangle and then join these dots. The result is the frequency polygon shown in Figure 2.10.

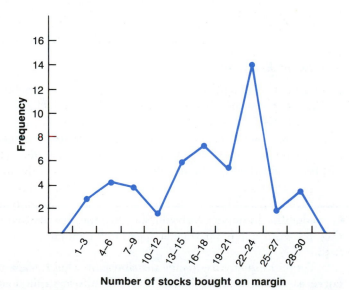

Figure 2.10

EXAMPLE 4

The Bookerville Red Cross Chapter compiles statistics on the number of pints of blood that it collects during the year. For the first six months of 1990, it has constructed the line graph shown in Figure 2.11.

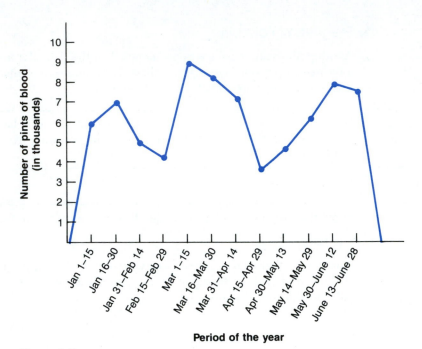

Figure 2.11

Answer the following:

a. During which months of the year were there at least 8000 pints of blood collected?
b. During which month(s) was the most blood collected?

COMMENT   Since frequency polygons emphasize changes (rise and fall) in frequency more clearly than any other graphical representation, they are often used to display business and economic data.

EXAMPLE 5

Consider the frequency polygon showing the distribution of the speeds of cars as they passed through a particular speed enforcement station (radar trap). See Figure 2.12.

There are many frequency distributions whose graphs resemble a bell-shaped curve as shown in Figure 2.13. Such graphs are called **normal curves** and their distributions are known as **normal distributions.** Since many things that occur in

Normal Curve
Normal Distribution

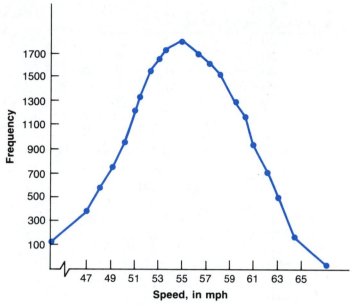

Figure 2.12

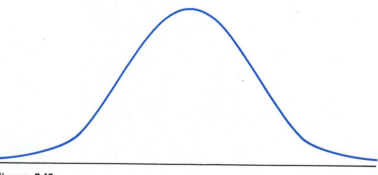

Figure 2.13
A normal curve.

nature are normally distributed, it is no surprise that they are studied in great detail by mathematicians. We will discuss this distribution in detail in a later chapter.

Cumulative Frequency Histograms and Cumulative Frequency Polygons

Suppose we analyze the data presented in the following table that represents the weights of 250 army recruits admitted to a particular training base.

Weight (rounded to the nearest lb)	Frequency (number)
131–140	23
141–150	43
151–160	59
161–170	69
171–180	56
	250

A histogram representing the grouped data is shown below:

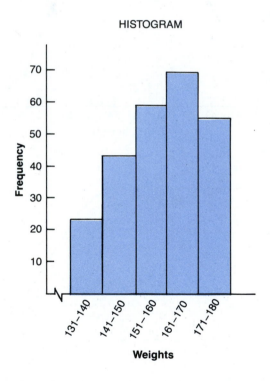

HISTOGRAM

Based on the frequency distribution and the histogram, we can say that the weights of 23 of the recruits were in the interval 131 to 140 pounds, that the weights of 43 of the recruits were in the interval 141 to 150 pounds, and so forth.

Often we are interested in answering questions of the type, "How many of the army recruits weighed less than or equal to a certain weight?" For example, suppose we wanted to know, "How many of the recruits weighed less than or equal to 170 pounds?" We can answer this question by adding or "accumulating" the frequencies in the grouped data. Thus, by adding the frequencies for the four lowest intervals, 23 + 43 + 59 + 69, we find that 194 of the recruits weighed 170 pounds or less.

Cumulative Frequency Histogram

A histogram that displays these "accumulated" figures is called a **cumulative frequency histogram.** For our example, the cumulative frequency histogram is given here.

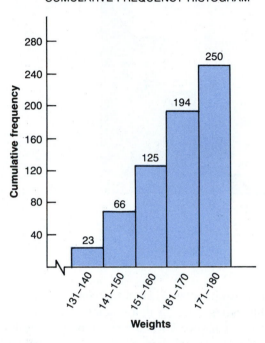

CUMULATIVE FREQUENCY HISTOGRAM

COMMENT   The frequency scale for our cumulative frequency histogram will go from 0 to 250, which represents the total frequency for our data.

Cumulative Frequency Polygon

We can also draw a cumulative frequency polygon for the preceding data. A **cumulative frequency polygon** is simply a line graph connecting a series of points that answer the question mentioned earlier, namely, "How many of the army recruits weighed less than or equal to a certain weight?" Such a graph is shown at the top of the next page.

We construct the cumulative frequency polygon as follows: For the interval 171 to 180, a point is placed at the upper right of the bar to show that 250 of the recruits weighed 180 pounds or less. For the interval 161 to 170, a point is placed at the upper right of the bar to show that 194 of the recruits weighed 170 pounds or less.

We continue this process of placing dots for all five intervals. You will notice that the last dot is placed at the bottom left of the interval 131 to 140 since "0 of the recruits weighed 130 pounds or less." The line graph connecting these six points represents the cumulative frequency polygon.

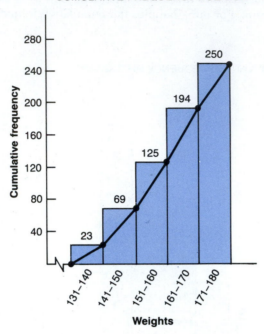

CUMULATIVE FREQUENCY POLYGON

COMMENT It is also possible to label the vertical scale of a cumulative frequency histogram (or polygon) in a slightly different manner, namely, one involving percents. This will be done in Chapter 3 when we discuss percentiles.

Ogives

Cumulative frequency polygons are also known as **ogives** (pronounced "oh-jive"). Any frequency polygon can be converted into an ogive by simply replacing the frequency wiith cumulative frequency. As indicated earlier, the cumulative frequency for any given class is obtained by adding the frequency for that class with the frequencies for *all* classes of smaller values. (A **cumulative relative frequency distribution** combines the cumulative frequency with the relative frequency.) The vertical scale is used to indicate frequency. However, we must adjust the scale so that the total of all individual frequencies will fit. The horizontal scale gives the class boundaries. Ogives are often used when we want to know how many scores are above or below some level.

Cumulative Relative Frequency Distribution

Circle Charts or Pie Charts

Circle Chart
Pie Chart

A common method for graphically describing qualitative data is the **circle chart** or **pie chart.** A circle (which contains 360°, or 360 degrees) is broken up into various

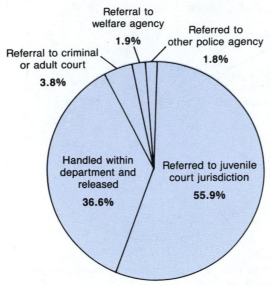

Figure 2.14

Percent distribution of juveniles taken into police custody by method of disposition. United States, 1984.

Source: U.S. Department of Justice, Federal Bureau of Investigation, *Uniform Crime Reports for the United States.* 1988 (Washington, D.C.) U.S. Government Printing Office, 1989.

categories of interest in the same way as one might slice a pie. Each category is assigned a certain percentage of the 360° of the total circle, depending on the data.

The magazine clipping shown in Figure 2.14 is an example of a pie chart that reveals interesting statistics. Since crimes committed by juveniles are on the increase, many people are interested in knowing how the courts are handling such cases. Are the courts too liberal? According to the chart, 36.6% of all juveniles taken into police custody in 1988 were handled within the department and released. Also, 55.9% of the juveniles were referred to juvenile court jurisdiction for appropriate action. Similarly, the pie chart given in Figure 2.15 indicates the reason that people gave for possessing a handgun or a pistol.

The information contained in both of these charts has important implications for our criminal justice system in particular and/or our society in general.

To illustrate the procedure for drawing pie charts, consider the data given in Table 2.5, which indicates the monthly living expenses of a doctoral student at a state university during 1989.

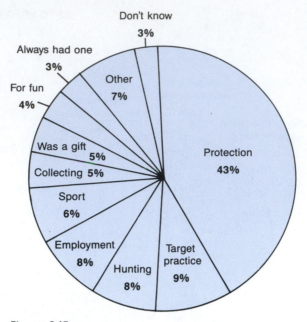

Figure 2.15

Gun owners' reasons for possessing a handgun or pistol in the United States, 1988.

Source: Sourcebook of Criminal Justice Statistics. U.S. Government Printing Office, Washington, D.C., p. 204.

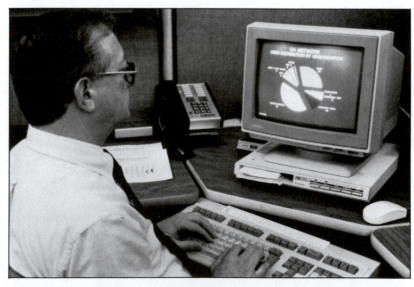

With modern computer graphics technology, pie charts can easily be drawn using many of the available statistical packages. Pictured above is a person analyzing a computer drawn pie chart. (*Courtesy of UNISYS*)

TABLE 2.5 Monthly Living Expenses in 1989 for a Doctoral Student at a State University

Item	Amount (in dollars)
Food	100
Apartment	75
Car and transportation	40
Entertainment	85
Laundry	20
Miscellaneous	40
	Total = 360

Since the total expenditure was $360 and there are 360° in a circle, we can construct the pie chart directly (without any conversions). We draw a circle and partition it in such a way that each category will contain the appropriate number of degrees, as shown in Figure 2.16.

Since it is more convenient to work with percentages than with amounts of money, we may want to convert the amount of money spent in each category into percentages. Thus, the student's money was spent as follows:

Relative Frequency

$$\frac{100}{360} = 0.2778, \text{ or } 27.78\%, \text{ for food}$$

$$\frac{75}{360} = 0.2083, \text{ or } 20.83\%, \text{ for the apartment}$$

$$\frac{40}{360} = 0.1111, \text{ or } 11.11\%, \text{ for the car}$$

$$\frac{85}{360} = 0.2361, \text{ or } 23.61\%, \text{ for entertainment}$$

$$\frac{20}{360} = 0.0556, \text{ or } 5.56\%, \quad \text{for laundry}$$

$$\frac{40}{360} = 0.1111, \text{ or } 11.11\%, \text{ for miscellaneous items}$$

We have indicated these percentages in the pie chart of Figure 2.16.

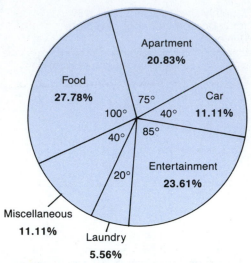

Figure 2.16
Pie chart showing the monthly living expenses in 1989 for a doctoral student at a state university.

COMMENT We convert a fraction into a decimal by dividing the denominator, that is, the bottom number, into the numerator, that is, the top number. Thus, $\frac{100}{360}$ becomes

$$
\begin{array}{r}
0.27777 \\
360\overline{)100.00000} \\
\underline{720} \\
2800 \\
\underline{2520} \\
2800 \\
\underline{2520} \\
2800
\end{array}
$$

When rounded, this becomes 0.2778. This number is written in percentage form as 27.78%.

To round decimals, we use the following rule.

RULE Rounding Decimals

1. Underline the digit that appears in the position to which the member is to be rounded.
2. Examine the first digit to the right of the underlined position.

> **a.** If the digit is 0, 1, 2, 3, or 4, replace all digits to the right of the underlined position by zeros.
> **b.** If the digit is 5, 6, 7, 8, or 9, add 1 to the digit in the underlined position and replace all digits to the right of the underlined position by zeros.
> **3.** If any of these zeros (from step 2) are to the right of the decimal point, omit them.

Let us see how this rule is used.

EXAMPLE 6

1. Round 61.379 to the nearest hundredth.

We underline the digit that appears in the position to which the number is to be rounded:

61.3<u>7</u>9

Since the digit to the right of the underlined position is 5 or more, we add 1 to the digit in the underlined postion. Thus, 61.379 rounded to the nearest hundredth is 61.38.

2. 0.0792 rounded to the nearest thousandth is 0.079.
3. 36.746 rounded to the nearest whole number is 37.

COMMENT The sum of the percentages may not necessarily be 100%. This discrepancy is due to the rounding off of numbers. However, in Example 5, the sum is 100%.

We further illustrate the technique of drawing pie charts by working several examples.

EXAMPLE 7

During 1990, a nationwide auto-leasing company sold 15,000 cars to the individuals who had originally leased them. The types of cars involved are shown in the following frequency distribution.

Type of Car	Number Sold
Manufactured by Japanese companies	3100
Manufactured by General Motors	4800
Manufactured by Chrysler Corp.	2000
Manufactured by Ford Motor Co.	1150
Manufactured by American Motors	850
Manufactured by German companies	2330
Manufactured by other companies	770
	15,000 = *Total sold*

Draw the pie chart for these data.

Solution

We first convert the numbers into percentages by dividing each number by the total 15,000. Thus, we have they following results.

Type of Car	Number Sold	Percentage of Total
Japanese companies	3100	$\dfrac{3100}{15000} = 0.2067$, or 20.67%
General Motors	4800	$\dfrac{4800}{15000} = 0.32$, or 32%
Chrysler Corp.	2000	$\dfrac{2000}{15000} = 0.1333$, or 13.33%
Ford Motor Co.	1150	$\dfrac{1150}{15000} = 0.0767$, or 7.67%
American Motors	850	$\dfrac{850}{15000} = 0.0567$, or 5.67%
German companies	2330	$\dfrac{2330}{15000} = 0.1553$, or 15.53%
Other companies	770	$\dfrac{770}{15000} = 0.0513$, or 5.13%

Now we multiply each percentage by 360° (the number of degrees in a circle) to determine the number of degrees to assign to each part. We get

$$0.2067 \times 360° = 74.41°, \quad \text{or } 74°, \quad \text{for Japanese companies}$$
$$0.32 \times 360° = 115.20°, \quad \text{or } 115°, \quad \text{for General Motors}$$

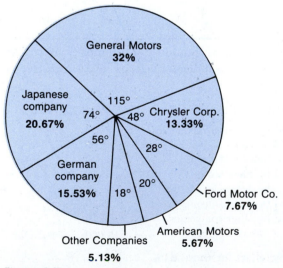

Figure 2.17

$0.1333 \times 360° = 47.99°$,   or 48°,   for Chrysler Corp.
$0.0767 \times 360° = 27.61°$,   or 28°,   for Ford Motor Co.
$0.0567 \times 360° = 20.41°$,   or 20°,   for American Motors
$0.1553 \times 360° = 55.91°$,   or 56°,   for German companies
$0.0513 \times 360° = 18.47°$,   or 18°,   for other companies

Then we use a protractor and compass to draw each part in order, using the appropriate number of degrees. In our case we obtain the pie chart represented in Figure 2.17.

EXAMPLE 8   Oil Reserves

It has been estimated that the number of billions of barrels of oil in the Western Hemisphere, excluding Alaska, is given by the pie chart shown in Figure 2.18. Assuming that there are 130 billion barrels of oil in reserve, answer the following:

a. How many barrels are in reserve in the United States?
b. How many barrels are there in Mexico?
c. How many barrels are there in Canada?

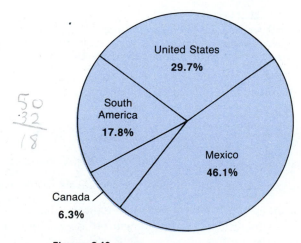

Figure 2.18
Source: Department of the Interior.

Solution
a. The United States has 29.7% of the 130 billion barrels. Thus, since

$$0.297 \times 130 = 38.61$$

the United States has 38.61 billion barrels of oil. (In decimal form 29.7% is written as 0.297.)

b. Mexico has 59.93 billion barrels of oil since

$0.461 \times 130 = 59.93$

c. Canada has 8.19 billion barrels of oil since

$0.063 \times 130 = 8.19$

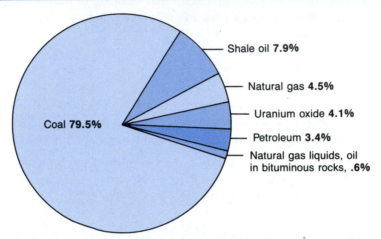

Shale oil **7.9%**

Natural gas **4.5%**

Uranium oxide **4.1%**

Petroleum **3.4%**

Natural gas liquids, oil in bituminous rocks, **.6%**

Coal **79.5%**

Statistics show that America has been using its natural resources at an increasing rate. As a result, government officials are looking for alternate sources of energy. The circle graph indicates some possible sourcles.

We summarize the procedure to be used in constructing pie charts:

RULES FOR CONSTRUCTING A PIE CHART OR CIRCLE GRAPH

1. Determine all the categories that are of interest from the data.

2. For each category determined in step 1, calculate its relative frequency.

3. Draw a circle and assign a slice of the circle to each category. The size of each slice should be proportional to the fraction of observations in that category. Also, the central angle should be an angle whose measure is 360° times the relative frequency for that category. The sum of the measures of all the central angles should be 360° (except for possible rounding errors).

4. Place an appropriate label in each category and indicate the percentage of the total number of observations in the category. This can be found by using the fact that for each category,

Percentage = relative frequency × 100

The sum of all the percentages must always be 100% (except for possible rounding errors).

COMMENT   Again, we wish to point out that pie charts are commonly used to summarize qualitative (or categorical) data. Some people believe that pie charts are more difficult to read than other graphs; however, such charts are particularly useful when discussing distributions of money.

EXERCISES FOR SECTION 2.3

1. In a survey conducted by Young and Hodges* healthy adults were asked whether they would be willing to be cared for in a nursing home should the need arise. Their findings are given below:

Age of Adult (In Years)	In Favor %	Not Sure %	Opposed %
21–30	10	21	69
31–40	13	20	67
41–50	24	18	58
51–60	35	14	51
61 and older	46	9	45

Draw a bar graph to picture this information.

2. According to the Department of Commerce, the amount of recycling of different items in the United States during 1989 was as follows:

Item	Percentage Recycled
Copper	61%
Aluminum	58%
Nickel	40%
Iron	26%
Paper	25%
Textiles	17%
Zinc	14%

Draw a bar graph to picture the preceding data.

3. Each of the world's six largest suspension bridges is longer than 1000 meters. Use the information given in the graph in the following figure to answer the questions.

    a. Find the length of each bridge to the nearest 100 meters.

*Young and Hodges, New York, 1990.

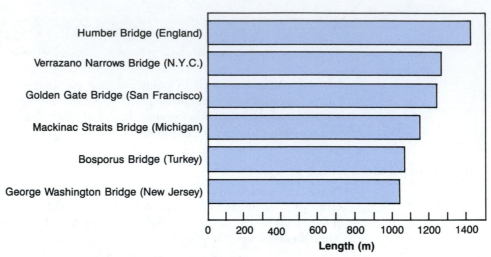

The six longest suspension bridges in the world.

b. What is the approximate difference in length between the Humber Bridge and the Verrazano Narrows Bridge?

c. What is the approximate difference in length between the Golden Gate Bridge and the Bosporus Bridge?

4. The following line graph shows the motor vehicle traffic fatalities per year per one million vehicle miles for the state of California.

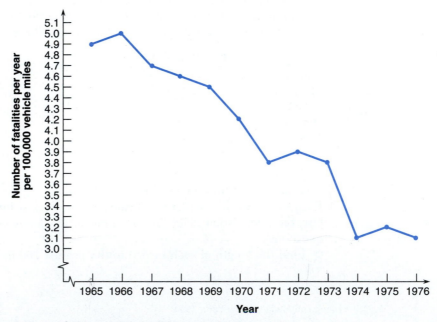

Source: Accident Facts, National Safety Council, Chicago, IL., 1981.

a. In what year was the number of fatalities the lowest?
b. Between what 2 years did the number of fatalities increase the most?
c. Between what two years did the number of fatalities decrease the most sharply?
d. What is the range of the number of fatalities over the 12-year period from 1965 to 1976?

FLOOD WATERS BEGIN TO RECEDE AS RAIN STOPS

May 28: After two days of continuous rainfall, the skies finally cleared yesterday afternoon and the sun reappeared. The flood waters began to slowly recede. The weekend rainfall dumped 6.1 inches of rain into an already overburdened river system. So far, this year's rainfall is running far ahead of last year's total.

Monday—May 28, 1990

---

5. Consider the accompanying newspaper article. The following list gives the total monthly rainfall (in inches) for the region over the past year. Draw a line graph for these data:

Jan	11.4	April	18.8	July	19.9	Oct	15.6
Feb	17.6	May	14.6	Aug	10.3	Nov	13.1
Mar	19.3	June	12.1	Sept	18.9	Dec	12.7

6. The National Basketball Association's (NBA) gross revenue for the 1985 to 1986 season was $196 million. The source of this revenue is shown in the following pie chart.

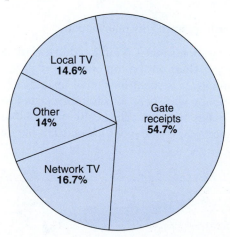

a. How much revenue was obtained from gate receipts?

b. How much revenue was obtained from TV (local or network)?

7. A 1990 survey of 4000 applicants to medical schools indicated that these applicants would most likely practice medicine in the following areas of specialization.

Area of Specialization	Number of Applicants
Neurology	775
Cardiology	827
Pediatrics	769
Obstetrics	584
Urology	428
Orthopedics	337
Opthamology	178
Podiatry	102

Draw a pie chart to picture this information.

8. According to the National Solid Waste Management Association, the average American generates about 3 to 4 pounds of trash each day. This translates into approximately 700 million pounds per day for the whole country. The type of trash thrown away daily consists of the following:

Type of Trash	Percent Thrown Away Daily
Paper	40
Food	17
Yard waste	13
Glass	9
Metals	9
Wood	3
Textiles	2
Plastics	2
Rubber and Leather	2
Miscellaneous	3

Draw a pie chart to picture this information.

9. The Moore Corporation spent 1.5 million dollars last year for employee benefits as shown in the pie chart on the top of the next page.

a. How much money was spent for dental insurance?

b. How much money was spent for disability insurance?

c. How much money was spent for health or dental insurance?

d. How much money was *not* spent for pension benefits or life insurance?

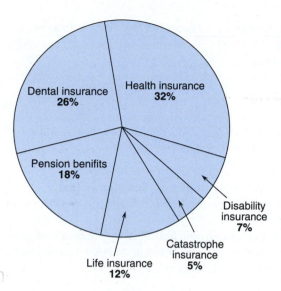

10. *How do we pay for things?* On January 2, 1991 a major U.S. airline reported that customers paid for their plane tickets as follows.

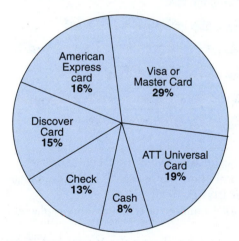

If the airline received payment for 1276 tickets on that day, how many customers paid for their tickets
a. by using the American Express card?
b. by using the Discover card or the ATT Universal card?
c. by not using a check or cash?
11. During 1980 American companies generated 2350 billion kilowatt hours of electricity. The sources of energy used to produce this electricity is as follows.

Source of Energy	Percent of Electricity Produced
Coal	51.1
Natural Gas	15.1
Hydro	12.1
Nuclear	11.0
Oil	10.7

Source: Statistical Abstract of the United States, Bureau of the Census, U.S. Department of Commerce, 1986.

a. Draw a bar graph to picture this information.
b. Draw a pie chart to picture this information.
c. Which graph (bar graph or pie chart) presents the information in a more useful form?

12. The following pictograph indicates the number of automatic transmissions replaced by Bay Auto Transmission Service Corp. over the past 7 years. (Each symbol represents 150 cars.)

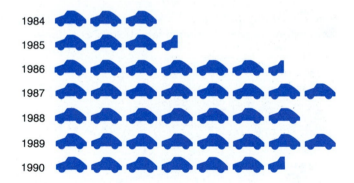

a. How many transmissions were replaced in 1984?
b. By approximately how many cars did the number of transmissions replaced increase from 1984 to 1990?
c. What is the total number of transmissions replaced by this company during the past 7 years?

13. The following pictograph indicates the 1980 population of the world's most populated countries. (Each symbol represents 100 million people.)
a. Approximately how many people did each of these countries have in 1980?
b. Approximately how many more people did China have than did the Soviet Union in 1980?

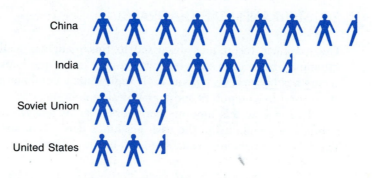

China	
India	
Soviet Union	
United States	

14. Draw a cumulative frequency histogram and a cumulative frequency polygon for the data given in each of the following frequency distributions.

a. The distribution of IQ scores in a particular school.

IQ Range	Frequency
71–80	57
81–90	160
91–100	360
101–110	462
111–120	298
121–130	179
Above 130	32
	Total = 1548

b. The weights (rounded to the nearest pound) of 600 female members of a health club.

Weight	Frequency
100–110	110
111–120	226
121–130	89
131–140	74
141–150	62
Over 150	39
	Total = 600

15. Based on your own experience, which of the following do you think is (are) likely to be normally distributed?
a. Height of individuals in your college
b. IQ scores of college students
c. Intelligence of math teachers
d. Savings account balances for depositors in a large bank
e. Blood pressure of students in a statistics class
f. Age at which women in New York City marry for the first time

☆ 2.4 STEM-AND-LEAF DIAGRAMS

The graphical techniques discussed to this point are well suited to handle most situations. In recent years, however, a new technique known as **stem-and-leaf diagrams** has become very popular. It represents a combination of sorting techniques often used by computers and a graphical technique.

To see how this new method works, let us analyze some information on the number of people using the cash machines daily at an automated banking facility. The following data are available for 30 business days:

162	146	110	219	174	165
128	159	197	205	153	166
151	142	188	212	162	123
203	137	167	178	183	152
178	198	143	179	189	138

Using stem-and-leaf diagrams, we can group the data and at the same time obtain a display that looks like a histogram. This is done as follows. The first number on the list is 162. We designate the first two leading digits (16) as its **stem.** We call the last (or trailing) digit its **leaf** as illustrated here.

Stem (first or leading digits)	Leaf (last or trailing digit)
16	2

= # 162

The stem and leaf of the number 128 are 12 and 8, respectively. Also, the stem and leaf of the number 151 are 15 and 1, respectively.

To form a stem-and-leaf display for the preceding data, we first list all stem possibilities in a column starting with the smallest stem (11, which corresponds to the number 110) and ending with the largest stem (21, which corresponds to the number 219). Then we place the leaf of each number from the original data in the row of the display corresponding to the number's stem. This is accomplished by placing the last (or trailing) digit on the right side of the vertical line opposite its corresponding leading digit or stem. For example, our first data value is 162. The leaf 2 is placed in the stem row 16. Similarly, for the number 128, the leaf 8 is placed in the stem row 12. We continue in this manner until each of the leaves is placed in the appropriate stem rows. The completed stem-and-leaf display will appear as shown in Figure 2.19.

The stem-and-leaf diagram arranges the data in a convenient form since we can now count the number of leaves for each stem. We then obtain the frequency distribution. From this it is very easy to draw the histogram or bar graph. If we turn the preceding stem-and-leaf display on its side, we obtain the same type of bar graph provided by the frequency distribution. This is shown in Figure 2.20.

Stem	Leaves
11	0
12	8 3
13	7 8
14	6 2 3
15	1 9 2 3
16	2 7 2 5 6
17	8 8 9 4
18	8 3 9
19	8 7
20	3 5
21	9 2

Figure 2.19

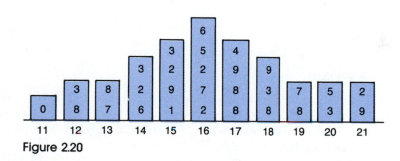

Figure 2.20

COMMENT One major advantage of a stem-and-leaf diagram over a frequency distribution is that the original data are preserved. The stem-and-leaf diagram displays the value of each individual score as well as the size of each data class. This is not possible with a bar graph.

COMMENT A stem-and-leaf diagram presents the data in a more convenient form. This will make it easy to perform various arithmetic calculations to be studied in the next chapter.

We summarize the procedure to be used in constructing stem-and-leaf diagrams in the following rules.

> **RULES FOR CONSTRUCTING STEM-AND-LEAF DIAGRAMS**
>
> To construct a stem-and-leaf diagram, proceed as follows:
>
> 1. Determine how the stems and leaves will be identified.
> 2. Arrange the stems in order in a vertical column, starting with the smallest stem and ending with the largest.
> 3. Go through the original data and place a leaf for each observation in the appropriate stem row.
> 4. If the display looks too cramped and narrow, we can stretch the display by using two lines (or more) per stem so that we can place leaf digits 0, 1, 2, 3, and 4 on one line of the stem and leaf digits 5, 6, 7, 8, and 9 on the other line of the stem.

Let us illustrate the preceding rule with another example.

EXAMPLE 1

A new drug treatment clinic recently opened in the East and the number of addicts treated with methadone per day during the first month was as follows:

37	95	34	26	45
88	89	24	61	28
42	78	67	32	79
67	29	28	24	35
68	72	91	86	78

Draw a stem-and-leaf diagram for these data.

Solution

Let us use the first digit of each of the numbers as the stem and the second digit as the leaf. Then we arrange the stems in order in a vertical column. Although they can be arranged horizontally, the stems are usually arranged vertically. We get the following table.

Stem	
2	
3	
4	
5	
6	
7	
8	
9	

Stem	Leaves
2	9 4 8 6 4 8
3	7 4 2 5
4	2 5
5	
6	7 8 7 1
7	8 2 9 8
8	8 9 6
9	5 1

Figure 2.21

Now we go through the original data and place a leaf for each observation in the appropriate stem row. The stem-and-leaf diagram is shown in Figure 2.21.

Dotplot

Another commonly used graphical display of numerical data is the *dotplot*. **Dotplots** are very helpful in showing the relative positions of numbers in a set of data or for comparing several sets of data. We illustrate the construction of dotplots in the following example.

EXAMPLE 2

Consider the newspaper article here. The following list gives the rainfall (in inches) for the region over the past year. Construct a dotplot for the data.

RAIN EASES DROUGHT SLIGHTLY

Nov 20: The weekend rainfall of 1.1 inches of rain should help ease our region's drought condition somewhat. Over 400 million gallons of water were added to our reservoirs. Further conservation is still needed.

Tuesday—Nov. 20, 1989

Jan.	7	Apr.	8	Jul.	4	Oct.	11
Feb.	6	May	6	Aug.	4	Nov.	8
Mar.	10	Jun.	5	Sept.	3	Dec.	9

Solution

We first draw a horizontal line that indicates the possible number of inches of rain. Then we go through the data and place a dot over the appropriate value on the horizontal axis. Thus, for example, we place a dot over the "7" on the horizontal axis to correspond to the rainfall in January. The complete dotplot for these data is pictured below.

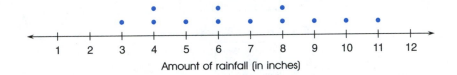

Amount of rainfall (in inches)

COMMENT Although dotplots are quite similar to histograms, they are generally more convenient to work with when dealing with single-value grouped data that involve decimals.

EXERCISES FOR SECTION 2.4

For each of Exercises 1–5, draw a stem-and-leaf diagram for the data.

1. The number of income tax audits completed daily by agents of the Internal Revenue Service at one regional office over a 30-day period is as follows:

45	51	37	16	29	31
36	43	39	28	34	16
28	28	37	36	39	28
19	24	18	33	29	28
25	38	45	47	49	40

2. The following bowling scores were obtained by 100 young campers from a day camp:

111	117	93	110	83	137	109	128	144	99
87	122	127	114	97	139	108	139	141	132
104	93	104	109	101	76	83	142	158	87
83	118	87	121	107	94	141	115	156	99
96	113	127	91	83	79	118	153	121	122
129	101	105	92	91	111	91	129	143	134
131	111	119	107	94	103	138	139	137	136

115	102	118	102	97	111	76	79	153	78
103	116	97	107	101	121	82	91	149	84
119	101	83	111	103	91	148	85	84	144

3. Blood samples were taken from 50 random shoppers in the Bayview Shopping Mall and then analyzed to determine their blood serum cholesterol levels. The following results were obtained:

200	227	257	211	284	241	189	287	227	241
188	256	283	204	269	205	199	251	238	219
197	189	272	218	222	221	223	239	249	216
184	201	187	232	238	203	211	221	205	227
198	297	199	221	245	232	201	208	208	196

4. A scientist from the Environmental Protection Agency took samples of the toxic substance polychlorinated biphenyl (PCB) levels from the soil at 100 different waste disposal facilities located throughout the United States. The following results (in 0.0001 grams per kilogram of soil) were obtained:

38.8	35.6	31.8	32.8	36.3	40.2	39.7	33.9	34.4	33.1	39.3	34.8
35.3	38.1	35.7	39.1	37.8	39.5	36.4	38.6	37.6	37.8	31.7	35.7
39.1	35.8	38.4	34.5	37.9	38.2	38.3	40.1	38.8	33.9	30.8	37.6
31.8	32.4	35.9	36.1	38.1	37.6	36.7	30.8	37.8	35.5	39.8	36.9
40.2	33.8	34.7	39.0	36.0	37.3	31.4	31.7	32.9	30.7	37.5	31.8

5. Bill is interested in selling his car. In a recent auction the following sealed bids were made. Group the numbers so that 50 values fall on each line. Thus, the stems will be 3900–3949, 3950–3999, and so on.

$4000	$4149	$4050	$3975	$4000
4198	4249	3945	4249	3998
3945	3901	3989	3976	3987
3998	4150	4149	4004	4049
4000	3945	4200	3984	4145

6. Draw a dotplot for the following data:

68	69	60	76	68	69
67	65	65	77	65	65
74	71	59	69	71	67
73	70	72	70	73	70
71	69	75	69	75	72
74	70	69	65	72	68

2.5 HOW TO LIE WITH GRAPHS

In the previous sections we indicated how to analyze a list of numbers by graphical techniques. Nevertheless, it is possible to use these techniques to present the data in a misleading way. The following examples indicate some misuses of statistics and the reason for the incorrect use.

Since frequency polygons emphasize changes (rise and fall) in frequency more clearly than any other graphical representation, they are used often to display business and economic data. However, this must be done with great care. Figures 2.22 and 2.23 both represent the same idea, that is, the number of cases of malpractice insurance filed against neurosurgeons in a large northeastern city in the United States during the years 1986 to 1990. Figure 2.23 seems to indicate that the number of malpractice insurance cases filed increased significantly over this 5-year period. Figure 2.22 also indicates that the number of such cases increased, although not so dramatically. How can we have two *different* graphs representing the same situation? Which graph better displays the situation?

Truncated Graphs

Notice that the vertical scales in the two graphs are not the same. The vertical scale of Figure 2.23 has been truncated or cut off. **Truncated graphs** often tend to distort the information presented. Consider the graphs on a drought situation shown in Figures 2.24 and 2.25. Again, notice the truncated graph. How bad is the drought situation?

Scaling

Another misuse of statistics involves **scaling.** Consider the graphs given in Fig-

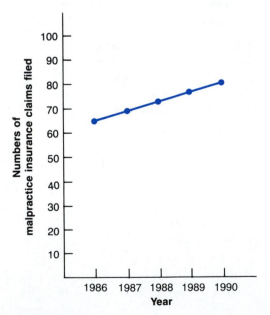

Figure 2.22

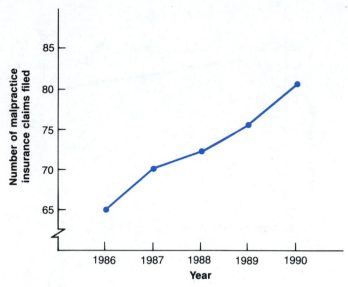

Figure 2.23

ures 2.26 and 2.27 to show how the circulation of a computer magazine has doubled. Anything wrong? The graph in Figure 2.27 is twice as tall as the graph in Figure 2.26. But it is also twice as wide, so it is four times as large. In Figure 2.27 each of the four rectangles is exactly the same size as the one in Figure 2.26. Do people draw graphs in such a misleading way? Unfortunately, the answer is yes.

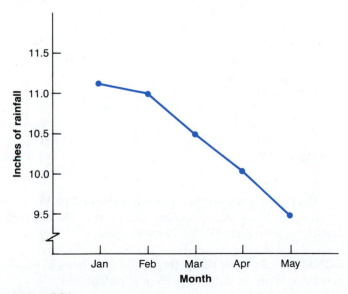

Figure 2.24

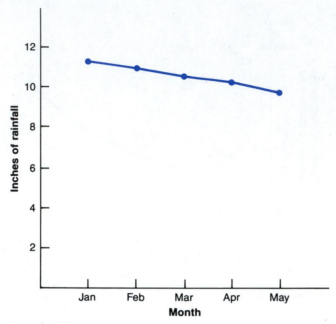

Figure 2.25

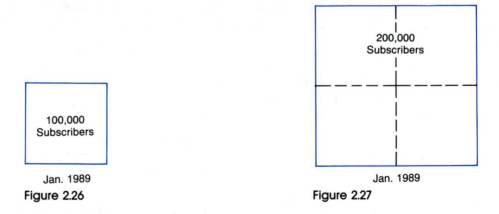

Jan. 1989
Figure 2.26

Jan. 1989
Figure 2.27

Up to this point we have merely indicated two ways that statistical graphs can be misused. Many other ways are discussed in Darrell Huff's book *How to Lie with Statistics* (New York: W. W. Norton, 1954).

COMMENT   By now the point should be obvious. Read and construct statistical graphs carefully to avoid the common pitfalls.

EXERCISES FOR SECTION 2.5

1. The number of arrests for drug abuse on a particular college campus over a 5-year period was presented graphically in the college student newspaper and in the local city newspaper. Which graph is misleading? Explain your answer.

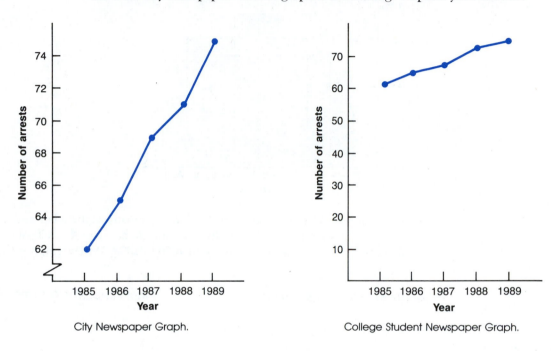

City Newspaper Graph.   College Student Newspaper Graph.

2. Consider the following advertisement promoted by a swimming pool construction company. Anything wrong with the claim?

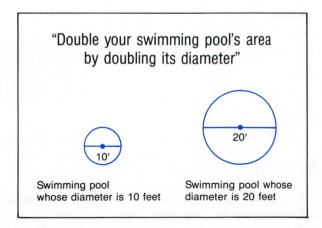

"Double your swimming pool's area by doubling its diameter"

Swimming pool whose diameter is 10 feet

Swimming pool whose diameter is 20 feet

3. The following graph concerning the unemployment rate in a large metropolitan city recently appeared in the city's newspaper indicating that unemployment was declining considerably. Anything wrong with this graph?

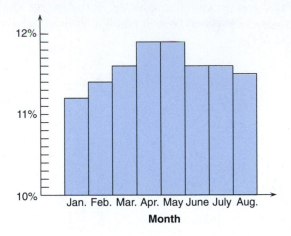

4. The following graphs, one prepared by a company statistician and the other prepared by an independent statistician, indicate how the Arjon Tool Company's share of the market has increased from 1986 to 1990. Which graph is the correct one?

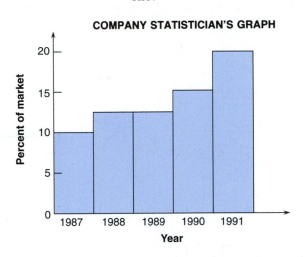

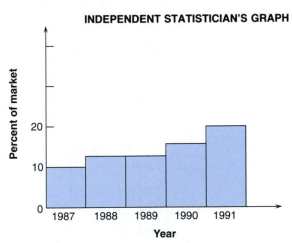

2.6 INDEX NUMBERS

Often when analyzing data we are interested in obtaining a clear picture of trend. We could then use index numbers that give us a good analysis of changes over a period of time. Index numbers allow us to compare such things as prices, production

Base Period

figures, sales figures, and so on for a given period of time with corresponding values in some earlier period of time. The earlier period is usually referred to as the **base period.** We have the following definition.

DEFINITION 2.4
Index Number

An **index number** is a special form of ratio that is used to show percentage changes over a period of time.

Consumer Price Index (CPI)

The best known index number is the **Consumer Price Index (CPI)** published by the U.S. Bureau of Labor Statistics. In the media this is often called the cost-of-living index. It reflects the changes in prices of goods and services purchased by a typical wage earner in a large city. Since the labor contracts negotiated by some unions often tie increases in salary to changes in the CPI through cost-of-living escalation clauses, a knowledge of how to compute index numbers is important.

The CPI is published each month and includes the results of surveys for the prices of over 400 goods and services including among other things food, clothing, medical costs, automobile repair, college tuition, and so on. Thus, it can be used as a measure of inflation. When the CPI goes up, wages, pensions, and Social Security payments go up.

COMMENT Great care must be exercised when using the CPI. The prices of the items included in the computation must be weighted according to their importance.

To calculate an index number, we use the following formula.

FORMULA 2.1

$$\text{Index number} = \frac{\text{given year's values}}{\text{base year's value}} \times 100$$

We illustrate the use of this formula with several examples.

EXAMPLE 1

The Musicktone Corporation sells various stereo components. The annual sales for these components over the years 1984 to 1989 is as follows.

Year	Sales
1984	$400,000
1985	425,000
1986	491,000
1987	350,000
1988	410,000
1989	380,000

Using 1984 as a base year, calculate the sales index for each year.

Solution

Since we are using 1984 as our base year, we must divide each year's value by the base year's value and multiply the result by 100. We have the following table.

Year	Ratio	Ratio $\times$ 100 = Index Number
1984	$\dfrac{400,000}{400,000} = 1$	$1 \times 100 = 100$
1985	$\dfrac{425,000}{400,000} = 1.0625$	$1.0625 \times 100 = 106.25$
1986	$\dfrac{491,000}{400,000} = 1.2275$	$1.2275 \times 100 = 122.75$
1987	$\dfrac{350,000}{400,000} = 0.875$	$0.875 \times 100 = 87.5$
1988	$\dfrac{410,000}{400,000} = 1.025$	$1.025 \times 100 = 102.5$
1989	$\dfrac{380,000}{400,000} = 0.95$	$0.95 \times 100 = 95$

The index numbers calculated in the previous example show us the percent change from the base year. This percent change is the difference from 100. We have

a. The sales index for 1985 indicates an increase of 6.25% when compared to 1984 since $106.25 - 100 = 6.25$.
b. The sales index for 1986 indicates an increase of 22.75% when compared to 1984 since $122.75 - 100 = 22.75$.
c. The sales index for 1987 indicates a decrease of 12.5% when compared to 1984 since $87.5 - 100 = -12.5$.

Similar conclusions can be arrived at for the other years.

EXAMPLE 2   Oil Dependence of the United States

The number of thousands of barrels of oil imported by the United States from OPEC (Organization of Petroleum Exporting Countries) sources for the years 1981 to 1984 is as follows.

Year	Oil Imported
1981	1,849,017
1982	2,260,482
1983	2,057,468
1984	2,023,341

Source: American Petroleum Institute, 1987

Using 1981 as a base year, compute the import index for each of the years given and interpret the results.

Solution

Using the same tabular arrangement as we did in the previous example, we have the following table.

Year	Ratio	Ratio × 100 = Index Number
1981	$\dfrac{1,849,017}{1,849,017} = 1$	$1 \times 100 = 100$
1982	$\dfrac{2,260,482}{1,849,017} = 1.22$	$1.22 \times 100 = 122$
1983	$\dfrac{2,057,468}{1,849,017} = 1.11$	$1.11 \times 100 = 111$
1984	$\dfrac{2,023,341}{1,849,017} = 1.09$	$1.09 \times 100 = 109$

The import indexes for these years indicate

a. for 1982 an increase of 22% when compared to 1981.
b. for 1983 an increase of 11% when compared to 1981.
c. for 1984 an increase of 9% when compared to 1981.

COMMENT In the previous example we can use a different base year and arrive at different conclusions. Thus, if 1983 is used as a base year, we get index numbers of 90, 110, 100, and 98, respectively, for the years 1981, 1982, 1983, and 1984. These index numbers give us different interpretations than when we used 1981 as a base year.

COMMENT Often when using index numbers an important decision involves deciding which base year is best to use.

EXERCISES FOR SECTION 2.6

For Exercises 1–4, use the following information. The price of a cup of coffee (including tax) at Joe's Hiway Mart on State Highway No. 3 over a period of 7 years was as follows.

Year	Price of Cup of Coffee
1986	$0.43
1987	0.49
1988	0.60
1989	0.65
1990	0.71
1991	0.75

1. Using 1986 as a base year, what does the price index for 1991 tell us?
2. Using 1986 as a base year, what does the price index for 1989 tell us?
3. Using 1991 as a base year, what does the price index for 1986 tell us?
4. Using 1991 as a base year, what does the price index for 1989 tell us?
5. The price of a general admission ticket to a particular baseball stadium was as follows.

Year	Price of Ticket
1985	$2.75
1986	3.25
1987	4.00
1988	4.25
1989	5.00
1990	6.00

Using 1985 as a base year, compute cost indexes for each of the years given and interpret the results.

6. Mr. Johnson is analyzing his heating bills for several years. He has accumulated the following data.

Year	Total Heating Bill for Season ($)
1984	1420
1985	1605
1986	1840
1987	1992
1988	2079
1989	2250
1990	2408

Using 1984 as a base year, compute cost indexes for each of the years given and interpret the results.

7. Refer back to Exercise 6. Using 1987 as a base year, compute cost indexes for each of the years given and interpret the results. Compare the results obtained for Exercises 6 and 7. Comment.

8. An administrator of a lunch program has computed the following price indexes for several items used in the lunch program.

Index of Prices (January 1988 = 100)

Month	Eggs	Vegetables	Paper Supplies
January 1989	102	112	103
February 1989	105	109	99
March 1989	107	111	105

a. Which item showed the smallest increase between January and March 1989?

b. Which item showed the greatest increase between January and March 1989?

9. Refer back to Exercise 8. For the month of February 1989, paper supplies had a price index of 99. Explain this price index.

10. Refer back to the newspaper article at the beginning of this chapter (on page 24). In 1988 Ellen Ingram earned $55,000 and in 1989 she earned $56,000. According to government records, the CPI for Ellen's region was 106.7 in 1988 and 108.1 in 1989. Did Ellen's purchasing power increase or decrease over the year?

2.7 USING COMPUTER PACKAGES

To illustrate how we can use the MINITAB computer package to draw a histogram, let us consider the following data representing the number of credit card sales reported by a large department store chain at all of its branches over a 50-day period.

Day	Number of Sales	Day	Number of Sales	Day	Number of Sales
Mon.	52	Mon.	60	Mon.	53
Tues.	68	Tues.	62	Tues.	58
Wed.	51	Wed.	58	Wed.	62
Thurs.	62	Thurs.	48	Thurs.	61
Fri.	49	Fri.	61	Fri.	60
Mon.	37	Mon.	45	Mon.	45
Tues.	74	Tues.	49	Tues.	61
Wed.	64	Wed.	53	Wed.	53
Thurs.	48	Thurs.	67	Thurs.	42
Fri.	63	Fri.	49	Fri.	37

Day	Number of Sales	Day	Number of Sales
Mon.	34	Mon.	47
Tues.	68	Tues.	59
Wed.	51	Wed.	38
Thurs.	47	Thurs.	32
Fri.	56	Fri.	76
Mon.	59	Mon.	57
Tues.	69	Tues.	41
Wed.	42	Wed.	39
Thurs.	38	Thurs.	31
Fri.	36	Fri.	37

After logging in on the computer, we can construct the histogram for the preceding data by typing in the following instructions:

```
MTB  >   SET THE FOLLOWING DATA INTO C1
DATA >   52  68  51  62  49  37  74  64  48  63
DATA >   60  62  58  48  61  45  49  53  67  49
DATA >   53  58  62  61  60  45  61  53  42  37
DATA >   34  68  51  47  56  59  69  42  38  36
DATA >   47  59  38  32  76  57  41  39  31  37
DATA >   END

MTB  >   HISTOGRAM OF C1
```

The first line tells the computer that the data should be placed in column C1 of the worksheet that MINITAB maintains in the computer. The next few lines contain the data themselves. The last line instructs the computer to construct the histogram.

After typing the above information, the computer will automatically print out the frequency distribution and histogram shown below. Notice that the histogram is printed sideways with the bars (actually they are asterisks) drawn horizontally rather than vertically.

```
Histogram of C1    N = 50
Midpoint    Count
      30        2   **
      35        5   *****
      40        6   ******
      45        4   ****
      50        8   ********
      55        5   *****
      60       12   ************
      65        3   ***
      70        3   ***
      75        2   **
```

We can use MINITAB to obtain a dotplot of the data in C1. We simply type the command DOTPLOT followed by the storage location of the data (C1). In our case we have

```
MTB > DOTPLOT OF C1
```

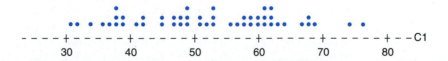

We can also have MINITAB draw a stem-and-leaf diagram for the data in C1.

We have the following:

```
MTB > STEM-AND-LEAF OF C1

Stem-and-leaf of C1          N = 50
Leaf Unit = 1.0

    3     3 124
   10     3 6777889
   13     4 122
   22     4 557788999
   (6)    5 112333
   22     5 678899
   16     6 0011122234
    6     6 7889
    2     7 4
    1     7 6
```

```
MTB > STOP
```

The first line of this output gives us a verbal description of what MINITAB is doing—in our case "stem-and-leaf of C1." It also tells us that N = 50 or that we have 50 pieces of data. The second line of the output (Leaf Unit = 1.0) tells us where the decimal point goes—in our case directly after each leaf digit.

There are three columns in this stem-and-leaf printout. The second column of numbers gives us the stems. The third column of numbers gives us the leaves. The leaves are ordered so that we now have an ordered stem-and leaf diagram. The first column of numbers gives us **depth**. The depths are used to display cumulative frequencies. Starting from the top, the depths tell us the number of leaves (pieces of data) that lie on a given row or on earlier rows. Thus, the "22" in the fourth row tells us that there is a total of 22 leaves in the first four rows.

When we reach the row in which the middle observations lie, the cumulative frequency is replaced by the number of leaves in that row enclosed by parentheses. In our case the middle observations lie in the 50 to 54 interval and there are six data values in this interval. Finally, the depths following the 50 to 54 interval indicate the number of leaves that lie in the given row and in subsequent rows. Thus, the "16" in the seventh row indicates that there is a total of 16 leaves in the last four rows.

Depth

COMMENT Our objective here is not to teach you how to use the MINITAB statistical package but merely to familiarize you with its general nature and usefulness. The details can be obtained from manuals. For a manual specifically dealing with MINITAB, see T. Ryan, B. Joiner, and B. Ryan, *MINITAB STUDENT HANDBOOK* (Boston: Duxbury Press, 1985).

2.8 SUMMARY

In this chapter we discussed the different graphical methods that one can use to picture a mass of data so that meaningful statements can be made. When given a large quantity of numbers to analyze, it is recommended that frequency tables be constructed. Some forms of graphical representation should then be used to serve as visual aids for thinking about and discussing statistical problems in a clear and easily understood manner. We discussed the different graphical techniques that one can use and also pointed out in Figures 2.22 and 2.24 on pages 74 and 75 how they can be misused. We studied an alternate and relatively new way of analyzing data. This is by means of stem-and-leaf diagrams or dotplots. We also analyzed index numbers, which are special ratios that can be used to study percentage changes over a period of time. We pointed out that great care must be exercised when interpreting such numbers. The best known of these index numbers is the CPI published by the U.S. Bureau of Labor Statistics. There are many other indexes that can be computed. Some of these measure changes in regional and national retail and wholesale prices, industrial and agricultural production, and so on. Each of these is computed in a manner similar to the way we computed index numbers.

Study Guide

The following is a chapter outline in capsule form. You should now be able to demonstrate your knowledge of the ideas mentioned by giving definitions, descriptions, or specific examples. Page references are given in parentheses.

All sample data in this book are the results of **random samples**. This means that each individual of the population must have an equally likely chance of being selected. (page 26)

A **frequency distribution** is a convenient way of grouping data in order to find meaningful patterns. The word **frequency** means how often some number occurs. (page 26)

Data are often arranged by groups called **classes** or **intervals**. (page 27)

When data are arranged so that we have a series of intervals with the corresponding frequency for each interval, we have **grouped data**. (page 28)

A **class mark** represents the point that is midway between the limits of a class; that is, it is the midpoint of a class. (page 28)

The **class interval width** is determined by

$$\frac{\text{Largest measurement} - \text{smallest measurement}}{\text{Number of intervals}}$$

We usually choose between 5 and 15 measurement classes. (page 28)

The tally for each class is called the **class frequency**. The class frequency for class i is denoted by the symbol f_i. (page 29)

The **relative frequency** for a given class is defined as the frequency of that class divided by the total number of measurements (the total frequency). (page 29)

A graphical portrayal of a frequency distribution is called a **histogram.** The height of each rectangle of a histogram represents its frequency. (page 29)

A **relative frequency histogram** is drawn by constructing a histogram in which the rectangle heights are relative frequencies of each class. (page 33)

The **bar graph** is commonly used to describe data graphically. The height of each bar represents the number of members, that is, the frequency of that class. (page 40)

When symbols and pictures are used in place of bars, the resulting graph is called a **pictograph**. (page 43)

If the midpoints (class marks) of the tops of the bars in a bar graph or histogram are joined together by straight lines, then the resulting figure (without the bars) is called a **frequency polygon**. (page 47)

A frequency distribution whose graph resembles a bell-shaped curve is known as a **normal distribution** and its graph is called a **normal curve.** (page 48)

A histogram that displays "accumulated" frequencies which are obtained by adding frequencies for the lower intervals is called a **cumulative frequency histogram**. (page 51)

A **cumulative frequency polygon** is simply a line graph connecting a series of points that answers the question, "How many of the scores are less than or equal to a given score?" (page 51)

Cumulative frequency polygons are also known as **ogives**. (page 52)

A **cumulative relative frequency distribution** combines the cumulative frequency with the relative frequency. (page 52)

Circle charts or **pie charts** are commonly used to describe qualitative data. A circle is broken up into various categories of interest in the same way that a pie is sliced, where each category is assigned a certain percentage of the 360° of the total circle. (page 52)

A useful graphical description of quantitative date is the **stem-and-leaf diagram**, where each measurement is partitioned into two components—a **stem** and a **leaf**. (page 68)

Dotplots are also used to display numerical data. They are helpful in showing the relative position of numbers in a set of data or for comparing several sets of data. (page 71)

Statistical graphs can be used to misrepresent the true picture. This is done by using **truncated graphs**, where the vertical scales of two "comparable" graphs are not the same. Sometimes the vertical scale is truncated or cut off. (page 74)

Another misuse of statistics involves **scaling**, where the scales of two comparable graphs are not the same. (page 74)

An **index number** is a special form of ratio that is used to show percentage changes over a period of time. The earlier period is usually referred to as the **base period**. The best known index number is the **Consumer Price Index (CPI)** published by the U.S. Bureau of Labor Statistics. (page 79)

When the MINITAB stem-and-leaf program is used, the **depths** indicate the number of leaves (pieces of data) that lie on a given row or on earlier rows. (page 85)

Formulas to Remember

The following list summarizes the formulas given in this chapter:

1. Rules for grouping data (page 30)
2. Class interval width $= \dfrac{\text{largest measurement} - \text{smallest measurement}}{\text{number of intervals}}$ (page 28)
3. Relative frequency for class $i = \dfrac{f_i}{n}$ (page 29)
4. Rules for drawing histograms (page 30)
5. Rules for rounding decimals (page 56)
6. Rules for drawing bar graphs, frequency polygons, pie charts (pages 47, 60)
7. Rules for constructing stem-and-leaf diagrams (page 70)
8. Index number $= \dfrac{\text{given year's values}}{\text{base year's values}} \times 100$ (page 79)

Testing Your Understanding of This Chapter's Concepts

1. After surveying numerous families in Breyerville to determine their monthly telephone bills, several students set up the following class intervals. Is anything wrong?

Monthly Bill	Frequency
$10–15	22
15–25	39
25–35	57
35–45	61
Over $45	89

2. When frequency tables and histograms are constructed, must we include a class if its frequency is zero?

3. *True or false.* Every ogive starts on the left with a relative frequency of zero at the lower class boundary of the first class and ends on the right with a relative frequency of 100% at the upper class boundary of the last class. Explain your answer.

4. An FAA official is interested in determining the class interval widths for constructing a frequency distribution on the number of late arriving planes. The following information is available:

Number of Minutes Late

Largest value	112 minutes
Smallest value	0 minutes
No. of intervals	12

What should the class width of each interval be?

THINKING CRITICALLY

1. An inexperienced statistician constructed the following frequency distribution of the salary of 6229 people. Is there anything wrong with the designations for class intervals?

Interval	Frequency
Under $10,000	269
$11,000–$15,000	1384
$16,000–$25,000	2469
$26,000–$35,000	1201
$36,000–$45,000	861
$46,000–$55,000	45

2. The MINITAB statistical package was used to generate the following computer printout of a histogram:

Midpoint	Count	
30	4	****
35	6	******
40	9	*********
45	8	********
50	4	****
55	3	***
60	7	*******
65	5	*****
71	1	*

a. How many pieces of data are described by this histogram?
b. What type of histogram is displayed—relative, frequency, or cumulative histogram? Explain your answer.
3. A university in Texas has three female faculty members in the mathematics de-

partment. Recently, one of them married one of her students. The student news-paper then printed an article with the following headline:

"33 ⅓% OF OUR FEMALE MATHEMATICS FACULTY MEMBERS MARRY THEIR STUDENTS"

What, if anything, is wrong with this headline?

4. Refer back to the newspaper article given at the beginning of this chapter concerning American cars and fuel economy.
   a. Is the picture presented by this graph fair? Explain your answer.
   b. What should the height of the 1987 bar be as compared to the height of the 1979 bar?
   c. From these data, would you conclude that American car makers have doubled their average fuel economy? Explain.

Review Exercises for Chapter 2

1. The following data were collected from 30 customers regarding their preferences on a scale from 1 to 100 about a Chevrolet Impala. The data are listed below. Fill in the table. Use six classes. Draw the histogram and frequency polygon.

Data		Class No.	Interval	Midpoint	Tally	Frequency
58	87					
29	19					
2	11					
73	23					
51	35					
17	80					
62	53					
74	33					
19	17					
23	79					
87	21					
89	42					
72	77					
38	28					
71	12					

2. Despite oil spills, sewage discharges, and continuing development, many waters around the New York–New Jersey harbors are suddenly beginning to show signs of life. The water quality for several regions in the harbor is shown in the following graph. Draw a bar graph to show the improvement in oxygen standards for the Lower East River over the years indicated.

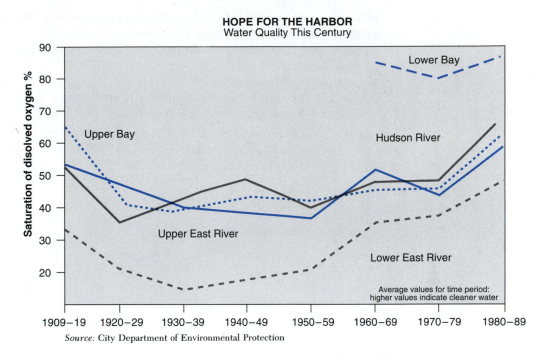

HOPE FOR THE HARBOR
Water Quality This Century

Source: City Department of Environmental Protection

3. According to the U.S. Burea of the Census, in 1985 there were more than 29 million children under age 15 who had mothers that worked. Furthermore, 19 million of these children had mothers who worked full-time. The method by

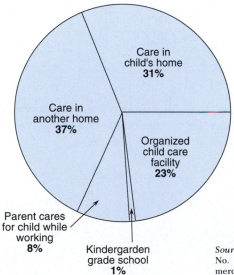

Source: Whose Minding the Kids? *Data User News,* Vol. 22, No. 6, Bureau of the Census, U.S. Department of Commerce, June 1967.

which the estimated 8 million preschool children were cared for while their mothers worked is shown in the pie chart on page 91.

Approximately how many preschoolers with working mothers were cared for in
a. their own home?
b. organized child-care facilities?
c. neither their own home nor in another home?

For questions 4–7, use the following information.

Automobile manufacturers in the United States today are trying to design cars that give the greatest mileage per gallon of gas. The following mileage results were obtained by 16 test versions of a particular model:

43   41   38   39   45   40   36   43
42   44   43   37   38   38   36   41

4. Complete the following table.

Mileage	Frequency	Relative Frequency
36–37		
38–39		
40–41		
42–43		
44–45		

5. Draw the frequency histogram and relative frequency histogram based on the data.
6. Draw a stem-and-leaf diagram for the data.
7. Draw a dotplot for the data.
8. The number of new housing permits issued for each day in February is as follows:

32   34   31   46   27   31
17   19   16   21   39   36
14   21   17   19   32   28
29   18   16   37   22   30
18   29   26   24   31

Construct a frequency distribution using 12 intervals and then draw its histogram.

9. The average cost for the same type of auto insurance coverage over the years is as follows.

Year	Cost ($)
1982	822
1983	864
1984	912
1985	939
1986	1141
1987	1202
1988	1288
1989	1301
1990	1463

    a. Draw a bar graph to picture this information.

    b. Draw a frequency polygon to picture this information.

10. Refer to Question 9. Using 1982 as a base year, compute price indexes for each of the years 1983 to 1990. Comment.

Chapter Test

1. The number of paramedics enrolled in the CPR program at Washington Hospital during July 1990 was as follows.

Period	Number of Paramedics Enrolled
July 1–5	91
6–10	80
11–15	76
16–20	62
21–25	73
26–30	40

Source: Washington Hospital Medical Records for 1990

Draw a bar graph to picture this information.

2. The population of the world was over 4 billion people in 1984 as shown here.

Location	Population (in millions)
Africa	513
Asia	2730
Europe	489
Latin America	390
North America	259
Oceania	24
USSR	272

Draw a pie chart to picture these data.

3. Several students in a probability class tossed ten coins many times and obtained the following results.

Number of Heads Obtained	Frequency
0	1
1	2
2	5
3	10
4	19
5	26
6	19
7	10
8	5
9	2
10	1

Draw the histogram for the distribution of the number of heads obtained and then draw the frequency polygon by using a line graph connecting the midpoints of the upper parts of each bar. What type of curve does the frequency polygon resemble?

4. The number of rooms rented by a large motel chain at 50 of its locations on August 2, 1990 was as follows:

63, 104, 98, 84, 107, 73, 58, 86, 83, 91, 54, 76, 80, 53, 88, 92, 85, 105, 79, 80, 92, 71, 89, 74, 69, 104, 97, 85, 55, 88, 74, 82, 93, 91, 104, 83, 101, 71, 73, 96, 106, 76, 85, 102, 63, 64, 105, 98, 86, 109

a. Draw a stem-and-leaf diagram for these data.
b. Complete the following frequency distribution and then draw the ogive (cumulative frequency distribution) for the data.

Interval	Tally	Cumulative Frequency
50–59		
60–69		
70–79		
80–89		
90–99		
100–109		

5. The cost for installing a particular type of auto alarm system in one northwestern city has been increasing as shown here.

Year	Cost for Alarm System
1984	$341
1985	363
1986	387
1987	399
1988	407
1989	419
1990	439

Find index numbers for the price charged over the years. Use 1984 as a base year.

6. The bar graph below shows how the population of the United States has changed from 1790 to 1990.
   a. How many more people were there in the United States in 1940 than in 1880?
   b. How many more people were there in the United States in 1970 than in 1950?

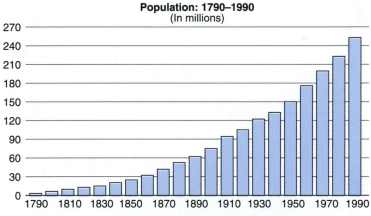

Population: 1790–1990
(In millions)

Source: U.S. Department of Commerce, Bureau of the Census, *Statistical Abstract of the United States*, 1991.

7. Draw a bar graph to picture the following data on the percent change of the U.S. population by region for 1980 to 1990.

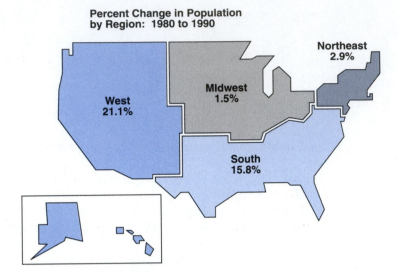

Percent Change in Population by Region: 1980 to 1990

8. Although the number of housing units in the United States has been increasing over the years, the average number of persons per household has been decreasing as can be seen from the following graph. How do you explain these trends?

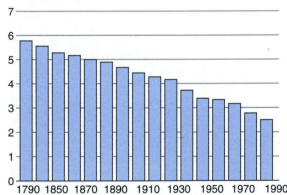

Average Number of Persons per Household for Selected Years

9. The following pictographs indicate that the total amount of sales tax collected in Boyerville during 1989 is double that collected in 1980. How should it be modified so that it will convey a fair impression of the actual change?

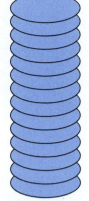

7 million dollars
1980

14 million dollars
1989

10. Which of the following is (are) likely to be normally distributed?
    a. The weight of joggers
    b. The useful life of an auto battery
    c. The drying time for paint

11. Consider the following frequency distribution. Find the class mark for interval 5.

Interval	Score	Frequency
1	20–29	12
2	30–39	16
3	40–49	28
4	50–59	13
5	60–69	11
6	70–79	6
		86

    a. 60   b. 69   c. 64.5   d. 11   e. 11/86

12. When constructing frequency polygons, the relative frequency of a class is defined as the frequency of that class divided by
    a. the total number of measurements
    b. the class mark
    c. its probability
    d. the class number
    e. none of these

13. Consider the following frequency distribution. What is the relative frequency of the score of 61?

Score	Frequency
12	11
17	10
19	16
41	14
58	31
61	82
82	69
	233

   a. 82   **b.** 82/233   **c.** 82/61   **d.** 61/82   **e.** 61/233

14. Refer back to question 13. What is the relative frequency of the score of 31?
    a. 58/233   **b.** 31/233   **c.** 233/31   **d.** 233/58   **e.** 0

15. If the midpoints of the tops of the bars of a bar graph or histogram are joined by straight lines, then the resulting figure, without the bars, is called a
    a. normal curve
    b. circle graph
    c. frequency polygon
    d. pictograph
    e. none of these

Suggested Further Reading

1. Folks, J. Leroy. *Ideas of Statistics*. New York: John Wiley, 1981.
2. Hamburg, Morris. *Statistical Analysis for Decision Making*, 4th ed. Orlando, FL: Harcourt, Brace, Jovanovich, 1987.
3. Hooke, Robert. *How To Tell Liars From Statisticians*. New York: Marcel Dekker, Inc., 1983.
4. Huff, D. *How to Lie with Statistics*. New York: W.W. Norton, 1954.
5. McClave, J. T. and P. G. Benson. *Statistics for Business and Economics*, 4th ed. San Francisco: Dellen, 1988.
6. Newmark, Joseph, and Frances Lake. *Mathematics as a Second Language*, 4th ed. Reading, MA: Addison-Wesley, 1987.
7. Tanur, J. M., F. Mosteller, W. H. Kruskul, R. F. Link, R. S. Pieters, and G. R. Rising. *Statistics: A Guide to the Unknown*. San Francisco: Holden-Day, 1978.
8. Tukey, J. W. *Exploratory Data Analysis*. Reading, MA: Addison-Wesley, 1977.

3 Numerical Methods for Analyzing Data

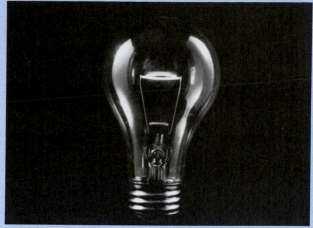

Often to compare brands, we must calculate the mean life of a light bulb. See discussion on p. 108. (© *Herb C. Ohlmeyer, Fran Heyl Associates*)

In this chapter we discuss arithmetic methods of summarizing data. We compute various numbers such as the mean, standard deviation, and so on. In turn, these descriptive measures enable us to analyze a data set adequately.

CHAPTER OBJECTIVES

After studying the material in this chapter, you should be able:

- **To work with** summation notation, which is used to denote the addition of numbers. (Section 3.2)

- **To discuss** the mean, median, and mode as three different ways of measuring some general trend or location of the data. These will be called measures of central tendency. (Section 3.3)

- **To calculate** the range, standard deviation, variance, and average deviation. Even when we know the mean, median, or mode, we might still want to know whether the numbers are close to each other or whether they are spread out. These are called measures of dispersion or measures of variation. (Section 3.4)

- **To apply** several shortcut formulas for calculating the mean, variance, and standard deviation for grouped or ungrouped data. (Section 3.5)

- **To find** an interpretation for the standard deviation. (Section 3.6)

- **To understand** Chebyshev's theorem, which gives us some information about where many terms of a distribution will lie. (Section 3.6)

- **To draw** the graphs of cumulative frequency distributions and the graphs of cumulative relative frequency distributions. These are called ogives. (Section 3.7)

- **To analyze** the idea of percentiles, which give us the relative standing of one score when compared to the rest. Specifically, percentiles tell us what percent of the scores are above or below a given score. (Section 3.7)

- **To compute** z-scores. These give us the relative position of a score with respect to the mean and are expressed in terms of standard deviations. (Section 3.8)

Statistics in the News

MUST WE SPEED?

Nov. 10—A new study conducted by the Department of Transportation found that more and more drivers are ignoring the states' legal 55 mph speed limit on the highways and freeways. A survey of 500 cars found them travelling at an average speed of 63 mph. Stricter enforcement of existing speed limits was urged.

November 10, 1989

CREDIT CARD INTEREST STILL AN AVERAGE 18%

New York—(Dec. 2) Despite the recent sharp drop in interest rates, many of our country's major lending institutions have still not lowered the annual interest rates charged to their credit card customers. A survey of 100 banks nationwide disclosed that the average annual interest rate charged was 18%. Several U.S. senators have called this legal usury.

December 2, 1988

Note the use of the word "average" in both newspaper articles. Is the word "average" used in the same manner in both instances? How do we calculate such averages?

Now consider the last newspaper article shown below. In this case the word "median" is used. Exactly what is meant by a median age? Is median age the same as average age? In this chapter we analyze the meaning of these words.

CENSUS SAYS U.S. MEDIAN AGE A RECORD 30.8

Washington—(Dec. 28) Americans are older than ever. Their median age rose to a record 30.8 years as of 1990 the Census Bureau reported in a population profile of the nation released yesterday. While record-high at 30.8 years, the median U.S. age is exceeded by other countries. In 1980, for instance, the median was 36.6 in West Germany and 34.4 in Great Britain. It was 29.4 in the Soviet Union, 23 in China, 19.9 in Brazil, 17.4 in Mexico and only 14.7 in Kenya.

December 28, 1990

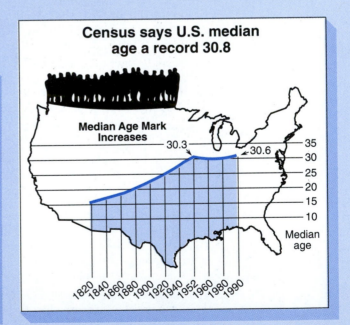

Census says U.S. median age a record 30.8

3.1 INTRODUCTION

In the preceding chapter we learned how to summarize data and present it graphically. Although the techniques discussed there are quite useful in describing the features of a distribution, statistical inference usually requires more precise analysis of the data. In particular, we will discuss many measures that locate the *center* of a distribution of a set of data and analyze how dispersed or spread out the distribution is. In this chapter we examine measures of central tendency and measures of variation (dispersion).

Most analyses involve various arithmetic computations that must be performed on the data. In each case the operation of addition plays a key role in these calculations. It is for this reason that we introduce a special shorthand notation called **summation notation.**

Now consider the card from the Emerson Medical Laboratory shown below.

Summation Notation

Emerson Medical Laboratory **Staten Island, N.Y.**			
Smith, Mary Ellen	F	32	Nov 1, 1986
patient's name	sex	age	date

	Triglycerides	Cholesterol
Results	109	256
Percentile rank	79	68

The card gives Mary Ellen Smith her percentile rank for triglycerides and cholesterol. How should she interpret this result? Is it better to have a higher or lower percentile rank? It should be apparent that we cannot answer this unless we understand the meaning of percentiles and learn how to calculate them. This will be done in this chapter.

After the data have been analyzed numerically, the techniques of statistical inference can be applied. These will be discussed in later chapters.

3.2 SUMMATION NOTATION

Often when working with a distribution of many numbers, we use letters with subscripts, that is, small numbers attached to them. Thus, we write x_1, x_2,

$x_3, \ldots, x_n$, which is read "x sub one, x sub two, x sub three, $\ldots$, x sub n." To be specific, consider the following set of numbers:

7, 15, 5, 3, 9, 8, 14, 21, 10

Here we let x_1 denote the first number; that is, $x_1 = 7$, and let x_2 denote the second number; that is, $x_2 = 15$. Similarly, $x_3 = 5$, $x_4 = 3$, $x_5 = 9$, $x_6 = 8$, $x_7 = 14$, $x_8 = 21$, and $x_9 = 10$.

To indicate the operation of taking the sum of a sequence of numbers, we use the Greek symbol Σ, which is read as sigma. To add all the x's, we write

$$\sum_{i=1}^{9} x_i$$

This tells us to add all consecutive values of x starting with x_1 and proceeding to x_9. Thus,

$$\sum_{i=1}^{9} x_i = x_1 + x_2 + x_3 + x_4 + x_5 + x_6 + x_7 + x_8 + x_9$$
$$= 7 + 15 + 5 + 3 + 9 + 8 + 14 + 21 + 10$$
$$= 92$$

If we only wanted to add $x_1 + x_2 + x_3 + x_4 + x_5$, we would write $\sum_{i=1}^{5} x_i$. The i

Index

(or j) in the summation symbol $\sum_{i=1}^{n}$ is referred to as the **index.**

Lower Limit of Summation

Upper Limit of Summation

The **lower limit of summation** is the value of the index placed below the summation symbol and the **upper limit of summation** is the value of the index placed above the summation symbol. In $\sum_{i=1}^{4} i$ the lower limit is 1 and the upper limit is 4.

Of course, $\sum_{i=1}^{4} i$ means the sum of the integers designated by i, from $i = 1$ to $i = 4$, so that $\sum_{i=1}^{4} i = 1 + 2 + 3 + 4 = 10$.

Throughout our study of statistics we will have need to work with various applications of the summation Σ symbol. Great care must be exercised when using this symbol. For example, using the Σ notation, we have the following:

$$\sum_{i=1}^{n} x_i^2 = x_1^2 + x_2^2 + x_3^2 + \cdots + x_n^2$$

$$\sum_{i=1}^{n} x_i y_i = x_1 y_1 + x_2 y_2 + \cdots + x_n y_n$$

$$\sum_{i=1}^{n} x_i^2 f_i = x_1^2 f_1 + x_2^2 f_2 + \cdots + x_n^2 f_n$$

COMMENT The symbols $\sum_{i=1}^{n} x_i^2$ and $\left(\sum_{i=1}^{n} x_i\right)^2$ are quite different. The symbol $\sum_{i=1}^{n} x_i^2$ means that we first square the numbers and add them together, whereas the symbol $\left(\sum_{i=1}^{n} x_i\right)^2$ means that we add all the x_i's together to obtain a sum and then square the sum.

Let us illustrate the use of the Σ symbols with several examples.

EXAMPLE 1

Find $\sum_{i=1}^{5} x_i^2$ and $\left(\sum_{i=1}^{5} x_i\right)^2$ for the following data:

$$x_1 = 3, \qquad x_2 = 7, \qquad x_3 = 6, \qquad x_4 = 3, \qquad x_5 = 9$$

Solution

$$\sum_{i=1}^{5} x_i^2 \text{ means } x_1^2 + x_2^2 + x_3^2 + x_4^2 + x_5^2$$

Thus,

$$\sum_{i=1}^{5} x_i^2 = 3^2 + 7^2 + 6^2 + 3^2 + 9^2$$
$$= 9 + 49 + 36 + 9 + 81$$
$$= 184$$

Also $\left(\sum_{i=1}^{5} x_i\right)^2$ means $(x_1 + x_2 + x_3 + x_4 + x_5)^2$.

Thus,

$$\left(\sum_{i=1}^{5} x_i\right)^2 = (3 + 7 + 6 + 3 + 9)^2$$
$$= 28^2$$
$$= 784$$

Therefore, $\sum_{i=1}^{5} x_i^2$ and $\left(\sum_{i=1}^{n} x_i\right)^2$ are quite different.

NOTATION Throughout this text, whenever we use the Σ notation, we will want to add *all* available data. Therefore, to simplify the formulas, we will sometimes write the summation symbol without any index. *Thus, when no index is indicated, it is understood that all the data are being used.*

COMMENT In Example 1 we can write $\sum_{i=1}^{5} x_i^2$ and $\left(\sum_{i=1}^{5} x_i\right)^2$ as $\sum x^2$ and $\left(\sum x\right)^2$ since we are using all available data.

EXAMPLE 2

Using the data given below, find (a) $\sum x$, (b) $\sum y$, (c) $\sum xy$, (d) $\sum x \cdot \sum y$.

x	3	5	6	8	11
y	8	7	9	10	12

a. $\sum x$ means $3 + 5 + 6 + 8 + 11 = 33$. Thus, $\sum x = 33$.
b. $\sum y$ means $8 + 7 + 9 + 10 + 12 = 46$. Thus, $\sum y = 46$.
c. To find $\sum xy$, we must first find all the products of the corresponding x and y values and then add these products together. Thus, we have

$$\sum xy = 3(8) + 5(7) + 6(9) + 8(10) + 11(12)$$
$$= 24 + 35 + 54 + 80 + 132$$
$$= 325$$

Therefore, $\sum xy = 325$
d. The symbol $\sum x \cdot \sum y$ means the product of the two summations $\sum x$ and $\sum y$. From parts (a) and (b) we already know that $\sum x = 33$ and $\sum y = 46$ so that

$$\sum x \cdot \sum y = 33(46) = 1518$$

EXAMPLE 3

If $x_1 = 7$, $x_2 = 3$, $x_3 = 9$, $x_4 = 4$, $f_1 = 5$, $f_2 = 1$, $f_3 = 8$, and $f_4 = 2$, find

a. $\sum_{i=1}^{4} x_i$ b. $\sum_{i=1}^{4} f_i$ c. $\sum_{i=1}^{4} x_i f_i$.

Solution

a. $\sum_{i=1}^{4} x_i$ means $x_1 + x_2 + x_3 + x_4$ so that

$$\sum_{i=1}^{4} x_i = 7 + 3 + 9 + 4 = 23$$

b. $\sum_{i=1}^{4} f_i = f_1 + f_2 + f_3 + f_4 = 5 + 1 + 8 + 2 = 16$

c. $\displaystyle\sum_{i=1}^{4} x_i f_i = x_1 f_1 + x_2 f_2 + x_3 f_3 + x_4 f_4$

$$= 7(5) + 3(1) + 9(8) + 4(2) = 118$$

When we work with summations, there are certain rules that we use. These rules are easily verified by simply writing out in full what each of the summations represents.

RULE 1

$$\sum_{i=1}^{n} (x_i + y_i) = \sum_{i=1}^{n} x_i + \sum_{i=1}^{n} y_i$$

RULE 2

$$\sum_{i=1}^{n} kx_i = k \cdot \sum_{i=1}^{n} x_i$$

RULE 3

$$\sum_{i=1}^{n} k = n \cdot k$$

In words, the first rule says that the summation of a sum of two terms equals the sum of the individual summations. The second rule says that we can "factor" a constant out from under the operation of a summation. Thus,

$$\sum_{i=1}^{n} kx_i = kx_1 + kx_2 + kx_3 + \cdots + kx_n$$

$$= k(x_1 + x_2 + \cdots + x_n)$$

$$= k \cdot \sum_{i=1}^{n} x_i$$

The third rule says that the summation of a constant is simply n times that constant or the constant times the number of indicated terms in the summation.

These rules are easy to use and will be applied throughout the book.

EXAMPLE 4

If $x_1 = 4$, $x_2 = 5$, $x_3 = 7$, and $x_4 = 9$, find $\displaystyle\sum_{i=1}^{4} (3x_i - 1)$.

Solution

$$\sum_{i=1}^{4} (3x_i - 1) = (3x_1 - 1) + (3x_2 - 1) + (3x_3 - 1) + (3x_4 - 1)$$

$$= (3 \cdot 4 - 1) + (3 \cdot 5 - 1) + (3 \cdot 7 - 1) + (3 \cdot 9 - 1)$$
$$= 11 + 14 + 20 + 26$$
$$= 71$$

Thus, $\sum_{i=1}^{4} (3x_i - 1) = 71$ for the previous values of x_i.

COMMENT   Example 4 can also be evaluated by using Rules 1 to 3. Can you see how?

EXERCISES FOR SECTION 3.2

1. Rewrite each of the following using summation notation:
   a. $y_1 + y_2 + y_3 + y_4 + y_5 + y_6 + y_7 + y_8$
   b. $y_1^2 + y_2^2 + y_3^2 + y_4^2 + y_5^2 + y_6^2 + y_7^2 + y_8^2$
   c. $y_1 f_1 + y_2 f_2 + y_3 f_3 + y_4 f_4 + y_5 f_5 + y_6 f_6 + y_7 f_7 + y_8 f_8$
   d. $19y_1 + 19y_2 + 19y_3 + 19y_4 + 19y_5 + 19y_6 + 19y_7$
   e. $(3y_1 + x_1) + (3y_2 + x_2) + (3y_3 + x_3) + \cdots + (3y_n + x_n)$

2. Rewrite each of the following without summation notation:
   a. $\sum_{i=1}^{7} y_i$
   b. $\sum_{i=4}^{9} x_i y_i$
   c. $\sum_{i=4}^{11} (5x_i + 2)$

3. If $x_1 = 5$, $x_2 = 19$, $x_3 = 8$, $x_4 = 29$, and $x_5 = 25$, find each of the following:
   a. Σx
   b. $(\Sigma x)^2$
   c. Σx^2
   d. $\Sigma(x + 1)$
   e. $\Sigma(x + 1)^2$
   f. $\Sigma(2x + 3)$
   g. $\Sigma(2x + 3)^2$

4. If $x_1 = 5$, $x_2 = 8$, $x_3 = 14$, $x_4 = 34$, $f_1 = 3$, $f_2 = 7$, $f_3 = 9$, $f_4 = 12$, $y_1 = 2$, $y_2 = 4$, $y_3 = -3$ and $y_4 = -6$, find each of the following:
   a. Σx^2
   b. $\Sigma x \cdot f$
   c. $\Sigma x \cdot \Sigma y$
   d. Σxy
   e. $\Sigma x \cdot y \cdot f$
   f. $\Sigma(x - y)f$

5. If $\sum_{i=1}^{10} x_i = 27$ and $\sum_{i=1}^{10} x_i^2 = 50$, find each of the following:
   a. $\sum_{i=1}^{10} (x_i + 9)$
   b. $\sum_{i=1}^{10} (x_i + 9)^2$

6. Show that each of the following are true:

a. $\displaystyle\sum_{i=1}^{n} (x_i - y_i) = \sum_{i=1}^{n} x_i - \sum_{i=1}^{n} y_i$

b. $\displaystyle\sum_{i=1}^{n} (x_i + k) = \sum_{i=1}^{n} x_i + nk$

7. Professor Lorenz requires each of the students in her English composition course to submit five different essays during the semester. In the following frequency distribution she has recorded the number of students who have completed 1, 2, 3, 4, or 5 of these assignments.

Number of Essays Completed x	Frequency f	xf
1	14	14
2	6	12
3	8	24
4	9	36
5	10	50
Total =	47	136

a. The sum $\Sigma f = 47$. What does this sum represent?
b. The sum $\Sigma xf = 136$. What does this sum represent?

3.3 MEASURES OF CENTRAL TENDENCY

To help us understand what we mean by measures of central tendency, consider the Metropolis Police Department, which recently purchased tires from two different manufacturers. The police department is interested in determining which tire is superior and has compiled a list on the number of miles each set of tires lasted before replacement was needed for 14 of its identical police cars. Seven cars were fitted with Brand X tires and seven with Brand Y. The number of miles each set lasted before replacement is indicated in the following chart.

	Brand X	Brand Y
	14,000	10,000
	12,000	8,000
	12,000	14,000
	14,000	10,000
	14,000	8,000
	11,000	40,000
	14,000	8,000
Total =	91,000	98,000

It would appear that Brand Y is the better tire since the seven police cars were driven a combined total of 98,000 miles using Brand Y tires but only 91,000 miles with Brand X tires. Let us, however, analyze the data by computing the average number of miles driven with each brand of tires. We will divide each total by the number of police cars used for each brand. We have the following.

Average

Brand X	Brand Y
$\dfrac{91,000}{7} = 13,000$	$\dfrac{98,000}{7} = 14,000$

Since Brand Y tires lasted on the average 14,000 miles and Brand X on the average 13,000 miles, it again appears that Brand Y is superior. If we look at the data more carefully, however, we find that Brand X tires consistently lasted around 13,000 miles. As a matter of fact, they lasted 14,000 miles most often, four times. Brand Y, on the other hand, lasted 14,000 miles only once. They lasted 8000 miles most often. Thus, in terms of consistency of performance one might say that Brand X is more consistent. Let us arrange the data for each brand in order from smallest to largest. We have the following.

Brand X	Brand Y
11,000	8,000
12,000	8,000
12,000	8,000
(14,000)	(10,000)
14,000	10,000
14,000	14,000
14,000	40,000

In this chart we have circled two numbers. These are the numbers that are in the middle for each brand: 14,000 for Brand X and 10,000 for Brand Y.

All the preceding ideas are summarized in the following definitions.

DEFINITION 3.1
Mean or Average

The **mean,** or **average,** of a set of numbers is obtained by adding the numbers together and dividing the sum by the number of numbers added. We denote the mean by the symbol $\bar{x}$, which is read x bar, or by the Greek letter μ, which is read mu, depending on the situation. The use of each symbol will be explained shortly.

DEFINITION 3.2
Mode

The **mode** of a set of numbers is the number (or numbers) that occurs most often. If no number occurs more than once, there is no mode.

| DEFINITION 3.3 Median | The **median** of a set of numbers is the number that is in the middle when the numbers are arranged in order from the smallest to the largest. The median is easy to calculate when we have an odd number of numbers. If we have an even number of them, the median is defined as the average of the middle two numbers when arranged in increasing order of size. We denote the median by the symbol $\tilde{x}$, read "x tilde." |

COMMENT A set of numbers may have more than one mode. (See Example 1 of this section.)

When the preceding definitions are applied to our example, we get the following table.

	Brand X	Brand Y
Mean	13,000	14,000
Median	14,000	10,000
Mode	14,000	8,000

Which number should the police department use to determine the superior tire: the mean, median, or mode? You might say that the mean is not particularly helpful since one set of Brand Y tires lasted 40,000 miles and this instance had the effect of increasing the average for Brand Y considerably. In terms of consistency Brand X appears to be superior to Brand Y.

Notice that in calculating the mean we had to add numbers. Since the operation of addition plays a key role in our calculations, we use the summation notation discussed in the last section.

Consider the following set of numbers:

7, 15, 5, 3, 9, 8, 14, 21, 10

We have

$$\Sigma x = x_1 + x_2 + x_3 + x_4 + x_5 + x_6 + x_7 + x_8 + x_9$$
$$\Sigma x = 7 + 15 + 5 + 3 + 9 + 8 + 14 + 21 + 10$$
$$\Sigma x = 92$$

If we divide this sum by n, the number of terms added, the result will be the mean for these numbers.

COMMENT Remember that usually no indexes will be shown in the summation formulas in this book: The summations are understood to be over all available data.

FORMULA 3.1 Sample Mean

The mean of a set of sample values $x_1, x_2, x_3, \ldots, x_n$ is given by

Sample Mean

$$\text{Sample mean} = \bar{x} = \frac{\Sigma x}{n}$$

$$= \frac{x_1 + x_2 + x_3 + \cdots + x_n}{n} \quad \begin{array}{l} \leftarrow \text{sum of the data} \\ \leftarrow \text{number of pieces of data} \end{array}$$

For our set of numbers the mean is

$$\bar{x} = \frac{\Sigma x}{n} = \frac{92}{9} = 10.22$$

EXAMPLE 1 The Better Worker

The district office of a state unemployment insurance department recently hired two new employees, Rochelle and Sharon, to interview prospective aid recipients. Their supervisor is interested in determining who is the better worker. The following chart indicates the number of clients interviewed daily by each on seven randomly selected days.

Rochelle	Sharon
54	38
67	51
46	46
52	49
45	46
39	38
41	44
344	312

Calculate the sample mean, median, and mode for each employee.

Solution

Let x represent the number of clients interviewed daily by Rochelle and let y represent the number of clients interviewed daily by Sharon. For Rochelle the sample mean is

$$\bar{x} = \frac{\Sigma x}{n} = \frac{x_1 + x_2 + x_3 + x_4 + x_5 + x_6 + x_7}{7}$$

$$= \frac{54 + 67 + 46 + 52 + 45 + 39 + 41}{7}$$

$$= \frac{344}{7} = 49.14$$

For Sharon the sample mean is

$$\bar{y} = \frac{\Sigma y}{n} = \frac{y_1 + y_2 + y_3 + y_4 + y_5 + y_6 + y_7}{7}$$

$$= \frac{38 + 51 + 46 + 49 + 46 + 38 + 44}{7}$$

$$= \frac{312}{7} = 44.57$$

Let us now arrange the numbers for each employee in order from the lowest to the highest. We get the following.

Rochelle	39 41 45 (46) 52 54 67
Sharon	38 38 44 (46) 46 49 51

Notice that one number has been circled for each worker. This number is in the middle and it represents the median. For both workers the median is 46.

The mode is the number that occurs most often. For Rochelle there is no mode since no number occurs more than once. For Sharon there are two modes, 38 and 46.

Which statistic, the mean, mode, or median, should the supervisor use in determining who is the better worker?

EXAMPLE 2 The personnel manager of the Manhattan Detective Agency has compiled the following list on the number of years several former employees worked before retiring.

Worker	Number of Years of Service Before Retiring
Fay	43
Renée	38
Trudy	47
Hilda	35
Jack	42
Pedro	41
José	39
Beebabats	31

Find the mean, median, and mode.

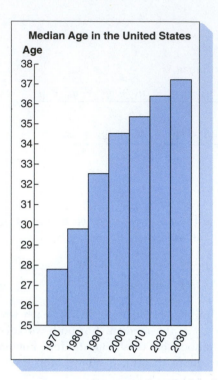

Note the use of the word "median" in this newspaper clipping. The information contained in it has important implications for our future.

Solution

We first arrange the numbers in order from the lowest to the highest. The median is the number that is in the middle. Since there are an even number of terms, the median is between the two circled numbers, 39 and 41.

31,   35,   38,   ③⑨,   ④①,   42,   43,   47

Definition 3.3 tells us that it is the average of 39 and 41. Thus, we have

$$\bar{x} = \text{median} = \frac{39 + 41}{2} = 40$$

Since no number occurred more than once, there is no mode. The mean is obtained by dividing the total sum, which is 316, by the number of terms, 8:

$$\text{Mean} = \frac{316}{8} = 39.5$$

Summarizing, we have

Mean: 39.5 years
Median: 40 years
Mode: none

Which is more useful to the personnel manager?

EXAMPLE 3

Calculate the sample mean of the following numbers:

28	19	25	17	28	19	26	17	28	25
17	28	31	22	31	17	31	28	31	14
31	14	17	28	24	31	17	14	26	24
24	24	19	24	14	28	22	31	17	22
25	12	26	19	26	12	19	26	19	28

Solution

One way of calculating the mean is to add the numbers and divide the result by 50, which is the number of terms. This takes a considerable amount of time, but nevertheless we have

$$\bar{x} = \frac{\Sigma x}{n} = \frac{x_1 + x_2 + \cdots + x_{50}}{n}$$

$$= \frac{28 + 19 + 25 + \cdots + 28}{50}$$

$$= \frac{1145}{50} = 22.9$$

Thus, the sample mean is 22.9.
We can also group the data (not necessarily in the usual interval format) as follows.

Column 1	Column 2	Column 3	Column 4
Number (x)	Tally	Frequency (f)	x · f
12	||	2	24
14	||||	4	56
17	⫫ ||	7	119
19	⫫ |	6	114
22	|||	3	66
24	⫫	5	120
25	|||	3	75
26	⫫	5	130
28	⫫ |||	8	224
31	⫫ ||	7	217
	Total = 50		1145

Column 4 was obtained by multiplying each number by its frequency. If we now sum column 3 and column 4 individually and divide the column 4 total by the column 3 total, our answer will be 22.9, which is the mean. Although we get the same result as before, it was considerably easier to obtain it by grouping the data as we did.

FORMULA 3.2

Sample Mean of a Distribution of Grouped Data

The **sample mean of a distribution of grouped data** is given by

$$\bar{x} = \frac{\Sigma \, xf}{\Sigma \, f}$$

where xf represents the product of each class mark and its frequency and $\Sigma \, f$ represents the total number of items in the distribution.

COMMENT It may seem that Formula 3.2 is considerably different from Formula 3.1 for calculating the sample mean. In reality it is not. In ungrouped data the frequency of each observation is 1 and the total number of terms, $\Sigma \, f$, is n. So, Formulas 3.1 and 3.2 are really the same.

To illustrate further the use of Formula 3.2, we consider the following example (where we have grouped the data in interval form).

EXAMPLE 4

George is a maintenance person in the Auburn Shopping Center. He keeps accurate records on the life of the special security light bulbs that he services. He has recorded the life of 60 light bulbs in the following chart.

Life of Bulb (hours)	Class Mark x	Frequency f	Product $x \cdot f$
40–49	44.5	5	222.5
50–59	54.5	7	381.5
60–69	64.5	8	516.0
70–79	74.5	9	670.5
80–89	84.5	12	1014.0
90–99	94.5	9	850.5
100–109	104.5	6	627.0
110–119	114.5	4	458.0
		Total = 60	4740

Calculate the sample mean life of a light bulb.

Solution

In using Formula 3.2, we use the class mark (see page 28) for each interval. Thus, we assume that any bulb that lasted between 40 and 49 hours actually lasted 44.5 hours. Although this may introduce a slight error, the error will be minimal when the number of bulbs is large. Thus, we use the class mark. Applying Formula 3.2, we have

$$\bar{x} = \frac{\Sigma \, xf}{\Sigma \, f} = \frac{222.5 + 381.5 + \cdots + 458.0}{5 + 7 + \cdots + 4} = \frac{4740}{60} = 79$$

In the previous example we assumed that any bulb that lasted between 40 and 49 hours actually lasted 44.5 hours, which is the class mark. Some statisticians prefer to compute medians (and percentiles, to be discussed later) for grouped data by assuming that the entries in a class are evenly distributed in that class. For example, suppose we were interested in calculating the median of a distribution for grouped data. These statisticians would then use the formula

$$\tilde{x} = L + \frac{j}{f} \cdot c$$

where L is the lower boundary or lower limit of the category into which the median must fall, f is its frequency, c is its interval length, and j is the number of items still to be counted after reaching L. To illustrate, suppose we were interested in computing the median for the grouped data given below.

(from)	Interval (up to but not including)	Frequency
9.5	19.5	12
19.5	29.5	10
29.5	39.5	13
39.5	49.5	11
49.5	59.5	7
		53

When the numbers are arranged in order of size (from the smallest to the largest), the median will be the 27th term. This falls in the 29.5 to 39.5 interval. Since there are $12 + 10$ or 22 below this category, we need another five items in addition to the 22 that fall below this class. We add $\frac{5}{13}$ of the class interval to the lower bound of that class. Here we have $L = 29.5$, $f = 13$, $c = 10$, and $j = 5$ so that

$$\tilde{x} = 29.5 + \frac{5}{13} \cdot 10$$

$$= 33.35 \quad \text{when rounded}$$

Now let us consider the following examples in which all values are not of equal importance. To solve such examples we must use Formula 3.3.

FORMULA 3.3	If $w_1, w_2, \ldots, w_n$ are the weights assigned to the numbers $x_1, x_2, \ldots, x_n$, then the **weighted sample mean** denoted by the symbol $\bar{x}_w$ is given by
Weighted Sample Mean	$$\bar{x}_w = \frac{\Sigma\, xw}{\Sigma\, w} = \frac{x_1 w_1 + x_2 w_2 + \cdots + x_n w_n}{w_1 + w_2 + \cdots + w_n}$$ This formula indicates that we multiply each number by its weight and divide the sum of these products by the sum of the weights.

EXAMPLE 5 Calculating Term Grades

The grades that Liz received on her exams in a statistics course and the weight assigned to each are as follows.

	Grade	Weight Assigned
Test 1	84	1
Test 2	73	2
Test 3	62	5
Test 4	91	4
Final exam	96	3

Find Liz's average term grade.

Solution

Since each test did not have the same weight, that is, count as much, Formula 3.1 or 3.2 has to be modified. This change is necessary because Formula 3.1 assumes that all numbers are of equal importance, which is not the case in this example. To calculate a weighted mean, we use Formula 3.3. When this formula is applied to our example, we have

$$\bar{x}_w = \frac{x_1 w_1 + x_2 w_2 + \cdots + x_5 w_5}{w_1 + w_2 + \cdots + w_5}$$

$$= \frac{(84 \cdot 1) + (73 \cdot 2) + (62 \cdot 5) + (91 \cdot 4) + (96 \cdot 3)}{1 + 2 + 5 + 4 + 3}$$

$$= \frac{1192}{15} = 79.47 \quad \text{when rounded off.}$$

Thus, the weighted mean is 79.47, not 81.2, which is obtained by adding the numbers together and dividing the sum by n, which is 5.

EXAMPLE 6 Average Cost of Gasoline

On a recent vacation trip Bill Hunt kept the following record of his gasoline purchases:

	Price per Gallon	Number of Gallons Purchased
Town 1	$1.53	17
Town 2	1.46	21
Town 3	1.49	16
Town 4	1.35	11
Town 5	1.51	19

What is the average cost per gallon of gasoline for the entire trip?

Solution

Since Bill did not purchase an identical amount of gasoline in each town, we cannot use Formula 3.1. We must use Formula 3.3 instead. We have

$$\bar{x}_w = \frac{(1.53)(17) + (1.46)(21) + (1.49)(16) + (1.35)(11) + (1.51)(19)}{17 + 21 + 16 + 11 + 19}$$

$$= \frac{124.05}{84} = \$1.48 \quad \text{rounded off}$$

Thus, the weighted average cost per gallon of gasoline for the entire trip was $1.48.

Midrange

There are other measures that describe in some way the middle or center of a set of numbers. One of these is known as the **midrange,** which is found by taking the average of the lowest value L and the highest value H. Thus,

$$\text{Midrange} = \frac{L + H}{2}$$

For the sample 3, 7, 11, 15, 16, 18, 21, 22, and 23, the lowest value L is 3 and the highest value H is 23. Therefore, the midrange is

$$\frac{3 + 23}{2} = 13$$

Geometric Mean

Harmonic Mean

Other measures that are sometimes used are the **geometric mean** and the **harmonic mean.** (See Exercise 17 at the end of this section.)

We mentioned earlier that both $\bar{x}$ and μ will be used to represent the mean. The symbol $\bar{x}$ is used to represent the sample mean or the mean of a sample. Thus, the mean $\bar{x}$ of the sample values $x_1, x_2, x_3, \ldots, x_n$ is given by the formula

$$\text{Sample mean } \bar{x} = \frac{\Sigma x}{n}$$

The symbol μ is used to represent the mean of the entire population. Thus the mean of a population of N values, $x_1, x_2, x_3, \ldots, x_N$ is given by the formula

$$\text{Population mean } \mu = \frac{\sum x}{N}$$

where the N values of x constitute the entire population.

COMMENT　Generally speaking, Greek letters are used when we are referring to a description of the population as opposed to English letters, which are used when we are referring to a description of a sample.

COMMENT　Throughout this section we used $\bar{x}$ since we were calculating sample means. From a given problem we can almost always tell whether we are referring to only part of the population or to the entire population.

Measures of Central Tendency

COMMENT　The mean, median, and mode are known as **measures of central tendency** since each measures some central or general trend, that is, location of the data. In any particular situation one measure will usually be more helpful than the others.

mode - # that occurs the most

EXERCISES FOR SECTION 3.3

mean - average $\bar{x}$
median - $\bar{x}$ = middle

1. *Average rainfall*. According to weather bureau records, the total amount of rainfall in Big Falls during 1990 was as follows.

Month	Rainfall (in inches)
January	39
February	42
March	38
April	39
May	36
June	25
July	25
August	19
September	23
October	27
November	31
December	35

19 23 25 25 27 31 35 36 38 39 42

31.58

379

Find the mean, median, and mode for the amount of rainfall. Which is more useful?

mean - 31.58
mode - 25, 39
median - 31, 35

2. *Home mortgages.* On February 8, 1991, 21 people applied to the National Commercial bank for a home mortgage loan. The amount of money applied for was

$68,000 $ 58,000 $61,000 $53,000 $48,500 $65,000
 51,000   49,000   60,000   55,000   43,000
 34,500   239,000   47,000   64,000   61,500   *35 19,000*
 79,000   61,000   49,500   70,000   69,500

Calculate the mean, median, and mode for these data.

3. Refer back to Exercise 2. Neglecting the loan for $239,000, calculate the mean, median, and mode for the remaining data. Compare the answers obtained with the answers obtained in Exercise 2. Comment.

 4. Mr. Potter is analyzing his telephone bills for the past nine months. They are as follows.

Month	Amount of Telephone Bill
January	$41.12
February	48.58
March	46.19
April	47.28
May	39.35
June	41.12
July	56.28
August	39.17
September	48.58

Calculate the mean, median, and mode for the amount of the telephone bill. Which is more useful?

 5. Sonya owns a car rental agency. Last year she leased six cars and paid $6800, $5700, $6300, $5995, $6100, and $5850 for these cars. Find the mean, median, and modal price for leasing a car.

6. Refer back to Exercise 5. This year Sonya again intends to lease six new cars. She is informed by the dealers that each car's price has been increased by $400. Find the mean, median, and modal price for leasing a new car under these new conditions.

 7. Mary is the manager of a dress store. Last week she sold 64 dresses. Their sizes were

 8 14 12 14 24 12 12 10
 8   8   8 18 20   8 10 12
 6   6   6 16 12 12   8 10
10 10 12 12 16 10 12 10
12 16 16 16 14 10 10 10
12 18 14 14   8 12 10 12
12 14 18 10 10 10 10 12
18 12 20   8   8   8 10 12

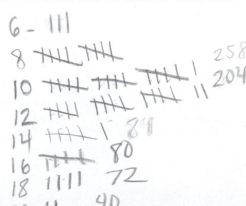

By grouping the data, find the mean, median, and mode. Which is more important to the management of the store?

8. The IQ of the 50 students in the fifth grade was measured. The results are as follows.

IQ score	Frequency
71–80	3
81–90	6
91–100	12
101–110	14
111–120	8
121–130	5
131–140	2

Find the mean, median, and mode for the data.

9. The Consumer's Affairs Department purchased a pack of a particular brand of cigarettes in 112 different stores in town to determine the average price charged. The results are as follows.

Number of Stores	Price Charged by These Stores
38	$1.70
64	1.75
7	1.80
3	1.85

Find the average price charged by all stores.

10. Consider the newspaper article on the next page. Find the average price of an admission ticket paid by these 100 people.

STATE TO INVESTIGATE
TICKET SCALPING

Dover: The Attorney General announced yesterday that he would launch an immediate investigation into the widespread practice of ticket scalping for tickets to rock concerts. A survey of 100 ticket holders for last night's rock concert indicated that these people paid exorbitant prices for the tickets. The prices paid for the same general admission ticket were as follows:

Number of people	Price paid
36	$30
28	35
19	40
17	45

The consumer is being victimized by this unlawful practice.

December 1, 1985

11. According to a survey, the starting salary of 50 graduates of a prestigious college was as follows.

Starting Salary	Frequency
$25,000–$29,999	3
30,000–34,999	6
35,000–39,999	12
40,000–44,999	14
45,000–49,999	8
50,000–54,999	5
55,000–59,999	2

Find the mean, median, and mode for the data.

12. An important number to hay fever sufferers is the pollen count. Health officials monitor the situation by taking daily pollen counts. The following data for Stevensville is available concerning the 1990 hay fever season.

Pollen Count Number	Frequency
1–5	6
6–10	13
11–15	14
16–20	17
21–25	19
26–30	15
31–35	8
36–40	6
41–45	3

Source: Lutheran Medical Center, Stevensville, 1990.

Find the mean, median, and mode for the data.

13. Professor Gonzales is teaching two sections of a statistics course. In one class the average midterm grade was 70 and in the second class the average midterm grade was 66. Is it safe to assume that the average midterm grade for both classes combined is 68? Explain your answer.

14. The combined weight of the 12 trucks on a bridge is 68,000 pounds. Find the average weight of a truck on the bridge.

15. There are nine checkout counters at a local supermarket. The number of people waiting to be checked out at these lines is 7, 3, 5, 4, 8, 4, 3, 3, and 6. Calculate what you consider to be the best "measure of central tendency." Which measure did you calculate? Why? Which is more important to management?

16. Three insurance companies claim that the average time that it takes their company to process a claim is five days. A consumer's group decides to test each insurance company's claim. It obtains the following list on the number of days needed by each company to process a claim.

Insurance Company *A* No. of Days Needed to Process 9 Different Claims	Insurance Company *B* No. of Days Needed to Process 11 Different Claims	Insurance Company *C* No. of Days Needed to Process 8 Different Claims
1 day	4 days	2 days
2 days	4 days	3 days
3 days	5 days	4 days
4 days	5 days	4 days
5 days	5 days	6 days
7 days	6 days	13 days
8 days	11 days	14 days
10 days	13 days	15 days
	14 days	
	15 days	
	16 days	

a. Which measurement of average was each insurance company using to support its claim?

b. From which insurance company would you buy insurance?

17. The **harmonic mean** of n numbers $x_1, x_2, \ldots, x_n$ is defined as n divided by the sum of the reciprocals of the numbers; that is,

$$\text{Harmonic mean} = \frac{n}{\sum\limits_{i=1}^{n} 1/x_i}$$

Also, the **geometric mean** of a set of n positive numbers $x_1, x_2, \ldots, x_n$ is defined as the nth root of their product; that is,

$$\text{Geometric mean} = \sqrt[n]{x_1 \cdot x_2 \cdots x_n}$$

Find the harmonic mean and the geometric mean of the numbers 8, 5, 9, and 6. (Both the harmonic mean and the geometric mean are used in certain applications.)

18. *Bank failures.* To insure a depositor's money, many banks belong to the Federal Deposit Insurance Corporation (FDIC). In the event of a bank failure the FDIC will insure each customer's deposits (currently) up to $100,000. According to government records, the number of banks that failed and were subsequently taken over by the FDIC in a certain region over the years is as follows.

Year	Number of Bank Failures
1980	2
1981	7
1982	6
1983	4
1984	9
1985	8
1986	6
1987	7
1988	5
1989	14

Outliers

Trimmed Mean

We notice that the preceding set of numbers contains two extremes (or **outliers**) that really should not be included in the data set for one reason or another. For situations such as these, statisticians will compute a **trimmed mean** where high and low values are excluded or "TRIMMED OFF" before calculating the mean. In our case if we exclude *both* the top 10% (the number 14) and the bottom 10% (the number 2) and then calculate the mean of the remaining data, we get

$$10\% \text{ trimmed mean} = \frac{7 + 6 + 4 + 9 + 8 + 6 + 7 + 5}{8} = 6.5$$

The following are the scores of 20 students on the mathematics part of the Scholastic Aptitude Test (SAT):

560 482 555 528 541 623 583 591 577 561
539 553 538 586 537 589 592 601 599 543

a. Calculate the arithmetic mean for the data.
b. Calculate the 5% trimmed mean for the data.
c. Calculate the 10% trimmed mean for the data.
d. Compare the three means that you obtained in parts (a) to (c). Which of the three means do you think is the best measure of central tendency for the data? Explain your answer.

19. Refer back to Exercise 18. Calculate the midrange for
a. the number of bank failures.
b. the mathematics SAT scores.

3.4 MEASURES OF VARIATION

Although the mean, median, or mode are very useful in analyzing a distribution, there are some disadvantages in using them alone. These measures only locate the center of the distribution. In certain situations location of the center may not be adequate. We need some method of analyzing variation, that is, the difference among the terms of a distribution. In this section we discuss some of the most commonly used methods for analyzing variation.

First let us consider Christina, who is interested in determining the best route to drive to work. During one week she drove to work on the Brooks Expressway and during a second week she drove on the Kingston Expressway. The number of minutes needed to drive to work each day was

| Brooks Expressway | 15 | 26 | 30 | 39 | 45 |
| Kingston Expressway | 29 | 30 | 31 | 32 | 33 |

In each case the average time that it took her to drive to work was 31 minutes. Which way is better?

When she used the Brooks Expressway, the time varied from 15 to 45 minutes. We then say that for the Brooks Expressway the *range* is $45 - 15 = 30$ minutes.

On the Kingston Expressway the time varied from 29 to 33 minutes. Thus, the range is $33 - 29 = 4$ minutes.

DEFINITION 3.4	The **range** of a set of numbers is the difference between the largest number in the
Range	distribution and the smallest number in the distribution.

The range is frequently used by manufacturers as a measure of dispersion (spread) in specifying the variation in the quality of a product. So, although the average diameter of a drill bit may be $\frac{15}{32}$ inches, in reality the range in size may be enormous. The manufacturer usually specifies the range to prospective customers.

The range is also used frequently by stock brokers to describe the prices of certain stock. One often hears such statements as "Stock X had a price range of 15 to 75 dollars, or 60 dollars, during the year."

The range is by far the simplest measure of variation to calculate since only two numbers are needed to calculate it; however, it does not tell us anything about how the other terms vary. Furthermore, if there is one extreme value in a distribution, the dispersion or the range will appear very large. If we remove the extreme term, the dispersion may become quite small. Because of this, other measures of variation such as variance, standard deviation, or average deviation are used.

To calculate the sample variance of a set of numbers, we first calculate the sample mean of the numbers. We then subtract the sample mean from each number and square the result. Finally, we divide the result by $n - 1$, where n is the number of items in the sample. The result is called the **sample variance** of the numbers. If we now take the square root of the sample variance, we get the **sample standard deviation** for the numbers.* If instead of squaring the differences from the mean we take the absolute value (that is, we neglect any negative signs) of these differences and find the average of these absolute values, the resulting number is called the **average deviation.** The symbol for absolute value is two vertical lines. Thus, $|+8|$ is read as "the absolute value of $+8$."

Let us illustrate the preceding ideas by calculating the sample variance, sample standard deviation, and average deviation for the two routes that Christina uses to drive to work. Since the sample mean, $\bar{x}$, is 31 we can arrange our calculations as shown in the following chart.

	Time (x)	Difference from Mean $(x - \bar{x})$	Square of Difference $(x - \bar{x})^2$	Absolute Value of Difference $|x - \bar{x}|$
	15	$15 - 31 = -16$	$(-16)^2 = 256$	16
By Way of	26	$26 - 31 = -5$	$(-5)^2 = 25$	5
Brooks	30	$30 - 31 = -1$	$(-1)^2 = 1$	1
Expressway	39	$39 - 31 = 8$	$8^2 = 64$	8
	45	$45 - 31 = \underline{14}$	$14^2 = \underline{196}$	$\underline{14}$
		$Sum = 0$	$Sum = 542$	$Sum = 44$

Therefore, if Christina travels to work by way of the Brooks Expressway, the sample variance is $\frac{542}{5 - 1}$, or 135.5; the sample standard deviation is $\sqrt{135.5}$ or approximately 11.64; and the average deviation is $\frac{44}{5}$, or 8.8.

---

*A knowledge of how to compute square roots is not needed. Such values can be obtained by using a calculator or a square root table.

Notice that in computing these measures of variation, we used symbols. Thus, $\bar{x}$ represents the sample mean, $x - \bar{x}$ represents the difference of any number from the mean, $(x - \bar{x})^2$ represents the square of the difference, and $|x - \bar{x}|$ represents the absolute value of the difference from the sample mean. Furthermore, the sum of the differences from the sample mean, $\Sigma(x - \bar{x})$, is 0. Can you see why?

Let us now compute the sample variance, sample standard deviation, and average deviation for traveling to work by way of the Kingston Expressway. Again we will use symbols.

	Time (x)	Difference from Mean (x − x̄)	Square of Difference (x − x̄)²	Absolute Value of Difference |x − x̄|
	29	29 − 31 = −2	(−2)² = 4	2
By Way of	30	30 − 31 = −1	(−1)² = 1	1
Kingston	31	31 − 31 = 0	0² = 0	0
Expressway	32	32 − 31 = 1	1² = 1	1
	33	33 − 31 = 2	2² = 4	2
		Sum = 0	Sum = 10	Sum = 6

In this case the sample variance is $\dfrac{10}{5 - 1}$, or 2.5; the sample standard deviation is $\sqrt{2.5}$, or approximately 1.58; and the average deviation is $\dfrac{6}{5}$, or 1.2. Here again, the sum of the differences from the mean, $\Sigma(x - \bar{x})$, is 0. This is always the case.

We now formally define sample variance, sample standard deviation, and average deviation.

DEFINITION 3.5 Sample Variance

The sample variance of a sample of n numbers is a measure of the spread of the numbers about the sample mean and is given by

Sample Variance

$$\textbf{Sample variance} = s^2 = \frac{\Sigma(x - \bar{x})^2}{n - 1}$$

DEFINITION 3.6 Sample Standard Deviation

The sample standard deviation of a sample of n numbers is the positive square root of the sample variance. Thus,

Sample Standard Deviation

$$\textbf{Sample standard deviation} = s = \sqrt{s^2} = \sqrt{\frac{\Sigma(x - \bar{x})^2}{n - 1}}$$

DEFINITION 3.7	The **average deviation** of a sample of numbers is the average of the absolute value of the differences from the sample mean. Symbolically,
Average Deviation	$$\text{Average deviation} = \frac{\Sigma|(x - \bar{x})|}{n}$$

If we are working with the entire population rather than with a sample, then we have the following:

$$\text{Population variance} = \sigma^2 = \frac{\Sigma(x - \mu)^2}{N}$$

$$\text{Population standard deviation} = \sigma = \sqrt{\frac{\Sigma(x - \mu)^2}{N}}$$

COMMENT When calculating the sample variance (or sample standard deviation), we use $n - 1$ instead of n in the denominators of the two formulas given in Definitions 3.5 and 3.6. However, when calculating the population variance (or standard deviation), we use N instead of $N - 1$ in the denominators of the formulas. There is a sound statistical reason for doing this, but we will not concern ourselves with the reason at this point. To summarize, the difference between σ and s is whether we divide by N or by $n - 1$ and whether we use the population mean, μ or the sample mean, $\bar{x}$. When calculating the population standard deviation, we use μ and divide by N and denote our result by σ, whereas when calculating a sample standard deviation, we use $\bar{x}$ and divide by $n - 1$ and denote our result by s. Thus, s is really an estimate of σ, the population standard deviation. Very often statisticians will refer to s as *the* standard deviation, even though it is only an estimate.

COMMENT It may seem that the standard deviation is a complicated and useless number to calculate. At the moment let us say that it is a useful number to the statistician. Just as the measures of central tendency help us locate the "center" of a relative frequency distribution, the standard deviation helps us measure its "spread." It tells us how much the observations differ from the mean. Notice that most of the observations in Figure 3.1 deviate very little from the mean of the distribution. As opposed to this, most of the observations in Figure 3.2 deviate substantially from the mean of the distribution. When we discuss the normal distribution in later chapters, you will understand the significance and usefulness of the standard deviation. Statisticians prefer to work with the standard deviation as opposed to the variance, since the standard deviation is usually smaller than the variance.

In most statistical problems we do not have all the data for the population. Instead, we have only a small part, that is, a sample, of the population. It is for this reason that we use the sample variance and sample standard deviation, which we restate below using a formula number for later reference.

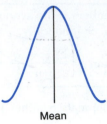

Figure 3.1
Most of the observations
deviate very little from the
mean.

Figure 3.2
Most of the observations deviate substantially from the mean.

| FORMULA 3.4 | Sample Variance | Sample Standard Deviation |

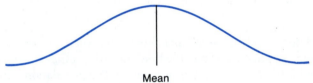

$$s^2 = \frac{\Sigma(x - \bar{x})^2}{n - 1} \qquad s = \sqrt{\frac{\Sigma(x - \bar{x})^2}{n - 1}}$$

We will illustrate the preceding ideas with another example.

EXAMPLE 1   The number of hours per day that several technicians spent adjusting the timing mechanism on CAT scan machines during the past week is 5, 3, 2, 6, 4, 2, and 6. Find the sample variance, sample standard deviation, and average deviation.

Solution

We arrange the data in order as shown in the following chart and perform the indicated calculations. The sample mean is

$$\bar{x} = \frac{\Sigma x}{n} = \frac{2 + 2 + 3 + 4 + 5 + 6 + 6}{7}$$

$$= \frac{28}{7} = 4$$

Number of Hours x	Difference from Mean $(x - \bar{x})$	Square of Difference $(x - \bar{x})^2$	Absolute Value of Difference $|x - \bar{x}|$
2	$2 - 4 = -2$	$(-2)^2 = 4$	2
2	$2 - 4 = -2$	$(-2)^2 = 4$	2
3	$3 - 4 = -1$	$(-1)^2 = 1$	1
4	$4 - 4 = 0$	$0^2 = 0$	0
5	$5 - 4 = 1$	$1^2 = 1$	1
6	$6 - 4 = 2$	$2^2 = 4$	2
6	$6 - 4 = 2$	$2^2 = 4$	2
	$Sum = 0$	$Sum = 18$	$Sum = 10$

Our answers then are

$$\text{Sample variance} = \frac{\Sigma(x - \bar{x})^2}{n - 1} = \frac{18}{7 - 1} = 3$$

$$\text{Sample standard deviation} = \sqrt{\text{variance}} = \sqrt{3}, \quad \text{or approximately } 1.73$$

$$\text{Average deviation} = \frac{\Sigma|x - \bar{x}|}{n} = \frac{10}{7} = 1.43 \quad \text{when rounded}$$

3.5 COMPUTATIONAL FORMULA FOR CALCULATING THE VARIANCE

Although the formulas in Definitions 3.5 and 3.6 of the preceding section can always be used for calculating the sample variance, in practice, it turns out that the calculations become quite tedious. For this reason we use more convenient formulas, which follow.

FORMULA 3.5 Computational Formulas

$$\text{Population variance} = \sigma^2 = \frac{\Sigma x^2}{N} - \frac{(\Sigma x)^2}{N^2} \quad \text{or} \quad \frac{N \Sigma x^2 - (\Sigma x)^2}{N^2}$$

$$\text{Sample variance} = s^2 = \frac{n \Sigma x^2 - (\Sigma x)^2}{n(n - 1)}$$

$$\text{Population standard deviation} = \sigma = \sqrt{\frac{\Sigma x^2}{N} - \frac{(\Sigma x)^2}{N^2}}$$

$$\text{Sample standard deviation} = s = \sqrt{\frac{n \Sigma x^2 - (\Sigma x)^2}{n(n - 1)}}$$

In Formula 3.5, Σx^2 means that we square each number and add the squares together, whereas $(\Sigma x)^2$ means we first sum the numbers and then square the sum.

Using summation notation for the data in Table 3.1, we have

$$\Sigma x = 1 + 2 + 3 + 4 + 5 = 15$$
$$\Sigma x^2 = 1^2 + 2^2 + 3^2 + 4^2 + 5^2$$
$$= 1 + 4 + 9 + 16 + 25 = 55$$

We now use Formula 3.5 to calculate the sample variance for the data of Table 3.1:

TABLE 3.1 Squares of Integers
(x^2 Means x Times x)

x	x^2
1	1
2	4
3	9
4	16
5	25
15	55
$\Sigma x = 15$	$\Sigma x^2 = 55$

$$s^2 = \frac{n \, \Sigma x^2 - (\Sigma x)^2}{n(n-1)}$$

$$= \frac{5(55) - (15)^2}{5(5-1)}$$

$$= \frac{275 - 225}{5(4)}$$

$$= \frac{50}{20} = 2.5$$

If you now calculate the sample variance by using the formula in Definition 3.5 of the preceding section and compare the results, your answer will be the same. It is considerably simpler, however, to get the answer by using Formula 3.5. The sample standard deviation is obtained by taking the square root of the variance. Thus,

$$\text{Standard deviation} = \sqrt{\text{variance}}$$

$$= \sqrt{2.5} \approx 1.58$$

(The symbol $\approx$ stands for approximately.)

COMMENT The advantage of computing the sample variance by Formula 3.5 is that we do not have to subtract the sample mean from each term of the distribution.

BEWARE   Do not confuse the symbols $\Sigma\ x^2$ and $(\Sigma\ x)^2$. The symbol $\Sigma\ x^2$ represents the sum of the squares of each number, whereas the symbol $(\Sigma\ x)^2$ represents the square of the sum of the numbers. If your calculation of the variance results in a negative number, you probably have confused the two symbols.

How are the mean, variance, and standard deviation affected if we *multiply* each term of a distribution by some number? To see what happens, multiply each number of the distribution given in Table 3.1 by, say, 10, and compute the mean, variance, and standard deviation of the new distribution. (You will be asked to do this in one of the exercises.)

How are the mean, variance, and standard deviation affected if we *add* the same constant to each term of a distribution? (Again, you will be asked to do this in one of the exercises.)

To calculate the variance and standard deviation of sample data that is presented in frequency distribution form, as is often the case with published data, we need a computational formula for determining the sample variance and sample standard deviation for grouped data. We can always use the definitions. Thus, we have

Sample Variance for Grouped Data

$$s^2 = \frac{\Sigma(x - \bar{x})^2 \cdot f}{n - 1}$$

Sample Standard Deviation for Grouped Data

$$s = \sqrt{\frac{\Sigma(x - \bar{x})^2 \cdot f}{n - 1}}$$

However, the computations involved in using these formulas can be very tedious. Formula 3.6 is a shortcut formula that gives the same results as if we had used the definition.

FORMULA 3.6   Computational Formulas for Grouped Data

Sample Variance for Grouped Data

$$s^2 = \frac{n(\Sigma\ x^2 \cdot f) - (\Sigma\ x \cdot f)^2}{n(n - 1)}$$

Sample Standard Deviation for Grouped Data

$$s = \sqrt{\frac{n(\Sigma\ x^2 \cdot f) - (\Sigma\ x \cdot f)^2}{n(n - 1)}}$$

EXAMPLE 1

The IQ of 50 students in the fifth grade of a special school was measured. The results are as follows:

IQ Score	Frequency
71–79	3
80–88	6
89–97	12
98–106	14
107–115	8
116–124	5
125–133	2

Compute the sample variance and sample standard deviation.

Solution

We rearrange the data as shown in the following table. Note that we use the class marks for each category. In grouped data we assume that all the numbers in each group are at the class mark for that group.

IQ Scores	Class Marks x	Frequencies f	$x \cdot f$	$x^2 \cdot f$
71–79	75	3	225	16,875
80–88	84	6	504	42,336
89–97	93	12	1116	103,788
98–106	102	14	1428	145,656
107–115	111	8	888	98,568
116–124	120	5	600	72,000
125–133	129	2	258	33,282
		50	5019	512,505

Here $n = \Sigma f = 50$ since the sample consisted of 50 students. Then using Formula 3.6, we get

$$s^2 = \frac{n(\Sigma x^2 \cdot f) - (\Sigma x \cdot f)^2}{n(n-1)}$$

$$= \frac{50(512,505) - (5019)^2}{50(50-1)}$$

$$= 177.5057$$

and

$$s = \sqrt{s^2} = \sqrt{177.5057} \approx 13.32.$$

Thus, the sample variance is 177.51 and the sample standard deviation is approximately 13.32. (Do not use a rounded variance to get a standard deviation. Round all answers at the end of all calculations.)

EXERCISES FOR SECTION 3.5

1. The ages of the executive officers of ten large companies doing business in Greensberg are 41, 38, 44, 49, 52, 58, 63, 39, 31, and 69 years.
   a. Find the range in ages for these people.
   b. Calculate the sample variance and sample standard deviation.

2. The annual salaries of the executive officers mentioned in Exercise 1 are as follows:

   $65,000 $58,000 $79,000 $97,000 $121,000
   $88,000 $145,000 $121,000 $104,000 $134,500

   Find the range of salaries for these people.

3. The 1990 Federal tax returns of 15 randomly selected politicians were analyzed to determine how much money each politician had deducted for charity contributions. The results are as follows:

   $2578 $1579 $1210 $2398 $4100 $3100 $1995 $1400
   $1800 $2500 $2650 $3600 $2900 $3055 $3175

   Find the average deviation, sample variance, and standard deviation for the charity deductions.

4. Often psychologists perform experiments with mice to analyze behavior and learning patterns. In one such experiment a psychologist found that 12 mice needed 13, 19, 10, 8, 26, 17, 19, 7, 12, 13, 5, and 31 minutes to find their way through a complicated maze. Find the sample variance, sample standard deviation, and average deviation for these numbers.

5. *Physical fitness.* To maintain their physical fitness, many Americans jog or exercise regularly each day. Gregory keeps accurate records on the number of pushups that he does each morning. For a two-week period he has recorded the following:

   28   56   49   42   40   52   53
   48   63   55   60   57   64   61

   Calculate the population variance, population standard deviation, and average deviation for the number of pushups that Gregory can do.

6. *Drugs and the senior citizen.* Are senior citizens in nursing homes taking too many medications? In a random sample of nine senior citizens in the Palmos Gardens Nursing Home, a government fact-finding commission found that these individuals were taking 5, 6, 4, 3, 1, 6, 4, 3, and 4 different kinds of medications. Find the range, sample mean, sample variance, sample standard deviation, and average deviation for the number of medications taken by these senior citizens.

7. Multiply each number in the preceding exercise by 3 and then compute the sample mean, sample variance, sample standard deviation, and average deviation for the new distribution. How do these results compare with those of the preceding exercise? Can you generalize?

8. *Jogging marathon.* The number of people entered in the annual jogging marathon is as follows:

Years	Number of Men Entered	Number of Women Entered
1984	31	26
1985	37	21
1986	36	23
1987	41	22
1988	33	35
1989	37	39
1990	38	43

a. Find the population standard deviation for the number of men entered in the marathon over the years.

b. Find the population standard deviation for the number of women entered in the marathon over the years.

c. Compare the answer obtained in part (a) with the answer obtained in part (b). Comment.

9. *Fire insurance.* The average cost for fire insurance for a two-story commercial building (on an 80 × 300 lot) in a certain state is $3400 per year with a (population) standard deviation of $375. Next year insurace costs will be increased by $500. What will the new population mean and population standard deviation be?

10. Refer back to Exercise 9. If the insurance costs are increased next year by 20% instead of $500, what will the new population mean and population standard deviation be?

11. The price of a gallon of home heating oil charged by five dealers in New York City on January 1, 1991 was $1.41, $1.38, $1.45, $1.36, and $1.29.

a. Calculate the population mean and population standard deviation by subtracting $1.29 from each price.

b. Calculate the population mean and population standard deviation by subtracting $1.36 from each price.

c. How are the (population) mean and standard deviation affected if we subtract a different number from each price?

12. Researchers at the Avery Research Corp. have developed two different kinds of long-lasting 9-volt batteries. The useful life of these two different batteries was calculated and is as follows:

	Battery A	Battery B
Average life	180 hours	178 hours
Standard deviation	7 hours	4 hours

Which of these batteries would you buy? Why?

13. Each year New York City conducts a 10-kilometer running marathon. A sample of the finishing time of 50 runners in the 1990 marathon produced the following frequency distribution.

Finishing Time (in minutes)	Frequency
30–under 40	2
40–under 50	4
50–under 60	16
60–under 70	14
70–under 80	9
80–under 90	5

Compute the sample mean, sample variance, and sample standard deviation for these runners.

14. **Case Study** In 1990 there were approximately 25 million Americans, mostly women, who had osteoporosis. This is a disorder in which the bones deteriorate due to the excessive loss of bone tissue. Abnormally low bone mass results in an increase in susceptibility to bone fractures. While the precise cause of osteoporosis is not fully understood, some researchers believe that calcium is a key to preventing osteoporosis. The recommended daily allowance (RDA) of calcium varies with age and gender. For women of child-bearing age, the RDA is 800 milligrams daily. A research team* randomly sampled 100 women suffering from osteoporosis to determine their daily intakes of calcium. The following frequency distribution gives the results.

Daily Intake of Calcium (in milligrams)	Frequency
Under 200	6
200–under 400	7
400–under 600	16
600–under 800	32
800–under 1000	21
1000–under 1200	15
1200–under 1400	3

Calculate the sample mean, sample variance, and sample standard deviation for these women.

15. **Case Study** The attorney general of a particular state believes that two oil companies operating gas stations within the state discriminate against poor people by charging more for a gallon of unleaded premium gas in poor neighborhoods than in middle-class neighborhoods. The oil companies deny the charge. Both sides agree to sample five company-operated gas stations in each area and to calculate the average cost per gallon of gas. When this is done the following data are obtained.

* Todd & Rodgers, New York, 1990.

Cost per Gallon of Unleaded Premium Gas

Poorer Neighborhoods	Middle-class Neighborhoods
$1.52	$1.61
1.53	1.45
1.51	1.47
1.54	1.49
1.40	1.48

The average price per gallon of gas in both neighborhoods turns out to $1.50. Thus, the oil companies claim that they are not discriminating against people who live in poorer neighborhoods. Analyze the data to determine if there is anything wrong with this reasoning. Comment.

3.6 INTERPRETATION OF THE STANDARD DEVIATION: CHEBYSHEV'S THEOREM

Up to this point we have been discussing formulas for calculating the standard deviation of a set of numbers. If the standard deviation turns out to be small, then we can conclude that all the data values are concentrated around the mean. On the other hand, if the standard deviation is large, then the data values will be widely scattered about the mean.

In a later chapter when we discuss the normal distribution, we will notice that a substantial number of the data is bunched within 1, 2, or 3 standard deviations above or below the mean. Nevertheless, a more general result, which is true for any set of measurements—population or sample—and regardless of the shape of the frequency distribution, is known as **Chebyshev's Theorem.** We state it now.

Chebyshev's Theorem

CHEBYSHEV'S THEOREM:

Let k be any number equal to or greater than 1. Then the proportion of any distribution that lies within k standard deviations of the mean is at least $1 - \dfrac{1}{k^2}$.

To see what Chebyshev's Theorem means, let us compute some values of $1 - \dfrac{1}{k^2}$ as shown here.

Value of k	Value of $1 - \dfrac{1}{k^2}$
1	0
2	$\dfrac{3}{4}$ or 75%
3	$\dfrac{8}{9}$ or 89%

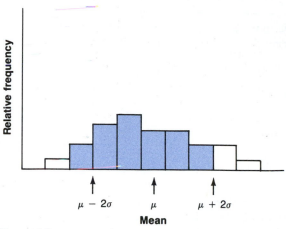

Figure 3.3
At least $\frac{3}{4}$ of the measurements will fall within the shaded portion.

When $k = 2$, then if μ is the population mean and σ the population standard deviation, Chebyshev's Theorem says that you will always find at least $\frac{3}{4}$, that is, 75% or more, of the measurements will lie within the interval $\mu - 2\sigma$ and $\mu + 2\sigma$, that is, within 2 standard deviations of the mean on either side. This can be seen in Figure 3.3. Similarly, when $k = 3$, Chebyshev's Theorem says that at least $\frac{8}{9}$ of the measurements, or 89% of the data, will lie within the interval $\mu - 3\sigma$ and $\mu + 3\sigma$, that is, within 3 standard deviations of the mean on either side. This can be seen in Figure 3.4.

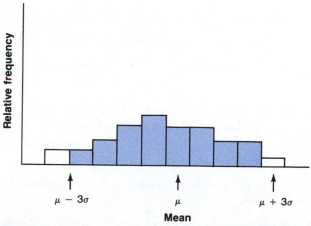

Figure 3.4
At least $\frac{8}{9}$ of the measurements will fall within the shaded portion.

EXAMPLE 1

A statistician is analyzing the claims filed with an auto insurance company. A sample of 100 claims discloses that the average claim filed was $831, with a standard deviation of $150. If we let $k = 2$, then at least $1 - \dfrac{1}{2^2}$ or $\dfrac{3}{4}$, that is, 75% or more, of the claims will be within

$831 - 2(150)$ and $831 + 2(150)$
or between $531 and $1131

Similarly, at least $1 - \dfrac{1}{3^2}$ or $\dfrac{8}{9}$, or 89% or more, of the claims will be within

$831 - 3(150)$ and $831 + 3(150)$
or between $381 and $1281

EXAMPLE 2

If all the light bulbs manufactured by a certain company have a mean life of 1600 hours with a standard deviation of 100 hours, at least what percentage of the bulbs will have a mean life of between 1450 and 1750 hours?

Solution

We first find $1750 - 1600 = 1600 - 1450 = 150$. Using Chebyshev's Theorem, we know that k standard deviations or $k(100)$ will equal 150. That is,

$$100k = 150$$

so that

$$k = 1.5$$

Thus, at least $1 - \dfrac{1}{(1.5)^2} = 1 - \dfrac{1}{2.25} = 0.5556$, or at least 55.56% of the light bulbs will have a mean life between 1450 and 1750 hours.

For a specific set of data, we can always answer the question, "How many measurements are within 1, 2, or 3 standard deviations of the mean?" We can simply count the number of measurements in each of the intervals. Based on Chebyshev's Theorem, we can say the following:

FACT I   For any sample of measurements (regardless of the shape of the frequency distribution):

1. Possibly none of the measurements fall within 1 standard deviation of the mean, that is, within $\bar{x} \pm s$.

2. At least $\frac{3}{4}$ of the measurements will fall within 2 standard deviations of the mean, that is, within $\bar{x} \pm 2s$.

3. At least $\frac{8}{9}$ of the measurements will fall within 3 standard deviations of the mean, that is, within $\bar{x} \pm 3s$.

4. At least $(1 - 1/k^2)$ will fall within k standard deviations of the mean, that is, within $\bar{x} \pm ks$, where $k > 1$.

FACT II   If the frequency distribution of a set of sample data is mound-shaped (normally distributed), then

1. Approximately 68% of the measurements will fall within 1 standard deviation of the mean, that is, within $\bar{x} \pm s$.

2. Approximately 95% of the measurements will fall within 2 standard deviations of the mean, that is, within $\bar{x} \pm 2s$.

3. Approximately 99.7% of the measurements will fall within 3 standard deviations of the mean, that is, within $\bar{x} \pm 3s$.

Empirical Rule

Fact I is often called Chebyshev's Theorem and Fact II is often called the **Empirical Rule**. We will discuss the normal distribution in considerable detail in a later chapter.

EXERCISES FOR SECTION 3.6

1. The daily number of calls for emergency road service to the local American Automobile Association (AAA) involving cars with dead batteries during the recent ten-day cold spell was as follows:

   69 58 68 37 76 53 41 38 39 37

   Verify Chebyshev's Theorem with $k = 2$ and $k = 3$ for these data.

2. Although many companies have backup computers, nevertheless, computer "downtime" can be very costly to businesses. In the last quarter of 1990 The Baxter Payroll Processing Company experienced the following weekly number of hours of computer downtime:

   8.3 5.1 6.9 2.7 5.1 3.8
   8.1 5.5 4.3 3.5 2.9 2.2

   Verify Chebyshev's theorem with $k = 2$ and $k = 3$ for these data.

3. The average charge for all merchandise purchased by a credit card at Cindy's Boutique is $54 with a standard deviation of $8. At least what percentage of credit card sales will be between $42 and $66?

4. For a variety of health reasons, many people must restrict their intake of sodium (salt). Often when eating in a restaurant, diners find it difficult to control sodium intake. The average sodium content of all the dinners at Luigi's Restaurant is

985 milligrams (mg). with a standard deviation of 25 mg. At least what percentage of the dinners will contain between 935 and 1035 mg of sodium?

5. The Biology Department at Salvia University gave a final exam in January. The grades of 20 students who took the exam are

85 74 66 53 90 92 84 71 68 87
87 64 56 72 62 50 82 68 78 59

a. Find the mean grade, $\bar{x}$, for this sample of students. Also, find s^2 and s.
b. According to Chebyshev's Theorem, what percentage of the measurements would you expect to find in the interval $\bar{x} \pm 2.5s$ and $\bar{x} \pm 4s$?
c. What percentage of measurements actually fall in the intervals of part (b)?
d. Compare the results obtained in part (c) with the results of part (b). Comment.

3.7 PERCENTILES AND PERCENTILE RANK

Consider the following newspaper article. The article indicates that percentile ranks are used by many colleges in determining which students will be admitted. Also, if we analyze the card from the Emerson Medical Laboratory shown in Figure 3.5, we again notice the use of percentiles. How are percentiles calculated? How do we interpret them?

NEW ADMISSIONS GUIDELINES ADOPTED

Trudy Hoffman

March 20: As a result of further budget cutbacks the Chancellor's office announced new admission guidelines. Effective this fall, no students will be admitted to a senior college unless his or her high school average is 80% or higher or is in at least the 75th percentile in his or her graduating class. To be admitted to a community college, a student must have a 70% high school average or have a percentile rank of 40 or higher.

Sunday, March 20, 1987

Age (in years)	Frequency
21–30	1
31–40	1
41–50	6
51–60	8
61–70	11
71–80	3

Now consider the table shown above, which shows the distribution of the ages of 30 runners in a Florida jogging marathon. The histogram for the data is shown in Figure 3.6.

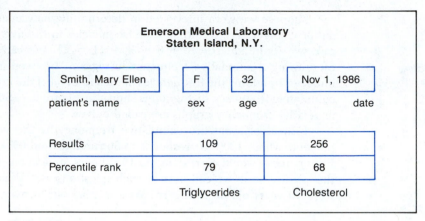

Figure 3.5

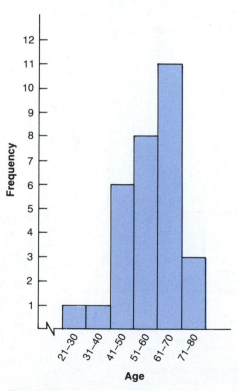

Figure 3.6

Suppose we were interested in determining how many runners were 70 years old or younger. We would add the frequencies in the five lowest categories. In our case we would add $1 + 1 + 6 + 8 + 11 = 27$. We can obtain the same answers by drawing a **cumulative frequency histogram.** In essence, we accumulate the frequencies by adding the frequencies in each interval of the grouped data. The resulting cumulative frequency histogram is shown in Figure 3.7. Such cumulative frequency or relative frequency graphs are called **ogives.**

Instead of using the cumulative frequency as the vertical scale, we can use percents where 100% corresponds to 30 runners, and 0% corresponds to 0 runners. Fifty percent would correspond to 15 runners, and so on. By having the percent on the vertical scale we can answer such questions as, "What percent of the runners were 70 years old or younger?" We can read our answer from the graph, called a

Cumulative Frequency Histogram

Ogives

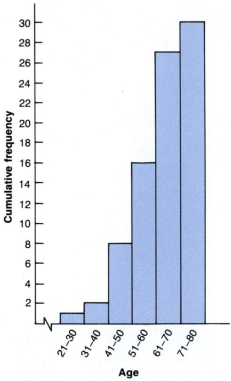

Figure 3.7
A cumulative frequency histogram.

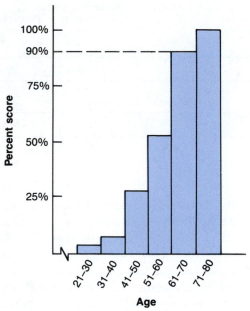

Figure 3.8
A cumulative relative frequency histogram.

Cumulative Relative
Frequency
Histogram

cumulative relative frequency histogram, given in Figure 3.8. The answer is 90%. More generally, a score that tells us what percent of the total population scored at or below that measure is called the *percentile*. How do we determine such percentiles?

To get us started, let us consider Lorraine, who received a 76 on her midterm psychology examination. There are 150 students, including her, in the class. She knows that 60% of the class got below 76, 10% of the class got 76, and the remaining 30% got above 76.

Since 60% of the class got below her grade of 76 and 30% got above her grade, her percentile rank should be between 60 and 70. We will use 65, which is midway between 60 and 70. What we do is find the percent of scores that are below the given score and add one-half the percent of the scores that are the same as the given score. In our case 60% of the class grades were below Lorraine's and 10% were the same as Lorraine's. Thus, the percentile rank of Lorraine's grade is

$$60 + \left(\frac{1}{2}\right)(10) = 65$$

Figure 3.9 illustrates the situation.

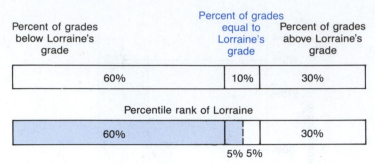

Figure 3.9

We now say that Lorraine's percentile rank is 65. This means that approximately 35% of the class did better than Lorraine on the exam and that she did better than 65% of the class. Essentially, the percentile rank of a score tells us the percentage of the distribution that is below that score. Formally, we have the following definition.

DEFINITION 3.8
Percentile Rank
Percentile

The **percentile rank** or **percentile** of a term in a distribution is found by adding the percentage of terms below it with one-half the percentage of terms equal to the given term.

Let X be a given score, let B represent the *number* of terms below the given score X, and let E represent the *number* of terms equal to the given score X. If there are N terms altogether (that is, if the entire population consists of N terms together), then the percentile rank of X is given by Formula 3.7.

FORMULA 3.7

$$\text{Percentile rank of } X = \frac{B + \dfrac{1}{2}E}{N} \cdot 100$$

We now illustrate the use of Formula 3.7.

EXAMPLE 1

Bill and Jill are twins, but both are in different classes. Recently, they both got 80 on a math test. The grades of the other students in their class were as follows:

Jill's class
64   67   73   73   73   74   77   77   78   78   79   80   80   82   91   94   100

Bill's class
43   65   68   73   75   76   76   77   79   80   80   80   80   85
86   87   88   90   92   96

Find the percentile rank of each student.

Solution

We use Formula 3.7. Jill's grade is 80. There were two 80s (including Jill's) in the class, so $E = 2$. There were 11 grades below 80, so $B = 11$. Since there are 17 students in the class altogether, $N = 17$. Thus,

$$\text{Jill's percentile rank} = \frac{11 + \dfrac{1}{2}(2)}{17} \cdot 100$$

$$= \frac{11 + 1}{17} \cdot 100 = \frac{12}{17} \cdot 100$$

$$= \frac{1200}{17} = 70.59$$

Jill's percentile rank is 70.59. Using a similar procedure for Bill's class, we find that $B = 9$, $E = 4$, and $N = 20$. Thus,

$$\text{Bill's percentile rank} = \frac{9 + \dfrac{1}{2}(4)}{20} \cdot 100$$

$$= \frac{9 + 2}{20} \cdot 100 = \frac{11}{20} \cdot 100$$

$$= \frac{1100}{20} = 55$$

Bill's percentile rank is 55.

COMMENT   The percentile rank of an individual score is often more helpful than the particular score value. Although both Bill and Jill had grades of 80, Jill's percentile rank is considerably higher. If we assume that the levels of competition are equivalent in both classes, this may indicate that Jill's performance is superior to Bill's performance when compared to the rest of their respective classes.

We often use the word **percentile** to refer directly to a score in a distribution. So, instead of saying that the percentile rank of Lorraine's grade is 65, we say that her grade is in the 65th percentile. Similarly, if a term has a percentile rank of 40, we say that it is in the 40th percentile.

Percentiles are used quite frequently to describe the results of achievement tests and the subsequent ranking of people taking those tests. This is especially true when applying for many civil service jobs. If there are more applicants than available jobs, candidates are often ranked according to percentiles. Many colleges use only percentile ranks, rather than the numerical high school average, to determine which candidates to admit. The reason is that percentile ranks of a student's high school average reflect how they did with respect to their classmates, whereas numerical averages only indicate an individual student's performance.

Since percentiles are numbers that divide the set of data into 100 equal parts, we can easily compare percentiles. Thus, in Example 1 we were able to find the percentile rank of Jill and Bill, even though they both were in different classes.

During World War II the United States Army administered the Army General Classification Tests (AGCT) to thousands of enlisted men. The results showed important differences in the average IQ of men in various jobs, ranging from 93 for miners and farmhands to around 120 for accountants, lawyers, and engineers. Figure 3.10 shows the IQ range between the 10th and 90th percentiles for workers in various

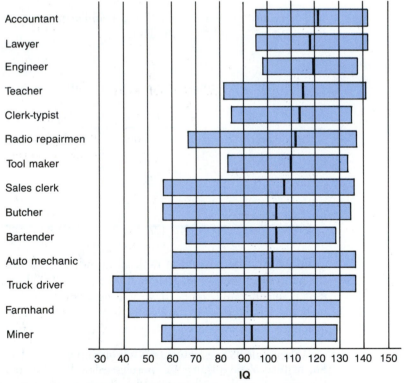

Figure 3.10

Each bar shows the IQ range between the 10th and 90th percentiles for men in that occupation. The vertical bars represent the 50th percentiles. Note that although the average IQ score of accountants was 121 and that of miners 93, some miners had higher IQ scores than some accountants.

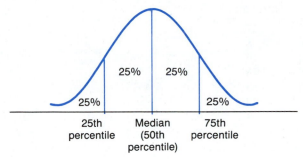

Figure 3.11
Some of the frequently used percentiles.

occupations. Furthermore, the vertical bars represent the 50th percentile or median scores. Very often, when such tests are administered to large groups of people, the results are given in terms of percentile bands, as shown in Figure 3.11. Since percentiles are used quite often, special names are given to the 25th and 75th percentiles of a distribution. Of course, the *50th percentile is the median of the distribution*. See Figure 3.11.

DEFINITION 3.9 Lower Quartile	The 25th percentile of a distribution is called the **lower quartile.** It is denoted by P_{25} or Q_1. Thus, 25% of the terms are below the lower quartile and 75% of the terms are above it.

DEFINITION 3.10 Upper Quartile	The 75th percentile of a distribution is called the **upper quartile.** It is denoted by P_{75} or Q_3. Thus, 75% of the terms are below the upper quartile and 25% of the terms are above it. These percentiles are pictured in Figure 3.11 for a normal distribution.

NOTATION   The various percentiles are denoted by the letter P with the appropriate subscript. Hence, P_{37} denotes the 37th percentile, P_{99} denotes the 99th percentile, and so on.

When calculating percentiles involving grouped data, we use the class mark. This is illustrated in the following example.

EXAMPLE 2   One hundred candidates have applied for a high-paying acting job. The candidates were tested and then rated according to their dancing ability, poise, and overall acting skill. The following ratings were obtained.

Rating	Frequency
10–19	17
20–29	8
30–39	6
40–49	18
50–59	28
60–69	16
70–79	4
80–89	3
	Total = 100

Calculate the percentile rank of Heather McAllister, who scored 44.5 in the ratings.

Solution

Since Heather scored 44.5 in the ratings, she is in the 40 to 49 category. We know that there are $17 + 8 + 6$ or 31 below the 40 to 49 category. We now assume that all the 18 candidates who scored in the 40 to 49 category scored 44.5, which is the class mark. Thus, there are 18 people including Heather who scored 44.5. Using Formula 3.6 with $N = 100$, $B = 31$, and $E = 18$, we have

$$\text{Heather's percentile rank} = \frac{31 + \frac{1}{2}(18)}{100} \times 100$$

$$= \frac{31 + 9}{100} \times 100$$

$$= 40$$

Heather is in the 40th percentile.

COMMENT   In the previous example we assumed that all of the candidates who scored in the 40 to 49 category scored 44.5, which is the class mark. Some statisticians prefer to compute percentiles for grouped data by assuming that the entries in a class are evenly distributed in that class. These statisticians then use the formula $\tilde{x} = L + \frac{j}{f} \cdot c$. Refer back to the discussion of this approach given on page 115.

Box Plots
Box-and-Whisker
Diagram
Interquartile Range

A relatively new concept in the descriptive analysis of data involves **box plots** or **box-and-whisker diagrams**. This method, invented by Professor John Tukey, is based on the quartiles of a data set and the **interquartile range**, which is the distance between the upper and lower quartile. We illustrate the procedure for constructing box plots in the following example.

EXAMPLE 3

Mary Aquilla is the telephone switchboard operator for the Hewlit Corporation. The number of "wrong number" telephone calls that she received over the past 20 days is as follows:

3   4   13   31   16   24   9   17   13   33
7   12   8   23   21   18   14   19   21   35

Construct a box plot for these data.

Solution

Step 1. We first find the smallest and the largest data value. In our case we have

smallest value = 3   and   largest value = 35

Step 2. We arrange the data in increasing order to find the median (middle quartile). In our case when the data are arranged in order, we get the following:

3 4 7 8 9 12 13 13 14 16 17 18 19 21 21 23 24 31 33 35

so that the median is between 16 and 17 or that

$$\text{Median or middle quartile} = \frac{16 + 17}{2} = 16.5$$

Step 3. Now we find the median of the top half of the data and also the median of the bottom half of the data. In our case we get the following:

Lower half of the data: 3   4   7   8   9   12   13   13   14   16
   Median of lower half or lower quartile = 10.5
Upper half of the data: 17   18   19   21   21   23   24   31   33   35
   Median of upper half or upper quartile = 22

Hinges

The lower and upper quartiles are called the **hinges.**

Step 4. Locate the values obtained in steps 1 to 3 on a horizontal axis. Now we draw a rectangle (the box) with the ends of the rectangle (the hinges) drawn at the lower and upper quartiles. We connect the hinges to each other to get a box. Now we mark the median (middle quartile), the low value, and the high value on the graph above the axis as shown in Figure 3.12. Finally, we connect the low and high values to the hinges by means of lines called **whiskers.** The resulting graph is called the box-and-whisker diagram.

Whiskers

In Figure 3.12 we have labeled everything on the graph so that we can interpret correctly the information that it provides. We note that the two boxes in the box-and-whisker diagram tell us how spread out the two middle quartiles are. Also, the two whiskers indicate how spread out the lower and upper quartiles are. By visually examining the lengths of the whiskers, we can safely say that if one whisker is clearly longer, then the distribution of the data is probably skewed in the direction of the longer whisker.

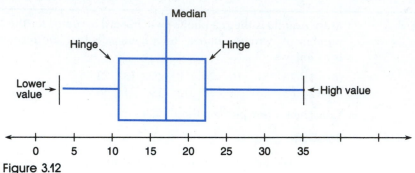

Figure 3.12
Box plot or box-and-whisker diagram.

COMMENT   From the preceding box-and-whisker diagram we can conclude that since the right whisker is longer than the left whisker, the data is skewed to the right. There is more variation in the upper quarter than in any other quarter.

EXERCISES FOR SECTION 3.7

1. The 24 graduating seniors of Village High School took the Scholastic Aptitude Test during May. Their results on the mathematics part are as follows:

Maurice—387	Arline—623	Alice—512	Harley—691	Sandy—589
Richard—399	Helen—401	Boris—482	Dan—555	Doris—701
Mabel—600	Seymour—493	Jason—507	Mitchel—401	Bruce—399
Jane—593	Alfred—482	Heather—705	Fred—683	Lucille—704
Bob—605	John—576	Gwen—601	Hollie—456	

Find the percentile rank of Alfred and the percentile rank of Bruce.

2. One hundred people are at an amusement park. One of the games requries each player to throw a ball at a target. The velocity (speed) of the ball as it is thrown is measured. The following results were obtained for these people.

Velocity of Ball Thrown (in miles per hour)	Frequency
80–89	3
70–79	9
60–69	18
50–59	24
40–49	19
30–39	12
20–29	9
10–19	6

a. Find the percentile rank of Steven who scored 74.5 in the contest.

b. Find the percentile rank of Leina who scored 54.5 in the contest.

3. On January 17, 1991 the Emerson Medical Laboratories received blood samples from 84 patients. Ed and five other patients had a blood serum cholesterol level of 268, whereas 59 others had a cholesterol level below 268. Find Ed's percentile rank.

4. In a recent jogging marathon the pulse rate of the joggers was taken. The following statistics are available:

$$P_{25} = 185 \text{ beats/minute}$$
$$P_{50} = 188 \text{ beats/minute}$$
$$P_{75} = 194 \text{ beats/minute}$$
$$P_{85} = 196 \text{ beats/minute}$$

What percent of the joggers had a pulse rate

a. less than 185 beats/minute?

b. more than 196 beats/minute?

c. more than 188 beats/minute?

d. between 185 and 196 beats/minute?

5. The following grades were obtained by the students of Pablo Aviation School on an endurance test:

68	98	83	55	93
59	69	96	86	84
78	80	65	76	80
85	95	50	90	73
80	85	90	80	97

Find the percentile rank of Andrea who scored 90 on this exam.

6. One hundred prisoners of the state correctional facility at Newport volunteered to be subjects for an endurance test performed by a panel of psychologists. The following scores were obtained.

Score	Frequency
10–19	6
20–29	9
30–39	12
40–49	19
50–59	24
60–69	18
70–79	9
80–89	3
Total =	100

a. Draw the cumulative frequency histogram for these data.
b. Draw the cumulative relative frequency histogram (in percent form) for these data.

7. Refer back to Exercise 5.

a. Find the lower quartile for the data.
b. Find the upper quartile for the data.
c. Draw the box-and-whisker diagram for the data.

8. Some people often make reservations in a hotel but fail to show up. The number of no-shows for a random sample of 25 days at the 600-room Galaxy Hotel is as follows:

11 23 31 17 21 12 10 15 22 28 25 13 14
 9 16 19   8 19   3 15 17 24 32 10 18

a. Find the lower quartile for the data.
b. Find the upper quartile for the data.
c. Find the interquartile range.
d. Draw the box-and-whisker diagram for the data.

9. The number of nurses calling in sick at the 1000 nurse Hollow Hill Hospital for a random sample of 20 days is as follows:

12 18 13 15 19 18 14 19 9 20
17 16   8 10 14 21 17 11 7 17

a. Find the lower quartile for the data.
b. Find the upper quartile for the data.
c. Find the interquartile range.
d. Draw the box-and-whisker diagram for the data.

10. It has been definitely proven that alcohol consumption affects a person's ability to operate a motor vehicle safely. The following relative frequencies were obtained by experienced drivers who were given a special obstacle avoidance test after consuming limited amounts of alcohol.

Score	Relative Frequency
40–49	0.01
50–59	0.02
60–69	0.06
70–79	0.10
80–89	0.17
90–99	0.19
100–109	0.18
110–119	0.15
120–129	0.08
130–139	0.03
140–149	0.01
150–159	0.00

Approximately what score must a driver have to be in the
a. 70th percentile?   b. 95th percentile?

3.8 z-SCORES

As we saw in the preceding section, one way of measuring the performance of an individual score in a population is by determining its percentile rank. Using percentile rank alone, however, can sometimes be misleading. For example, two students in different classes may have the same percentile rank. Yet one student may be far superior to his or her competitors, whereas the second student may only slightly surpass the others in his or her class.

z-Score

Statisticians have another very important way of measuring the performance of an individual score in a population. This measure is called the **z-score.** The z-score measures how many standard deviations an individual score is away from the mean. We define it formally as follows:

DEFINITION 3.11 FORMULA 3.8	The z-score of any number x in a distribution whose mean is μ and whose standard deviation is σ is given by

$$z = \frac{x - \mu}{\sigma} \quad \text{or} \quad z = \frac{x - \bar{x}}{s} \quad \rightarrow sample$$

where x = value of number in original units
μ = population mean
σ = population standard deviation
$\bar{x}$ = sample mean
s = sample standard deviation

z-Value

Measurement in
Standard Units

COMMENT The z-score of a number in a population is sometimes called the **z-value** or **measurement in standard units.**

COMMENT Since σ is always a positive number, z will be a negative number whenever x is less than μ, as $x - \mu$ is then a negative number. A z-score of 0 implies that the term has the same value as the mean.

We now illustrate how to calculate z-scores with several examples.

EXAMPLE 1

A certain brand of flashlight battery has a mean life, μ, of 40 hours and a standard deviation of 5 hours. Find the z-score of a battery that lasts

a. 50 hours   b. 35 hours   c. 40 hours

Solution

Since $\mu = 40$ and $\sigma = 5$, we use Formula 3.8.

a. The z-score of 50 is

$$\frac{50 - 40}{5} = \frac{10}{5} = 2$$

b. The z-score of 35 is

$$\frac{35 - 40}{5} = \frac{-5}{5} = -1$$

c. The z-score of 40 is

$$\frac{40 - 40}{5} = 0$$

EXAMPLE 2

Testing Tuna Fish

Two consumer's groups, one in New York and one in California, recently tested at numerous local colleges a number of different brands of canned tuna fish for taste appeal. Each consumer group used a different rating system. The following results were obtained:

Brand	New York Rating		Brand	California Rating
A	1		M	25
B	10		N	35
C	15		P	45
D	21		Q	50
E	28		R	70

Which brand has the greatest taste appeal?

Solution

At first glance it would appear that Brand R is superior since its California rating was 70. However, we see from the rating and from the given information that the two consumer's groups awarded their points differently so that the point value alone is not enough of a basis for deciding among the different brands. We therefore convert each of the ratings into standard scores. These calculations are shown in Tables 3.2 and 3.3.

We can now use the z-scores as a basis for comparison of the different brands. Clearly, Brand R for which $z = 1.65$ is superior to Brand E for which $z = 1.41$.

Notice that the sum of the z-scores for the New York ratings of the brands is 0:

$$(-1.52) + (-0.54) + (0) + (0.65) + (1.41) = 0$$

Table 3.2 New York Rating of Tuna Fish ($\mu = 15$, $\sigma = 9.23$)

	Rating x	Mean μ	Difference From Mean $(x - \mu)$	z-Score $z = \dfrac{x - \mu}{\sigma}$
A	1	15	$1 - 15 = -14$	$\dfrac{-14}{9.23} = -1.52$
B	10	15	$ac10 - 15 = -5$	$\dfrac{-5}{9.23} = -0.54$
C	15	15	$15 - 15 = 0$	$\dfrac{0}{9.23} = 0$
D	21	15	$21 - 15 = 6$	$\dfrac{6}{9.23} = 0.65$
E	28	15	$28 - 15 = 13$	$\dfrac{13}{9.23} = 1.41$

Table 3.3 California Rating of Tuna Fish ($\mu = 45$, $\sigma = 15.17$)

Brand	Rating x	Mean μ	Difference from Mean $(x - \mu)$	z-Score $z = \dfrac{x - \mu}{\sigma}$
M	25	45	$25 - 45 = -20$	$\dfrac{-20}{15.17} = -1.32$
N	35	45	$35 - 45 = -10$	$\dfrac{-10}{15.17} = -0.66$
P	45	45	$45 - 45 = 0$	$\dfrac{0}{15.17} = 0$
Q	50	45	$50 - 45 = 5$	$\dfrac{5}{15.17} = 0.33$
R	70	45	$70 - 45 = 25$	$\dfrac{25}{15.17} = 1.65$

This means that the average of the z-scores is 0 since 0 divided by 5, the number of z-scores, is 0. Also, the sum of the z-scores for the California rating of the brands is 0:

$$(-1.32) + (-0.66) + (0) + (0.33) + (1.65) = 0$$

Therefore, the mean is 0. If you now compute the standard deviations of the z-scores in Tables 3.2 and 3.3, you will find that the standard deviation in each case is 1. We summarize these facts in the following rule.

> **RULE**
>
> In any distribution the mean of the z-scores is 0 and the standard deviation of the z-scores is 1.

Formula 3.8 can be changed so that if we are given a particular z-score, we can calculate the corresponding original score. The changed formula is as follows.

FORMULA 3.9	$x = \mu + z\sigma$

EXAMPLE 3

In a recent swimming contest the mean score was 40 and the standard deviation was 4. If Carlos had a z-score of -1.2, how many points did he score?

Solution

Since $\mu = 40$, $\sigma = 4$, and $z = -1.2$, we can use Formula 3.9. Thus, we have

$$x = 40 + (-1.2)(4)$$
$$= 40 - 4.8$$
$$= 35.2$$

Carlos's score was 35.2.

EXERCISES FOR SECTION 3.8

1. The average bowling score of a member of the Yorkville Bowling Club in a recent contest was 185 with a standard deviation of 15. Find the z-score of
   a. Howie who scored 200 points.
   b. Sheila who scored 120 points.
   c. Paul who scored 185 points.
   d. Carolyn who scored 240 points.

2. Refer back to Exercise 1. The z-scores of five players participating in the contest were as follows:

   Hilda -2 Ike -1 Susan $+1.5$ Sherry -0.2 Marilyn $+1.8$

   a. Rank these people from lowest to highest.
   b. Which of these players were above average?
   c. Which of these players were below average?

3. *Pollution.* There are 6 particular polluted lakes in New York and 5 particular polluted lakes in neighboring New Jersey that were analyzed by environmen-

Garbage and industrial pollutants dumped into our nation's waterways are adversely affecting our environment.

talists from both states. (Each state has its own method for measuring the pollution level of its waterways.) The following results on the amount of pollution were obtained by these two groups.

New York Lake	Rating
A	22
B	15
C	29
D	33
E	24
F	17

New Jersey Lake	Rating
Q	68
R	75
S	61
T	70
U	70

a. Which of these 11 lakes tested has the greatest amount of pollution?

b. Which of these 11 lakes tested has the least amount of pollution?

4. The average age of a paramedic in a particular city is 36 years with a standard deviation of 2.4 years. Find the ages of Joe, Drew, and Frances if their z-scores are −1.6, 2.1, and −0.12, respectively.

5. Leona has two twin boys, Alan and Derek, who recently took two different kinds of math aptitude tests. The average score on the exam that Alan took was 82 with a standard deviation of 4. Alan scored 75 on the exam. The average score on the exam that Derek took was 137 with a standard deviation of 12. Derek

$X = 36 + (2.4)(z)$

$X = 82 + z$

-82

$75 = 82 + z(4)$

$-7 = z4$

scored 120 on the exam. On the basis of these two tests only, who has a higher math aptitude?

6. The following information concerning fines imposed for speeding convictions is posted on the bulletin board of the statistician's office at the Motor Vehicle Office.

Fine Imposed	z-score	Percentile Rank
$40	−2	2
$50	−1	16
$60	0	30
$70	1	50
$80	2	85

Note: $\sigma = 10$.

a. What was the amount of the average fine imposed?
b. What percent of the fines were between $50 and $70 inclusive?
c. On the basis of information given in the preceding chart, what was the fine imposed if its z-score is − 1.38?

7. *Deciding on a major.* Marjorie has decided to become a business major. However, she cannot decide in which particular aspect of business she should major. She takes an aptitude test given to all business majors. Her results, as well as the results of the other candidates, are as follows.

Specialty Within Business	Marjorie's Score on Exam	Mean Score	Standard Deviation
Accounting	49	45	4.7
Finance	81	80	8.3
Marketing/management	62	79	11.2
Real estate/insurance	19	20	3.1
Retailing/sales	40	41	2.6

a. Transform each of Marjorie's scores into a z-score.
b. In which area of business does she have the most talent?
c. In which area of business does she have the least talent?

8. Verify that the standard deviation of the z-scores of Tables 3.2 and 3.3 is 1.

9. The distribution shown on top of the next page indicates the range of grades that can be expected on many intelligence tests. Notice that the scores are normally distributed.

a. If someone scores 600 on the ETS exam, what is the corresponding percentile rank?
b. What percent of the children taking the WISC exam will score higher than 85?
c. Find the percentile rank of a score that has a z-value of + 1.

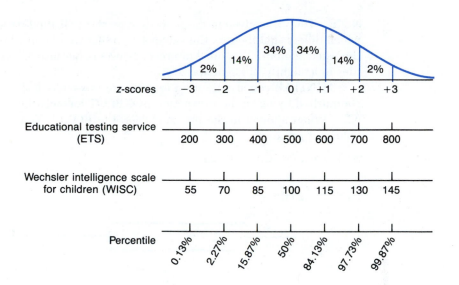

3.9 USING COMPUTER PACKAGES

To illustrate how to use the MINITAB computer package to compute the various statistical measures discussed in this chapter, let us consider the following data on the ages of ten patients who were admitted to the 1990 Rainville Drug Detoxification Program because of their addiction to "Crack." Their ages were 16, 24, 32, 19, 21, 14, 18, 19, 17, and 25. The following MINITAB program gets the data into column C1:

```
MTB   > SET THE FOLLOWING DATA INTO C1
DATA  > 16   24   32   19   21
DATA  > 14   18   19   17   25
DATA  > END
```

The DESCRIBE statement in MINITAB yields the population size (N), the mean, median, trimmed mean (TRMEAN); the standard deviation; the standard error of the mean (to be discussed in a later chapter); the maximum and minimum values; the upper (75th) percentile Q_3; and the lower (25th) percentile Q_1. The DESCRIBE statement for the preceding data produces the following computer printout:

```
MTB   >   DESCRIBE DATA IN C1

       N     MEAN    MEDIAN    TRMEAN    STDEV    SEMEAN
C1    10    20.50    19.00     19.88     5.28      1.67

      MIN      MAX        Q1        Q3
C1  14.00    32.00     16.75     24.25
```

NOTE The value given in the preceding printout for the first quartile (Q1) is 16.75, which differs slightly from the value of 16.5 that we obtain when we calculate Q1 in the usual manner. This small discrepancy occurs because MINITAB computes quartiles in a slightly different way.

MINITAB can also be used to generate box-and-whisker diagrams. This is accomplished by using the instruction BOXPLOT followed by the location of the data (C1). When applied to the previous data, the BOXPLOT instruction produces the following printout:

```
MTB  > BOXPLOT OF C1


MTB  > STOP
```

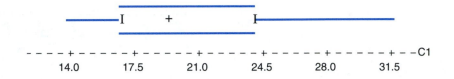

In this printout the median is indicated by the + symbol. Note the length of the whiskers. There is less variation in the bottom two quarters of the drug-age data than in the top two quarters, and the top quarter has the greatest variation of all. MINITAB's box-and-whisker diagram is slightly different from the following boxplot for the same data that we obtain when constructing it by hand. You should compare both diagrams.

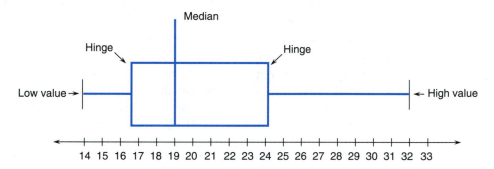

3.10 SUMMARY

In this chapter we discussed various numerical methods for analyzing data. In particular, we calculated and compared three measures of central tendency: the mean, median, and mode. We pointed out that each has its advantages and disadvantages. In addition, various properties of each measure were discussed. Thus, we mentioned that the mean is affected by extreme values and that the sum of the differences from the mean is zero. The mean is the most frequently used measure of central tendency. We also demonstrated how to calculate a weighted mean when the terms of a distribution are not of equal weight. In the process we introduced summation notation.

We then discussed four measures of variation that tell us how dispersed, that is, how spread out, the terms of the distribution are around the *center* of the distribution. These were the range, variance, standard deviation, and average deviation. Various shortcuts for computing the standard deviation and variance were introduced.

We also saw that an individual score is sometimes meaningless unless it is accompanied by a percentile rank or z-score. When scores are converted into percentile ranks or z-values, we can then make meaningful statements about them and compare them with other scores.

We discussed and demonstrated how to calculate percentile ranks as well as z-scores. The latter play an important role in the normal distribution to be discussed in a later chapter.

STUDY GUIDE

The following is a chapter outline in capsule form. You should now be able to demonstrate your knowledge of the ideas mentioned by giving definitions, descriptions, or specific examples. Page references are given in parentheses.

To indicate the operation of taking the sum of a sequence of numbers, we use **summation notation.** (page 101)

The i (or j) in the summation symbol $\sum_{i=1}^{n}$ is referred to as the **index.** (page 102)

The **lower limit of summation** is the value of the index placed below the summation symbol and the **upper limit of summation** is the value of the index placed above the summation symbol. (page 102)

The **mean** or **average** of a set of numbers, denoted by $\bar{x}$ or μ is obtained by adding the numbers together and dividing the sum by the number of numbers added. (page 108)

The **mode** of a set of numbers is the number (or numbers) that occurs most often. (page 108)

The **median** of a set of numbers, denoted by $\tilde{x}$, is the number that is in the middle when the numbers are arranged in order from the lowest to the highest. (page 109)

The mean of a sample of numbers is called the **sample mean.** (page 110)

When all values in a set of numbers are not of equal importance, we must multiply each number by its weight and divide the sum of these products by the sum of the weights to obtain the **weighted sample mean.** (page 116)

The **midrange** of a set of numbers is found by taking the average of the lowest value L and the highest value H. (page 117)

The **harmonic mean** of n numbers x_1, x_2, . . . , x_n is defined as n divided by the sum of the reciprocals. (page 117)

The **geometric mean** of a set of n positive numbers x_1, x_2, . . . , x_n is defined as the nth root of their product. (page 117)

The mean, median, and mode are known as **measures of central tendency** since each measures some central or general trend of the data. (page 118)

Any extremes in a set of numbers are called **outliers.** (page 123)

A **trimmed mean** is calculated by deleting high and low values from a set of numbers and then calculating the mean of the remaining numbers. (page 123)

The **range** of a set of numbers is the difference between the largest number in the distribution and the smallest number in the distribution. (page 124)

The **sample variance** of a sample of numbers is a measure of the spread of the numbers about the sample mean and is given by $s^2 = \dfrac{\Sigma (x - \bar{x})^2}{n - 1}$. (page 126)

The **sample standard deviation** of a sample of numbers, denoted by s, is the positive square root of the sample variance. (page 127)

The **average deviation** of a sample of numbers is the average of the absolute value of the differences from the mean. (page 136)

For any set of measurements—population or sample—and regardless of the shape of the distribution, **Chebyshev's Theorem** specifies the following: Let k be any number ≥ 1. Then the proportion of the distribution that lies within k standard deviations of the mean is at least $1 - \dfrac{1}{k^2}$. (page 139)

If the frequency distribution of a set of sample data is mound-shaped (normally distributed), then the **Empirical Rule** states that approximately 68% of the measurements will fall within 1 standard deviation of the mean, 95% of the measurements will fall within 2 standard deviations of the mean, and 99.7% of the measurements will fall within 3 standard deviations of the mean. (page 142)

If we add together all the frequencies in each interval of grouped data that are less than or equal to a given interval and graph, the results are a **cumulative frequency histogram** or **ogive.** (page 143)

A score that indicates what percent of the population scored at or below a given measure is called the **percentile.** (page 144)

The **percentile rank** of a term in a distribution is found by adding the percentage of terms below it with one-half the percentage of terms equal to the given term. (page 144)

The 25th percentile of a distribution, denoted as P_{25} or Q_{25}, is called the **lower** or **first quartile.** (page 147)

The 50th percentile or middle quartile of a distribution is its **median.** (page 147)

The 75th percentile of a distribution, denoted as P_{75} or Q_{75}, is called the **upper** or **third quartile.** (page 147)

The **interquartile range** is the difference between the upper and lower quartiles. (page 148)

A **boxplot** or **box-and-whisker diagram** is a way of analyzing data that is based on the quartiles of the data set and the interquartile range. The lower and upper quartiles are called **hinges.** They are connected to the low and high values by means of **whiskers.** (page 148)

The **z-score** of any number x in a distribution whose mean is μ and whose standard deviation is σ is given by $z = \dfrac{x - \mu}{\sigma}$ or $z = \dfrac{x - \bar{x}}{s}$. (page 153)

The **z-score** of a number in a population is sometimes called the **z-value** or **measurement in standard units.** (page 153)

FORMULAS TO REMEMBER

At this point you have learned some of the common terms used in statistical analysis and some of the graphic techniques. The formulas are important too.

The following list is a summary of all formulas given in the chapter. You should be able to identify each symbol, understand the relationships among the symbols expressed in each formula, understand the significance of each formula, and use the formulas in solving problems.

1. When using summation notation the following are true:

$$\sum_{i=1}^{n} (x_i + y_i) = \sum_{i=1}^{n} x_i + \sum_{i=1}^{n} y_i$$

$$\sum_{i=1}^{n} kx_i = k \cdot \sum_{i=1}^{n} x_i$$

$$\sum_{i=1}^{n} k = n \cdot k$$

2. Sample mean: $\bar{x} = \dfrac{\sum x}{n}$

Population mean: $\mu = \dfrac{\sum x}{N}$

3. Sample mean for grouped data: $x = \dfrac{\sum xf}{\sum f}$

4. Weighted sample mean: $\bar{x}_w = \dfrac{\Sigma\, xw}{\Sigma\, w}$

5. Sample variance: $s^2 = \dfrac{\Sigma(x - \bar{x})^2}{n - 1}$

6. Sample standard deviation: $s = \sqrt{\dfrac{\Sigma(x - \bar{x})^2}{n - 1}}$

7. Average deviation: $\dfrac{\Sigma|x - \bar{x}|}{n}$

8. Population variance: $\sigma^2 = \dfrac{\Sigma(x - \mu)^2}{N}$

9. Population standard deviation: $\sigma = \sqrt{\dfrac{\Sigma(x - \mu)^2}{N}}$

10. Computational formula for population variance: $\sigma^2 = \dfrac{\Sigma\, x^2}{N} - \dfrac{(\Sigma\, x)^2}{N^2}$

11. Computational formula for population standard deviation:
$$\sigma = \sqrt{\dfrac{\Sigma\, x^2}{N} - \dfrac{(\Sigma\, x)^2}{N^2}}$$

12. Computational formula for sample variance:
$$s^2 = \dfrac{n(\Sigma\, x^2) - (\Sigma\, x)^2}{n(n - 1)}$$

13. Computational formula for sample standard deviation:
$$s = \sqrt{\dfrac{n(\Sigma\, x^2) - (\Sigma\, x)^2}{n(n - 1)}}$$

14. Computational formulas for grouped data:

Sample variance: $s^2 = \dfrac{n(\Sigma\, x^2 \cdot f) - (\Sigma\, x \cdot f)^2}{n(n - 1)}$

Sample standard deviation: $s = \sqrt{\dfrac{n(\Sigma\, x^2 \cdot f) - (\Sigma\, x \cdot f)^2}{n(n - 1)}}$

15. *Chebyshev's Theorem:* At least $1 - \dfrac{1}{k^2}$ of a set of measurements will lie within k standard deviation units of the mean (assuming $k \geq 1$).

16. Percentile rank of $X = \dfrac{B + \dfrac{1}{2}E}{N} \cdot 100$

17. z-score: $z = \dfrac{x - \mu}{\sigma}$ or $z = \dfrac{x - \bar{x}}{s}$

18. Original score: $x = \mu + z\sigma$

Testing Your Understanding of This Chapter's Concepts

1. Joe has a high school average of 89. The college that he wishes to attend will not accept any applicant with a percentile rank below 85. Is Joe sure he will be accepted by this college or is it possible that he will be denied admission? Explain your answer.

2. Can a percentile rank of 56 have a negative z-score? Explain your answer.

3. Verify, by expanding, that $\Sigma (x - \mu) = 0$

4. Verify that the following two formulas for calculating the sample standard deviation are the same:

$$s = \sqrt{\frac{\Sigma(x - \bar{x})^2}{n - 1}} = \sqrt{\frac{n(\Sigma x^2) - (\Sigma x)^2}{n(n - 1)}}$$

5. Locate the approximate location of the mean, median, and mode for the frequency distributions given in parts (b) and (c). We have already done it for the diagram of part (a).

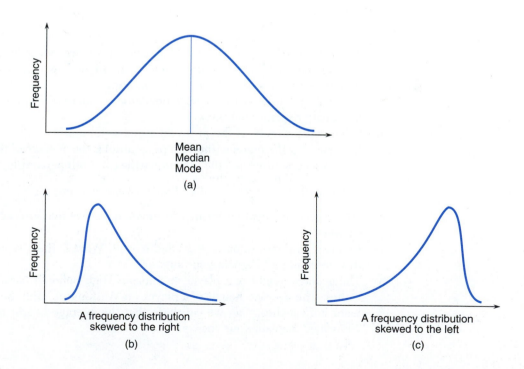

Mean
Median
Mode

(a)

A frequency distribution
skewed to the right

(b)

A frequency distribution
skewed to the left

(c)

6. When working with a normal distribution, z-scores will usually be between -3 and $+3$. Can you explain why?

THINKING CRITICALLY

1. A certain set of 10 measurements has a mean equal to 75 and a variance equal to 64. Find   (a) Σx (b) Σx^2.
2. Trudy's two twin daughters Heather and Cindy, who are graduating from high school in June, plan to go on to college and major in business. However, they cannot decide in which particular aspect of business they should major. The twins take a series of aptitude tests given to all prospective majors. The following results are available.

Specialty Within Business	Heather's z-score	Cindy's Percentile rank
Accounting	2.00	99
Finance	1.00	72
Marketing/management	1.00	87
Real estate/insurance	0	50
Retailing/sales	-1.00	33

Assume that the results of these aptitude tests are normally distributed. For each of the specialties (within business) mentioned, which of the girls has the higher relative talent? Explain your answer.

3. Is it ever possible for the average deviation, standard deviation, and variance to be equal? Explain your answer.
4. What is the z-score of the mean (in any distribution)?
5. Is it true that the median (the 50th percentile) is the average of the lower quartile (the 25th percentile) and the upper quartile (the 75th percentile)? In other words, is it true that $\dfrac{P_{25} + P_{75}}{2} = P_{50}$. Explain your answer.
6. It is true that negative terms can never have positive z-scores? Explain your answer.
7. In a certain distribution $\mu = 85$ and $\sigma = 0$. What is the z-score of any term in this distribution? Explain your answer.
8. The average height of a player on Lincoln High School's basketball team is 68 inches. The average height of a player on Washington High School's basketball team is 72 inches. Can we conclude that the average height for both teams is 70 inches? Explain your answer.

Review Exercises for Chapter 3

1. A consumer's group tested two different brands of beer for flavor retention. The following data are available on the number of minutes that glasses of beer, using each of the following brands, maintained their flavor once opened:

$$\text{Brand A: mean} = 10 \qquad \text{standard deviation} = 2$$

$$\text{Brand B: mean} = 12 \qquad \text{standard deviation} = 4$$

   Which brand of beer would you choose? Why?

2. If $x_1 = 19$, $x_2 = 23$, $x_3 = 11$, $x_4 = 16$, and $x_5 = 13$, find   (a) Σx (b) Σx^2.

3. In an attempt to conserve energy many homeowners install storm windows or tape up existing windows with plastic to prevent heat loss during the winter. A random sample of four houses on Trafalagar Avenue disclosed that these houses had 3, 5, 6, and 8 storm or taped windows. A similar sample of four houses on Billings Lane disclosed that these houses had 3, 10, 17, and 21 storm or taped windows. Without actually doing the calculation, how does the standard deviation for the number of windows on Trafalagar Ave compare with the standard deviation for the number of windows on Billings Lane?

4. Is it safe for a man who is 6 feet tall and a nonswimmer to walk in a pool that has an *average* depth of 3 feet? Explain your answer.

5. Mr. and Mrs. Spector recently celebrated their 50th wedding anniversary. Their four boys contributed an average of $750 to pay for a round-the-world trip, whereas their seven daughters contributed an average of $650 for the trip. Find the average amount contributed by each child for the trip.

6. John, who is 23 years old, wants to date a girl who is approximately 20 years old. He is considering two computer dating services in the Los Angeles area, both of which claim that the average age of the girls they have on file is 20 years. John does not know which dating service to select for he believes that both are equally good. His friend persuades him to look at the individual ages of the girls and not at the average. The individual ages are as follows.

Dating Service A	Dating Service B
15	18
17	19
18	20
23	21
27	22

   Is his friend right? Explain your answer.

7. The Aura Chemical Company claims to have developed a cheap 9-volt rechargeable battery that lasts many hours and requires very little time to charge. The following frequency distribution indicates the number of minutes required to fully charge 100 batteries.

Number of Minutes	Frequency
100–199	12
200–299	28
300–399	20
400–499	18
500–599	14
600–699	8
	100

Calculate the sample mean and sample standard deviation for the data.

8. Refer back to Exercise 7. Find P_{55}.

9. A certain type of toner for a laser jet printer will give an average of 6000 copies with a standard deviation of 185. Find the z-score of a toner that gives 6200 copies.

10. The following relative frequencies were obtained by entering college students from Roosevelt High School on the Scholastic Aptitude Test administered by the Educational Testing Service:

Test Score	Relative Frequency
200–249	0.01
250–299	0.02
300–349	0.06
350–399	0.10
400–449	0.17
450–499	0.19
500–549	0.18
550–599	0.15
600–649	0.08
650–699	0.03
700–749	0.01
750–799	0

Approximately what score must a student have in order to be in the   (a) 70th percentile,   (b) 90th percentile.

11. If the third quartile for scores on a test is 83, what percent of the scores are above 83?

Chapter Test

1. Given the followng measurements:

$$9, 13, 15, 15, 15, 24, 35, 42, 64, 75, 82, 96$$

a. The mode is _____

b. The median is _____

c. The mean is _____

d. The lower quartile is _____

e. The upper quartile is _____

2. Marlene's grade on the calculus midterm and on the calculus final as well as information on the grades of the entire class are as follows.

	Midterm	Final
Marlene's grade	80	60
Class mean	85	68
Class standard deviation	8	12

When compared to the rest of the class, on which exam did she do better?

3. In any distribution, what proportion of terms lies within 1.5 standard deviations of the mean?

4. A survey was conducted at a major U.S. airport to determine how many minutes planes arrived late after their scheduled arrival time. The following results were obtained.

Minutes Late	Frequency
5–9	2
10–14	16
15–19	55
20–24	111
25–29	65
30–34	9

Find the mean and standard deviation for this distribution.

5. A survey was conducted in Yorkville to determine the number of telephones per household. The following table lists the results of the survey.

Number of Telephones per Household	Frequency
1	30
2	45
3	120
4	80
5	60
6	20
7	5

Find   (a) Q_1, (b) Q_3 (c) P_{80} (d) P_{95}.

6. MINITAB was used to generate the following boxplot.
    a. Find the median of the data set (approximately).
    b. Find the upper and lower quartile for the data set (approximately).
    c. Find the interquartile range for the data set.

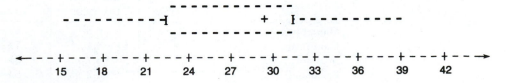

7. The closing price (to the nearest dollar) of a share of stock of the Afgax Corp. for 30 randomly selected days is as follows:

$43 $41 $35 $38 $33 $42 $39 $29 $31 $40
39 32 28 29 43 45 31 30 44 29
40 33 34 32 29 28 34 35 37 38

Contruct a relative frequency histogram for these data.

8. The average time required for firefighters to race up the stairs of a high-rise building from street level to the top floor is 48 seconds, with a standard deviation of 9.1 seconds. Find the time required by Bob, Mark, and Chris to race up the stairs if their z-scores are 3.3, -0.45, and 1.43, respectively.

9. The average age of 25 women who work for Mack's Department Store is 28 years. The average age of the 36 men who work for the store is 32 years. What is the average age for the entire group?

10. If $x_1 = 16$, $x_2 = 12$, $x_3 = 19$, $x_4 = 26$, and $x_5 = 23$, find $\Sigma(x - \bar{x})$.

11. Fifty people have applied to be contestants on a TV variety show. Maureen O'Brien is one of the applicants. She scored better than 32 of the applicants on the qualifying exam and 6 other people scored the same as she did. Find her percentile rank.

ACME CLOTHING TO CLOSE

Dec. 5: Acme Clothing Company officials announced yesterday that they would close the Patchogue plant on Feb. 1, 1991, with the resulting loss of 1400 jobs. The closing is a direct result of rising foreign imports with which the company cannot compete.

Tuesday—Dec. 4, 1990

12. Refer to the accompanying newspaper article. The company agreed to retrain some of the workers so that they could find other jobs. Each of the workers was given an aptitude test to determine suitability for particular jobs. Laura Snyder's results as well as the results of the other workers who took the test are shown here.

Skill	Average Test Score	Standard Deviation	Laura's Score
Marketing/sales	58	6.8	59
Plant maintenance	69	4.1	72
Inventory/shipping	27	3.7	31
Personnel	83	5.2	84

a. Transform each of Laura's scores into a z-score.

b. In which skill does she have the most talent?

c. In which skill does she have the least talent?

13. *Case Study* Despite governmental supervision, many companies often dispose of industrial wastes by dumping them into nearby rivers, lakes, or streams, thereby polluting the water supply for the animals and fish that feed in these lakes. In one such study an environmental group documented the devastating effect of such activities for a particular river.* They collected data on the number of fish and deer that died (based on pathological studies) as a result of industrial dumping. The data that follow are for a six-month period.

Month	Number of Dead Deer Found
April	12
May	14
June	6
July	9
August	8
September	11

Find the (sample) mean and standard deviation for the preceding data.

14. If $x_1 = 4$, $x_2 = 6$, $x_3 = 9$, $x_4 = 11$, and $x_5 = 15$, find $(\Sigma\ x)^2$.

    a. 2025   **b.** 479   **c.** 45   **d.** 25   **e.** none of these

For questions 15–19 refer to the following data which gives the ages of some of the Board members at the Bayview Senior Citizens Center:

Sally	68	Don	69	Caldwell	67	Stan	70
Bob	64	Gregg	68	Juan	65		
Ellen	73	Stuart	72	Miguel	77		

15. Find the (sample) mean age of these Board members.

    a. 69.3   **b.** 69   **c.** 68   **d.** 68.5   **e.** none of these

16. Find the median age of these Board members.

    a. 69.3   **b.** 69   **c.** 68   **d.** 68.5   **e.** none of these

17. Find the modal age of these Board members.

    a. 69.3   **b.** 69   **c.** 68   **d.** 68.5   **e.** none of these

18. Find Sally's percentile rank.

    a. 30th percentile   **b.** 35th percentile   **c.** 40th percentile

    b. 70th percentile   **e.** none of these

19. Find the sample standard deviation for these Board members.

    a. 136.1   **b.** 13.61   **c.** 3.689   **d.** 3.889   **e.** none of these

20. On September 14, the 21 sales representatives at the downtown office of a Wall Street brokerage firm and the 17 sales representatives at the uptown office of the brokerage firm executed an average of 89 and 93 transactions. What was the average number of transactions for the entire group on that day?

    a. 91   **b.** 90.789   **c.** 89   **d.** 93   **e.** none of these

*Boggs and Jones, New York, 1986.

Suggested Further Reading

1. Harrell, T. W., and M. S. Harrell. "Army General Classification Test Scores for Civilian Occupations." *Educational and Psychological Measurements* (1945), pp. 222–239.
2. Mendelhall, W. *Introduction to Probability and Statistics*, 7th ed. Boston: Duxbury Press, 1987, Chapter 3.
3. Neter, J., W. Wasserman, and G. A. Whitmore, *Applied Statistics*. Boston: Allyn & Bacon, 1982, Chapter 3.
4. Newmark, J., and F. Lake. *Mathematics as a Second Language*, 4th ed. Reading, MA: Addison-Wesley, 1987.
5. Newmark, J. *Using Finite Mathematics*. New York: Harper & Row, 1982.
6. Tanur, J. M., F. Mosteller, W. H. Kruskal, R. G. Link, R. S. Pieters, and G. R. Rising. *Statistics: A Guide to the Unknown.* (E. L. Lehman, Special editor.) San Francisco: Holden-Day, 1978.

4 Probability

Gamblers frequently speak of odds when betting on horses. See discussion on p. 222. (© Bancroft, Fran Heyl Associates)

Until now we have been discussing methods for organizing and summarizing data, that is, descriptive statistics. Since we often use results of a sample to make inferences about the unknown population from which the sample was drawn, we can never be certain that our inferences are correct. Thus, before studying the techniques of statistical inference, we need to be familiar with probability theory that involves the science of uncertainty.

In this chapter we introduce some of the basic ideas of probability that will enable us to evaluate the likelihood that our statistical inferences are correct.

CHAPTER OBJECTIVES

After studying the material in this chapter, you should be able:

- **To define** what is meant by the word "probability." (Section 4.2)

- **To explain** the use of the word "probability" in such statements as "the probability of a particular thing happening is 0.78." (Section 4.2)

- **To discuss** a formula for determining the number of possible outcomes when an experiment is performed. This is done so that we can find the total number of possible outcomes, which we use for probability calculations. (Section 4.3)

- **To analyze** the number of different ways of arranging things, depending on whether or not order counts. Thus, we will analyze permutations and combinations. (Sections 4.4 and 4.5)

- **To work** with a convenient notation used to represent a special type of multiplication. This is the factorial notation. (Section 4.4)

- **To apply** a computational device for calculating the number of possible combinations. This is Pascal's triangle. (Section 4.5)

- **To talk** about "odds" and "mathematical expectation." These words, often used by gamblers, represent the payoff for a situation and the likelihood of obtaining it. (Section 4.6)

GOVERNORS ASK FOR MORE STATE AID

May 10: It was decided yesterday at a governor's meeting that a committee of four governors would be selected to go to Washington to plead for more state aid. Many governors expressed the view that the current level of federal aid to the states was appalling, and that it was the obligation of the federal government to share the state's financial burden

May 10, 1990

The newspaper article indicates that a committee of four governors would be selected to go to Washington to plead for state aid. How do we select such committees? How many four-governor committees can possibly be formed from among all the governors?

SMOKING AND YOUR HEALTH— THE LATEST STATISTICS

Washington: Smoking is a powerful addiction. The latest statistics released by the American Cancer Society indicate that young women under age 23 make up the fastest growing group of new smokers in the United States. One study of high school seniors found that an average 31.6% of girls smoked compared to 28% for boys. Most of the kids began the habit before the age of 19. According to a report from the Surgeon General Antonia Novello, after just one year off cigarettes the probability of the added cancer risk caused by previous smoking is reduced by 50%. Furthermore, after about 15 years of nonsmoking, the probability of lung cancer is about the same as that of individuals who never smoked.

Friday—January 4, 1991

Note the use of the word "probability" in the accompanying newspaper article. How are such probabilities determined? Are they reliable?

In this chapter we analyze what we mean by probability and how we determine such probabilities.

4.1 INTRODUCTION

Although the word "probability" may sound strange to you, it is not so unfamiliar as you may think. In everyday situations we frequently make decisions and take action as a result of the probability of certain events. Thus, if the weather forecaster predicts rain with a probability of 80%, we undoubtedly would prepare ourselves accordingly.

Let us, however, analyze the weather forecaster's prediction. What the forecaster really means is that based on past records, 80% of the time when the weather conditions have been the same as they are today, rain has followed. Thus, the probability calculations and resultant forecasts are based on past records. They are based on the assumption that since in the *past* rain has occurred a certain percentage of the time, it will occur the same percentage of times in the *future*. This is but one usage of probability. It is based on **relative frequency.** We will explain this idea in greater detail shortly.

Probability is also used in statements that express a personal judgment or conviction. This can be best illustrated by the following statements: "If the United States had not dropped the atomic bomb on Japan, World War II would *probably* have lasted several more years" or "If all the New York Mets players had been healthy the entire season, they *probably* would have won the pennant last year."

Probability can also be used in other situations. For example, if a fair coin* is tossed, we would all agree that the probability is $\frac{1}{2}$ that heads comes up. This is because there are only two possible outcomes when we flip a coin, heads or tails.

Now consider the following conversation overheard in a student cafeteria.

Bill: I am going to cut math today.
Eric: Why?
Bill: I didn't do my homework.
Eric: So what? I didn't either.
Bill: So, since the teacher calls on at least half of the class each day for answers, he will *probably* call on me today and find out that I am not prepared.
Eric: The teacher called on me yesterday, so he *probably* won't get me today. I am going to class.

In the preceding situation each student is making a decision based on probability.

Basically, the theory of probability deals with the study of uncertainties. Thus, it has been found to have wide applications in the following situations:

1. It is used by insurance companies when they calculate insurance premiums and the *probable* life expectancies of their policyholders.

---

*A fair coin is a coin that has the same chance of landing on heads as on tails. Throughout this book we will always assume that we have fair coins unless told otherwise.

2. It is used (formally or informally) by a gambler who decides to bet 10 to 1 on a particular horse.
3. It is used by industry officials in determining the reliability of certain equipment.
4. It is used by medical researchers who claim that smoking increases your *chance* of getting lung cancer.
5. It is used by biologists in their study of genetics.
6. It is used by pollsters in such polls as the Harris poll, the Gallup poll, and the Nielsen ratings to determine the reliability of their polls.

HISTORICAL NOTE

Historically, probability had its origin in the gambling room. The Chevalier de Méré, a professional French gambler, had asked his friend Blaise Pascal (1623–1662) to solve the following problem: In what proportion should two players of equal skill divide the stakes remaining on the gambling table if they are forced to stop before finishing the game? Pascal wrote to and began an active correspondence with Pierre Fermat (1601–1665) concerning the problem. Although Pascal and Fermat agreed on the answer, they gave different proofs. It was in this series of correspondences during the year 1652 that they developed the modern theory of probability.

A century earlier the Italian mathematician and gambler Girolomo Cardan (1501–1576) wrote *The Book on Games of Chance*. For a further discussion of Cardan, see p. 11.

Another famous mathematician who contributed to the theory of probability was Abraham De Moivre (1667–1754). Like de Méré, De Moivre spent many hours with London gamblers and wrote a manual for gamblers entitled *Doctrine of Chances*. Like Cardan, De Moivre predicted the day of his death. A rather interesting story is told of De Moivre's death. De Moivre was ill and each day he noticed that he was sleeping 15 minutes longer than he did on the preceding day. Using progressions, he computed that he would die in his sleep on the day after he slept 23 hours and 45 minutes. On the day following a sleep of 24 hours, De Moivre died.

The French mathematician Pierre Simon de Laplace (1749–1827) also contributed much to the historical development of probability. In 1812 he published *Théorie Analytique des Probabilités*, in which he referred to probability as a science that began with games but that had wide-ranging applications. In particular, he applied probability theory not to gambling situations but as an aid in astronomy.

Over the course of many years probability theory has left the gambling rooms and has grown to be an important and ever-expanding branch of mathematics.

7. It is used by an investor who decides that a particular stock has a greater chance for future growth than any other stock.
8. It is used by business managers in determining which products to manufacture, which products to advertise, and through which media: television, radio, magazines, newspapers, subway and bus advertisements, and so on.
9. It is used by psychologists in predicting reactions or behavioral patterns under certain stimuli.
10. It is used by government economists in predicting that the inflation rate will increase or decrease in the future.

Since probability has so many possible meanings and uses, we will first analyze the nature of probability and how to calculate it. This will be done in this chapter. In the next chapter we will discuss various rules that allow us to calculate probabilities for many different situations.

4.2 DEFINITION OF PROBABILITY

Probability theory can be thought of as that branch of mathematics that is concerned with calculating the probability of outcomes of experiments.

Since many ideas of probability were derived from gambling situations, let us consider the following experiment. An honest die (the plural is dice) was rolled many times and the number of 1's that came up was recorded. The results are

Number of 1's that came up	1	11	18	99	1001	10,001
Number of rolls of the die	6	60	120	600	6000	60,000

Notice that in each case the number of 1's that appeared is approximately $\frac{1}{6}$ of the total number of tosses of the die. It would then be reasonable to conclude that the probability of a 1 appearing is $\frac{1}{6}$.

Although when a die is rolled there are six equally likely possible outcomes if it is an honest die (see Figure 4.1), we are concerned with the number of 1's appearing. Each time a 1 appears, we call it a **favorable outcome.** There are six possible outcomes of which only one is favorable. The probability is thus the number

Favorable Outcome

Figure 4.1
Possible outcomes when a die is rolled once.

of favorable outcomes divided by the total number of possible outcomes, which is 1 divided by 6, or $\frac{1}{6}$. The preceding chart indicates that our guess, that the probability is approximately $\frac{1}{6}$, is correct.

Possible outcomes when a coin is flipped.

Sample Space

Event

Similarly, if a coin is tossed once, we would say that the probability of getting heads is $\frac{1}{2}$ since there are two possible outcomes, heads and tails, and only one is favorable. These are the only two possible outcomes in this case. All the possible outcomes of an experiment are referred to as the **sample space** of the experiment. We will usually be interested in only some outcomes of the experiment. The outcomes that are of interest to us will be referred to as an **event.** Thus, in flipping a coin once the sample space is Heads or Tails, abbreviated as H,T. The event of interest is H.

In the rolling of a die the sample space consists of six possible outcomes: 1, 2, 3, 4, 5, and 6. We may be interested in the event "getting a 1."

If we toss a coin twice, the sample space is HH, HT, TH, and TT. There are four possibilities. In this abbreviated notation HT means heads on the first toss and tails on the second toss, whereas TH means tails on the first toss and heads on the second toss. The event "getting a head on both tosses" is denoted by HH. The event "no head" is TT.

To illustrate further the idea of sample space and event, consider the following examples.

EXAMPLE 1

Two dice are rolled at the same time. Find the sample space.

Solution

There are 36 possible outcomes as pictured in Figure 4.2. The possible outcomes are summarized as follows:

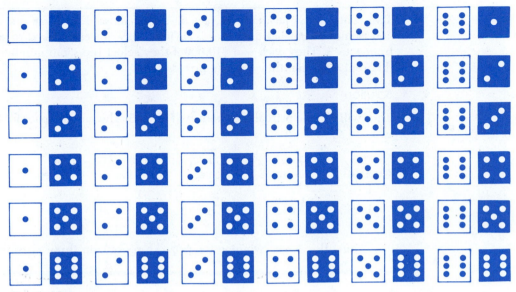

Figure 4.2
Thirty-six possible outcomes when two dice are rolled at the same time.

Die 1	Die 2	Die 1	Die 2	Die 1	Die 2
1	1	3	1	5	1
1	2	3	2	5	2
1	3	3	3	5	3
1	4	3	4	5	4
1	5	3	5	5	5
1	6	3	6	5	6
2	1	4	1	6	1
2	2	4	2	6	2
2	3	4	3	6	3
2	4	4	4	6	4
2	5	4	5	6	5
2	6	4	6	6	6

The event "sum of 7 on both dice together" can happen in six ways. These are

Die 1	Die 2	Die 1	Die 2
1	6	3	4
6	1	5	2
4	3	2	5

Similarly, the event "sum of 9 on both dice together" can happen in four ways. The event "sum of 13 together" can happen in zero ways.

EXAMPLE 2

Two contestants, Jack and Jill, are on a quiz show. Each is in a soundproof booth and cannot hear what the other is saying. They are each asked to select a number from 1 to 3. They each win $10,000 if they select the same number. In how many different ways can they win the prize?

Solution

They will win the prize if they both say 1, 2, or 3. The sample space for this experiment is

Jill Guesses	Jack Guesses	Jill Guesses	Jack Guesses	Jill Guesses	Jack Guesses
1	1	2	1	3	1
1	2	2	2	3	2
1	3	2	3	3	3

Thus, the event "winning the prize" can occur in three possible ways.

We now define formally what we mean by probability.

DEFINITION 4.1
Probability

If an event can occur in n equally likely ways and if f of these ways are considered favorable, then the **probability** of getting a favorable outcome is

$$\frac{\text{Number of favorable outcomes}}{\text{Total number of outcomes}} = \frac{f}{n}$$

Thus, the probability of any event equals the number of favorable outcomes divided by the total number of possible outcomes.

We use the symbol $p(A)$ to stand for "the probability of event A."

Classical Interpretation of Probability

Definition 4.1 is often called the **classical interpretation of probability.** It assumes that all outcomes of an experiment are equally likely. Thus, if we say that the probability of getting heads when flipping an honest coin is $\frac{1}{2}$, we are basing this on the following facts: When a coin is flipped once, there are 2 possible outcomes, heads and tails. Therefore, the probability of getting heads is $\frac{1}{2}$ (1 out of a possible 2 outcomes.) Similarly, the probability of getting a 1 when an honest die is rolled is $\frac{1}{6}$ since there are 6 possible outcomes, only 1 of which is favorable.

Relative Frequency Concept of Probability

An alternate interpretation of probability is called the **relative frequency concept of probability.** Suppose we tossed a coin 100 times and it landed heads 50 of the times. It would seem reasonable to claim that the probability of landing heads is approximately $\dfrac{50}{100}$ or $\dfrac{1}{2}$. We can think of the probability as the relative frequency of the event. Of course, to convince ourselves that our answer is reasonable, we could toss the coin many, many additional times. Since it is much easier to determine relative frequencies, this interpretation is very easy to understand and is commonly used. The relative-frequency interpretation represents the percentage of times that the event will happen in repeated experiments.

Probability can also be defined from a strictly mathematical, that is, axiomatic point of view; however, this is beyond the scope of this text. Let us now illustrate the concept of probability with several examples.

EXAMPLE 3

A family plans to have three children. What is the probability that all three children will be girls? (Assume that the probability of a girl being born in a given instance is $\dfrac{1}{2}$).

Solution

We will use Definition 4.1. We first find the total number of ways of having three children, that is, the sample space. There are eight possibilities as shown in the following table:

Child 1	Child 2	Child 3
Boy	Boy	Boy
Boy	Boy	Girl
Boy	Girl	Boy
Boy	Girl	Girl
Girl	Boy	Boy
Girl	Girl	Boy
Girl	Boy	Girl
Girl	Girl	Girl

Of these, only one is favorable, namely, the outcome Girl, Girl, Girl. Thus,

$$p(3 \text{ girls}) = \frac{1}{8}$$

EXAMPLE 4

Playing Cards

A card is selected from an ordinary deck of 52 cards. What is the probability of getting

a. a queen?
b. a diamond?

c. a black card?

d. a picture card?

e. the king of clubs?

Solution

A deck of playing cards consists of 52 cards, as shown in Figure 4.3. When we perform the experiment of randomly selecting a card from the deck, there are 52 possible outcomes, namely, the 52 cards pictured in Figure 4.3.

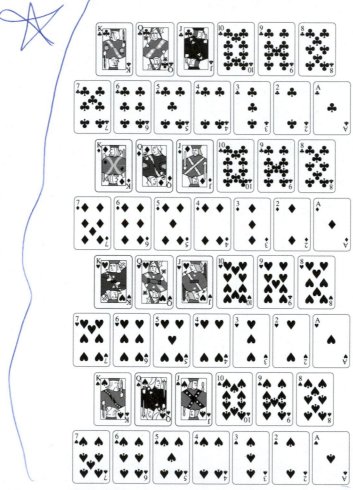

Figure 4.3

The sample space when a card is drawn from an ordinary deck of 52 cards.

a. As shown in Figure 4.3, there are 4 queens in the deck, so there are 4 favorable outcomes. Definition 4.1 tells us that

$$p(\text{queen}) = \frac{4}{52} = \frac{1}{13}$$

b. There are 13 diamonds in the deck, so there are 13 favorable outcomes. Therefore,

$$p(\text{diamonds}) = \frac{13}{52} = \frac{1}{4}$$

c. Since a black card can be either a spade or a club, there are 26 black cards in the deck as shown in Figure 4.3. Therefore,

$$p(\text{black card}) = \frac{26}{52} = \frac{1}{2}$$

d. There are 12 picture cards (4 jacks, 4 queens, and 4 kings), so there are 12 favorable outcomes. Therefore,

$$p(\text{picture card}) = \frac{12}{52} = \frac{3}{13}$$

e. There is only one king of clubs in a deck of 52 cards. Thus,

$$p(\text{king of clubs}) = \frac{1}{52}$$

EXAMPLE 5

Mary and her friend Gwendolyn are visiting the Sears' Tower in Chicago. Each of them enter a different elevator in the main lobby that is going up and that can let them off on any floor from 1 to 6. Assuming that Mary and Gwendolyn are as likely to get off at one floor as another, what is the probability that they both get off on the same floor?

Solution

Since both Mary and Gwendolyn can get off at any floor between 1 and 6, there are 36 possible outcomes as shown in the following table. A favorable outcome occurs if both get off on floor 1, floor 2, and so on as shown. Thus, there are 6 favorable outcomes out of 36 possible outcomes, so that

Floor Where Mary Gets Off	Floor Where Gwendolyn Gets Off	Floor Where Mary Gets Off	Floor Where Gwendolyn Gets Off
1	1 ← Favorable outcome	4	1
1	2	4	2
1	3	4	3
1	4	4	4 ← Favorable outcome
1	5	4	5
1	6	4	6
2	1	5	1
2	2 ← Favorable outcome	5	2
2	3	5	3
2	4	5	4
2	5	5	5 ← Favorable outcome
2	6	5	6
3	1	6	1
3	2	6	2
3	3 ← Favorable outcome	6	3
3	4	6	4
3	5	6	5
3	6	6	6 ← Favorable outcome

$$p(\text{both get off on same floor}) = \frac{6}{36} = \frac{1}{6}$$

EXAMPLE 6   A wheel of fortune has the numbers 1 through 50 painted on it (see the diagram on top of page 185). Tickets numbered 1 through 50 have been sold. The wheel will be rotated and the number on which the pointer lands will be the winning number. If the pointer stops on the line between two numbers, the wheel must be turned again. A prize of a new car will be awarded to the person whose ticket number matches the winning number. What is the probability that

a. ticket number 42 wins?
b. a ticket between the numbers 26 and 39 wins?
c. ticket number 51 wins?

Solution

Since 50 tickets were sold, the total number of possible outcomes, that is, the sample space, is 50.

a. Only 1 ticket numbered 42 was sold. There is only one favorable outcome. Thus,

$$p(\text{ticket number 42 wins}) = \frac{1}{50}$$

b. There are 12 ticket numbers between 26 and 39, not including 26 and 39. Thus,

$$p(\text{a ticket between the numbers 26 and 39 wins}) = \frac{12}{50} = \frac{6}{25}$$

c. Since tickets numbered up to 50 were sold, ticket number 51 can never be a winning number. Thus,

$$p(\text{ticket number 51 wins}) = \frac{0}{50} = 0$$

Null Event   An event that can never happen is called a **null event** and its probability is 0.

EXAMPLE 7   Two fair dice are rolled at the same time and the number of dots appearing on both dice is counted. Find the probability that this sum

a. is 7.
b. is an odd number larger than 6.
c. is less than 2.
d. is more than 12.
e. is between 2 and 12, including these two numbers.

Solution

When two dice are rolled, there are 36 possible outcomes; that is, the sample space has 36 possibilities. These were listed in Example 1 on page 179.

a. The sum of 7 on both dice together can happen in 6 ways so that

$$p(\text{sum of } 7) = \frac{6}{36} \text{ or } \frac{1}{6}$$

b. The statement "a sum that is an odd number larger than 6" means a sum of 7, a sum of 9, or a sum of 11. A sum of 7 on both dice together can happen in 6 ways. Similarly, a sum of 9 on both dice together can happen in 4 ways, and a sum of 11 can happen in 2 ways. There are then 12 favorable outcomes out of 36 possibilities. Thus,

$$p(\text{a sum that is an odd number larger than } 6) = \frac{12}{36} = \frac{1}{3}$$

c. When two dice are rolled, the minimum sum is 2, and we cannot obtain a sum less than 2. There are *no* favorable events. This is the null event. Hence,

$$p(\text{a sum that is less than } 2) = 0$$

d. When two dice are rolled, the maximum sum is 12. We cannot obtain a sum that is more than 12. This is the null event. Thus,

$$p(\text{a sum that is more than } 12) = 0$$

e. When two dice are rolled we *must* obtain a sum that is between 2 and 12, including the numbers 2 and 12. There are 36 possible outcomes and *all* these are favorable. Thus,

$$p(\text{a sum between 2 and 12, including 2 and 12}) = \frac{36}{36} = 1$$

Therefore, a favorable outcome *must* occur in this case.

Certain Event or Definite Event

An event that is certain to occur is called the **certain event** or **definite event** and its probability is 1.

COMMENT Any event, call it A, may or may not occur. If it is sure to occur, we have the certain event and its probability is 1. If it will never occur, we have the null event and its probability is 0. Thus, if we are given any event A, then we know that its probability *must* be between 0 and 1 and possibly equal to 0 or 1. This is because the event may or may not occur. Probability can *never* be a negative number.

COMMENT We mentioned earlier that probability can be thought of as the fraction of times that an outcome will occur in a long series of repetitions of an experiment. However, there may be certain experiments that cannot be repeated. For example, if Gary's kidney has to be removed surgically, we cannot think of this as an experiment that can be repeated over and over again, at least as far as Gary is concerned. How

1970 Draft Lottery

January Birthday	Draft-priority number	February Birthday	Draft-priority number	March Birthday	Draft-priority number	April Birthday	Draft-priority number	May Birthday	Draft-priority number	June Birthday	Draft-priority number
1	•305	1	86•	1	108•	1	32•	1	•330	1	•249
2	159•	2	144•	2	29•	2	•271	2	•298	2	•228
3	•251	3	•297	3	•267	3	83•	3	40•	3	•301
4	•215	4	•210	4	•275	4	81•	4	•276	4	20•
5	101•	5	•214	5	•293	5	•269	5	•364	5	28•
6	•224	6	•347	6	139•	6	•253	6	155•	6	110•
7	•306	7	91•	7	122•	7	147•	7	35•	7	85•
8	•199	8	181•	8	•213	8	•312	8	•321	8	•366
9	•194	9	•338	9	•317	9	•219	9	•197	9	•335
10	•325	10	•216	10	•323	10	•218	10	65•	10	•206
11	•329	11	150•	11	136•	11	14•	11	37•	11	134•
12	•221	12	68•	12	•300	12	•346	12	133•	12	•272
13	•318	13	152•	13	•259	13	124•	13	•295	13	69•
14	•238	14	4•	14	•354	14	•231	14	178•	14	•356
15	17•	15	89•	15	169•	15	•273	15	130•	15	180•
16	121•	16	•212	16	166•	16	148•	16	55•	16	•274
17	•235	17	•189	17	33•	17	•260	17	112•	17	73•
18	140•	18	•292	18	•332	18	90•	18	•278	18	•341
19	58•	19	25•	19	•200	19	•336	19	75•	19	104•
20	•280	20	•302	20	•239	20	•345	20	183•	20	•360
21	•186	21	•363	21	•334	21	62•	21	•250	21	60•
22	•337	22	•290	22	•265	22	•316	22	•326	22	•247
23	118•	23	57•	23	•256	23	•252	23	•319	23	109•
24	59•	24	•236	24	•258	24	2•	24	31•	24	•358
25	52•	25	179•	25	•343	25	•351	25	•361	25	137•
26	92•	26	•365	26	170•	26	•340	26	•357	26	22•
27	•355	27	•205	27	•268	27	74•	27	•296	27	64•
28	77•	28	•299	28	•223	28	•262	28	•308	28	•222
29	•349	29	•285	29	•362	29	•191	29	•226	29	•353
30	164•			30	•217	30	•208	30	103•	30	•209
31	•211			31	30•			31	•313		

July Birthday	Draft-priority number	August Birthday	Draft-priority number	September Birthday	Draft-priority number	October Birthday	Draft-priority number	November Birthday	Draft-priority number	December Birthday	Draft-priority number
1	93•	1	111•	1	•225	1	•359	1	19•	1	129•
2	•350	2	45•	2	161•	2	125•	2	34•	2	•328
3	115•	3	•261	3	49•	3	•244	3	•348	3	157•
4	•279	4	145•	4	•322	4	•202	4	•266	4	165•
5	•188	5	54•	5	82•	5	24•	5	•310	5	56•
6	•327	6	114•	6	6•	6	87•	6	76•	6	10•
7	50•	7	168•	7	8•	7	•234	7	51•	7	12•
8	13•	8	48•	8	•184	8	•283	8	97•	8	105•
9	•277	9	106•	9	•263	9	•342	9	80•	9	43•
10	•284	10	21•	10	71•	10	•220	10	•282	10	41•
11	•248	11	•324	11	158•	11	•237	11	46•	11	39•
12	15•	12	142•	12	•242	12	72•	12	66•	12	•314
13	42•	13	•307	13	175•	13	138•	13	126•	13	163•
14	•331	14	•198	14	1•	14	•294	14	127•	14	26•
15	•322	15	102•	15	113•	15	171•	15	131•	15	•320
16	120•	16	44•	16	•207	16	•254	16	107•	16	96•
17	98•	17	154•	17	•255	17	•288	17	143•	17	•304
18	•190	18	141•	18	•246	18	5•	18	146•	18	128•
19	•227	19	•311	19	177•	19	•241	19	•203	19	•240
20	•187	20	•344	20	63•	20	•192	20	•185	20	135•
21	27•	21	•291	21	•204	21	•243	21	156•	21	70•
22	153•	22	•339	22	160•	22	117•	22	9•	22	53•
23	172•	23	116•	23	119•	23	•201	23	182•	23	162•
24	23•	24	36•	24	•195	24	•196	24	•230	24	95•
25	67•	25	•286	25	149•	25	176•	25	132•	25	84•
26	•303	26	•245	26	18•	26	7•	26	•309	26	173•
27	•289	27	•352	27	•233	27	•264	27	47•	27	78•
28	88•	28	167•	28	•257	28	94•	28	•281	28	123•
29	•270	29	61•	29	151•	29	•229	29	99•	29	16•
30	•287	30	•333	30	•315	30	38•	30	174•	30	3•
31	•193	31	11•			31	79•			31	100•

In December 1969 the United States Selective Service established a priority system for determining which young men would be drafted into the army (see chart above). Capsules representing each birth date were placed in a drum and selected at random. Those men whose numbers were selected first were almost certain that they would be drafted. The different birthdays were given draft priority numbers as indicated in the clipping on the draft lottery. Supposedly, each birthday had an equally likely probability of being selected. However, by analyzing the numbers carefully, we find that the majority of those birthdays that occurred later in the year had a higher priority number than those that occurred earlier in the year. Did each birthday have an equal probability of being selected?

do we assign probabilities in this case? This is not an easy task. Calculating the probability in such situations requires the judgment and experience of a doctor familiar with *many* experiments of a similar type. Thus, if doctors tell you that you have an 80% chance of surviving the operation, they mean that based upon their previous experiences with such situations, 80% of the patients with similar operations have survived. Usually an experienced surgeon can assign a fairly reasonable probability to the success of a nonrepeatable operation.

EXERCISES FOR SECTION 4.2

1. A survey was taken of 10,672 families to determine the number of televisions owned by each. The following results were obtained:

Number of Televisions Owned	Frequency
0	72
1	1,424
2	2,619
3	3,227
4	1,545
5	1,023
6	762
	Total: 10,672

What is the probability that a randomly selected family owns
a. two televisions? 2,619/10,672
b. between one and three televisions, inclusive?
c. seven televisions? 0
d. at most seven televisions (that is, seven or fewer)? 1

2. Consider the accompanying newspaper article.

HOW DO YOU GET TO WORK?

January 5: According to the latest U.S. Bureau of the Census statistics, most people prefer to use their automobiles or car pools to travel to work. Many Americans simply refuse to use public transportation to get to work. Among the reasons given for this refusal, are unreliability, lack of safety, and convenience.

January 5, 1991

234-6920

The following statistics are available on how 702,000 Americans get to work.

	By Car	By Public Transportation
Urban Worker	470,000	157,000
Rural Worker	70,000	5,000

Source: U.S. Bureau of the Census, 1991.

If one of these 702,000 American workers is selected at random, what is the probability that the individual
a. is a rural worker?
b. comes to work by means of public transportation?
c. is an urban worker who travels to work by car?

3. The following is a breakdown of the 274 faculty members in various ranks at Crescent University:

T = 274

Rank	Number of Males	Number of Females
Professor	17	12
Associate Professor	48	40
Assistant Professor	55	42
Instructor/Lecturer	31	29

What is the probability that a faculty member selected at random will be
a. a female professor? $12/274 = 6/137$
b. an assistant professor? $97/274$
c. an instructor? $60/274 = 30/137$
d. a female? $123/274$

4. A man and a woman who do not know each other enter an elevator and can get off at any floor from 1 to 4. Assume that the man and woman are as likely to get off at one floor as another.
a. What is the probability that they both get off on the same floor?
b. Is this assumption of equally likely outcomes reasonable?

5. A woman has 7 nickels, 4 pennies, 3 dimes, and 6 quarters in her purse. Her young child selects a coin at random from her purse. Find the probability that the coin selected is
a. a penny. $4/20 = 1/5$
b. a nickel. $7/20$
c. a dime or quarter. $9/20$ $16/20$
d. worth at least five cents.

6. Which of the following cannot be the probability of some event?

a. 0.796   b. 3.96   c. 0.0   d. $\dfrac{-5}{7}$ e. -0.001

7. Five college students have been nominated to receive the $7000 annual George

[handwritten: ten possible combos w/ those three in there]

[handwritten: 1/10]

Smith Scholarship. The students are Jeremy Rogers, Nancy Simmons, Mary McLave, Molene Tavoukjian, and Ahmad Spence. If only three of these students will receive the scholarship, find the probability that the winners are Jeremy Rogers, Mary McLave, and Ahamad Spence.

8. In the town of Bolton there are seven people with the name of Patrick Shoemaker. An announcement is heard on television saying that Patrick Shoemaker of Bolton has just won 2 million dollars in the state lottery. Find the probability that it is the Patrick Shoemaker who lives at 138 Main Street. (Assume that all seven live at different addresses in Bolton.)

9. The Smiths have four children. What is the probability that they have two boys and two girls? (*Hint*: Find the sample space.) *[handwritten: 6/16 = 3/8]*

[handwritten left margin: ?, 0, 4, 8, 0]

[handwritten list: BBBB, BBBG, BBGB, BGBB, GBBB, GGGG, GGGB, GGBG, GBGG, BGGG]

10. Bob is taking a true-false test and has no idea of the answers to the last three questions. He decides to guess at the answers.
   a. If c stands for a correct answer and w stands for a wrong answer, list all possible outcomes.
   b. Find the probability that he guesses all three answers correctly.
   c. Find the probability that at least one of his answers is correct.
   d. Find the probability that all three answers are wrong.

11. *Illegal aliens*:   A random survey of 1000 voters in California was taken. Each person was asked to indicate his or her opinion on granting American citizenship to the illegal aliens in this country. The following results were obtained:

	In favor of granting citizenship under all circumstances	In favor of granting citizenship under certain circumstances only	Opposed to granting citizenship under any circumstances	Had no opinion
Male	226	153	79	17
Female	307	112	97	9

[handwritten: T=1000   533   265   176   26]

Source: Crescent and Bowes, Los Angeles, 1990.

Find the probability that an individual selected at random from this group
   a. favors granting citizenship unconditionally. *[handwritten: 533/1000]*
   b. is a male who had no opinion. *[handwritten: 17/1000]*
   c. is a female. *[handwritten: 525/1000 = 21/40]*
   d. is a female who opposes granting citizenship under all circumstances. *[handwritten: 97/1000]*

12. A nurse has collected 200 pints of blood from student volunteers and stores them in 200 identical pint bottles. The bottles are not labeled. A bottle is selected at random. If it is known that three of the bottles contain contaminated blood, find the probability that the selected bottle is not contaminated.

13. The following is a breakdown of the different types of credit sales of a large department store for a particular day.

Type of Credit Card Used

	Master Card	VISA	American Express	Discover Card
Under $25	88	95	31	56
Between $25 and $75	65	107	19	49
Over $75	39	63	11	33

What is the probability that a credit card sale selected at random will be
a. a Master Card sale. 192/656
b. under $25. 270/656
c. a Master Card sale that is between $25 and $75. 65/256

14. A computer company manufactured 10,000 enhanced computer chip boards during one month, each stamped with a number from 1 to 10,000. A few months later the company decides to recall those chip boards whose last digit is 8 or 9. Charles Price owns one of these enhanced chip boards. What is the probability that he owns one of the chip boards being recalled?

15. A computer programmer prepares three different computer disks (with different programs on each) and three different identifying labels. Before attaching each label to the appropriate disk, the programmer drops everything on the floor. If the programmer picks up all the disks and labels and randomly attaches a label to a disk, what is the probability that each disk will be labeled correctly?

16. Due to a mechanical malfunction of a luggage carrel at the airport, four identification tags have been found on the floor next to four unidentified pieces of luggage. A security officer randomly attaches one identification tag to each piece of luggage. Find the probability that each identification tag is attached to the proper piece of luggage.

17. A slot machine in a gambling casino has three wheels, and each wheel has a picture of a lemon, a cherry, and an apple on it. When the appropriate amount of money is deposited and the button is pushed, each wheel rotates and then displays a picture of one of the items mentioned. Each wheel operates independently of the other. When all three wheels show the same item, then the player wins $5000.
a. List all possible outcomes for this machine.
b. Find the probability of a player winning $5000 when playing this slot machine.

4.3 COUNTING PROBLEMS

In determining the probability of an event, we must first know the total number of possible outcomes. In many situations it is a rather simple task to list all possible outcomes and then to determine how many of these are favorable. In other situations there may be so many possible outcomes that it would be too time consuming to list all of them. Thus, when two dice are rolled, there are 36 possible outcomes. These are listed on page 179. Exercise 17 of Section 4.2 has 27 possible outcomes. When there are too many possibilities to list, we can use rules, which will be given shortly, to determine the actual number.

Tree Diagram

One technique that is sometimes used to determine the number of possible outcomes is to construct a **tree diagram.** The following examples will illustrate how this is done.

EXAMPLE 1

By means of a tree diagram we can determine the number of possible outcomes when a coin is repeatedly tossed. If one coin is tossed, there are two possible outcomes, heads or tails, as shown in the tree diagram in Figure 4.4. If two coins are tossed, there are now four possible outcomes. These are HH, HT, TH, and TT, since each of the possible outcomes on the first toss can occur with each of the 2 possibilities on the second toss. So, if heads appeared on the first toss, we may get

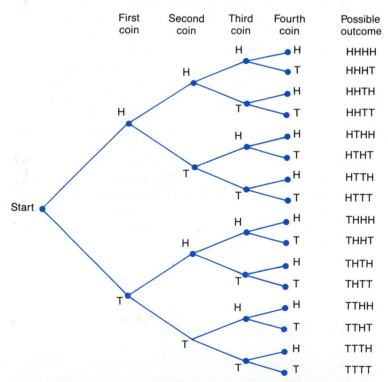

Figure 4.4
Tree diagram of the number of possible outcomes when a coin is repeatedly tossed.

heads or tails on the second toss. The same is true if tails appeared on the first toss. The tree diagram shows that there are 4 possibilities. When three coins are tossed, the diagram shows that there are 8 possible outcomes. Also, there are 16 possible outcomes when four coins are tossed.

| EXAMPLE 2 | Hazel Brown is about to order dinner in a restaurant. She can choose any one of three main courses and any one of four desserts. |

Main Course	Dessert
Hamburger	Hot-fudge sundae
Steak	Jello
Southern-fried chicken	Cake
	Fruit

Using a tree diagram, find all the possible dinners that Hazel can order.

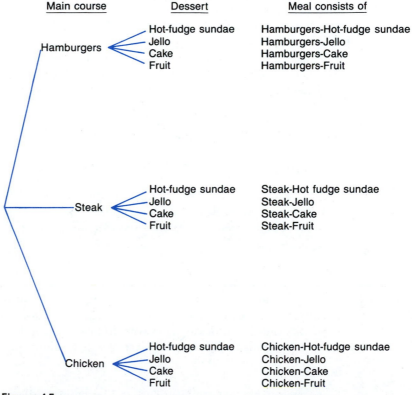

Figure 4.5
Tree diagram of the possible dinners that can be ordered from a choice of any one of three main courses and any one of four desserts.

Solution

Hazel may order any one of the three dishes as a main course. With each main course she may order any one of four desserts. These possibilities are pictured in Figure 4.5. The diagram shows that there are 12 possible meals that Hazel can order.

COMMENT Although counting the number of possible outcomes by using a tree diagram is not difficult when there are only several possibilities, it becomes very impractical to construct a tree when there are many possibilities. For example, if ten coins are tossed, there are 1024 different outcomes. Similarly, if a die is rolled four times, there are 1296 different outcomes. For situations such as these we need a rule to help us determine the number of possible outcomes.

Before stating the rule, however, let us analyze the following situations. When one die is rolled, there are six possible outcomes, 1, 2, 3, 4, 5, 6. When a second die is rolled, there are 6 × 6, or 36, possible outcomes for both dice together. These are listed on page 179.

If again we analyze Exercise 17 of Section 4.2, we find that the first wheel can show a lemon, cherry, or apple. This gives us 3 possibilities. Similarly, the second wheel can also show a lemon, cherry, or apple. The same is true for the third wheel. This gives us 3 possibilities for wheel 1, 3 possibilities for wheel 2, and 3 possibilities for wheel 3. We then have a total of 27 possible outcomes since

$$3 \times 3 \times 3 = 27$$

This leads us to the following rule.

COUNTING RULE

If one thing can be done in m ways, and if after this is done, something else can be done in n ways, then both things can be done in a total of $m \cdot n$ different ways in the stated order. (The same rule extends to more than two things done in sequence.)

EXAMPLE 3   A geology teacher plans to travel from New York to Florida and then on to Mexico to collect rock specimens for her class. From New York to Florida she can travel by train, airplane, boat, or car. However, from Florida to Mexico she cannot travel by train. In how many different ways can she make the trip?

Solution

Since she can travel from New York to Florida in 4 different ways and from Florida to Mexico in 3 different ways, she can make the trip in 4 × 3, or 12, different ways.

EXAMPLE 4

In a certain state license plates have three letters followed by two digits. If the first digit cannot be 0 how many different license plates can be made if

a. repetitions of letters or numbers are allowed?
b. repetitions of letters are not allowed?

Solution

There are 26 letters and 10 possible digits (0, 1, 2, . . . , 9).

a. If repetitions are allowed, the same letter can be used again. Since 0 cannot be used as the first digit, the total number of different license plates is

$$26 \times 26 \times 26 \times \textcircled{9} \times 10 = 1,581,840$$

Thus, 1,581,840 different license plates are possible. Note that the circled position has only 9 possibilities. Why?

b. If repetition of letters is not allowed, there are 26 possibilities for the first letter, but only 25 possibilities for the second letter since once a letter is used it may not be used again. For the third letter there are only 24 possibilities. There are then a total of 1,404,000 different license plates since

$$26 \times 25 \times 24 \times 9 \times 10 = 1,404,000$$

EXAMPLE 5

In Example 1 (see page 192) there are two possible outcomes for the first toss, two possibilities for the second toss, and two possibilities for the third toss. Thus, there are a total of eight possible outcomes when three coins are tossed or when one coin is tossed three times since

$$2 \times 2 \times 2 = 8$$

This is the same result we obtained using tree diagrams. It is considerably easier to do it this way.

EXERCISES FOR SECTION 4.3 1–13 odds

1. A certain model car comes with one of three possible engine sizes and with or without an AM/FM radio. Furthermore, it is equipped with automatic or standard transmission. In how many different ways can a buyer select a car? 12 ways
2. There are nine approach roads leading to an airport. Because of heavy traffic, a taxi driver decides to go to the airport by one road and to leave by another road. In how many different ways can this be done?

9-

3. In the waiting room of Memorial Hospital the following items can be purchased from the vending machines:

Juice	Sandwich	Beverage
1 Orange	1 Tunafish	1 Pepsi
2 Apple	2 Cheese	2 Coke
3 Tomato	3 Meat	3 Coffee
4 Pineapple		4 Orange soda
		5 Tea

4 x 3 x 5 = 60 possible

If a visitor selects one item from each category, how many meals can be obtained?

4. All liberal arts students at Amerville College must successfully complete one course chosen from each of the following categories in order to graduate:

Category A	Category B	Category C
(*Science and Technology*)	(*Social, Historical, Philosophical and Psychological Analysis*)	(*Literature, Language and the Arts*)
Math 1	Political Science 5	English 100
Physics 10	History 100	Music 13
Chemistry 100	Philosophy 10	Art 12
Astronomy 12	Sociology 25	
Geology 17		
Computer 5		

In how many different ways can a student satisfy the college requirement?

5. In how many different ways can the letters of the word "STAMP" be arranged if each arrangement is to consist of five letters and
   a. repetition is not allowed? *120*
   b. repetition is allowed? *3125*

6. In a certain state all domestic animals (dogs and cats) must have a coded identification tag around their neck. These identification tags are of the following type: The first place of each identification tag must be an 8, the second and third place must be letters, and the fourth and fifth places can be any numbers with repetition allowed. How many different identification tags can be made?

7. A jewelry salesman carries all of his samples in his attaché case. There is a combination lock on either side of the case. Each lock has three dials that have to be rotated independently so that any number from 0 to 9 inclusive shows on each dial. Both locks can be opened only if the correct number shows on each dial displayed. How many combinations could a thief have to try before the locks open and the thief can steal the sample jewels from the attaché case?

8. Consider the following newspaper article. Seven depositors are anxiously waiting in line to withdraw their money. In how many different ways can they stand in line?

POLICE QUELL DEPOSITOR UNREST

New York: Local police had to be called in yesterday to restore law and order at the Golden Pacific Bank following the announcement by state banking officials that the bank was insolvent and that depositors would be allowed to withdraw only $100 per person pending further action by the Federal Deposit Insurance Corporation.

June 10, 1985

9. How many different numbers greater than 3000 can be formed from the digits 2, 3, 5, and 9 if no repetitions are allowed?

*10. In the baseball world series two teams play against each other in a series of up to seven games. The first team that wins four games is the winner. Assume that the two teams are labeled A and B. Make a tree diagram showing all possible ways in which the World Series can end.

11. Using tree diagrams, determine the total number of possible ways (in terms of gender) that a family can consist of five children.

12. The U.S. Postal Service recommends that all pieces of mail have on it the correct nine-digit zip code number (five digits for the area to which the item is to be delivered and four digits for the block number). Assuming no restrictions, how many different possible zip codes are there?

13. In a certain state license plates consist of three numbers followed by three letters. How many different license plates can be formed if
    a. no digits can be repeated but letters may be repeated?
    b. there are no restrictions at all?

14. There are three cables connected to a disassembled computer. One is to be used for connecting the computer to a printer, one is for connection to a monitor, and one is for connection to a modem. After the computer is reassembled, the operator must reconnect the cables to the appropriate device. If there are no identifying marks on either cable, in how many different ways can these cables be reconnected?

15. There are seven members on the Board of Directors of the Ross Chemical Corp. as follows: Alice Johnson, Steve Bachtel, George Klangman, Jeremy Billingsley, Fern Washington, Wallace McKenzie, and Eugene Brooks. At any stockholder's meeting these members are seated at a (straight) head table.
    a. In how many different ways can the Board members be seated?
    *b. If Jeremy Billingsley and Wallace McKenzie are bitter enemies and cannot be seated together, how many different seating arrangements are possible?

*c. Find the probability that the two female board members (Alice Johnson and Fern Washington) are seated together.

16. A jewelry manufacturer wishes to make cuff links for men. He will engrave the first letter of the first name, the middle initial, and the first letter of the last name on cuff links. How many possible pairs of cuff links can be formed using this scheme and the letters of the alphabet?

17. Mabel is getting dressed. She can select any one of four pairs of slacks, any one of three blouses, and any one of two pairs of shoes depending on their color as shown below.

Slacks	Blouses	Shoes
Black	White print	White
White	Red print	Black
Blue	Black	
Red		

Neglecting any consideration as to what colors match with one another, make a tree diagram showing all possible outfits that she may select.

18. Marilyn is getting dressed. She can select any one of six blouses, any one of four skirts, any one of three pairs of shoes, any one of six handbags, and any one of three scarves to wear for an important business meeting. In how many ways can Marilyn select an outfit?

4.4 PERMUTATIONS

Consider the following situation: Three vacationing students, Mel, Carl, and Rhoda, have purchased standby tickets on a transatlantic flight that will take them from London back to New York. There are three aisle seats remaining vacant in the plane; one in the forward section, one in the midsection, and one in the tail section of the airplane. In how many different ways can the three travelers line up to board the plane and pick the best of the remaining seats?

There are six ways as shown in the accompanying table. In this case you will notice that in each arrangement order is important.

First Person to Board Plane	Second Person to Board Plane	Last Person to Board Plane
Mel	Carl	Rhoda
Mel	Rhoda	Carl
Carl	Mel	Rhoda
Carl	Rhoda	Mel
Rhoda	Carl	Mel
Rhoda	Mel	Carl

Thus, the arrangement Mel, Carl, Rhoda means that Mel boards first, then Carl, and finally Rhoda. If the number of seats is limited, then the person who boards first has first choice of selecting the best of the remaining seats as compared with the person who boards last.

This important idea in mathematics, in which a number of objects can be arranged in a particular order, is known as a *permutation*.

DEFINITION 4.2	A **permutation** is any arrangement of distinct objects in a particular order.
Permutation	

Let us now examine another such problem.

EXAMPLE 1

Four young ladies, Stephanie Marie Gallagher, Liz Armstrong, Ann Sullivan, and Patricia Beth O'Connell, are finalists in a state beauty contest. The judges must select a winner and an alternate from among these four contestants. In how many different ways can this be done?

Solution

The winner and the alternate can be selected in 12 different ways. We list these possibilities in a table.

Winner	Alternate	Winner	Alternate
Stephanie	Liz	Ann	Stephanie
Stephanie	Ann	Ann	Liz
Stephanie	Patricia	Ann	Patricia
Liz	Stephanie	Patricia	Stephanie
Liz	Ann	Patricia	Liz
Liz	Patricia	Patricia	Ann

$$_nP_r = \frac{n!}{(n-r)!}$$

COMMENT Perhaps you are wondering whether there are any formulas that can be used to determine the number of possible permutations. The answer is yes and this will be done shortly.

EXAMPLE 2

A doctor has five examination rooms. There are five patients in the waiting room. In how many different ways can the patients be assigned to the examination rooms?

Solution

We can solve this problem by numbering the waiting rooms 1 through 5 as shown here and then considering them in sequence.

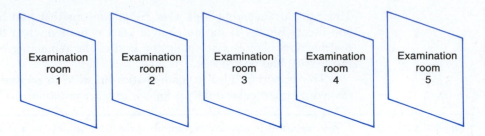

Examination room 1 can be used by any one of the five patients. Once a patient has been assigned to examination room 1, there are four patients who can use examination room 2, so there are (using the counting rule)

$$5 \cdot 4 = 20 \text{ ways}$$

of using examination room 1 and examination room 2. Similarly, once examination rooms 1 and 2 are occupied, there are three patients remaining who can use examination room 3, so there are

$$5 \cdot 4 \cdot 3 = 60 \text{ ways}$$

of using examination rooms 1, 2, and 3. Continuing in this manner we find that there are

$$5 \cdot 4 \cdot 3 \cdot 2 \cdot 1 = 120 \text{ ways}$$

of using examination rooms 1, 2, 3, 4, and 5. Thus, there are 120 different ways in which the five patients can be assigned to the five examination rooms.

Notice in Example 2 that we had to multiply $5 \cdot 4 \cdot 3 \cdot 2 \cdot 1$. Often in mathematics we have to multiply a series of numbers together starting from a given whole number and multiplying this by the number that is 1 less than that, and so on until we get to the number 1, which is where we stop. For this type of multiplication we introduce **factorial notation** and use the symbol $n!$ read as "n factorial." In this case we would write the product as $5!$. Thus,

Factorial Notation

$$5! = 5 \cdot 4 \cdot 3 \cdot 2 \cdot 1 = 120.$$

Accordingly,

$$1! = 1$$
$$2! = 2 \cdot 1 = 2$$
$$3! = 3 \cdot 2 \cdot 1 = 6$$
$$4! = 4 \cdot 3 \cdot 2 \cdot 1 = 24$$
$$5! = 5 \cdot 4 \cdot 3 \cdot 2 \cdot 1 = 120$$
$$6! = 6 \cdot 5 \cdot 4 \cdot 3 \cdot 2 \cdot 1 = 720$$
$$7! = 7 \cdot 6 \cdot 5 \cdot 4 \cdot 3 \cdot 2 \cdot 1 = 5{,}040$$
$$8! = 8 \cdot 7 \cdot 6 \cdot 5 \cdot 4 \cdot 3 \cdot 2 \cdot 1 = 40{,}320$$

$$9! = 9 \cdot 8 \cdot 7 \cdot 6 \cdot 5 \cdot 4 \cdot 3 \cdot 2 \cdot 1 = 362,880$$
$$10! = 10 \cdot 9 \cdot 8 \cdot 7 \cdot 6 \cdot 5 \cdot 4 \cdot 3 \cdot 2 \cdot 1 = 3,628,800$$

Also, we define 0! to be 1. This makes formulas involving 0! meaningful.

Let us now return to the beauty contest example (Example 1) discussed earlier. In that example the judges were interested in selecting two winners from among the four finalists. This actually represents the number of permutations of four things taken two at a time, or the number of possible permutations of two things that can be formed out of a possible four things. We abbreviate this by writing $_4P_2$. This is read as the number of permutations of four things taken two at a time.

NOTATION $_nP_r$ represents the number of permutations of n things taken r at a time, and $_nP_n$ represents the number of permutations of n things taken n at a time.

To determine the number of possible permutations, we use the following formula:

FORMULA 4.1

$$_nP_r = \frac{n!}{(n - r)!}$$

EXAMPLE 3

a. Find $_6P_4$.
b. Find $_7P_5$.
c. Find $_5P_5$.

Solution

a. $_6P_4$ means the number of possible permutations of six things taken four at a time. Using Formula 4.1 we have $n = 6$, $r = 4$, and $n - r = 6 - 4 = 2$. Thus,

$$_6P_4 = \frac{6!}{(6 - 4)!} = \frac{6!}{2!} = \frac{6 \cdot 5 \cdot 4 \cdot 3 \cdot 2 \cdot 1}{2 \cdot 1} = \frac{6 \cdot 5 \cdot 4 \cdot 3 \cdot \cancel{2} \cdot \cancel{1}}{\cancel{2} \cdot \cancel{1}}$$
$$= 6 \cdot 5 \cdot 4 \cdot 3 = 360$$

Thus, $_6P_4 = 360$.

b. $_7P_5$ means the number of permutations of seven things taken five at a time, so that $n = 7$ and $r = 5$. Using Formula 4.1 we have

$$_7P_5 = \frac{7!}{(7 - 5)!} = \frac{7!}{2!} = \frac{7 \cdot 6 \cdot 5 \cdot 4 \cdot 3 \cdot 2 \cdot 1}{2 \cdot 1}$$
$$= \frac{7 \cdot 6 \cdot 5 \cdot 4 \cdot 3 \cdot \cancel{2} \cdot \cancel{1}}{\cancel{2} \cdot \cancel{1}}$$
$$= 7 \cdot 6 \cdot 5 \cdot 4 \cdot 3 = 2520$$

Thus, $_7P_5 = 2520$.

c. $_5P_5$ means the number of permutations of five things taken five at a time, so that $n = 5$ and $r = 5$. Using Formula 4.1 we have

$$_5P_5 = \frac{5!}{(5 - 5)!} = \frac{5!}{0!}$$

$$= \frac{5!}{1} \quad \text{(Since 0! is 1)}$$

$$= 5! = 5 \cdot 4 \cdot 3 \cdot 2 \cdot 1$$

$$= 120$$

Thus, $_5P_5 = 5! = 120$.

EXAMPLE 4

A baseball scout has received a list of 15 promising prospects for the team's consideration. He is asked to list, in order of preference, the 5 most outstanding of these prospects. In how many different ways can he select the 5 best players?

Solution

As there are 15 promising prospects, $n = 15$. Also only the top 5 are to be considered, so that $r = 5$ and $n - r = 15 - 5 = 10$. Using Formula 4.1, we have

$$_nP_r = \frac{n!}{(n - r)!}$$

$$= \frac{15!}{10!}$$

$$= \frac{15 \cdot 14 \cdot 13 \cdot 12 \cdot 11 \cdot \cancel{10} \cdot \cancel{9} \cdot \cancel{8} \cdot \cancel{7} \cdot \cancel{6} \cdot \cancel{5} \cdot \cancel{4} \cdot \cancel{3} \cdot \cancel{2} \cdot \cancel{1}}{\cancel{10} \cdot \cancel{9} \cdot \cancel{8} \cdot \cancel{7} \cdot \cancel{6} \cdot \cancel{5} \cdot \cancel{4} \cdot \cancel{3} \cdot \cancel{2} \cdot \cancel{1}}$$

$$= 15 \cdot 14 \cdot 13 \cdot 12 \cdot 11 = 360,360$$

Thus, the scout can select the top 5 players in 360,360 different ways.

COMMENT   In Example 3 we calculated $_5P_5$, which represents the number of permutations of five things taken five at a time. This really means the number of different ways of arranging five things, where order counts. Our answer turned out to be 5! This leads us to Formula 4.2.

FORMULA 4.2

The number of possible permutations of n things taken n at a time, denoted as $_nP_n$, is

$$_nP_n = n!$$

EXAMPLE 5 How many different permutations can be formed from the letters of the word CAT?

Solution

There are six permutations. These are

CAT CTA TAC TCA ACT ATC

This actually represents the number of possible permutations of three letters taken three at a time. Formula 4.2 tells us that there are 3!, or 6, possible permutations since

$$3! = 3 \cdot 2 \cdot 1 = 6.$$

These are listed above.

Suppose a librarian has two identical algebra books and three identical geometry books to be shelved. In how many different ways can this be done? Since there are five books altogether and all have to be shelved, we are tempted to use Formula 4.2 or Formula 4.1. Unfortunately, since the two algebra books are identical, we cannot tell them apart. There are actually ten possible permutations. These are:

algebra	algebra	geometry	geometry	geometry
algebra	geometry	geometry	geometry	algebra
algebra	geometry	algebra	geometry	geometry
algebra	geometry	geometry	algebra	geometry
geometry	geometry	geometry	algebra	algebra
geometry	geometry	algebra	algebra	geometry
geometry	geometry	algebra	geometry	algebra
geometry	algebra	algebra	geometry	geometry
geometry	algebra	geometry	algebra	geometry
geometry	algebra	geometry	geometry	algebra

Of course, if we label the algebra books as copy 1 and copy 2, which is sometimes done by some libraries, we can use Formula 4.1. Thus,

Algebra copy 1	Algebra copy 2	Geometry copy 1	Geometry copy 2	Geometry copy 3

would be a different permutation than

Algebra copy 2	Algebra copy 1	Geometry copy 1	Geometry copy 2	Geometry copy 3

However, this is usually not done. Thus, these two permutations have to be counted as the same or as only one permutation. Formulas 4.1 and 4.2 have to be revised somewhat to allow for the possibility of repetitions. This leads us to the following formula:

FORMULA 4.3

The number of different permutations of n things of which p are alike, q are alike, or r are alike, and so on, is

$$\frac{n!}{p!\,q!\,r!}\cdots$$

It is understood that $p + q + r + \cdots = n$

In our example we have five books to be shelved, so $n = 5$. Since the two algebra books are identical and the three geometry books are also identical, $p = 2$ and $q = 3$. Formula 4.3 tells us that the number of permutations is

$$\frac{5!}{2! \cdot 3!}$$

$$= \frac{5 \cdot 4 \cdot 3 \cdot 2 \cdot 1}{2 \cdot 1 \cdot 3 \cdot 2 \cdot 1}$$

$$= \frac{5 \cdot \overset{2}{\cancel{4}} \cdot \cancel{3} \cdot \cancel{2} \cdot \cancel{1}}{\cancel{2} \cdot \cancel{1} \cdot \cancel{3} \cdot \cancel{2} \cdot \cancel{1}}$$

$$= 5 \cdot 2 = 10$$

Thus, there are ten permutations, which were listed previously.

EXAMPLE 6

How many different permutations are there of the letters in the word

a. "IDIOT?"
b. "STATISTICS?"

Solution

a. The word "IDIOT" has five letters, so $n = 5$. There are 2 I's, so $p = 2$. Formula 4.3 tells us that the number of permutations is

$$\frac{5!}{2!\,1!\,1!\,1!}$$

$$= \frac{5 \cdot 4 \cdot 3 \cdot \cancel{2} \cdot \cancel{1}}{\cancel{2} \cdot \cancel{1} \cdot 1 \cdot 1 \cdot 1}$$

$$= 5 \cdot 4 \cdot 3 = 60$$

STAMP
n=5
5! 5/20

There are 60 permutations.

b. Since the word "STATISTICS" has ten letters, $n = 10$. The letter "S" is repeated three times, "T" is repeated three times, and "I" is repeated twice, so $p = 3$, $q = 3$, and $r = 2$. Formula 4.3 tells us that the number of permutations is

$$\frac{10!}{3!\,3!\,2!\,1!\,1!}$$

$$= \frac{10 \cdot 9 \cdot 8 \cdot 7 \cdot \cancel{6} \cdot 5 \cdot \cancel{4} \cdot \cancel{3} \cdot \overset{2}{\cancel{2}} \cdot \cancel{1}}{\cancel{3} \cdot \cancel{2} \cdot \cancel{1} \cdot \cancel{3} \cdot \cancel{2} \cdot \cancel{1} \cdot \cancel{2} \cdot \cancel{1} \cdot 1 \cdot 1}$$

$$= 10 \cdot 9 \cdot 8 \cdot 7 \cdot 5 \cdot 2 = 50,400$$

There are 50,400 permutations.

Calculations involving factorials can easily be done using a hand-held calculator. However, if you are familiar with the MINITAB statistical computer package, then you can use MINITAB to evaluate factorials quite easily. For example, to evaluate 10! using MINITAB, we do the following:

```
MTB>SET THE FOLLOWING INTO C1
DATA>1:10
DATA>END
MTB>PARPRODUCTS OF C1 PUT INTO C2
MTB>PRINT C1   C2
```

ROW	C1	C2
1	1	1
2	2	2
3	3	6
4	4	24
5	5	120
6	6	720
7	7	5040
8	8	40320
9	9	362880
10	10	3628800

```
MTB>STOP
```

In a similar manner, we can have MINITAB evaluate any factorial.

formula

EXERCISES FOR SECTION 4.4

1. Evaluate each of the following symbols:

40,320

a. $8!$ **b.** $9!$ **c.** $3!$ **d.** $\dfrac{6!}{5!} = 6$

e. $\dfrac{0!}{4}$ **f.** $\dfrac{5!}{3!\,2!}$ **g.** $\dfrac{6!}{4!\,2!}$ **h.** $\dfrac{7!}{5!\,2!}$

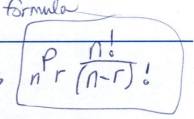

$${}_nP_r \quad \frac{n!}{(n-r)!}$$

i. $_8P_5$ j. $_7P_4$ k. $_9P_5$ l. $_0P_0$
m. $_7P_0$ n. $_7P_7$ o. $_6P_4$ p. $_3P_2$

2. In a supermarket there is a long line at the checkout counter. The manager notices this and decides to open an additional checkout counter. Seven people rush over to the new checkout counter. In how many different ways can these seven people line up to be checked out?

3. Susan Rao is an IRS agent. She has made appointments with eight taxpayers to review their 1040 tax forms on May 2 on a first-come, first-served basis. However, due to a computer malfunction, she finds that she has time to meet with only five taxpayers to review their forms. Assuming order counts, in how many different ways can this be done? *6720*

4. Each year movie-goers in a certain city are asked to rank the 5 best movies from among a list of 14 movies. In how many different ways can this be done?

5. There are eight members on the Board of Directors of the Brack Corporation. A new executive officer, chairperson, comptroller, and recording secretary are to be elected from among these eight members.
   a. How many permutations of the list of these officers is possible from the eight-board members? *1680*
   b. If Manya is a member of the Board of Directors, what is the probability that she will be elected executive officer? *1/8*
   c. What is the probability that she will be elected to any office? *1/2*

6. Recently, a wealthy man donated nine famous paintings to a museum. In how many different ways can three of these paintings be exhibited in the first floor of the museum if one is to be placed by the main entrance, one by the elevators, and one by the information desk?

7. In how many different ways can the letters of the word "ECONOMICS" be arranged if a vowel must be the first letter of each permutation?

8. How many different permutations are there of the letters in the words
   a. ELEMENTARY   b. ANALYSIS
   c. MANAGEMENT   d. ORGANIZATION

9. In how many different ways can the letters of the word "PROBABILITY" be arranged if the letter "P" must be the first letter of each arrangement?

10. Five math books, four sociology books, and three psychology books are to be arranged on a bookshelf. None of the books is identical.
    a. How many different permutations of these books are there?
    b. How many different permutations of these books are there if books on the same subject must be grouped together?

11. At a recent university retirement luncheon five professors came with their spouses. In how many ways can these ten people be seated at the head (straight) table if
    a. any person can sit in any seat?
    b. men are to be seated on one side of the president and women on the other side of the president?
    c. each husband will sit with his wife? (*Note:* The president will not be seated.)

12. Bill Carter knows that the first five digits of his hospital insurance identification

number are 376-49. However, he does not recall the sequence of the remaining four digits 5, 0, 2, and 8. How many possibilities are there?

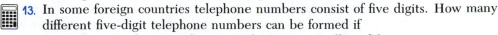 13. In some foreign countries telephone numbers consist of five digits. How many different five-digit telephone numbers can be formed if

    a. the first number must be a 3 (with repetitions allowed)?

    b. no digit can be repeated?

    c. there are no restrictions at all?

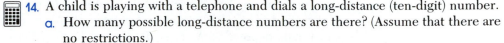 14. A child is playing with a telephone and dials a long-distance (ten-digit) number.

    a. How many possible long-distance numbers are there? (Assume that there are no restrictions.)

    b. What is the probability that the child dials his or her own number?

15. The number of different ways in which n distinct objects can be arranged in a rcle is $(n - 1)!$

    a. Explain why this formula is valid.

    b. In how many distinct ways can a baker display six different cakes in a circular arrangement in a showcase window?

16. Six people are seated around a circular table playing cards. One of the players is annoyed by the smoking at an adjoining table. It is decided that the players should change seats. In how many different ways can the players rearrange themselves?

17. Many companies solicit employee safety suggestions by using a suggestion box. Bonuses are usually rewarded for any suggestions that are implemented. Recently, workers of the Carter Paste Company submitted ten different suggestions for improvements in safety. Company officials plan to rank these suggestions and to reward the top four suggestions with prizes of $1000, $500, $100, and $50, respectively. In how many different ways can the winning suggestions be selected?

18. A baseball manager has selected his nine starting players.

    a. How many different batting orders are possible?

    b. Assuming that all different batting orders are equally likely, what is the probability that the pitcher bats last?

    c. Is the assumption given in part (b) reasonable?

19. In how many different ways can seven swimmers be assigned to seven lanes in a swimming olympic tryout?

20. Each week Noel does the dishes on three days, Kelly does the dishes on two days, and Adel does them on the remaining two days. In how many different orders could they choose to do the dishes? (*Hint:* This is an arrangement of seven things with repetition.)

4.5 COMBINATIONS

Imagine that a six-person rescue party is climbing a mountain in search of survivors of an airplane crash. They suddenly spot the plane on a ledge but the passageway to the ledge is very narrow and only four people can proceed. The remaining two people will have to return. How many different four-person rescue groups can be formed to reach the ledge?

In this situation we are obviously interested in selecting four out of six people. However, since the order in which the selection is to be made is not important, Formula 4.1 of Section 4.4 (see page 201) has to be changed.

Any selection of things in which the order is not important is called a *combination*. Let us determine how many possible combinations there actually are. If the names of the people of the six-person rescue party are Alice, Betty, Calvin, Drew, Ellen, and Frank, denoted as A, B, C, D, E, and F, then Formula 4.1 of Section 4.4 tells us that there are $_6P_4$ possible ways of selecting the four people out of a possible six. This yields

$$_6P_4 = \frac{6!}{(6-4)!}$$

$$= \frac{6!}{2!}$$

$$= 360 \text{ possibilities}$$

However, we know that permutations take order into account. Since we are not interested in order, there cannot be 360 possible combinations. Thus, if the four-person rescue party consists of A, B, C, and D, there would be the following 24 different permutations:

A B C D	B A C D	C A B D	D A B C
A B D C	B A D C	C A D B	D A C B
A D C B	B D A C	C B A D	D B A C
A D B C	B D C A	C B D A	D B C A
A C B D	B C A D	C D A B	D C B A
A C D B	B C D A	C D B A	D C A B

Since these permutations consist of the same four people, A, B, C, and D, we consider them as only one combination of these people.

Similarly, for any other combination of four people there are 24 permutations of these people. Thus, it would seem reasonable to divide the 360 by 24 getting 15 and to conclude that there are only 15 different combinations.

Notice that $24 = 4!$. Therefore, the number of combinations of six things taken four at a time is

$$\frac{_6P_4}{4!} = \frac{6!}{4!2!}$$

$$= 15$$

$$_nC_r = \frac{n!}{r!\cdot(n-r)!}$$

In general, consider the problem of selecting r objects from a possible n objects. We have the following definitions and formula.

DEFINITION 4.3
Combination

A **combination** is a selection from a collection of distinct objects where order is not important.

DEFINITION 4.4

The number of different ways of selecting r objects from a possible n distinct objects, where the order is not important, is called the **number of combinations of n things taken r at a time** and is denoted as $_nC_r$. Some books use the symbol $\binom{n}{r}$ instead

Binomial Coefficient

of $_nC_r$. The symbol $\binom{n}{r}$ is called a **binomial coefficient.**

FORMULA 4.4

The number of combinations of n things taken r at a time is

$$\binom{n}{r} = {_nC_r} = \frac{n!}{r!(n - r)!}$$

Formula 4.4 is especially useful when calculating the probability of certain events. We illustrate the use of this formula with several examples.

EXAMPLE 1

Consider any five people, whom we shall name A, B, C, D, and E.

a. In how many ways can a committee of three be selected from among them?
b. What is the probability of selecting the three-person committee consisting of A, B, and C? (Assume that each committee is equally likely to be selected.)
c. In how many ways can a committee of five be selected from among them?

Solution

a. We must select any three people from a possible five, and order does not matter. Since this is the number of combinations of five things taken three at a time, we want $_5C_3$. Using Formula 4.4 with $n = 5$ and $r = 3$, we have

$$\binom{5}{3} = {_5C_3} = \frac{5!}{3!(5 - 3)!} = \frac{5!}{3!2!} = \frac{5 \cdot \overset{2}{\cancel{4}} \cdot \cancel{3} \cdot \cancel{2} \cdot \cancel{1}}{\cancel{3} \cdot \cancel{2} \cdot \cancel{1} \cdot \cancel{2} \cdot \cancel{1}} = 10$$

There are ten possible three-person committees that can be formed. We can verify this answer by listing them:

```
A B C   A B E   A C E   B C D   B D E
A B D   A C D   A D E   B C E   C D E
```

b. There are ten possible three-person committees that can be formed, as listed in part (a). Of these only one consists of A, B, and C. Thus,

$$p(\text{committee consists of A, B, and C}) = \frac{1}{10}$$

c. We are interested in selecting any five people from a possible five, and order does not matter. This is $_5C_5$. Thus,

$$\binom{5}{5} = {_5C_5} = \frac{5!}{5!(5-5)!}$$

$$= \frac{5!}{5!0!} \quad (\text{Remember } 0! = 1)$$

$$= \frac{5 \cdot 4 \cdot 3 \cdot 2 \cdot 1}{5 \cdot 4 \cdot 3 \cdot 2 \cdot 1 \cdot 1}$$

$$= 1$$

So, there is only one combination containing all five people.

EXAMPLE 2   Nuclear Accident

On Wednesday, March 27, 1979, the nuclear generating facility on Three Mile Island near Middletown, Pennsylvania, malfunctioned and began discharging radiation into the air. In an attempt to prevent a nuclear "meltdown" ten nuclear physicists were contacted to shed some light on the problem. (Later, additional physicists were called in.) Because the ten scientists were busily engaged on another project, they agreed that only seven of them would come to the crippled nuclear plant. (Any three of them could remain to oversee their other existing project.) In how many different ways could the seven scientists have been selected?

Solution

Order was not important, so the seven scientists had to be selected out of a possible ten. This could be done in $_{10}C_7$ ways. Using Formula 4.4, we have

$$\binom{10}{7} = {_{10}C_7} = \frac{10!}{7!(10-7)!} = \frac{10!}{7!3!} = 120$$

Thus, the 7 scientists could have been selected in 120 different ways.

The four cooling towers at the Three Mile Island Nuclear Power Plant.
(*Photo:* UPI/Bettmann Newsphotos)

$4 \times 3 \times 2 \times 1 = 24$

EXAMPLE 3

Football Teams

How many different 11-member football teams can be formed from a possible 20 players if any player can play any position?

Solution

We are interested in the number of combinations of 20 players taken 11 at a time. So, $n = 20$ and $r = 11$. Thus,

$$\binom{20}{11} = {}_{20}C_{11} = \frac{20!}{11!(20 - 11)!}$$

$$= \frac{20!}{11!9!}$$

$$= 167,960$$

There are 167,960 possible 11-member football teams.

EXAMPLE 4

a. How many different poker hands consisting of 5 cards can be dealt from a deck of 52 cards?

b. What is the probability of being dealt a royal flush in five-card poker? (A royal flush consists of the ten, jack, queen, king, and ace of the same suit.)

Solution

a. Since order is not important, we are interested in the number of combinations of 5 things out of a possible 52. So, $n = 52$ and $r = 5$. Thus,

$$\binom{52}{5} = {}_{52}C_5 = \frac{52!}{5!(52-5)!}$$

$$= \frac{52!}{5!47!}$$

$$= \frac{52 \cdot 51 \cdot 50 \cdot 49 \cdot 48 \cdot \cancel{47!}}{5 \cdot 4 \cdot 3 \cdot 2 \cdot 1 \cdot \cancel{47!}}$$

$$= 2,598,960$$

There are 2,598,960 possible poker hands.

b. Out of the possible 2,598,960 different poker hands only four are favorable. These are the ten, jack, queen, king, and ace of hearts; the ten, jack, queen, king, and ace of clubs; the ten, jack, queen, king, and ace of diamonds; and the ten, jack, queen, king, and ace of spades. Thus,

$$p(\text{royal flush}) = \frac{4}{2,598,960} = \frac{1}{649,740}$$

COMMENT   For those familiar with poker, we present certain probabilities in Table 4.1.

TABLE 4.1   Different Poker Hands and Their Probabilities

Type of Hand	Probability of It Being Dealt to You
Royal flush (ace, king, queen, jack, 10 in the same suit)	0.0000015
Four of a kind (four of a kind, like four 4's or four queens)	0.00024
Flush (five cards in a single suit but not straight)	0.0020
Two pairs	0.0475
Nothing of interest	0.5012

EXAMPLE 5

John has ten single dollar bills of which three are counterfeit. If he selects four of them at random, what is the probability of getting two good bills and two counterfeit bills?

Solution

We are interested in selecting four bills from a possible ten. Thus, the number of possible outcomes is $_{10}C_4$. The two good bills to be drawn must be drawn from the seven good ones. This can happen in $_7C_2$ ways. Also, the two counterfeit bills to be drawn must be drawn from the three counterfeit ones. This can happen in $_3C_2$ ways. Thus,

$$p(\text{two good bills and two counterfeit bills}) = \frac{_7C_2 \cdot {_3C_2}}{_{10}C_4}$$

Now

$$_7C_2 = \frac{7!}{2!(7-2)!}$$

$$= \frac{7!}{2!5!} = 21$$

and

$$_3C_2 = \frac{3!}{2!(3-2)!}$$

$$= \frac{3!}{2!1!} = 3$$

Also,

$$_{10}C_4 = \frac{10!}{4!(10-4)!}$$

$$= \frac{10!}{4!6!} = 210$$

Therefore,

$$p(\text{two good bills and two counterfeit bills}) = \frac{21 \cdot 3}{210}$$

$$= \frac{3}{10}$$

EXAMPLE 6   Find the probability of selecting all good bills in the preceding example.

Solution

Since we are interested in selecting only good bills, we will select four good bills from a possible seven and no counterfeit bills. Thus,

$$p(\text{all good bills}) = \frac{{}_7C_4 \cdot {}_3C_0}{{}_{10}C_4}$$

$$= \frac{35 \cdot 1}{210} = \frac{1}{6}$$

The probability of selecting all good bills is $\frac{1}{6}$.

There is an alternate method for computing the number of possible combinations of n things taken r at a time, ${}_nC_r$, which completely avoids the factorial notation. This can be accomplished by using *Pascal's triangle*. Such a triangle is shown in Figure 4.6. How do we construct such a triangle?

Each row has a 1 on either end. All in-between entries are obtained by adding the numbers immediately above and directly to the right and left of them as shown by the arrows in the diagram of Figure 4.7. For example, to obtain the entries for the eighth row we first put a 1 on each end (moving over slightly). Then we add the 1 and 7 from row 7, getting 8 as shown. We then add 7 and 21, getting 28; then we

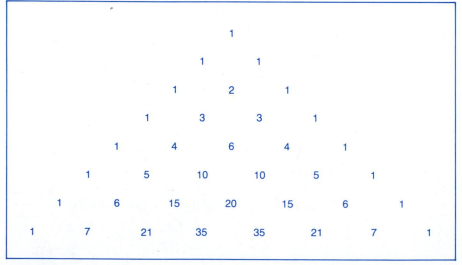

Figure 4.6
Pascal's triangle.

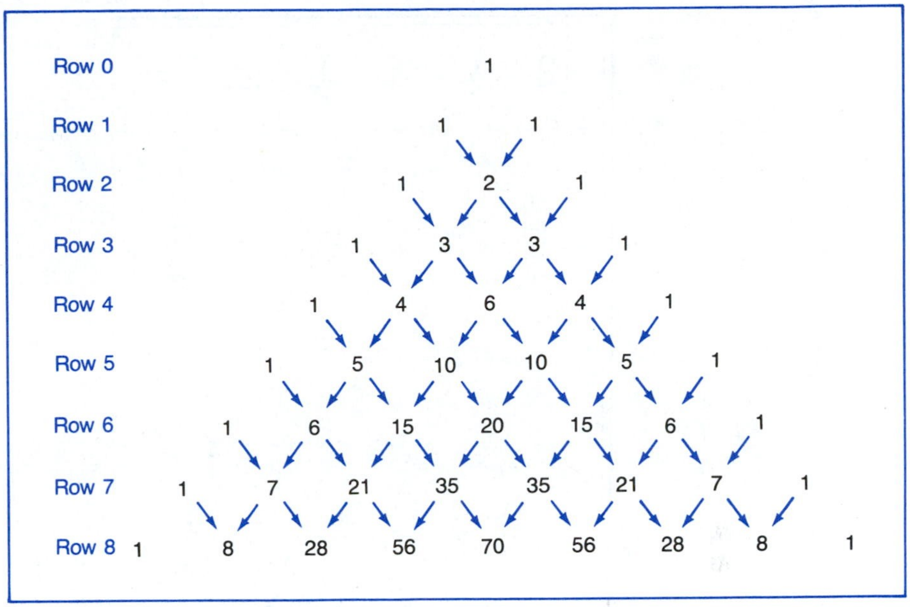

Figure 4.7
Construction of Pascal's triangle.

add 21 to 35, getting 56, and so on. Remember to place the 1's on each end. The numbers must be lined up as shown in the diagram. Such a triangle of numbers is known as **Pascal's triangle** in honor of the French mathematician who found many applications for it. Although the Chinese knew of this triangle several centuries earlier (see Figure 4.8), it is named for Pascal.

Pascal's Triangle

Let us now apply Pascal's triangle to solve problems in combinations.

EXAMPLE 7

Professor Matthews tells four jokes each semester. His policy is never to repeat any combination of four jokes once they are used. How many semesters will eight different jokes last the professor?

Solution

Since order is not important, we are interested in the number of combinations of eight things taken four at a time. So, $n = 8$ and $r = 4$. Therefore, we must evaluate $_8C_4$. We will find $_8C_4$ first using Formula 4.4 and then using Pascal's triangle. Using Formula 4.4, we have

$$_8C_4 = \frac{8!}{4!(8-4)!} = \frac{8!}{4!4!} = 70$$

Thus, the eight jokes will last him for 70 semesters.

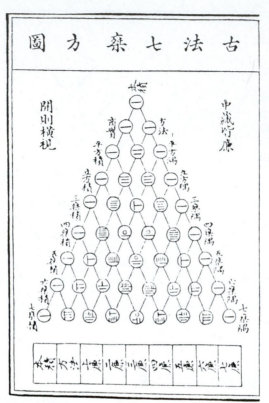

Figure 4.8
The Pascal triangle as used in 1303.
(Redrawn from Joseph Needham, *Science and Civilization in China, III*, Cambridge University Press, 135.)

Let us now evaluate $_8C_4$ using Pascal's triangle. Since $n = 8$, we must write down the first nine rows of Pascal's triangle as shown in Figure 4.9.

Now we look at row 8. Since we are interested in selecting four things from a possible eight things, we skip the 1 and move to the fourth entry appearing after the 1. It is 70. This number represents the value of $_8C_4$. Thus, $_8C_4 = 70$. Similarly, to find $_8C_6$ we move to the sixth entry after the end 1 on row 8. It is 28. Thus, $_8C_6 = 28$. Also, the first entry after the end 1 is 8 so that $_8C_1 = 8$. If we wanted $_8C_8$, we would move to the eighth entry after the end 1. It is 1, so that $_8C_8 = 1$. To find $_8C_0$ we must move to the zeroth entry after the end 1. This means we do not go anywhere; we just stay at the 1. Thus, $_8C_0 = 1$.

COMMENT When you use Pascal's triangle, remember that the rows are numbered from 0 rather than from 1 so that the rows are labeled row 0, row 1, row 2, and so on.

Row 0						1			
Row 1					1		1		
Row 2				1		2		1	
Row 3			1		3		3		1
Row 4		1		4		6		4	1
Row 5	1		5		10		10	5	1
Row 6	1	6		15		20	15	6	1
Row 7	1	7	21		35	35	21	7	1
Row 8	1	8	28	56	70	56	28	8	1

Figure 4.9
First nine rows of Pascal's triangle.

EXAMPLE 8

Using Pascal's triangle find the following:
a. $_5C_2$ **b.** $_5C_3$ **c.** $_6C_6$ **d.** $_7C_0$

Solution

We will use the Pascal triangle shown in Figure 4.9.

a. To find $_5C_2$, go to row 5 and move across to the second number from the left after the end 1. The entry is 10. Thus, $_5C_2 = 10$.
b. To find $_5C_3$, go to row 5 and move across to the third number from the left after the end 1. The entry is 10. Thus, $_5C_3 = 10$.
c. To find $_6C_6$, go to row 6 and move across to the sixth number from the left after the end 1. The entry is 1. Thus, $_6C_6 = 1$.
d. To find $_7C_0$, go to row 7 and do not move anywhere after the end 1. Remain there. Thus, $_7C_0 = 1$.

EXERCISES FOR SECTION 4.5 $1-9-odds$ formula $_NC_r = \dfrac{n!}{r!(N-r)!}$

  1. Evaluate each of the following:

a. $_7C_3$ **b.** $_6C_5$ **c.** $_8C_5$ **d.** $_7C_0$
35 $\dfrac{720=6}{120}$

e. $_7C_1$ **f.** $_8C_4$ **g.** $\binom{7}{4}$ **h.** $\binom{9}{5}$

i. $\binom{5}{5}$ **j.** $\binom{8}{9}$

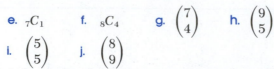

2. A disc jockey has been given 12 records to play on the air. However, the disc jockey has time to play only 9 of them. In how many different ways can the disc jockey select the 9 records to be played?

3. *Medical volunteers:* Medical researchers are testing a new drug for treating one form of a neurological disorder. It is decided to set up a control group of 18 people and then to select 8 of these people to be given the new drug. The remaining 10 people will be given a placebo. In how many different ways can the 8 subjects be selected?

$_{18}C_8$

4. There are 12 members on the Board of Directors of the Barry Corporation. A committee consisting of 7 people is to be selected from the 12 board members to look into the feasibility of establishing a new pension plan. In how many different ways can this be done?

$_{10}C_4$

5. There are ten nurses who work on the night shift on the tenth floor of General Hospital. In an effort to save money, the hospital administrator decides to fire four nurses. In how many different ways can the administrator select the four nurses to be fired? 210

17280

3628800

2432902 18

156920?2 14

6. *Paraplegics:* The campus newspaper recently ran an article on how the college mistreats paraplegics. To investigate the accuracy of these charges, the administration decides to select 7 male paraplegics from a possible 15 male paraplegics and 9 female paraplegics from a possible 13 paraplegics and to interview them in great detail. In how many different ways can the administration select these people?

$_{20}C_5$

7. An inspector from the health department arrives at a local restaurant to check the food for possible contamination. There are 20 portions prepared. The inspector decides to test 5 of these portions. In how many different ways can the inspector select the 5 portions to be tested?

8. A small airline company employs 11 male flight attendants and 13 female flight attendants. As an economy move, the management decides to fire two workers. Find the probability that
   a. both fired workers will be male flight attendants.
   b. both fired workers will be female flight attendants.
   c. one of the fired workers will be a male flight attendant and the other will be a female flight attendant.

9. The Alpha Beta student club consists of 8 men and 9 women members. A committee of 6 is to be selected. In how many different ways can this be done if
   a. the committee must have 3 men and 3 women on it?
   b. the committee must have at least 1 woman on it?
   c. the committee must have at least 1 man and 1 woman on it?

d. the committee must have more women than men on it?

10. Each year a state tax department audits the books of 5 of the 15 largest companies doing business within the state. In how many different ways can the tax department select the 5 companies to be audited?

11. There are 11 people with standby tickets hoping to get on a flight to Europe. Airline officials decide that they will be able to allow 4 people to board the plane. The remaining 7 passengers will have to wait for the next flight. Assuming order does not count,

a. in how many different ways can the 4 people allowed to board the plane be selected?

b. in how many different ways can the 7 passengers that will have to wait for the next flight be selected?

c. how do the answers in parts (a) and (b) compare? Comment.

*12. Verify that the probabilities given in Table 4.1 are accurate.

*13. A committee of four is to be set up to investigate charges of medical incompetence at Kings Hospital. The members of the committee are to be selected from the eight cardiologists, five surgeons, four anesthesiologists, and two obstetricians on the hospital staff. In how many different ways can the committee be selected, if

a. one member must be from each of the aforementioned specialties?

b. at least one cardiologist and one surgeon must be on the committee?

*14. Explain how the identity $\binom{n+1}{r} = \binom{n}{r} + \binom{n}{r-1}$ enables us to construct Pascal's triangle. (*Hint:* First verify the identity.)

*15. Using Pascal's triangle, show that $\binom{n}{r} = \binom{n}{n-r}$; that is, numbers in a particular row read from left to right are the same as read from right to left.

*16. Using Pascal's triangle, show that the sum of the numbers in any row is 2^n where n is the row number.

4.6 ODDS AND MATHEMATICAL EXPECTATION

Gamblers are frequently interested in determining which games are profitable to them. What they would really like to know is how much money can be earned in the long run from a particular game. If in the long run nothing can be won, that is, if as much money that is won will be lost, then why play at all? Similarly, if in the long run no money can be won but money can be lost, then the gambler will not play such a game. He or she will play any game only if money can be won. When applied to gambling or business situations, the **mathematical expectation** of an event is the amount of money to be won or lost in the long run.

COMMENT   In Chapter 6, when we discuss random variables and their probability distributions in detail, we will present a more general definition of mathematical expectation.

Since many applied examples of mathematical expectations involve gambling or business situations, where money can be won or lost in the long run, we have the following convenient formula for such cases.

FORMULA 4.5

Mathematical Expectation

Consider an event that has probability p_1 of occurring and that has a payoff, that is, the amount won, m_1. Consider also a second event with probability p_2 and payoff m_2, a third event with probability p_3 and payoff m_3, and so on. The **mathematical expectation** of the events is

$$m_1 p_1 + m_2 p_2 + m_3 p_3 + \cdots + m_n p_n$$

where the event has n different payoffs.

(The numbers attached to the letters are called subscripts. They have no special significance. We use them only to avoid using too many different letters.)

We illustrate these ideas with several examples.

EXAMPLE 1

A large company is considering opening two new factories in different towns. If it opens in town A, it can expect to make $63,000 profit per year with a probability of $\frac{4}{7}$. However, if it opens in town B, it can expect to make a profit of $77,000 with a probability of only $\frac{3}{7}$. What is the company's mathematical expectation?

Solution

We use Formula 4.5. We have

$$(63,000) \left(\frac{4}{7} \right) + (77,000) \left(\frac{3}{7} \right) = 36,000 + 33,000 = 69,000$$

Thus, the company's mathematical expectation is $69,000.

EXAMPLE 2

A contractor is bidding on a road construction job that promises a profit of $200,000 with a probability of $\frac{7}{10}$ and a loss, due to strikes, weather conditions, late arrival of building materials, and so on, of $40,000 with a probability of $\frac{3}{10}$. What is the contractor's mathematical expectation?

Solution

A "$+$" sign will denote a gain and a "$-$" sign will denote a loss. Using Formula 4.5, we find that the contractor's mathematical expectation is

$$(+\$200,000)\left(\frac{7}{10}\right) + (-\$40,000)\left(\frac{3}{10}\right)$$
$$= \$140,000 - \$12,000 = \$128,000$$

The contractor's mathematical expectation is \$128,000.

CAN YOUR CHANCES OF WINNING IN A LOTTERY BE IMPROVED?

Sept. 17: Because winning lottery numbers are generated randomly, no set of numbers has a higher probability of winning than any other, according to Dr. Jim Maxwell of the American Mathematics Society. "It is my understanding that lotteries are designed so that no one will have an advantage over anyone else," he said. "You could play the same number over and over again and have the same chance of winning as you would if you changed the number each time." Nevertheless, lottery players can decrease the number of players with whom they share their winnings by decreasing the odds that other players will choose the same numbers. "This is done by staying away from common number combinations such as months of the year, birthdays, holidays, et cetera," Dr. Maxwell said, since numbers drawn from the calendar are usually 31 or less. Because many lotteries call for numbers up to 48, players may increase their winnings, though not their chances of winning, by selecting some numbers above 31.

September 17, 1988

This newspaper article indicates that we can often make decisions (and select numbers accordingly) that affect our mathematical expectations.

EXAMPLE 3

Peter selects one card from a deck of 52 cards. If it is an ace, he wins $5. If it is a club, he wins only $1. However, if it is the ace of clubs, then he wins an extra $10. What is his mathematical expectation?

Solution

When one card is selected from a deck of cards, we have the following probabilities:

$$p(\text{ace}) = \frac{4}{52}$$

$$p(\text{clubs}) = \frac{13}{52}$$

$$p(\text{ace of clubs}) = \frac{1}{52}$$

Thus, Peter's mathematical expectation is

$$5\left(\frac{4}{52}\right) + 1\left(\frac{13}{52}\right) + 10\left(\frac{1}{52}\right) = \frac{20}{52} + \frac{13}{52} + \frac{10}{52}$$

$$= \frac{20 + 13 + 10}{52} = \frac{43}{52} = 0.83$$

His mathematical expectation is 83 cents.

COMMENT   83¢ is the fair price to pay to play the game, or is the break-even point. If he pays more than 83¢ per game to play, then in the long run he will lose money. If he pays less, then in the long run he will win.

Another interesting application of probability is concerned with betting. Gamblers frequently speak of the odds of a game. To understand this idea best, consider George, who believes that whenever he washes his car, it usually rains the following day. If George has just washed his car and if the probability of it raining tomorrow is $\frac{3}{10}$, gamblers would say that the odds in favor of it raining are 3 to 7 and the odds against it raining are 7 to 3. The 7 represents the 7 chances of it not raining. Formally we have the following definitions.

DEFINITION 4.5
Odds in Favor of
an Event

The **odds in favor** of an event occurring are p to q, where p is the number of favorable outcomes and q is the number of unfavorable outcomes.

favorable to unfavorable

DEFINITION 4.6 Odds Against an Event	If p and q are the same as in Definition 4.5, the **odds against** an event happening are q to p.

We now illustrate these definitions with several examples.

EXAMPLE 4

What are the odds in favor of the New York Mets winning the World Series if the probability of their winning is $\frac{4}{7}$ and the probability of their losing is $\frac{3}{7}$?

Solution

Since the probability of their winning is $\frac{4}{7}$, this means that out of 7 possibilities 4 are favorable and 3 are unfavorable. Thus, the odds in favor of the New York Mets winning the series are 4 to 3.

> ## METS ARE FAVORED TO WIN SERIES
>
> *New York (Sept. 20)*—The New York Mets are favored to win the upcoming baseball World Series. Local oddsmakers were betting 7-to-4 odds that the Mets would win the series in six games.
>
> September 20, 1990

EXAMPLE 5

Leon is in a restaurant. He decides to give the waiter a tip consisting of only 1 coin selected randomly from among the 6 that he has in his pocket. What are the odds against him giving the waiter a penny tip, if he has a penny, a nickel, a dime, a quarter, a half-dollar, or a dollar piece?

Solution

There are 6 coins; 5 are favorable and 1 is unfavorable. Thus, the odds against giving the waiter a penny tip are 5 to 1.

EXAMPLE 6

What are the odds in favor of getting a face card when selecting a card at random from a deck of 52 cards?

[handwritten: 180,000 +-30,000]

Solution

There are 52 possibilities; 12 are favorable and 40 are unfavorable. Thus, the odds in favor of getting a face card are 12 to 40.

EXERCISES FOR SECTION 4.6

[handwritten: 3/5   300,000 $\frac{3}{5}$ (300000) $\frac{3(50,000)}{5}$]

[handwritten in left margin: $\frac{3}{5}$ 300000 $\frac{3}{5}$(300,000)]

1. A publisher has invested \$50,000 to print a math book. If the book sells more than 100,000 copies over a 3-year period, the publisher will make a \$300,000 profit. Otherwise the publisher will lose \$50,000. If the probability that the book sells more than 100,000 copies is $\frac{3}{5}$, what is the mathematical expectation?

2. If graduation ceremonies at a certain college are conducted outside, \$5800 can be raised for alumni activities. If it rains, graduation ceremonies have to be conducted in the school auditorium and only \$2000 can be raised. If the probability of rain is $\frac{3}{5}$ and the probability of nice weather is $\frac{2}{5}$, how much money can the college expect to raise?

3. A coin is tossed three times. If heads appears on all 3 tosses, Mary will win \$2. If heads appears on 2 of the tosses, she will win \$16. Otherwise she loses \$8. What is her mathematical expectation?

4. A box contains 8 green marbles, 7 yellow marbles, and 5 black marbles. One marble is to be selected from the box. You get \$10 if the marble selected is black, but you lose either \$3 if the marble selected is green or \$5 if it is yellow. What is your mathematical expectation?

5. In a coin collection there are 9 pennies, 8 nickels, 12 dimes, 16 quarters, 11 half-dollars, and 4 dollar pieces. A coin is randomly selected from the collection by a blindfolded boy who gets to keep the coin picked. What is the boy's mathematical expectation?

6. A pair of dice is rolled. Depending on the sum of the dots appearing on both dice, Jake can win or lose money as shown in the following chart.

Sum of dots	Outcome
6 or 8 or 9	win \$6
2 or 4 or 5	lose \$4
10	win \$7
3 or 12	win \$5
7 or 11	lose \$1

What is Jake's mathematical expectation?

7. A survey in Denver of 1000 people found that 628 were in favor of tighter gun controls, whereas 372 believed that the current laws were adequate. If an in-

dividual from this group is randomly selected, what are the odds that this individual is against new gun control legislation?

8. Marlowe parks his car by a defective parking meter. The probability that he will get a summons is $\frac{4}{9}$. What are the odds against his getting a summons?

9. Rachel has just taken a new medicine that claims to be 95% effective in relieving arthritic pain. What are the odds against the medicine being effective?

10. Consider the following newspaper clipping. What are the odds in favor of a particular business being checked?

STATE TO CHECK ON SALES TAX CHEATS

Marlowe: (April 17)—The State Sales Tax Bureau announced yesterday that it would begin checking on merchants who consistently charge the wrong sales tax. Said a spokesman for the bureau, the State is losing an estimated 5 million dollars annually from merchants who do not collect sales tax.

It is expected that approximately 15% of all business establishments within the state will be checked.

April 17, 1990

11. George Salington is a salesperson for a medical supplies company. He is analyzing his five best accounts and comes up with the following information.

Account	Weekly Sales	Estimated Probability That Sales Will Be Realized
I	$3975	0.61
II	$3300	0.82
III	$4100	0.59
IV	$4880	0.51
V	$5705	0.46

Find the expected sales from each account.

12. Refer to Exercise 11. George's car has broken down and as a result George finds that he has enough time to visit only two of the accounts. Using mathematical expectation, help George decide which accounts to visit.

13. Consider the following newspaper clipping. Based on the information contained in it,

a. find the oil companies' mathematical expectation.
b. what are the odds in favor of finding oil in the area?

EXPLORATION FOR OIL TO BEGIN TODAY

Prudhoe Bay (March 17): A conglomeration of oil companies will begin drilling for oil in an area which is 70 miles north of area A. Geologists estimate that the probability of finding oil in the area is 0.85. If oil is discovered in large enough quantities to make it commercially feasible to drill for, the oil companies are expected to make a profit of $7 million. On the other hand, if the quantity of oil discovered is not large enough, then the oil companies will lose 2.5 million dollars in exploration costs.

Thursday, March 17, 1981

4.7 SUMMARY

In this chapter we discussed various aspects of probability. We noticed that probability is concerned with the total number of possible outcomes and experiments. Among the different ways of defining probability, we mentioned both the classical and the relative-frequency interpretation of probability. In the classical definition we have

$$p = \frac{\text{number of outcomes favoring an event}}{\text{total number of equally possible outcomes}}$$

We noticed that the probability of an event was between 0 and 1, the null event and the definite event, respectively.

To enable us to determine the total number of possible outcomes, we analyzed various counting techniques. Tree diagrams, permutations, and combinations were introduced and discussed in detail. Permutations represent arrangements of objects where order *is* important, whereas combinations represent selections of objects where order is *not* important. Applications of permutations and combinations to many different situations were given, in addition to the usual gambling problems.

Finally, probability was applied to determine the amount of money to be won in the long run in various situations. This was called the mathematical expectation of the event. We also discussed what is meant by statements such as odds in favor of an event and odds against an event. Definitions were given that allow us to calculate these odds.

| Study Guide | The following is a chapter outline in capsule form. You should now be able to demonstrate your knowledge of the ideas mentioned by giving definitions, descriptions, or specific examples. Page references are given in parentheses. |

The outcomes that are of interest to us represent a **favorable event**. (page 177)

All possible outcomes of an experiment are referred to as the **sample space** of the experiment. (page 178)

In the **classical interpretation of probability** we define the probability of any event as the number of favorable outcomes divided by the total number of possible outcomes, assuming that all outcomes of an experiment are equally likely. (page 180)

The **relative frequency interpretation of probability** represents the percentage of times that an event will happen in repeated experiments. (page 180)

An event that can never happen is called a **null event** and its probability is 0. (page 185)

An event that is certain to occur is called the **certain event** or **defininte event** and its probability is 1. (page 186)

The probability of any event must be between 0 and 1. (page 186)

One technique that is sometimes used to determine the number of possible outcomes for an experiment is to construct a **tree diagram** where each branch represents a different possible outcome. (page 192)

The **counting rule** states that if one thing can be done in m ways, and if after this is done, something else can be done in n ways, then both things can be done in a total of $m \cdot n$ different ways in the stated order. (page 194)

A **permutation** is any arrangement of distinct objects in a particular order. (page 199)

We use **factorial notation** to represent a special type of multiplication. Thus, $n!$ means that we start with the whole number n and multiply this by the number that is 1 less than n, and so on until we get to the number 1, which is where we stop. (page 200)

$_nP_r$ represents the **number of permutations of n things taken r at a time**. (page 201)

The **number of permutations of n things taken n at a time** is denoted by $_nP_n$ and is equal to $n!$ (page 202)

A **combination** is a selection from a collection of distinct objects where order is not important. (page 209)

The **number of combinations of n things taken r at a time** is denoted by $_nC_r$ or $\binom{n}{r}$. The symbol $\binom{n}{r}$ or $_nC_r$ is also called a **binomial coefficient**. (page 209)

Pascal's triangle can be used for computing the number of possible combinations of n things taken r at a time. This technique avoids the factorial notation. (page 215)

If one event has probability p_1 of occurring with a payoff of m_1, a second event has probability p_2 of occurring with payoff of m_2, a third event has probability p_3 of occurring with payoff m_3, and so on, then the **mathematical expectation** of the events is $m_1p_1 + m_2p_2 + m_3p_3 + \cdots + m_np_n$. When applied to gambling or business situations, the mathematical expectation of an event is the amount of money to be won or lost in the long run. (page 220)

The **odds in favor** of an event occurring are p to q, where p is the number of favorable outcomes and q is the number of unfavorable outcomes. (page 222)

The **odds against** an event happening are q to p, where p and q are the same as in the previous definition. (page 223)

Formulas to Remember

You should be able to identify each symbol in the following formulas, understand the relationships among the symbols expressed in each formula, understand the significance of each formula, and use the formulas in solving problems.

1. $p(A) = \dfrac{\text{number of favorable outcomes}}{\text{total number of equally possible outcomes}} = \dfrac{f}{n}$

2. $_nP_r = \dfrac{n!}{(n-r)!}$ The number of permutations of n things taken r at a time

3. $_nP_n = n!$ The number of permutations of n things taken n at a time

4. $\dfrac{n!}{p!\,q!\,r!} \cdots$ The number of permutations of n things where p are alike, q are alike, and so on

5. $_nC_r = \dfrac{n!}{r!(n-r)!}$ The number of combinations of n things taken r at a time

6. $m_1p_1 + m_2p_2 + m_3p_3 + \cdots$ Mathematical expectation of an event.

7. Odds in favor of an event are p to q, and odds against an event are q to p,

$$\text{where} \begin{cases} p = \text{the number of favorable outcomes} \\ q = \text{the number of unfavorable outcomes.} \end{cases}$$

Testing Your Understanding of This Chapter's Concepts

1. Construct a tree diagram to determine the number of possible ways that a coin can be flipped five times in succession so that throughout the flips there are always at least as many heads as tails.

2. Each of the 17,225 students at a state university commutes to school by bicycle. In an attempt to discourage theft, the administration is considering a plan to carve identification codes on each bike.

    a. Will a three-digit code carved on each bike be sufficient for all bikes? Explain your answer.

    b. Will a three-letter code carved on each bike be sufficient for all the bikes? Explain your answer.

 3. The Social Security Administration never reuses any Social Security number; that is, after an individual dies, that individual's Social Security number is no longer used.

    a. How many different nine-digit Social Security numbers are there?

    b. If a tenth digit is added, how many different Social Security numbers will there be?

    c. If a letter is added to everybody's Social Security number, how many different Social Security numbers will there be?

4. In an effort to protect the consumer, the agriculture department of a certain state requires that all canned goods manufactured within the state be stamped with numbers or letters as follows:

    i. The company numbers as assigned by the department: There are nine companies within the state.

    ii. The plant location within the state: The state is divided into eight geographical areas labeled A, B, C, D, E, F, G, or H.

    iii. The date of manufacture: This consists of the month and last two digits of the year. For example, a can with the code 1A0288 on it means that it was manufactured by company 1 located in area A during the month of February 1988.

    Using the preceding scheme, how many different codes are possible for cans manufactured during the years 1986 through 1991?

THINKING CRITICALLY

1. José tosses a fair coin twice and observes the number of heads obtained. He reasons that since the number of heads obtained can be either 0, 1, or 2, the probability of obtaining no heads is $\frac{1}{3}$. Do you agree with this conclusion? Explain your answer.

2. There are three cards on a table as shown in the following figure. Under one of these cards there is a one thousand dollar bill. Under the second card there is a one dollar bill and under the third card there is nothing. A player must guess under which card the $1000 bill appears.

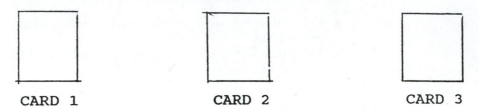

CARD 1   CARD 2   CARD 3

If the player guesses correctly, then the player keeps the $1000. After the player selects a card, the card dealer reveals one of the remaining cards under which there is nothing. The dealer now offers the player the opportunity of switching choices among the remaining cards since the probability of winning $1000 is now $\frac{1}{2}$ instead of $\frac{1}{3}$. Do you agree with this reasoning? Explain your answer.

3. A pair of dice is rolled and the sum of the number of dots appearing on both dice is recorded. Since the sum is either 2, 3, 4, 5, 6, 7, 8, 9, 10, 11, or 12, the probability of obtaining a sum of 12 is $\frac{1}{11}$. Do you agree with this reasoning? Explain your answer.

Review Exercises for Chapter 4

1. If the letters of the word "AXIOM" are rearranged at random, what is the probability that the first letter of the new "word" is a vowel?

2. Bob and Bill are gas station attendants. Each week, Bob works four days and Bill works three days. In how many different orders could they choose to work at the gas station? (*Hint:* This is an arrangement of seven things with repetition.)

3. A license plate consists of five letters of the alphabet selected at random with repetition allowed. What is the probability that the license plate number has at least four X's?

4. Mr. and Mrs. Boyer have four children. What is the probability that their oldest child and their youngest child are both girls?

5. There are nine players on a baseball team. The manager is arranging a batting order for these players. The pitcher will bat last. How many different batting orders are possible for these players?

6. How many three-digit numbers larger than 600 can be formed from the digits 4, 5, 6, and 7 if repetition is not allowed?

7. How many three-letter words can be formed from the letters of the word "VARIOUS" if each letter is used only once in a word?

8. How many different outfits can Joe put together if he has 4 pairs of pants, 6 shirts, and 2 pairs of shoes?

9. If the probability of the New York Mets winning a particular game is $\frac{3}{7}$, what are the odds in favor of it winning the game?

10. A coin is tossed six times, what is the probability of getting all heads?

11. Two dice are rolled. If you win $4 when you get a sum greater than 9, and lose $1 otherwise, which is your mathematical expectation?

Chapter Test

1. Which of the following cannot be the probability of an event?

   a. 0.3625   b. −0.5   c. $\frac{1}{3}$ d. 0   e. 1.00

 2. In how many different ways can seven people line up to buy a ticket for a football game?

 3. How many different committees of 6 students and 5 faculty members can be formed if each committee is to consist of 4 students and 3 faculty members?

4. How many four-digit numbers larger than 5000 can be formed using the digits 1, 2, 5, 6, 8, and 9? (No repetition allowed.)

 5. How many different permutations are there of the letters in the word "ELE-MENTARY"?

6. Refer back to the newspaper article given at the beginning of this chapter, on page 174. Assume that there are 38 governors that are Democrats and 12 governors that are Republicans. Find the probability that the committee will consist of

   a. all Republicans.
   b. at least two Republicans.
   c. no Republicans.

 7. A bakery is considering using one of two identification codes to place on each package of sliced bread that it manufactures; four letters followed by three numbers or four numbers followed by three letters. Which of these schemes will result in more identification codes?

8. An employee in a supermarket discovers three cans of food on the floor from which the identifying labels have been removed. One can contains carrots, one can contains corn, and one can contains peas. The labels are nearby on the floor. It is definitely known that the labels are for these cans. If the employee randomly attaches a label to a can, what is the probability that each can gets labeled properly?

9. Construct a tree diagram to determine the number of possible outcomes when a die is rolled twice.

10. Eleven patrons enter a crowded movie theater that has five vacant seats. In how many different ways can the five vacant seats be occupied?

11. There are 11 trucks and 13 cars waiting to be serviced at Bay's Transmission Service. Several mechanics have called in that they are sick. Management decides that it can service only 6 trucks and 7 cars. In how many different ways can management select the cars and trucks to be serviced?

12. Twelve cans of soda consisting of 4 Pepsi, 3 Coke, 2 Seven-Up, and 3 Sunkist are to be placed on a shelf. In how many different ways can the cans of soda be arranged if cans of the same brand are to stay together?

13. A wheel of fortune is equally divided into six colored areas: black, green, blue, yellow, red, and brown. If the wheel stops on black or blue after one spin, then the prize is $9. If it stops on green, the prize is $7. However, if it stops on red or brown, then $2 will be lost. There is no prize when the wheel stops on yellow.
    a. What are the odds of winning a prize in any one spin of the wheel?
    b. What is the mathematical expectation for someone who plays this game?

14. Last year the Internal Revenue Service (IRS) audited 2.3% of all tax returns filed by residents of a certain city. If a resident of this city is randomly selected, what are the odds in favor of this resident's tax return being audited?

15. Seven children are semifinalists in a beauty contest from which three children will be selected. In how many different ways can a winner and two runner-ups (runner-up 1 and runner-up 2) be selected?

16. A pair of dice is rolled. Find the probability that the sum of the number of dots appearing is 7.
    a. $\frac{1}{7}$ b. $\frac{6}{36}$ c. $\frac{7}{36}$ d. $\frac{5}{36}$ e. none of these

17. A family is known to have four children. Find the probability that the family consists of 2 boys and 2 girls.
    a. $\frac{1}{2}$ b. $\frac{5}{8}$ c. $\frac{3}{4}$ d. $\frac{3}{8}$ e. none of these

18. Which of the following cannot be the probability of some event?
    a. 0.00053   b. $-\frac{1}{7}$ c. 0   d. 0.99999   *e. none of these

19. A charity box contains 8 pennies, 5 nickels, 6 dimes, and 7 quarters. When the box is shaken, a coin falls out. Find the probability that the value of the coin is at least 10 cents.
    a. $\frac{7}{26}$ b. $\frac{13}{26}$ c. $\frac{6}{26}$ d. $\frac{8}{26}$ e. none of these

20. Each license plate in a certain state consists of 2 letters followed by 3 numbers with repetition permitted. How many different license plates are possible in this state?
    a. 676,000   b. 468,000   c. 492,804   d. 52,000   e. none of these

Suggested Further Reading

1. Bell, E. T. *Men of Mathematics*. New York: Simon & Schuster, 1961. Chapter 5 contains a bibliography of Pascal.

2. Bergamini, D. and eds. *Life. Mathematics* (Life-Science Library). New York: Time Life Books, 1970. Pages 126 to 147 discuss figuring the odds in an uncertain world.

3. Daniel, Wayne W., and James Terrell. *Business Statistics: Basic Concepts and Methodology*, 4th ed. Boston: Houghton-Mifflin, 1986.

4. Dixon, Wilfred J., and Frank J. Massey. *Introduction to Statistical Analysis*, 4th ed. New York: McGraw-Hill, 1982.

5. Epstein, R. A. *Theory of Gambling and Statistical Logic*. New York: Academic Press, 1967. Contains an interesting discussion on the fairness of coins.

6. Groebner, David, and Patrick Shannon. *Business Statistics*, 2nd ed. Columbus: Charles E. Merrill, 1985.

7. Havermann, E. "Wonderful Wizard of Odds" in *Life* 51, no. 14 (Oct. 6, 1961), p. 30 ff. contain a discussion of odds.

8. Huff, Darrell. *How to Take a Chance*. New York: Norton, 1959.

9. Kasner, E., and J. Newman. *Mathematics and the Imagination*. New York: Simon & Schuster, 1940. See the chapter on chance and probability.

10. *Mathematics in the Modern World* (Readings from *Scientific American*) San Francisco: W. H. Freeman, 1968. Article 22 discusses chance; Articles 23 and 24 discuss probability.

11. Naiman, Arnold, and Robert Rosenfeld. *Understanding Statistics*, 3rd ed. New York: McGraw-Hill, 1983.

12. Newmark, Joseph. *Using Finite Mathematics*. New York: Harper & Row, 1982.

13. Newmark, Joseph, and Frances Lake. *Mathematics as a Second Language*, 4th ed. Reading, MA: Addison-Wesley, 1987. Chapter 11 discusses probability and its applications.

14. Ore, Øystern. *Cardano, the Gambling Scholar*. New York: Dover, 1965.

15. Wonnacott, Ronald, and Thomas Wonnacott. *Introductory Statistics*, 4th ed. New York: John Wiley, 1985.

5 Rules of Probability

If a fair coin is tossed fifty times, how likely is it that it will land on "heads" fifty times? See discussion on p. 264. (© Rapho, Photo Researchers, Inc.)

Although probability was defined in the last chapter, nevertheless, we need rules for determining probabilities of events involving "and" or events involving "or." Also, we may be interested in determining probabilities given that other events have occurred. In this chapter we analyze various rules of probability.

BEWARE (A LITTLE) FALLING METEORITES

A falling meteorite may not be one of the major hazards of modern life, but a person can be struck by one. Scientists in Canada have even calculated the magnitude of the risk.

The scientists, at Herzberg Institute of Astrophysics in Ottawa, used a network of 60 cameras in Western Canada for the past nine years to study meteorite falls. From the observations, they said in a recent letter to *Nature*, it can be calculated that one human should be hit in North America every 180 years.

The researchers, I. Halliday, A. T. Blackwell and A. A. Griffin, based their calculations on the number of meteorite falls of a size large enough to be detected, the number of humans in the total Canadian and United States populations, the average human size, the time a person could be expected to be out of doors and other factors. On these assumptions they calculated an annual rate of human meteorite hits at 0.0055 per year, or one every 180 years.

Rare as these events should be, they noted, one such case actually occurred only 31 years ago. It is believed to be the only well-documented case of a collision between a meteorite and a human.

On Nov. 30, 1954, a nine-pound stony meteorite plunged through the roof of a home in Sylacauga, Ala., bounced off a large radio and struck Mrs. E. H. Hodges, inflicting painful bruises but causing her no serious injury.

"At first glance, it would appear unlikely that there would be even one known event only 31 years ago," the Canadian scientists said, "but the fact that there are no other verified cases elsewhere in the world indicates that the impacts on people are extremely rare."

Meteorite impacts on buildings are much less rare, they said, noting that there have been seven documented reports in North America during the past 20 years.

Worldwide, the Canadian scientists said, one could expect a human to be struck by a meteorite once in every nine years, while 16 buildings a year would be expected to sustain some meteorite damage.

November 14, 1986

What is the likelihood (probability) of a meteorite falling on a human being? The newspaper article indicates that given the average human size, the time that a person can be expected to be outdoors, and other factors, the probability of a human hit is still 0.0055 per year. How are such conditional probabilities calculated?

AIRPORTS TO TIGHTEN SECURITY

Washington (Jan. 10)—In the wake of the recent threats by international terrorists to bomb American facilities and the reports by many organizations on the easiness by which security can be breached at our airports, the Federal Aeronautics Administration (FAA) yesterday ordered all airports throughout the country to tighten security and to check all passengers thoroughly before they boarded planes. Several airports have already installed highly technological screening machines which can detect plastic explosives. Other airports have discontinued curbside check-ins.

January 10, 1991

As mentioned in the preceding article, many airports are increasing their security. What is the probability that someone can avoid being detected given all the additional security check? How do we calculate such conditional probabilities?

5.1 INTRODUCTION

In Chapter 4 we discussed the nature of probability and how to calculate its value. However, in order to determine the probability of an event in many situations, we must often first calculate the probability of other related events and then combine these probabilities. In this chapter we discuss several rules for combining probabilities, including rules for addition, multiplication, conditional probability, and Bayes' rule. Depending on the situation, these rules enable us to combine probabilities so that we can determine the probability of some event of interest.

5.2 ADDITION RULES

Mutually Exclusive Events

Let us look in on Charlie, who is playing cards. He is about to select one card from an ordinary deck of 52 playing cards. His opponent, Dick, will pay him $50 if the card selected is a face card (that is, a jack, queen, or king) *or* an ace. What is the probability that Charlie wins the $50?

Again, we note that a deck of playing cards consists of 52 playing cards as shown in Figure 5.1.

Now, to answer the question we first notice that a card selected cannot be a face card and an ace at the same time. Mathematically, we say that the events of "drawing a face card" and of "drawing an ace" are **mutually exclusive.**

Since there are 12 face cards in a deck (4 jacks, 4 queens, and 4 kings), the probability of getting a face card is $\frac{12}{52}$, or $\frac{3}{13}$. Similarly, the probability of getting an ace is $\frac{4}{52}$, or $\frac{1}{13}$, since there are 4 aces in the deck. There are then 12 face cards and 4 aces. Thus, there are 16 favorable outcomes out of a possible 52 cards in the deck. Applying the definition of probability, we get

$$p(\text{face card or ace}) = \frac{16}{52} = \frac{4}{13}$$

Let us now add the probability of getting a face card with the probability of getting an ace. We have

$$p(\text{face card}) + p(\text{ace}) = \frac{12}{52} + \frac{4}{52}$$

$$= \frac{12 + 4}{52} \qquad \text{(since the denominators are the same)}$$

$$= \frac{16}{52} = \frac{4}{13}$$

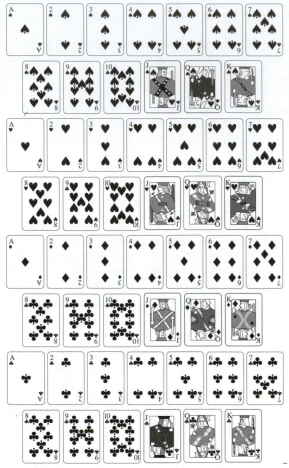

Figure 5.1
Deck of cards.

This indicates that

$$p(\text{face card or ace}) = p(\text{face card}) + p(\text{ace})$$

The same reasoning can be applied for any mutually exclusive events. First we have the following definition and formula.

DEFINITION 5.1
Mutually Exclusive Events

Consider any two events A and B. If both events cannot occur at the same time, we say that the events A and B are **mutually exclusive.**

FORMULA 5.1	Addition Rule

Addition Rule for Mutually Exclusive Events

If A and B are mutually exclusive, then

$$p(A \text{ or } B) = p(A) + p(B)$$

We illustrate the use of Formula 5.1 with several examples.

EXAMPLE 1

Louis has been shopping around for a calculator and has decided to buy the scientific model graphics calculator. The probability that he will buy the Casio model is $\frac{1}{9}$ and the probability that he will buy the Texas Instruments model is $\frac{4}{9}$. What is the probability that he will buy either of these two models?

Solution

Since Louis will buy only one calculator, the events "buys the Texas Instruments model" and "buys the Casio model" are mutually exclusive. Thus, Formula 5.1 can be used. We have

$$p(\text{buys either model}) = p(\text{buys Texas Instruments model}) \\ + p(\text{buys Casio model})$$

$$= \frac{4}{9} + \frac{1}{9}$$

$$= \frac{4 + 1}{9} = \frac{5}{9}$$

Therefore, $p(\text{Louis buys either model}) = \frac{5}{9}$.

EXAMPLE 2

Mary turns on the television set. The probability that the television is on Channel 2 is $\frac{1}{3}$, and the probability that it is on Channel 7 is $\frac{4}{13}$. What is the probability that it is either on Channel 2 or Channel 7?

Solution

Since the same television set cannot be on Channel 2 and on Channel 7 at the same time, the events "television set on Channel 2" and "television set on Channel 7" are mutually exclusive; thus, Formula 5.1 can be used. We have

$p(\text{television set on Channel 2 or 7}) = p(\text{television set on Channel 2})$

$+ \ p(\text{television set on Channel 7})$

$$= \frac{1}{3} + \frac{4}{13}$$

$$= \frac{13}{39} + \frac{12}{39} = \frac{25}{39}$$

Therefore,

$$p(\text{TV on Channel 2 or 7}) = \frac{25}{39}$$

$\left(\textit{Note:} \text{ We cannot add the fractions } \dfrac{1}{3} \text{ and } \dfrac{4}{13} \text{ together as they are since they do not}\right.$
have the same denominators. We must change both fractions to fractions with the
same denominators. Thus, $\dfrac{1}{3}$ becomes $\dfrac{13}{39}$ and $\dfrac{4}{13}$ becomes $\dfrac{12}{39}$. Hence, the probability
that the television is on Channel 2 or 7 is $\dfrac{25}{39}.\Big)$

EXAMPLE 3

Two dice are rolled. What is the probability that the sum of the dots appearing on both dice together is 9 or 11?

Solution

Since the events "getting a sum of 9" and "getting a sum of 11" are mutually exclusive, Formula 5.1 can be used. When 2 dice are rolled there are 36 possible outcomes. These were listed on page 179. There are 4 possible ways of getting a sum of 9. Thus

$$p(\text{sum of 9}) = \frac{4}{36}$$

Also, there are only two possible ways of getting a sum of 11. Thus,

$$p(\text{sum of 11}) = \frac{2}{36}$$

Therefore,

$$p(\text{sum of 9 or 11}) = p(\text{sum of 9}) + p(\text{sum of 11})$$

$$= \frac{4}{36} + \frac{2}{36}$$

$$= \frac{6}{36} = \frac{1}{6}$$

Hence, the probability that the sum is 9 or 11 is $\frac{1}{6}$.

EXAMPLE 4

Doris and her friends plan to travel to Florida during the winter intersession period. The probability that they go by car is $\frac{2}{3}$, and the probability that they go by plane is $\frac{1}{5}$. What is the probability that they travel to Florida by car or plane only?

Solution

Since they plan to travel to Florida either by car or by plane, not by both, we are dealing with mutually exclusive events. Formula 5.1 can be used. Therefore,

$$p(\text{go by car or plane}) = p(\text{go by car}) + p(\text{go by plane})$$

$$= \frac{2}{3} + \frac{1}{5}$$

$$= \frac{10}{15} + \frac{3}{15}$$

$$= \frac{13}{15}$$

Complementary Events

EXAMPLE 5

Rosemary buys a ticket in the state lottery. The probability that she will win the grand prize of one million dollars is $\frac{1}{50,000}$. What is the probability that she does not win the one million dollars?

Solution

Since the events "Rosemary wins the million dollars" and "Rosemary does not win the million dollars" are mutually exclusive, we can use Formula 5.1. One of those events must occur so that the event "Rosemary wins the million dollars or does not win the million dollars" is the definite event. We know that the definite event has probability 1 (see page 186). Thus,

$$p(\text{Rosemary wins million dollars or does not win}) = p(\text{wins million dollars})$$
$$+ \ p(\text{does not win million dollars})$$

$$1 = \frac{1}{50,000} + p(\text{does not win million dollars})$$

$$1 - \frac{1}{50,000} = p(\text{does not win million dollars})$$

$$\frac{50,000}{50,000} - \frac{1}{50,000} = p(\text{does not win million dollars})$$

$$\frac{49,999}{50,000} = p(\text{does not win million dollars})$$

Therefore, the probability that Rosemary does not win the million dollars is $\frac{49,999}{50,000}$.

More generally, consider any event A. Let $p(A)$ be the probability that A happens and let $p(A')$ read as the probability of A prime, be the probability that A does not happen. Since either A happens or does not happen, we can use Formula 5.1. Thus,

$$p(A \text{ happens or does not happen}) = p(A \text{ happens}) + p(A \text{ does not happen})$$
$$1 = p(A) + p(A')$$
$$1 - p(A) = p(A') \qquad \text{(We subtract } p(A) \text{ from both sides.)}$$

Complement of an
Event

Therefore, the probability of A not happening is $1 - p(A)$. (*Note:* Some books refer to the event A' as the **complement** of event A.)

Addition Rule—General Case

Now consider the following problem. One card is drawn from a deck of cards. What is the probability of getting a king or a red card? At first we might say that since there are 4 kings and 26 red cards, then

$$p(\text{king or red card}) = p(\text{king}) + p(\text{red card})$$

$$= \frac{4}{52} + \frac{26}{52}$$

$$= \frac{30}{52} = \frac{15}{26}$$

Thus, we would say that the probability of getting a king or a red card is $\frac{15}{26}$. Notice, however, that in arriving at this answer we have counted some cards twice. The 2 red kings have been counted as both kings and red cards. Obviously, we must count them only once in probability calculations. The events "getting a king" and "getting a red card" are not mutually exclusive. We therefore have to revise our original estimate of the total number of favorable outcomes by deducting the number of cards that have been counted twice. We will subtract 2. When this is done, we get

$$p(\text{king or red card}) = p(\text{king}) + p(\text{red card}) - p(\text{king also a red card})$$

$$= \frac{4}{52} + \frac{26}{52} - \frac{2}{52}$$

$$= \frac{4 + 26 - 2}{52}$$

$$= \frac{28}{52} = \frac{7}{13}$$

Thus, the probability of getting a king or a red card is $\frac{7}{13}$. This leads us to a more general formula.

Addition Rule
(general case)

ADDITION RULE (GENERAL CASE)

If A and B are events, the probability of obtaining either of them is equal to the probability of A plus the probability of B minus the probability of both occurring at the same time.

Symbolically, the addition rule is as follows.

FORMULA 5.2 — If A and B are any events, then

$$p(A \text{ or } B) = p(A) + p(B) - p(A \text{ and } B)$$

We now apply Formula 5.2 in several examples.

EXAMPLE 6

Sylvester is a member of The Hamilton Bay Ensemble. The probability that Sylvester plays a guitar is $\frac{1}{4}$, and the probability that he plays a clarinet is $\frac{5}{8}$. If the probability that he plays both instruments is $\frac{5}{24}$, what is the probability that he plays the guitar or that he plays the clarinet?

Solution

Since it is possible that Sylvester plays both these instruments, these events are not mutually exclusive. Thus, we must use Formula 5.2. We have

$$p(\text{plays guitar or clarinet}) = p(\text{plays guitar}) + p(\text{plays clarinet})$$
$$- p(\text{plays guitar and clarinet})$$

$$= \frac{1}{4} + \frac{5}{8} - \frac{5}{24}$$

$$= \frac{6}{24} + \frac{15}{24} - \frac{5}{24}$$

$$= \frac{6 + 15 - 5}{24}$$

$$= \frac{16}{24} = \frac{2}{3}$$

Thus, the probability that he plays either instrument is $\frac{2}{3}$.

EXAMPLE 7

Voting

Consider the newspaper article given on the top of the next page. What is the probability that either a husband will vote or his wife will vote in the coming mayoral election?

Solution

The events husband votes and wife votes are not mutually exclusive events, since both events can occur. Thus, we use Formula 5.2. We have

$$p(\text{husband or wife votes}) = p(\text{husband votes}) + p(\text{wife votes}) - p(\text{both vote})$$

Based on the newspaper article,

$$p(\text{husband votes}) = \frac{1}{11}$$

LOW VOTER TURNOUT EXPECTED

Greensburgh (Oct. 20)—According to a poll released yesterday, voter turnout for the coming mayoral election is expected to be at an all-time low. One out of eleven married men said that they would vote. For the women, the figure was one out of nine. Only one out of 28 couples said that they would both vote. Both candidates indicated that they were mobilizing their forces to get out more votes.

November 1, 1990

Reprinted by permission.

$$p(\text{wife votes}) = \frac{1}{9}$$

$$p(\text{both vote}) = \frac{1}{28}$$

Therefore,

$$p(\text{husband or wife votes}) = \frac{1}{11} + \frac{1}{9} - \frac{1}{28}$$

$$= \frac{252}{2772} + \frac{308}{2772} - \frac{99}{2772} = \frac{461}{2772}$$

Thus, $p(\text{husband or wife votes}) = \dfrac{461}{2772}$.

EXAMPLE 8

Controlling Pollution

Environmentalists have accused a large company in the eastern United States of dumping nuclear waste material into a local river. The probability that *either* the fish in the river or the animals that drink from the river will die is $\frac{11}{21}$. The probability that only the fish will die is $\frac{1}{3}$, and the probability that only the animals that drink from the river will die is $\frac{2}{7}$. What is the probability that both the fish *and* the animals that drink from the river will die?

Solution

Since both the fish and animals may die, the events are not mutually exclusive. We then use Formula 5.2. We have

$$p(\text{fish or animals die}) = p(\text{fish die}) + p(\text{animals die}) - p(\text{both fish and animals die})$$

$$\frac{11}{21} = \frac{1}{3} + \frac{2}{7} - p(\text{both die})$$

$$\frac{11}{21} = \frac{7}{21} + \frac{6}{21} - p(\text{both die})$$

$$\frac{11}{21} = \frac{13}{21} - p(\text{both die})$$

$$\frac{11}{21} + p(\text{both die}) = \frac{13}{21}$$

$$p(\text{both die}) = \frac{13}{21} - \frac{11}{21} = \frac{2}{21}$$

Thus, the probability that *both* the fish and animals that drink from the river will die is $\frac{2}{21}$.

COMMENT Although you may think that Formulas 5.1 and 5.2 are different, this is not the case. Formula 5.1 is just a special case of Formula 5.2. Formula 5.2 can always be used since if the events A and B are mutually exclusive, the probability of them happening together is 0. In this case Formula 5.2 becomes

$$p(A \text{ or } B) = p(A) + p(B) - 0$$

which is exactly Formula 5.1.

COMMENT Formula 5.2 can be used when we have only two events, A and B. For any three events A, B, and C, the probability of A or B or C is given by the formula

$$p(A \text{ or } B \text{ or } C) = p(A) + p(B) + p(C) - p(A \text{ and } B) - p(A \text{ and } C)$$
$$- p(B \text{ and } C) + p(A \text{ and } B \text{ and } C)$$

The use of this formula will be illustrated in the exercises.

EXERCISES FOR SECTION 5.2

1. Determine which of the following events are mutually exclusive.
   a. getting an *A* and getting a *B* in this course
   b. majoring in education and majoring in mathematics
   c. graduating from college in June and graduating from college in August
   d. smoking cigarettes and having lung cancer
   e. having a high incidence of air pollution and having an increase in upper respiratory infections
   f. coming to work by car and coming to work by public transportation
   g. going on vacation and quitting a job
   h. being a male and being a female

2. According to the American Heart Association, 31% of the people in Heightstown have high blood pressure, 40% of the people are overweight, and 16% of the people have high blood pressure and are also overweight. If a resident of Heightstown is randomly selected, what is the probability that this individual has high blood pressure *or* is overweight?

3. According to official 1990 police statistics, the probability of a person being mugged in Princetown is 0.19 and the probability of a person's house being burglarized in Princetown is 0.24. Furthermore, the probability of both these events happening is 0.02. If a resident of Princetown is randomly selected, what is the probability that this person was either mugged *or* that this person's house was burglarized?

4. A survey* of 10,000 children in the New York City area found that 65% of them lived in families owning a video cassette recorder (VCR). The survey further found that 76% of them lived in families that owned cars. Furthermore, 50% of them lived in families that owned both a VCR and a car. If a child from this group is randomly selected, what is the probability that the child's family owns either a VCR or a car?

5. Refer back to Exercise 4. The survey also found that 57% of the children had brothers or sisters that attended college. Also, 36% of these children needed remediation with their schoolwork. Furthermore, 10% of the children needed remediation *and* had brothers or sisters that attended college. If a child from this group is randomly selected, what is the probability that the child needs remediation *or* has a brother or sister that attended college?

6. Various studies indicate that the probability that a married man in a certain community smokes is $\frac{15}{28}$. The probability that his wife smokes is $\frac{2}{7}$ and the probability that both the husband and wife smoke is $\frac{1}{7}$. What is the probability

*Hodges and Kiddey, *Information About New York City Children*, New York, 1991.

that a randomly selected married man from this community *or* his wife (but not both) smokes?

7. 86% of the families in Monticello have either a pet dog or pet cat in their home. 64% of the families have a pet dog and 33% of the families have a pet cat. If a family from Monticello is randomly selected, what is the probability that the family has both a pet dog and a pet cat?

8. Whether purchases are made by cash, check, or credit card is of concern to vendors as they must pay a certain percentage of the sale value to the credit card issuer. Based on past experience, the owner of Atlas Travels knows that 88% of the travelers pay for their airline tickets by credit card or by personal check. Furthermore, 47% of the travelers pay by personal check. If a traveler that booked through Atlas Travels is randomly selected, what is the probability that the traveler paid for the airline ticket with a credit card? (Assume that nobody pays part by credit and the balance by check.)

9. It has been asserted that many of the trucks operating on our interstate highways are not safe. One study found that the probability of finding a truck operating on these highways with unsafe tires is 0.29, and the probability of finding a truck operating on these highways with defective brake systems is 0.15. Furthermore, the probability of finding a truck with either unsafe tires or defective brake systems is 0.36. A truck operating on the interstate highway system is randomly selected by a highway patrol officer. What is the probability that the truck has *both* unsafe tires and defective brakes?

10. Reggie is a pinch hitter in a baseball game. The probability that he gets a single is $\frac{1}{3}$ and the probability that he gets a home run is $\frac{1}{9}$. If Reggie will bat once in tomorrow's playoff game, what is the probability that he gets a single or a home run?

11. According to financial aid counselors, $\frac{1}{8}$ of all applications are returned because the applicant did not fill out the form completely, $\frac{1}{9}$ of all applications are returned because the wrong form was used, and $\frac{1}{16}$ of all applications are returned because the wrong form was used *and* also the applicant did not fill out the form completely. If a returned application is randomly selected, find the probability that it is either because the wrong form was used *or* because it was not filled out completely.

12. A hospital administrator is analyzing the division in which nurses on payroll line 16 work. The analysis shown on the top of the next page is available. If a nurse at the hospital is randomly selected, what is the probability that the nurse works in any of the three divisions mentioned?

Division in Which Nurse Works	Probability
Pediatrics	$\dfrac{63}{120}$
Cardiology	$\dfrac{70}{120}$
Urology	$\dfrac{64}{120}$
Pediatrics and Cardiology	$\dfrac{34}{120}$
Pediatrics and Urology	$\dfrac{30}{120}$
Cardiology and Urology	$\dfrac{38}{120}$
Pediatrics, Cardiology, and Urology	$\dfrac{8}{120}$

omit

13. All technology students at Jackmore College must complete the following three science courses:

Physics 100   –   Introduction to Physics
Biology 3   –   General Biology
Chemistry 120   –   General Chemistry

The probabilities that a student registers for one or more of these courses in the student's sophomore year are given in the accompanying table.

Science Course	Probability
Physics 100	$\dfrac{2}{3}$
Biology 3	$\dfrac{74}{120}$
Chemistry 120	$\dfrac{26}{60}$
Physics 100 and Biology 3	$\dfrac{38}{120}$
Biology 3 and Chemistry 120	$\dfrac{6}{20}$
Physics 100 and Chemistry 120	$\dfrac{2}{6}$
Either Physics 100, Biology 3, or Chemistry 120	1

Find the probability that a student registers for all three courses in his or her sophomore year.

14. A traffic enforcement official is analyzing the moving violation summonses issued by various police agencies throughout the state. The accompanying table gives the probability that a motorist was issued a summons for some moving violation.

Moving Violation	Probability
Speeding	$\dfrac{290}{470}$
Drunk driving	$\dfrac{250}{470}$
Driving recklessly	$\dfrac{170}{470}$
Speeding and drunk driving	$\dfrac{110}{470}$
Drunk driving and driving recklessly	$\dfrac{80}{470}$
Speeding and driving recklessly	$\dfrac{70}{470}$
Speeding, or drunk driving, or driving recklessly	1

Assuming that a driver was issued a summons for one of the moving violations mentioned, what is the probability that a driver was issued summonses for all three violations?

5.3 CONDITIONAL PROBABILITY

Although the addition rule given in Section 5.2 applies to many different situations, there are still other problems that cannot be solved by that formula. It is for this reason that we introduce conditional probability. Let us first consider the following problem.

EXAMPLE 1

Antismoking Campaign

In an effort to reduce the amount of smoking, the administration of Podunk University is considering the establishment of a smoking clinic to help students "break the habit." However, not all the students at the school favor the proposal. As a result,

	Against Smoking Clinic	For Smoking Clinic	No Opinion	Total
Freshmen	23	122	18	163
Sophomores	39	165	27	231
Juniors	58	238	46	342
Seniors	71	127	66	264
Total	191	652	157	1000

a survey of the 1000 students at the school was conducted to determine student opinion about the proposal. The table on the preceding page summarizes the results of the survey.

a. What is the probability that a student selected at random voted against the establishment of the smoking clinic?
b. If a student is a freshman, what is the probability that the student voted for the smoking clinic?
c. If a senior is selected at random, what is the probability that the senior has no opinion about the clinic?

Solution

a. Since there was a total of 191 students who voted against the establishment of the smoking clinic out of a possible 1000 students, we apply the definition of probability and get

$$p(\text{student voted against the smoking clinic}) = \frac{191}{1000}.$$

Thus, the probability that a student selected at random voted against the establishment of the smoking clinic is $\frac{191}{1000}$.

b. There are 163 freshmen in the school. One hundred twenty-two of them voted for the smoking clinic. Since we are concerned with freshmen only, the number of possible outcomes of interest to us is 163, not 1000. Out of these, 122 are favorable. Thus, the probability that a student voted for the smoking clinic given that the student is a freshman is $\frac{122}{163}$.

c. In this case the information given narrows the sample space to the 264 seniors, 66 of which had no opinion. Thus, the probability that a student has no opinion given that the student is a senior is $\frac{66}{264}$, or $\frac{1}{4}$.

Conditional Probability

The situation of part (b) or that of part (c) in Example 1 is called a **conditional probability** because we are interested in the probability of a student voting in favor of the establishment of the smoking clinic given that, or conditional upon the fact that, the student is a freshman. We express this condition mathematically by using a vertical line "|" to stand for the words "given that" or "if we know that." We then write

$$p(\text{student voted in favor of smoking clinic} \mid \text{student is a freshman}) = \frac{122}{163}$$

Similarly, for part (c) we write

$$p(\text{student had no opinion} \mid \text{student is a senior}) = \frac{66}{264} = \frac{1}{4}$$

EXAMPLE 2

Sherman is repairing his car. He has removed the six spark plugs. Four are good and two are defective. He now selects one plug and then, without replacing it, selects a second plug. What is the probability that both spark plugs selected are good?

Solution

We list the possible outcomes and then count all the favorable ones. To do this, we label the good spark plugs as g_1, g_2, g_3, and g_4, and the defective ones as d_1 and d_2. The possible outcomes are

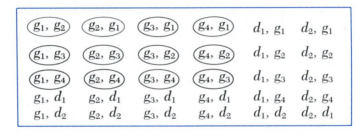

There are 30 possible outcomes. Twelve of these are favorable. These are the circled ones, which represent the outcome that both spark plugs are good. Thus,

$$p(\text{both spark plugs selected are good}) = \frac{12}{30} = \frac{2}{5}$$

EXAMPLE 3

In Example 2, what is the probability that both spark plugs selected are good if we know that the first plug selected is good?

Solution

Again, we list all possible outcomes and count the number of favorable ones.

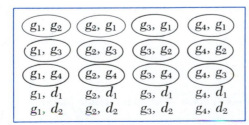

Since we know that the first plug selected is good, there are only 20 possible outcomes. Of these, 12 are favorable. These are the circled ones. Thus, the probability that both spark plugs are good if we know that the first plug is good is

$$\frac{12}{20} = \frac{3}{5}$$

Using the conditional probability notation, we can write this result as

$$p(\text{both spark plugs are good} \mid \text{first spark plug is good}) = \frac{3}{5}$$

COMMENT   Example 3 differs from Example 2 since in Example 3 we are interested in determining the probability of getting two good spark plugs once we know that the first one selected is good. On the other hand, in Example 2 we were interested in determining the probability of getting two good plugs without knowing whether or not the first plug is defective.

Let us analyze the problem discussed at the beginning of this section in detail. There are a total of 163 freshmen out of a possible 1000 students in the school. Thus,

$$p(\text{freshman}) = \frac{163}{1000}$$

Also, there were 122 freshmen who voted in favor of the clinic. Thus,

$$p(\text{freshman and voted in favor of clinic}) = \frac{122}{1000}$$

Summarizing these results we have

$$p(\text{freshman}) = \frac{163}{1000} \text{ and}$$

$$p(\text{freshman and voted in favor of clinic}) = \frac{122}{1000}$$

Let us now divide $p(\text{freshman }and\text{ voted in favor of clinic})$ by $p(\text{freshman})$. We get

$$\frac{p(\text{freshman }and\text{ voted in favor of clinic})}{p(\text{freshman})} = \frac{122/1000}{163/1000}$$

$$= \frac{122}{1000} \div \frac{163}{1000}$$

$$= \frac{122}{\cancel{1000}} \cdot \frac{\cancel{1000}}{163}$$

$$= \frac{122}{163}$$

This is the same result as p(student voted in favor of smoking clinic | student is a freshman). In both cases the answer is $\dfrac{122}{163}$.

If we let A stand for "student voted in favor of smoking clinic" and B stand for "student is a freshman," then the previous result suggests that

$$p(A \mid B) = \frac{p(A \text{ and } B)}{p(B)}$$

We can apply the same analysis for part (c) of the problem. We have

$$p(\text{senior}) = \frac{264}{1000} \quad \text{and} \quad p(\text{senior and no opinion}) = \frac{66}{1000}$$

If we divide p(senior and no opinion) by p(senior), we get

$$\frac{p(\text{senior and no opinion})}{p(\text{senior})} = \frac{66/1000}{264/1000}$$

$$= \frac{66}{1000} \div \frac{264}{1000}$$

$$= \frac{66}{1000} \cdot \frac{1000}{264} = \frac{66}{264} = \frac{1}{4}$$

Thus,

$$p(\text{student had no opinion} \mid \text{student is a senior}) = \frac{p(\text{senior and no opinion})}{p(\text{senior})}.$$

Conditional Probability Formula

We can generalize our discussion by using a formula that is called the **conditional probability formula.**

FORMULA 5.3   Conditional Probability Formula

If A and B are any events, then

$$p(A \mid B) = \frac{p(A \text{ and } B)}{p(B)}, \quad \text{provided } p(B) \neq 0$$

We illustrate the use of Formula 5.3 with several examples.

EXAMPLE 4   In Ashville the probability that a married man drives is 0.90. If the probability that a married man *and* his wife both drive is 0.85, what is the probability that his wife drives given that he drives?

Solution

We use Formula 5.3. We are told that

$$p(\text{husband drives}) = 0.90$$

and

$$p(\text{husband and wife drive}) = 0.85$$

Thus,

$$p(\text{wife drives} \mid \text{husband drives}) = \frac{p(\text{husband and wife drive})}{p(\text{husband drives})}$$

$$= \frac{0.85}{0.90}$$

$$= \frac{85}{90} \quad \text{(We multiply numerator and denominator by 100)}$$

$$= \frac{17}{18}$$

Thus, $p(\text{wife drives} \mid \text{husband drives}) = \dfrac{17}{18}$

EXAMPLE 5   Joe often speeds while driving to school in order to arrive on time. The probability that he will speed to school is 0.75. If the probability that he speeds and gets stopped by a police officer is 0.25, find the probability that he is stopped, given that he is speeding.

Solution

We use Formula 5.3. We are told that $p(\text{Joe speeds})$ is 0.75 and $p(\text{speeds and is stopped})$ is 0.25. Thus,

$$p(\text{he is stopped} \mid \text{he speeds}) = \frac{p(\text{speeds and is stopped})}{p(\text{speeds})}$$

$$= \frac{0.25}{0.75}$$

$$= \frac{25}{75}$$

$$= \frac{1}{3}$$

Thus, $p(\text{he is stopped} \mid \text{he speeds}) = \dfrac{1}{3}$

EXAMPLE 6

Janet likes to study. The probability that she studies *and* passes her math test is 0.80. If the probability that she studies is 0.83, what is the probability that she passes the math test, given that she has studied?

Solution

We use Formula 5.3. We have

$$p(\text{passes math test} \mid \text{she studied}) = \frac{p(\text{studies and passes math test})}{p(\text{she studies})}$$

$$= \frac{0.80}{0.83}$$

$$= \frac{80}{83}$$

Thus, $p(\text{she passes math test} \mid \text{she has studied}) = \dfrac{80}{83}$

EXERCISES FOR SECTION 5.3

1. There are 200 students receiving full scholarships at Island University as follows.

		Level	
		Undergraduate	*Graduate*
	Male	66	39
Sex			
	Female	54	41

If a student who is receiving a full scholarship is randomly selected, find the probability that the student is
a. an undergraduate.
b. a female.
c. a female undergraduate.
d. a male graduate.
e. a male graduate or a female undergraduate.

2. Seventy-two percent of all hospitalized senior citizen patients at Brook's Hospital have supplemental medicare insurance. Furthermore, the hospital medical records indicate that 32% of all hospitalized patients are females *and* that they have supplemental medicare insurance. If a patient is selected at random, what is the probability that the patient is a female given that the patient is receiving supplemental medicare insurance?

3. It is known that 9% of all people who have a particular enyzme in their blood and who are given a particular medication will develop a skin reaction. Marlene has a disease for which this drug could be used. Furthermore, the probability that her doctor prescribes this particular medication is 0.14. What is the probability that she develops a skin reaction when given this particular medication?

4. In one study it was found that 22% of all females between the age of 20 to 30 years jog to maintain their physical fitness. Twelve percent of the females jog *and* also exercise regularly at a health spa. If a female between the age of 20 to 30 years who is jogging is randomly selected, what is the probability that she exercises regularly at a health spa?

5. 92% of the campers at the Willowbrook Campgrounds are mountain climbers. Furthermore, 23% of the campers are both mountain climbers and licensed hunters. If a camper from the Willowbrook Campgrounds is seen climbing a mountain, find the probability that the camper is also a hunter.

6. Market research indicates that 41% of all people buy their computer and their computer software in a computer store as opposed to a department or general merchandise store. Furthermore, 77% of all people buy their computer in a computer store. Boris purchased his computer in a computer store and is now interested in buying software. What is the probability that he will buy it in a computer store?

7. 53% of the motels in a nationwide chain accept Master Card and Discover Card. 64% of the motels accept Master Card. If one motel of this chain that accepts Master Card is randomly selected, find the probability that it accepts Discover Card.

8. Due to the rising costs of medical insurance, many Americans simply cannot afford to have such coverage. The U.S. Bureau of the Census estimates that only about 65% of all American families have medical insurance. Moreover, the probability that an adult will be hospitalized and that he or she has no medical insurance is 0.13. If an adult American that has no medical insurance is randomly selected what is the probability that he or she will be hospitalized?

9. A sociologist interviewed 1000 individuals who had never flown in an airplane to determine the reason why these people had not flown. The following are the results of the interview.

REASON FOR NOT FLYING

	Afraid of hijacking	Financial considerations	No need to fly	Afraid of heights
Male	31	62	147	221
Female	151	94	118	176

Find the probability that an individual in the group
a. is a male that is afraid of heights.

b. is afraid of heights given that he is a male.

c. cited financial considerations as a reason for not flying.

10. The sociologist described in Exercise 9 then interviewed 1000 randomly selected individuals to determine which environmental issue concerned them the most. The results of the survey are shown in the accompanying table.

ENVIRONMENTAL PROBLEM

		Overpopulation	Water Pollution	Air pollution	Waste of natural resources
Age (Years)	Under 25	120	29	10	310
	25–50	69	360	13	6
	Over 50	8	11	60	4

Find the probability that a randomly selected individual in the group

a. is under 25 years of age.

b. is under 25 years of age given that he or she thought that water pollution was the main problem.

c. thought that overpopulation was the main problem.

d. thought that waste of natural resources was the main problem given that the individual was under 25 years old.

e. was under 25 years of age given that waste of natural resources was (is) the main problem.

11. Despite the numerous articles that have been written on the subject, many Americans have a blood serum cholsterol level that is considerably higher than the 200-milligram (mg) level recommended by the American Medical Association (AMA). Medical researchers tested 1412 residents of Patterson and reported the following cholesterol levels.

CHOLESTEROL LEVEL (in mg)

	Below 200	200–240	Above 240
Male	53	372	423
Female	79	297	188

Find the probability that an individual in the group

a. is a female.

b. is a female given that her cholesterol level is above 240 mg. 188/564

c. has a cholesterol level above 240 mg given that the individual is a female.

12. 89% of all office buildings in Newburg have a water-sprinkler system (in the event of fire). 67% of all office buildings in Newburg have smoke alarms *and* a water-sprinkler system. If an office building in Newburg with a water-sprinkler system is randomly selected, find the probability that the office building has *no* smoke alarms.

CASE STUDY Law

In June of 1964 an elderly woman was mugged in San Pedro, California. In the vicinity of the crime a bearded black man sat waiting in a yellow car. Shortly after the crime was committed, a young white woman, wearing her blonde hair in a ponytail, was seen running from the scene of the crime and getting into the car, which sped off. The police broadcast a description of the suspected muggers. Soon afterward, a couple fitting the description was arrested and convicted of the crime. Although the evidence in the case was largely circumstantial, the prosecutor based his case on probability and the unlikeliness of another couple having such characteristics. He assumed the following probabilities.

Characteristic	Assumed Probability
Drives yellow car	$\dfrac{1}{10}$
Black-white couple	$\dfrac{1}{1000}$
Black man	$\dfrac{1}{3}$
Man with beard	$\dfrac{1}{10}$
Blonde woman	$\dfrac{1}{4}$
Woman wears her hair in ponytail	$\dfrac{1}{10}$

He then multiplied the individual probabilities:

$$\frac{1}{10} \cdot \frac{1}{1000} \cdot \frac{1}{3} \cdot \frac{1}{10} \cdot \frac{1}{4} \cdot \frac{1}{10} = \frac{1}{12,000,000}$$

He claimed that the probability is $\dfrac{1}{12,000,000}$ that another couple has such characteristics. The jury agreed and convicted the couple. The conviction was overturned by the California Supreme Court in 1968. The defense attorneys got some professional advice on probability. Serious errors were found in the prosecutor's probability calculations. Some of these involved assumptions about independent events. As a matter of act it was demonstrated that the probability is 0.41 that another couple with the same characteristics existed in the area once it was known that there was at least one such couple.

For a complete discussion of this probability case, read "Trial by Mathematics," which appeared in *Time*, January 8, 1965, p. 42, and April 26, 1968, p. 41.

5.4 MULTIPLICATION RULES

In Section 5.3 we discussed the conditional probability formula and how it is used. In this section we discuss a variation of the conditional probability formula known as the multiplication rule.

Consider a large electric company in the northeastern United States. In recent years it has been unable to meet the demand for electricity. To prevent any cable damage and blackouts as a result of overload, that is, too much electrical demand, it has installed two special switching devices to shut off the flow of electricity automatically and thus prevent cable damage when an overload occurs. The probability that the first switch will not work properly, is 0.4, and the probability that the second switch will not work properly given that the first switch fails is 0.3. What is the probability that both switches will fail?

Let us look at Formula 5.3 in Section 5.3. It says that for any events A and B

$$p(A \mid B) = \frac{p(A \text{ and } B)}{p(B)}$$

If we multiply both sides of this equation by $p(B)$, we get

$$p(A \mid B) \cdot p(B) = p(A \text{ and } B)$$

Multiplication Rule   This equation is called the **multiplication rule.** We state this formally as follows:

FORMULA 5.4   Multiplication Rule

If A and B are any events, then

$$p(A \text{ and } B) = p(A \mid B) \cdot p(B)$$

If we now apply Formula 5.4 to our example, we get

$p(\text{both switches fail})$
$= p(\text{switch 2 fails} \mid \text{switch 1 has failed}) \cdot p(\text{switch 1 fails})$
$= (0.3)(0.4)$
$= 0.12$

Thus, the probability that both switches fail is 0.12.

EXAMPLE 1   In a certain community the probability that a man over 40 years old is overweight is 0.42. The probability that his blood pressure is high given that he is overweight is 0.67. If a man over 40 years of age is selected at random, what is the probability that he is overweight and that he has high blood pressure?

Solution

We use Formula 5.4. We have

p(overweight and high blood pressure)
= p(high blood pressure | overweight) · p(overweight)
= (0.67)(0.42)
= 0.2814

Thus, the probability that a man over 40 is overweight and has high blood pressure is approximately 0.28.

EXAMPLE 2

TV Commercials

A new cleansing product has recently been introduced and is being advertised on television as having remarkable cleansing qualities. The manufacturer believes that if a homemaker is selected at random, the probability that the homemaker watches television and sees the commercial between the hours of 12 noon and 4 P.M. is $\frac{4}{11}$. Furthermore, if the homemaker sees the commercial, then the probability that the homemaker buys the cleanser is $\frac{22}{36}$. What is the probability that a homemaker selected at random will watch television *and* buy the product?

Solution

We use Formula 5.4. We have

p(watches TV and buys product)
= p(buys product | watches TV) · p(watches TV)

$$= \frac{22}{36} \cdot \frac{4}{11}$$

$$= \frac{88}{396} = \frac{2}{9}$$

Thus, the probability that a homemaker selected at random watches television and buys the cleanser is $\frac{2}{9}$.

Independent Events

In many cases it turns out that whether or not one event happens does not affect whether another will happen. For example, if two cards are drawn from a deck and

Independent Events

the first card is replaced before the second card is drawn, the outcome on the first draw has nothing to do with the outcome on the second draw. Also, if two dice are rolled, the outcome for one die has nothing to do with the outcome for the second die. Such events are called **independent events.**

DEFINITION 5.2   Two events A and B are said to be **independent** if the likelihood of the occurrence of event B is in no way affected by the occurrence or non-occurrence of event A.

When dealing with the independent events, we can simplify Formula 5.4. The following example shows this.

EXAMPLE 3

Two cards are drawn from a deck of 52 cards. Find the probability that both cards drawn are aces if the first card

a. is *not* replaced before the second card is drawn.
b. is replaced before the second card is drawn.

Solution

a. Since the first card is not replaced, we use the multiplication rule. We have

$p(\text{both cards are aces}) = p(\text{2nd card is ace} \mid \text{1st card is ace}) \cdot p(\text{1st card is ace})$

Notice that since the first card is not replaced, there are only three aces remaining out of a possible 51 cards. This is because the first card removed was an ace. Thus,

$$p(\text{both cards are aces}) = \frac{3}{51} \cdot \frac{4}{52}$$

$$= \frac{12}{2652} = \frac{1}{221}$$

Thus, the probability that both cards are aces is $\frac{1}{221}$.

b. Since the first card is replaced before the second card is drawn, then whether or not an ace appeared on the first card in no way affects what happens on the second draw. The events "ace on second draw" and "ace on first draw" are independent. Thus, $p(\text{2nd card is ace} \mid \text{1st card is ace})$ is exactly the same as $p(\text{2nd card is ace})$. Therefore,

$p(\text{both cards are aces}) = p(\text{2nd card is ace} \mid \text{1st card is ace}) \cdot p(\text{1st card is ace})$

$$= p(\text{2nd card is ace}) \cdot p(\text{1st card is ace})$$

$$= \frac{4}{52} \cdot \frac{4}{52}$$

$$= \frac{16}{2704} = \frac{1}{169}$$

Hence, the probability that both cards are aces in this case is $\frac{1}{169}$.

Example 3 suggests that if two events A and B are independent, we can substitute $p(B)$ for $p(B \mid A)$ since B is in no way affected by what happens with A. We then get a special multiplication rule for independent events.

FORMULA 5.5	If A and B are independent events, then $$p(A \text{ and } B) = p(A) \cdot p(B)$$

EXAMPLE 4

Two randomly selected travelers, Carlos and Pedro, who do not know each other, are at the information desk at Kennedy International Airport. The probability that Carlos speaks Spanish is 0.86, and the probability that Pedro speaks Spanish is 0.73. What is the probability that they both speak Spanish?

Solution

Since both travelers do not know each other, the events "Carlos speaks Spanish" and "Pedro speaks Spanish" are independent. We therefore use Formula 5.5. We have

$$
\begin{aligned}
p(\text{both speak Spanish}) &= p(\text{Carlos speaks Spanish}) \cdot p(\text{Pedro speaks Spanish}) \\
&= (0.86)(0.73) \\
&= 0.6278
\end{aligned}
$$

Thus, the probability that they both speak Spanish is approximately 0.63.

EXAMPLE 5

If the probability of a skin diver in a certain community having untreated diabetes is 0.15, what is the probability that two totally unrelated skin divers from the community do *not* have untreated diabetes? (Assume independence.)

Solution

These are independent events, so we use Formula 5.5. The probability of a skin diver having untreated diabetes is 0.15. Thus, the probability that the diver does not have untreated diabetes is $1 - 0.15$, or 0.85. Therefore,

$$
\begin{aligned}
p(\text{both divers do not have diabetes}) &= p(\text{diver 1 does not have diabetes}) \\
&\quad\; p(\text{diver 2 does not have diabetes}) \\
&= (0.85)(0.85) \\
&= 0.7225
\end{aligned}
$$

Hence, the probability that neither of two totally unrelated skin divers have untreated diabetes is approximately 0.72.

COMMENT   The multiplication rule for independent events can be generalized for more than two independent events. We simply multiply all the respective probabilities. Thus, if event A has probability 0.7 of occurring, event B has probability 0.6 of occurring, and event C has probability 0.5 of occurring, and if these events are independent, then the probability that all three occur is

$$p(A \text{ and } B \text{ and } C) = p(A) \cdot p(B) \cdot p(C)$$
$$= (0.7)(0.6)(0.5)$$
$$= 0.21$$

Therefore, the probability that all three occur is 0.21.

EXERCISES FOR SECTION 5.4

1. According to the Marro Insurance Company, the probability that a married man on Roosevelt Island has as life insurance policy is 0.89. Furthermore, the probability that his wife has a life insurance policy given that her husband has a life insurance policy is 0.54. For a randomly selected married couple on Roosevelt Island, find the probability that both husband and wife have a life insurance policy.

2. Many colleges plan their course offerings well in advance. Such plans are often based on past experiences. At Island University the probability that a computer science major will register for Calculus II in the spring semester when the student registers for Calculus I in the fall semester is 0.86. Moreover, the probability that a computer science major registers for Calculus I in the fall semester is 0.47. Find the probability that a randomly selected computer science major registers for Calculus I in the fall semester and Calculus II in the spring semester.

3. Susan Howell is the executive officer of the Printex Corporation. She has two phones on her desk, each with a different number and neither of which is an extension of the other. Based on past experience, it is known that the probability that phone 1 rings during the hours of 9:00 A.M.–10:00 A.M. is 0.89 and the probability that phone 2 rings during these hours is 0.64. What is the probability that both phones ring during these hours?

4. The Board of Directors of the Jalem Transit Corporation is considering raising the prices of commuter tickets by 8%. The probability that it votes to raise the prices is 0.85. Past experience has shown that whenever the prices of commuter tickets are raised by 8%, ridership drops by approximately 6% with a probability of 0.97. Many commuters simply switch to their automobiles. Find the probability that the board votes to increase the prices *and* that the commuter ridership drops by 6%.

5. 76% of all business majors at Doyl University go on to get their M.B.A. degrees

after getting the B.A. degree. Furthermore, 53% of all business majors at Doyl University receive their B.A. degrees. Find the probability that a randomly selected business major at Doyl University is a business major who will get the B.A. and M.B.A. degrees.

6. Philip McGuire, Janet Rogers, and Fredric Mitchell have each applied to different banks for a home equity loan. The probability that Philip McGuire's application is approved is 0.85. The probability that Janet Rogers' application is approved is 0.92, and the probability that Fredric Mitchell's application is approved is 0.79. Assuming independence, find the probability that all three applications are approved.

7. Refer back to Exercise 6. Find the probability that only Janet Rogers' application will be approved.

8. Judy McGowan is driving on the highway when she notices two gas stations up ahead on the road. The probability that the first gas station has a functioning air pump is 0.57. The probability that the second gas station has a functioning air pump is 0.42. What is the probability that neither has a functioning air pump?

9. At Los Gables Airport two kinds of security checks are used to prevent any passenger from taking an explosive device and/or a gun on to an airplane. One is a visual check by a security guard coupled with an explosive sniff test by a specially trained dog and the second is a screening by a metal detector. The probability that the guard or dog stops a person carrying a gun or bomb is 0.74, and the probability that the person is caught by the metal detector given that he or she was not stopped by the guard is 0.99. What is the probability that a person will *not* be stopped by the guard nor caught by the metal detector? (See the newspaper article on page 236.)

10. Consider the following newspaper article. The probability that the printer's

NEWSPAPER WORKERS STILL ON STRIKE

New York (Jan. 7): The strike by workers of the *New York Daily News* enters its fourth month today with no end in sight. The workers are demanding more job security and increased contributions by management to the pension fund. Management claims that if all of the worker's demands are met, then the cost of a newspaper will rise to over a dollar or the company will be forced to go out of business.

No new talks were scheduled for this week.

January 7, 1991

demand will be met is 0.19. Also, the probability that the price of a newspaper will go up given that the printer's demands are met is 0.89. Find the probability that the printer's demands are met *and* that the price of a newspaper goes up.

11. A leading motel chain has been having bad experiences with customers who pay their bills by check. Based on past experience, the motel chain knows that the probability that a guest's check will "bounce" given that the bill was paid by check is 0.37. Furthermore, the probability that a guest will pay by check is 0.59. Bill is about to pay his bill. What is the probability that Bill pays the bill with a check that *will not* bounce?

12. The Bello Police Department operates a high-speed hydrofoil boat that is used for emergency purposes to rescue people from the adjacent rugged forests and swamplands. The boat is equipped with two motors that do not operate simultaneously. The second backup motor operates only when the first motor fails to operate. Due to the rugged terrain, the probability that the first motor will not operate is 0.20. Furthermore, the probability that the second backup motor will fail to operate given that the first failed to operate is 0.35.

   a. Find the probability that both motors will fail to operate and that the police department's hydrofoil boat will be unable to rescue someone.

   b. Find the boat's reliability. (The reliability of the police department's hydrofoil boat is the probability that it will be able to rescue someone.)

13. Smoke detectors save lives. Nevertheless, fire officials claim that people often do not replace the batteries in these alarms when they burn out, thus rendering the detectors useless. There are three smoke detectors in the Greene's home, each located in a different part of the house. The probabilities that these detectors are functioning properly are 0.92, 0.86, and 0.89, respectively. What is the probability that *none* of these is functioning properly and that a fire will go undetected?

14. There are seven families who live on the top floor of Irving Towers. The probability that any one of these families has a personal home computer is 0.47. Assuming independence, what is the probability that only one of these families has a personal home computer?

*5.5 BAYES' FORMULA

Let us look in on Dr. Carey, who has two bottles of sample pills on his desk for the treatment of arthritic pain. One day he gives Madeline a few pills from one of the bottles. (All other treatments have failed.) However, he does not remember from

---

*An asterisk indicates that the section requires more time and thought than other sections.

which bottle he took the pills. The pills in bottle B_1 are effective 70% of the time, with no known side effects. The pills in bottle B_2 are effective 90% of the time, with some possible side effects. Bottle B_1 is closer to Dr. Carey on his desk and the probability is $\frac{2}{3}$ that he selected the pills from this bottle. On the other hand, bottle B_2 is farther away from Dr. Carey and the probability is therefore $\frac{1}{3}$ that he selected the pills from this bottle. The problem is to determine the bottle from which the pills were taken.

In many problems we are given situations such as this one, where we know the outcome of the experiment and are interested in concluding that the outcome happened because of a particular event. Figure 5.2 is an example of this. For these situations we need a formula.

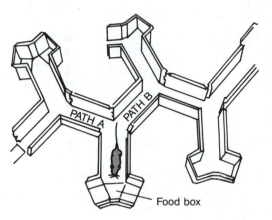

PATH A PATH B

Food box

Figure 5.2
Consider the situation pictured at the left. Often we read about psychologists experimenting with rats to determine how quickly they learn maze patterns. Mazes used to study human learning are similar, in principle, to those used with animals. We notice that the rat in the picture has already arrived at the food box. What is the probability that it came there from Path A and not from Path B? By using Bayes' formula, we will be able to answer this and similar questions.

EXAMPLE 1   Medicine

In the situation we are discussing, find the probability that the pills are effective in relieving Madeline's pain.

Solution

Madeline can be relieved of her pain by taking the pills from either bottle B_1 or bottle B_2. Let A represent the event *Madeline's pain is relieved*, let B_1 represent the event that *the pills were taken from bottle B_1*, and let B_2 represent the event that *the pills were taken from bottle B_2*. The tree diagram on the top of the next page illustrates how Madeline's pain can be relieved. From the given information

$$\text{Prob}(A \mid B_1) = 0.70 \quad \text{and} \quad \text{Prob}(B_1) = \frac{2}{3}$$

Result

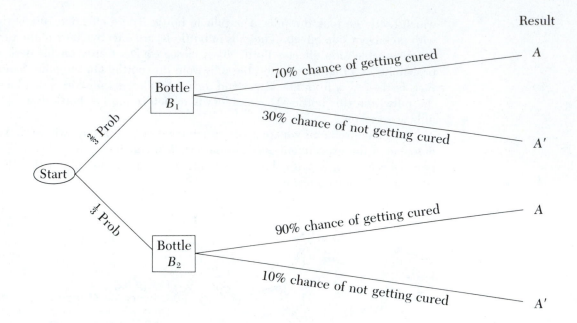

Applying the multiplication rule (Formula 5.4) gives

Prob(selects B_1 *and* pill from B_1 is effective)

= Prob(Madeline is cured | pill from B_1 was selected)
    · Prob(selecting B_1)

$$= (0.70)\frac{2}{3} = \frac{7}{10} \cdot \frac{2}{3}$$

$$= \frac{14}{30} = \frac{7}{15}$$

Symbolically,

$$\text{Prob}(A \text{ and } B_1) = \text{Prob}(A \mid B_1) \cdot \text{Prob}(B_1) = \frac{7}{15}$$

Also, from the given information

$$\text{Prob}(A \mid B_2) = \frac{9}{10} \quad \text{and} \quad \text{Prob}(B_2) = \frac{1}{3}$$

so that

$$\text{Prob}(A \text{ and } B_2) = \text{Prob}(A \mid B_2) \cdot \text{Prob}(B_2)$$

$$= \frac{9}{10} \cdot \frac{1}{3} = \frac{3}{10}$$

Madeline can be cured in one of two mutually exclusive ways:

1. Bottle B_1 is selected and a pill from it is effective.
2. Bottle B_2 is selected and a pill from it is effective.

Thus, using the addition rule for mutually exclusive events (Formula 5.1), we have

$$\text{Prob(pills are effective)} = \text{Prob(selects } B_1 \text{ and pills are effective)}$$
$$+ \text{Prob(selects } B_2 \text{ and pills are effective)}$$

Symbolically,

$$\text{Prob(pills are effective)} = \text{Prob}(A \text{ and } B_1) + \text{Prob}(A \text{ and } B_2)$$

$$\text{Prob}(A) = \text{Prob}(A \mid B_1) \cdot \text{Prob}(B_1) + \text{Prob}(A \mid B_2) \cdot \text{Prob}(B_2)$$

$$= \frac{7}{15} + \frac{3}{10}$$

$$= \frac{14}{30} + \frac{9}{30} = \frac{23}{30}$$

Therefore, the probability that the pills are effective in relieving Madeline of her pain is $\frac{23}{30}$.

EXAMPLE 2   Two weeks later Madeline returns to Dr. Carey and reports that the pills were extremely effective. Dr. Carey would now like to recommend the same medicine for his other patients who suffer from the same pain. What is the probability that the pills came from

a. Bottle B_1?
b. Bottle B_2?

Solution

a. Using the conditional probability formula (Formula 5.3), we have

$$\text{Prob(pills came from } B_1 \mid \text{pills are effective)}$$
$$= \frac{\text{Prob(selects } B_1 \text{ and pills are effective)}}{\text{Prob(pills are effective)}}$$

Symbolically,

$$\text{Prob}(B_1 \mid A) = \frac{\text{Prob}(A \text{ and } B_1)}{\text{Prob}(A)}$$

Substituting the values obtained in Example 1 gives

$$\text{Prob}(B_1 \mid A) = \frac{7/15}{23/30} = \frac{7}{15} \div \frac{23}{30}$$

$$= \frac{7}{15} \cdot \frac{30}{23} = \frac{14}{23}$$

Thus, the probability that the pills came from bottle B_1 is $\dfrac{14}{23}$.

b. To find the probability that the pills came from bottle B_2, we use a procedure similar to the one used in part (a). We have

$$\text{Prob}(\text{pills came from } B_2 \mid \text{pills are effective}) = \frac{\text{Prob}(\text{selects } B_2 \text{ and pills are effective})}{\text{Prob}(\text{pills are effective})}$$

In symbols,

$$\text{Prob}(B_2 \mid A) = \frac{\text{Prob}(A \text{ and } B_2)}{\text{Prob}(A)}$$

$$= \frac{3/10}{23/30} = \frac{3}{10} \div \frac{23}{30}$$

$$= \frac{3}{10} \cdot \frac{30}{23} = \frac{9}{23}$$

Thus, the probability that the pills came from bottle B_2 is $\dfrac{9}{23}$.

We can combine the results of the preceding two examples as follows:

$$\text{Prob}(B_1 \mid A) = \frac{\text{Prob}(A \text{ and } B_1)}{\text{Prob}(A)}$$

$$= \frac{\text{Prob}(A \text{ and } B_1)}{\text{Prob}(A \text{ and } B_1) + \text{Prob}(A \text{ and } B_2)}$$

$$= \frac{\text{Prob}(A \mid B_1) \cdot \text{Prob}(B_1)}{\text{Prob}(A \mid B_1) \cdot \text{Prob}(B_1) + \text{Prob}(A \mid B_2) \cdot \text{Prob}(B_2)}$$

and

$$\text{Prob}(B_2 \mid A) = \frac{\text{Prob}(A \text{ and } B_2)}{\text{Prob}(A)}$$

$$= \frac{\text{Prob}(A \text{ and } B_2)}{\text{Prob}(A \text{ and } B_1) + \text{Prob}(A \text{ and } B_2)}$$

$$= \frac{\text{Prob}(A \mid B_2) \cdot \text{Prob}(B_2)}{\text{Prob}(A \mid B_1) \cdot \text{Prob}(B_1) + \text{Prob}(A \mid B_2) \cdot \text{Prob}(B_2)}.$$

When these results are generalized we have Bayes' rule.

FORMULA 5.6   Bayes' Rule

Bayes' Rule   Consider a sample space that is composed of the mutually exclusive events A_1, A_2, A_3, . . . , A_n. Suppose each event has a nonzero probability of occurring and that one must definitely occur. If B is any event in the sample space, then

$$\text{Prob}(A_1 \mid B) = \frac{\text{Prob}(B \mid A_1) \cdot \text{Prob}(A_1)}{\text{Prob}(B \mid A_1) \cdot \text{Prob}(A_1) + \text{Prob}(B \mid A_2) \cdot \text{Prob}(A_2) + \cdots + \text{Prob}(B \mid A_n) \cdot \text{Prob}(A_n)}$$

$$\text{Prob}(A_2 \mid B) = \frac{\text{Prob}(B \mid A_2) \cdot \text{Prob}(A_2)}{\text{Prob}(B \mid A_1) \cdot \text{Prob}(A_1) + \text{Prob}(B \mid A_2) \cdot \text{Prob}(A_2) + \cdots + \text{Prob}(B \mid A_n) \cdot \text{Prob}(A_n)}$$

$$\vdots \qquad\qquad \vdots$$

$$\text{Prob}(A_n \mid B) = \frac{\text{Prob}(B \mid A_n) \cdot \text{Prob}(A_n)}{\text{Prob}(B \mid A_1) \cdot \text{Prob}(A_1) + \text{Prob}(B \mid A_2) \cdot \text{Prob}(A_2) + \cdots + \text{Prob}(B \mid A_n) \cdot \text{Prob}(A_n)}$$

Bayes' rule may seem rather complicated but it is easy to use, as the following examples illustrate.

EXAMPLE 3   If $p(A \mid B) = \frac{1}{5}$, $p(A \mid C) = \frac{2}{7}$, $p(B) = \frac{1}{2}$, and $p(C) = \frac{1}{2}$, find

a. $p(B \mid A)$.
b. $p(C \mid A)$.

Solution

a. Using Bayes' rule, we have

$$p(B \mid A) = \frac{p(A \mid B) \cdot p(B)}{p(A \mid B) \cdot p(B) + p(A \mid C) \cdot p(C)}$$

$$= \frac{\left(\frac{1}{5}\right)\left(\frac{1}{2}\right)}{\left(\frac{1}{5}\right)\left(\frac{1}{2}\right) + \left(\frac{2}{7}\right)\left(\frac{1}{2}\right)}$$

$$= \frac{\frac{1}{10}}{\left(\frac{1}{10}\right) + \left(\frac{1}{7}\right)}$$

$$= \frac{1/10}{17/70} = \frac{1}{10} \div \frac{17}{70}$$

$$= \frac{1}{10} \cdot \frac{70}{17} = \frac{7}{17}$$

Thus, $p(B \mid A) = \dfrac{7}{17}$.

b. Again, we use Bayes' formula. We have

$$p(C \mid A) = \frac{p(A \mid C) \cdot p(C)}{p(A \mid B) \cdot p(B) + p(A \mid C) \cdot p(C)}$$

$$= \frac{\left(\frac{2}{7}\right)\left(\frac{1}{2}\right)}{\left(\frac{1}{5}\right)\left(\frac{1}{2}\right) + \left(\frac{2}{7}\right)\left(\frac{1}{2}\right)} = \frac{\frac{1}{7}}{\left(\frac{1}{10}\right) + \left(\frac{1}{7}\right)} = \frac{\frac{1}{7}}{\frac{17}{70}}$$

$$= \frac{10}{17}$$

Thus, $p(C \mid A) = \dfrac{10}{17}$.

EXAMPLE 4

A prisoner has just escaped from jail. There are three roads leading away from the jail. If the prisoner selects road A to make good her escape, the probability that she succeeds is $\dfrac{1}{4}$. If she selects road B, the probability that she succeeds is $\dfrac{1}{5}$. If she selects road C, the probability that she succeeds is $\frac{1}{6}$. Furthermore, the probability that she selects each of these roads is the same. It is $\frac{1}{3}$. If the prisoner succeeds in her escape, what is the probability that she made good her escape by using road B?

Solution

We use Bayes' rule. We have

$$(p \text{ used road B} \mid \text{succeeded}) = \frac{p(\text{succeeds} \mid \text{uses road B}) \cdot p(\text{uses road B})}{p(\text{succeeds} \mid \text{uses road A}) \cdot p(\text{uses road A}) + p(\text{succeeds} \mid \text{uses road B}) \cdot p(\text{uses road B}) + p(\text{succeeds} \mid \text{uses road C}) \cdot p(\text{uses road C})}$$

$$= \frac{\left(\frac{1}{5}\right)\left(\frac{1}{3}\right)}{\left(\frac{1}{4}\right)\left(\frac{1}{3}\right) + \left(\frac{1}{5}\right)\left(\frac{1}{3}\right) + \left(\frac{1}{6}\right)\left(\frac{1}{3}\right)}$$

$$= \frac{\frac{1}{15}}{\left(\frac{1}{12}\right) + \left(\frac{1}{15}\right) + \left(\frac{1}{18}\right)}$$

$$= \frac{12}{37}$$

Thus, the probability that she made good her escape by using road B is $\frac{12}{37}$.

EXAMPLE 5   Smoke Detectors

A large real estate manager purchased 50,000 smoke detectors to comply with new city ordinances. She purchased 25,000 from company A, 15,000 from company B, and 10,000 from company C. It is known that some of the smoke detectors malfunction and go off spontaneously. It is also known that 4% of the detectors produced by company A are defective, 5% of the detectors produced by company B are defective, and 6% of the detectors produced by company C are defective. A call is received by the management office that one of the detectors is malfunctioning. Find the probability that it was produced by company A.

Solution

From the given information,

$$\text{Prob(produced by company A)} = \frac{25,000}{50,000} = 0.5$$

$$\text{Prob(produced by company B)} = \frac{15,000}{50,000} = 0.3$$

$$\text{Prob(produced by company C)} = \frac{10,000}{50,000} = 0.2$$

$$\text{Prob(defective} \mid \text{produced by company A)} = 0.04$$

$$\text{Prob(defective} \mid \text{produced by company B)} = 0.05$$

$$\text{Prob(defective} \mid \text{produced by company C)} = 0.06$$

Here we must determine Prob(produced by company A | defective). We apply Bayes' rule:

Prob(produced by A | defective)

$$= \frac{\text{Prob(defective} \mid \text{prod. by A)} \cdot \text{Prob(prod. by A)}}{\text{Prob(def.} \mid \text{prod. by A)} \cdot \text{Prob(prod. by A)} + \text{Prob(def.} \mid \text{prod. by B)} \cdot \text{Prob(prod. by B)} + \text{Prob(def.} \mid \text{prod. by C)} \cdot \text{Prob(prod. by C)}}$$

$$= \frac{(0.04)(0.5)}{(0.04)(0.5) + (0.05)(0.3) + (0.06)(0.2)}$$

$$= \frac{0.02}{0.02 + 0.015 + 0.012} = \frac{0.02}{0.047}$$

$$= \frac{20}{47}$$

Thus, the probability that the defective smoke detector was produced by company A is $\frac{20}{47}$.

EXAMPLE 6

There are four photocopying machines I, II, III, and IV on the third floor of a large office building. The probabilities that the copies produced from each of these machines will be blurred are 0.2, 0.5, 0.3, and 0.1, respectively. Furthermore, because of the location of these machines, management estimates that the probabilities that a worker will use any one of these machines are 0.6, 0.2, 0.1, and 0.1, respectively. The president of the company receives a blurred memo that was photocopied on one of these machines. What is the probability that it was photocopied on machine I?

Solution

We use Bayes' formula. Let A represent the event "copy is blurred," B_1, represent the event "Machine I is used," B_2 represent the event "Machine II is used," B_3 represent the event "Machine III is used," and B_4 represent the event "Machine IV is used." Then we are interested in $p(B_1 \mid A)$.

Using Bayes' rule, we have

$$p(B_1 \mid A) = \frac{p(A \mid B_1) \cdot p(B_1)}{p(A \mid B_1) \cdot p(B_1) + p(A \mid B_2) \cdot p(B_2) + p(A \mid B_3) \cdot p(B_3) + p(A \mid B_4) \cdot p(B_4)}$$

$$= \frac{(0.6)(0.2)}{(0.6)(0.2) + (0.2)(0.5) + (0.1)(0.3) + (0.1)(0.1)}$$

$$= \frac{0.12}{0.26} = \frac{12}{26} = \frac{6}{13}$$

Thus the probability that the blurred memo was photocopied on Machine I is $\frac{6}{13}$.

EXERCISES FOR SECTION 5.5

1. A manufacturer has just received two truckloads of microwave ovens. Truckload I contains 6 defective and 32 good microwave ovens. Truckload II contains 5 defective and 16 good microwave ovens. An inspector randomly selects a microwave oven from the unloading dock and notices that it is defective. Unfortunately, he does not know from which truckload the sample came. If the probability of selecting either of these truckloads is $\frac{1}{2}$, what is the probability that the defective microwave oven came from truckload II?

2. There are two drawers in a cash register. Drawer D_1 contains 25 good and 4 counterfeit dollar bills. Drawer D_2 contains 12 good and 5 counterfeit dollar bills. A counterfeit dollar bill is selected at random from one of these drawers. If the probability of selecting either of these drawers is $\frac{1}{2}$, what is the probability that the counterfeit dollar bill was selected from drawer D_2?

3. A hospital receives its blood supply from one of four possible blood banks A, B, C, or D, which it selects on a random basis. The probabilities that the blood received from these blood banks is contaminated are 0.11, 0.09, 0.13, and 0.10, respectively. The hospital randomly selects a blood bank. If the blood received is contaminated, what is the probability that the blood came from blood bank B?

4. Refer back to Exercise 3. If the blood received is *not* contaminated, find the probability that it came from blood bank C.

5. A computer store purchases computer chipboards from one of two distributors, with 75% of the chipboards purchased from distributor A. Past experience indicates that about 8% of distributor A's chipboards are defective, and 9% of distributor B's chipboards are defective. A computer chipboard is randomly

selected. If the chipboard turns out to be defective, what is the probability that it came from distributor B?

6. The Brier Health Club is experimenting with three different weight-reduction techniques for its members. Thus far, method A has been successful with 12 out of 16 members, method B has been effective with 10 out of 18 members, and method C has been successful with 7 out of 13 members. If a club member that was formerly overweight is randomly selected, what is the probability that the results were achieved by using method B? (Assume that each method can be selected with a probability of $\frac{1}{3}$.)

7. During a recent power blackout 250 people were arrested on suspicion of looting. Since each claimed to be innocent but did not have an alibi, they were given lie detector tests. (Many courts will not allow the results of polygraph (lie detector) tests to be used as evidence.) Past experience with such tests indicates that they are 98% reliable when given to an innocent person and 85% reliable when given to a guilty person. Witnesses are able to testify that only 25 of the 250 arrested people were actually involved in the looting. What is the probability that a given suspect is guilty given that the polygraph test says that he or she is innocent?

8. Many people install burglar alarms in their cars to prevent theft. However, police records indicate that these alarms actually go off many times for no apparent reason. The probability of the alarm sounding when someone is attempting to steal the car is 0.96. For one particular car, the alarm sounded "falsely" $\frac{7}{35}$ times in the last year. Furthermore, in this city the probability of a car theft is $\frac{1}{4}$. If the alarm for this car is ringing, what is the probability that someone is attempting to steal the car?

9. *Medicine:* The American Cancer Society as well as the medical profession recommends that people have themselves checked annually for any cancerous growths. If a person has cancer, then the probability is 0.97 that it will be detected by a test. Furthermore, the probability that the test results will be positive (meaning that cancer is possible) when no cancer actually exists is 0.08. Government records indicate that 7% of the population in the vicinity of a chemical corporation that produces asbestos have some form of cancer. Donald Williams takes the test and the results are positive. What is the probability that he does *not* have cancer?

10. *Ecology:* Environmentalists have developed a test for determining when the mercury level in fish is above the permissible levels. If the fish actually contain an excessive amount of mercury, then the test is 99% effective in determining this, and only 1% will escape undetected. On the other hand, if the mercury content is within permissible limits, then the test will correctly indicate this 96% of the time. Only 4% of the time will the test incorrectly indicate that the mercury

content is not within permissible limits. The test is to be used on fish from a river into which a chemical company has been dumping its wastes. It is estimated that 30% of the fish in the river contain excessive amounts of mercury. A fish is caught and tested by this procedure. The test indicates that the mercury level is within permissible limits. Find the probability that the mercury content is actually greater than the permissible level.

5.6 SUMMARY

In this chapter we discussed many different rules concerning the calculation of probabilities. Each formula given applies to different situations. Thus, the addition rule allows us to determine the probability of event A or event B or both events happening. We distinguished between mutually exclusive and non-mutually exclusive events and their effect on the addition rule.

We then discussed conditional probability and how the probability of one event is affected by the occurrence or nonoccurrence of a second event. This led us to the multiplication rule. When one event is in no way affected by the occurrence or nonoccurrence of a second event, we have independent events and a simplified multiplication rule.

Finally we discussed Bayes' formula, which is used when we know the outcome of some experiment and are interested in determining the probability that it was caused by or is the result of some other event. In each case many applications of all the formulas introduced were given.

Study Guide

The following is a chapter outline in capsule form. You should now be able to demonstrate your knowledge of the ideas mentioned by giving definitions, descriptions, or specific examples. Page references are given in parentheses.

Two events A and B that cannot occur at the same time are said to be **mutually exclusive**. (page 238)

If $p(A)$ is the probability of event A happening, then $p(A')$ is the probability that A will not happen. Thus, $p(A') = 1 - p(A)$. The event A' is called the **complement** of event A. (page 242)

If A and B are any events, then the **addition rule** states

$$p(A \text{ or } B) = p(A) + p(B) - p(A \text{ and } B). \text{ (page 243)}$$

We use the symbol $p(A|B)$ to represent the **conditional probability** of A given that B has occurred or conditional upon the fact that A has occurred. (page 251)

The **conditional probability formula** allows us to calculate $p(A|B)$ in terms of $p(A \text{ and } B)$ and $p(B)$. (page 254)

If A and B are any events, then the **multiplication rule** states that

$$p(A \text{ and } B) = p(A|B) \cdot p(B). \text{ (page 260)}$$

Two events are said to be **independent** if the likelihood of the occurrence of event B is in no way affected by the occurrence or nonoccurrence of event A. (page 262)

If we know the outcome of an experiment and are interested in determining the probability that the outcome happened because of a particularr event, then we use **Bayes' Formula**. (page 271)

Formulas to Remember

You should be able to identify each symbol in the following formulas, understand the relationships among the symbols expressed in each formula, understand the significance of each formula, and use the formulas in solving problems.

1. Addition rule, for mutually exclusive events:

   $p(A \text{ or } B) = p(A) + p(B)$

2. Addition rule, general case: $p(A \text{ or } B) = p(A) + p(B) - p(A \text{ and } B)$

3. Addition rule, for three events: $p(A \text{ or } B \text{ or } C) = p(A) + p(B) + p(C) - p(A \text{ and } B) - p(A \text{ and } C) - p(B \text{ and } C) + p(A \text{ and } B \text{ and } C)$

4. Complement of event A: $p(A') = 1 - p(A)$

5. Conditional probability formula: $p(A \mid B) = \dfrac{p(A \text{ and } B)}{p(B)}$

6. Multiplication rule, general: $p(A \text{ and } B) = p(A \mid B) \cdot p(B)$

7. Multiplication rule, for independent events: $p(A \text{ and } B) = p(A) \cdot p(B)$

8. Bayes' rule:

   $$p(A_i \mid B) = \frac{p(B \mid A_i) \cdot p(A_i)}{p(B \mid A_1) \cdot p(A_1) + p(B \mid A_2) \cdot p(A_2) + \cdots + p(B \mid A_n) \cdot p(A_n)}$$

Testing Your Understanding of This Chapter's Concepts

1. Explain the difference between mutually exclusive events and independent events.
2. Which of the following events are mutually exclusive?
   a. becoming pregnant and getting a headache.
   b. cheating on your math final and getting caught.
3. Which of the following events are independent?
   a. driving while under the influence of alcohol and getting into an accident.
   b. setting the alarm clock to wake you up in the morning and getting to work on time.

4. *True or false:* If two events are mutually exclusive, their probabilities must add up to 1. Explain your answer.

5. The probability that your math teacher will be absent tomorrow is 0.8. The probability that your psychology teacher will be absent tomorrow is 0.8. Therefore, the probability that both will be absent tomorrow is $(0.8)(0.8) = 0.64$, which would mean that the probability that both will be present is 0.36. Do you agree with this reasoning? Explain your answer.

6. In an attempt to attract new listeners, a radio station announces that it will randomly select a name from the local residential phone directory. The name will be announced "on the air." If the named person calls the station within 10 minutes, then that person will be awarded $1000. A curious listener whose last name begins with the letter "Z" calls the station and inquires about her chances of winning. She is informed that since there are 26 letters in the alphabet, then the probability of her winning is $\dfrac{1}{26}$. Do you agree? Explain your answer.

THINKING CRITICALLY

1. Mr. and Mrs. Pascal have two girls and would like to have a boy. After taking a course in probability they decide to have a third child, reasoning that the probability that the third child is a boy is $\dfrac{7}{8}$. There are eight possible ways in which a family can consist of three children, as shown in the following chart.

	Child 1	Child 2	Child 3
Possibility 1	Boy	Boy	Boy
Possibility 2	Boy	Boy	Girl
Possibility 3	Boy	Girl	Boy
Possibility 4	Boy	Girl	Girl
Possibility 5	Girl	Boy	Boy
Possibility 6	Girl	Boy	Girl
Possibility 7	Girl	Girl	Boy
Possibility 8	Girl	Girl	Girl

Of these, seven of them have at least one boy. Thus, they reason that the probability that the third child is a boy is $\dfrac{7}{8}$. Do you agree with this reasoning? Explain your answer.

2. If only two people are in an elevator, what is the probability that their birthdays are different?

3. If only three people are in an elevator, what is the probability that none of them has the same birthday?

4. Linda Greenleaf has taken the day off from her job to await delivery of her new washing machine and carpet. Both companies have promised to deliver the items

in the morning. The probability that the washing machine is delivered in the morning is 0.57. The probability that the carpet is delivered in the morning is 0.68. What is the probability that *only one* of the items will be delivered in the morning?

5. There are six families who live on the ground floor of Covington Estate. The probability that any one of the families subscribes to closed circuit cable television is 0.17. Find the probability that
   a. at least one of the families subscribes to closed circuit television.
   b. at most one of the families subscribes to closed circuit television.

6. If two events A and B, each having nonzero probability, are mutually exclusive, must they also be independent? Explain your answer.

Review Exercises for Chapter 5

1. Three women are independently attempting to swim around the island of Manhattan in New York City. Their probabilities of succeeding are 0.4, 0.5, and 0.35. Find the probability that
   a. all of them succeed.
   b. none of them succeeds.
   c. at least one of them succeeds.

2. A fair coin is tossed five times. If it is known that exactly one of these tosses showed heads, what is the probability that it was on the second toss?

3. In a certain community the probability that an individual has a certain type of blood is as follows.

Blood Type	Probability
A	0.45
B	0.15
AB	0.05
O	0.35

Robert is planning to marry Andrea. What is the probability that Robert has type A blood and Andrea has type O blood? (Assume independence.)

4. Refer back to question 3. What is the probabilty that Robert and Andrea both have the same blood type?

5. A card is drawn from an ordinary deck of cards. What is the probability that the card is a
   a. queen given that it is a spade?
   b. spade given that it is a queen?
   c. four if we know that it is not a picture (face) card?

6. Two fair dice are tossed. What is the probability that the sum equals 10 is we know that it exceeds 7?

7. Paul Masterson and Juliana Vasquez are two cardiologists on the staff of Maimonides Hospital. Each carries a pager that can be activated by personnel in the Emergency room. Paul is very conscientious and is within earshot of his

pager 75% of the time. Juliana, on the other hand, responds to her pager only 55% of the time. Each doctor operates independently of the other. What is the probability that at least one of them can be contacted in the event of a medical emergency?

8. The First National Bank recently acquired three mortgages from the failed Fargo Bank. It estimates that the chances of these mortgages being repaid are 0.6, 0.4, and 0.25. What is the probability that at least one of these mortgages will default?

9. The Maxtel Corp. receives a shipment of nine photocopying machines. It is known that two of these machines are defective. If two of these machines from the shipment are randomly selected without replacement, what is the probability that they both are defective?

10. A sociologist interviewed 3200 individuals from different age groups in New York to determine their reaction to the new "living will" law that was recently enacted. Under the new law, any individual can appoint another person to make medical decisions (such as disconnecting life-support systems) for the individual in case he or she is unable to make such decisions. The results of the survey are shown in the accompanying table.

Age Group	In Favor of New Law	Opposed to New Law	No Opinion
Under 30 years	604	366	322
31–50 years	404	424	286
Over 51 years	336	72	386

Find the probability that a randomly selected individual in the group
a. is over 51 years of age.
b. opposes the new law given that the individual is over 51 years of age.
c. is over 51 years of age given that the individual opposes the new law.

11. Three people who do not know each other are standing on a line at a movie theater waiting for the box office to open. What is the probability that their birthdays (month and day) are different?

Would you believe that we need only 23 people in a crowd to have a 50% probability that at least two of these people would have the same birthday? The probability increases to about 1, almost a certainty, when we have a crowd of 63 people.

12. Refer back to Question 11. If four people who do not know each other are standing on line, what is the probability that none of them has the same birthday (month and day)? Assume that no one was born on February 29.

Chapter Test

1. A family that is known to have three children, not twins, is selected at random from all families with three children. If it is known that there is a girl in the family, what is the probability that all the children are girls?

2. A large department store has just received a shipment of 50 food processors. Unknown to management, 8 of the food processors are defective. It is decided to sample the shipment by selecting 2 food processors without replacement. What is the probability that
   a. 2 good food processors are selected?
   b. 2 defective food processors are selected?
   c. 1 defective and 1 good food processor is selected?

3. There are three people in a doctor's waiting room. The nurse asks these patients to indicate their birthdays. What is the probability that at least two of their birthdays are the same?

4. Three students are in the student cafeteria discussing whether or not they should go to the weekend football game. The probability that Dennis decides to go to the game is 0.78, the probability that Chuck decides to go to the game is 0.65, and the probability that Victor decides to go to the game is 0.57. Assuming independence, what is the probability that all 3 decide to go to the game?

5. Refer back to Exercise 4. Find the probability that only Dennis decides to go to the game.

6. The State Insurance Department often receives complaints concerning delays in the processing of claims by medical insurance companies. According to an industry spokesperson, $\frac{1}{8}$ of the problems with delayed claim forms are caused by the patient, $\frac{1}{9}$ of the problems with delayed claim forms are caused by the doctor, and $\frac{1}{15}$ of the problems are caused by both the patient and doctor. If a claim form that was delayed is randomly selected, what is the probability that the problem was caused by either the patient *or* the doctor?

7. There are two particular slot machines into which a gambler in Las Vegas can insert money and possibly win the jackpot. The probability that the gambler selects the closer machine is $\frac{3}{5}$, and the probability that the gambler selects the machine that is farther away is $\frac{2}{5}$. It has been estimated that the probabilities of

winning at these machines are $\frac{1}{9}$ and $\frac{1}{15}$, respectively. If the gambler wins the jackpot, what is the probability that it was from the nearer machine?

8. A survey was conducted by Baker and Skibanez to determine how many high school students smoke cigarettes. The following results were obtained.

	Freshmen	Sophomores	Juniors	Seniors
Smoke	45	59	88	142
Don't Smoke	165	122	107	143

Source: Baker and Skibanez, New York, 1990.

If a student in this school is selected at random, what is the probability that the student
a. is a sophomore?
b. smokes given that the student is a sophomore?
c. is a sophomore given that the student smokes?

9. Martha is the manager of a local fast-food restaurant. Over the past few years she has determined the following probabilities on the items that a customer will order.

Item(s)	Probability
Fish	$\frac{4}{7}$
French fries	$\frac{1}{2}$
Salad	$\frac{5}{11}$
Fish and French fries	$\frac{2}{9}$
French fries and salad	$\frac{1}{7}$
Fish and salad	$\frac{1}{4}$
Fish, French fries, and salad	$\frac{1}{44}$

What is the probability that a randomly selected customer will order either fish, French fries, or a salad?

10. A recent survey of numerous obstetricians specializing in high risk pregnancies revealed some interesting facts about their patients and their recent deliveries. These doctors have determined the following probabilities for their patients:

Fact	Probability
Mother over 35 years of age	0.27
First child for mother	0.22
Mother had a well-paying career job	0.42
First child for mother and over 35 years of age	0.17
First child for mother and mother had a well-paying career job	0.09
Mother over 35 years of age and had a well-paying career job	0.16
First child for mother and mother over 35 years of age and had a well-paying career job	0.07

Source: Johnson and Baker, Chicago, IL 1989.

What is the probability that a randomly selected mother in this group is either over 35 years of age or had a well-paying career job or that the child is a first child for the mother?

11. A bag contains 4 white chips, 3 red chips, and 5 black chips. Two chips are randomly selected from this box without replacement. Find the probability that the 2 chips selected are of the same color.

12. The Calvington Corp. is considering the feasibility of opening an overseas branch office. It decides to canvass its employees to determine how many employees speak a foreign language. The following information is obtained: The probability that a male employee of the company speaks a foreign language is 0.83, and the probability that a female employee of the company speaks a foreign language is 0.66. Find the probability that two randomly selected and totally unrelated employees (one male and one female) both *do not* speak a foreign language.

13. Assume that we have three mutually exclusive events that partition a sample space. If $p(A|B) = 0.15$, $p(A|C) = 0.12$, $p(A|D) = 0.09$, $p(B) = 0.20$, $p(C) = 0.40$, and $p(D) = 0.40$, find
   a. $p(B|A)$
   b. $p(C|A)$
   c. $p(D|A)$

For questions 14–17 use the following information: Five hundred people were asked to rate a new television pilot show. Their results are shown below.

	Liked it Enthusiastically	Liked it Moderately	Hated it	Indifferent
Male	43	67	91	60
Female	57	84	28	70

14. One of these people is selected at random. Find the probability that the individual is *not* a male.

    a. $\dfrac{57}{500}$ b. $\dfrac{261}{500}$ c. $\dfrac{239}{500}$ d. $\dfrac{239}{261}$ e. none of these

15. One of these people is randomly selected. If it is known that this person liked it moderately, find the probability that the individual is a male.

    a. $\dfrac{67}{84}$ b. $\dfrac{67}{151}$ c. $\dfrac{67}{261}$ d. $\dfrac{84}{261}$ e. none of these

16. One of these people is randomly selected. If it is known that this person is a male, find the probability that the individual liked it moderately.

    a. $\dfrac{67}{84}$ b. $\dfrac{67}{151}$ c. $\dfrac{67}{261}$ d. $\dfrac{84}{261}$ e. none of these

17. Find the probability of selecting an individual from this group who is both a male and who liked it moderately.

    a. $\dfrac{67}{261}$ b. $\dfrac{67}{151}$ c. $\dfrac{67}{84}$ d. $\dfrac{67}{500}$ e. none of these

18. A card is drawn from an ordinary deck of cards. If it is known that it is a picture card, find the probability that it is the queen of diamonds.

    a. $\dfrac{2}{12}$ b. $\dfrac{2}{52}$ c. $\dfrac{1}{52}$ d. $\dfrac{1}{12}$ e. none of these

Suggested Further Reading

1. Freund, J. *Statistics, A First Course.* 5th ed. Englewood Cliffs, NJ: Prentice-Hall, 1991.
2. Groebner, David, and Patrick Shannon. *Business Statistics.* 2nd ed. Columbus: Charles E. Merrill, 1985.
3. Guilford, J. P., and B. Fruchter. *Fundamental Statistics in Psychology and Education.* New York: McGraw-Hill, 1973.
4. "Mathematics in the Modern World." Readings from *Scientific American*, article 22, San Francisco, CA: Freeman, 1968.
5. Newmark, Joseph. *Using Finite Mathematics.* New York: Harper & Row, 1982.
6. Wonnacott, Ronald, and Thomas Wonnacott, *Introductory Statistics*, 4th ed. New York: John Wiley, 1985.

6 Some Discrete Probability Distributions and Their Properties

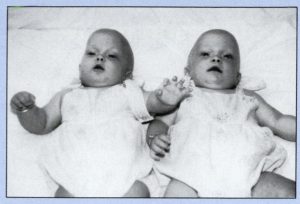

What is the probability of having identical twin babies? See discussion on p. 349. (© *Caroline Brown, Fran Heyl Associates*)

As random variables and probability distributions play a key role in our later studies of statistical inferences, in this chapter we analyze the nature of and applications involving discrete random variables and their associated probability distributions.

<table>
<tr><td>

CHAPTER OBJECTIVES

</td><td>

After studying the material in this chapter, you should be able:

- **To discuss** the concept of a random variable, where the outcome of some experiment is of interest to us. (Section 6.2)
- **To see** how probability functions assign probabilities to the different values of the random variable. (Section 6.2)
- **To understand** how the mean and variance of a probability function give us the expected value of a probability function as well as the spread of the distribution. (Sections 6.3 and 6.4)
- **To study** Bernoulli or binomial experiments, which are experiments having only two possible outcomes, success or failure. (Section 6.5)
- **To apply** the binomial probability distribution, which gives us the probability of obtaining a specified number of successes when an experiment is performed *n* times. (Section 6.5)
- **To work with** several important binomial distribution properties. (Section 6.6)
- **To analyze** the Poisson probability distribution, which can be applied in very specific cases. (Section 6.7)
- **To use** the hypergeometric probability distribution when the absence of certain assumptions makes the binomial distribution inappropriate. (Section 6.8)

</td></tr>
</table>

Statistics in the News

According to this article, about one out of every eight women in America is expected to develop breast cancer. For the American population as a whole, can we determine the expected number of women that will develop breast cancer? Such medical statistics have important implications for our society.

In this chapter we will discuss how such statistics are calculated.

6.1 INTRODUCTION

In Chapter 2 we discussed frequency distributions of sets of data. Using these distributions we were able to analyze data more intelligently to determine which outcomes occurred most often, least often, and so on. In Chapters 4 and 5 we discussed the various rules of probability and how they can be applied to many different situations. These rules enable us to predict how often something will happen in the long run. In this chapter we combine these ideas.

In any given experiment there may be many different things of interest. For example, if a scientist decides to mate a white rat with a black rat, she may be interested in the number of offspring that are white, black, gray, and so on. We will therefore have to define what is meant by a random variable and then discuss its probability function. Special emphasis will be given in this chapter to the binomial random variable and its probability distribution.

6.2 DISCRETE PROBABILITY FUNCTIONS

To understand what is meant by a discrete probability function, let us analyze the following examples.

Valerie is a dentist and keeps accurate records on the number of cavities of each of her patients. Her records indicate that each patient has anywhere from zero to five cavities. Based on past experience, she has compiled the data given in Table 6.1.

TABLE 6.1 The Number of Cavities and Their Probabilities

Number of Cavities	Probability
0	$\dfrac{1}{16}$
1	$\dfrac{4}{16}$
2	$\dfrac{5}{16}$
3	$\dfrac{3}{16}$
4	$\dfrac{2}{16}$
5	$\dfrac{1}{16}$
	$Total = \dfrac{16}{16} = 1$

Notice that each patient has anywhere from zero to five cavities. Thus, the number of cavities that each patient has is somehow dependent on chance, as the probabilities in the table indicate.

Now consider Eric who is an elevator operator. The number of people who enter the elevator at exactly 9:00 A.M. varies from zero to ten. The capacity of the elevator is ten people. From past experience Eric has been able to set up the chart shown in Table 6.2.

In each of the two previous examples the values assumed by the item of interest, that is, the number of cavities or the number of people entering the elevator, were whole numbers and were somehow dependent on chance. We refer to such a quantity as a **discrete random variable.**

TABLE 6.2   The Number of People Entering the Elevator and Their Probabilities

Number of People Entering Elevator	Probability	Number of People Entering Elevator	Probability
0	$\dfrac{1}{50}$	6	$\dfrac{10}{50}$
1	$\dfrac{3}{50}$	7	$\dfrac{6}{50}$
2	$\dfrac{4}{50}$	8	$\dfrac{3}{50}$
3	$\dfrac{5}{50}$	9	$\dfrac{2}{50}$
4	$\dfrac{7}{50}$	10	$\dfrac{1}{50}$
5	$\dfrac{8}{50}$		$Total = \dfrac{50}{50} = 1$

The term discrete random variable applies to many different situations. Thus, it may represent the number of people buying tickets to a movie, the number of mistakes made by a secretary in typing a letter, the number of telephone calls received by the school switchboard during the month of September, the number of games that the Green Bay Packers will win next season, or the number of students that will enroll in a particular course, Music 161, to be offered for the first time in the spring.

Basically, if an experiment is performed and some quantitative variable, denoted by x, is measured or observed, then the quantitative variable x is called a **random variable** since the values that x may assume in the given experiment depend on chance. It is a random outcome. Whenever all possible values that a random variable may assume can be listed (or counted), then the random variable is said to be **discrete.** As opposed to this, the time required to complete a transaction at a bank is a

Random Variable

Discrete Random Variable

Continuous Random Variable

continuous random variable because it could theoretically assume any one of an infinite number of values—namely, any value 0 seconds or more. Random variables that can assume values corresponding to any of the points contained in one or more intervals on a line are called **continuous random variables.**

Other examples of continuous random variables are:

1. the time it takes for a drug to take effect.
2. the height (in cm) of a player on a basketball team.
3. the blood serum cholesterol level of a person.
4. the weight of a bag of sugar or of a large jar of coffee.
5. the length of time between births in the maternity ward of a hospital.

In this chapter we discuss discrete random variables and their probability distributions. In the next chapter we will discuss continuous random variables and their distributions. *To summarize:* A **random variable** is a *numerical* quantity, the *value* of which is determined by an experiment. In other words, its value is determined by chance.

The following examples further illustrate the idea of a random variable.

EXAMPLE 1

Three cards are selected, without replacement, from a deck of 52 cards. The random variable may be the number of aces obtained. It would then have values of 0, 1, 2, or 3, depending on the number of aces actually obtained. This is a discrete random variable.

EXAMPLE 2

Calvin drives his car over some nails. The random variable is the number of flat tires that Calvin gets. The values of the random variable are 0, 1, 2, 3, 4, corresponding to zero flats, one flat, two flats, three flats, and four flats, respectively. This is a discrete random variable.

EXAMPLE 3

Let the number of people who will attend the next concert at the Hollywood Bowl be the random variable of interest. Then this random variable can assume values ranging from 0 to the seating capacity of the Hollywood Bowl. This is a discrete random variable.

EXAMPLE 4

A nurse is taking Chuck's blood pressure. Let the random variable be Chuck's systolic blood pressure. What values can the random variable assume? This represents a continuous random variable.

EXAMPLE 5

Get on a scale and weigh yourself. Let the random variable be your weight. What values can the random variable assume? This represents a continuous random variable.

Let us now return to the two examples discussed at the beginning of this section. You will notice that the probabilities associated with the different values of the random variable are indicated. Thus, Table 6.1 tells us that the probability of having one cavity is $\frac{4}{16}$ and that the probability of having three cavities is $\frac{3}{16}$. Similarly, Table 6.2 tells us that the probability of three people entering the elevator is $\frac{5}{50}$.

When discussing a random variable, we are almost always interested in assigning probabilities to the various values of the random variable. For this reason we now discuss probability functions.

DEFINITION 6.1
Probability Function
Probability
Distribution

A **probability function** or **probability distribution** is a correspondence that assigns probabilities to the values of a random variable.

EXAMPLE 6

If a pair of dice is rolled, the random variable that may be of interest to us is the number of dots appearing on both dice together. When a pair of dice is rolled, there are 36 possible outcomes. These are shown in Figure 6.1.

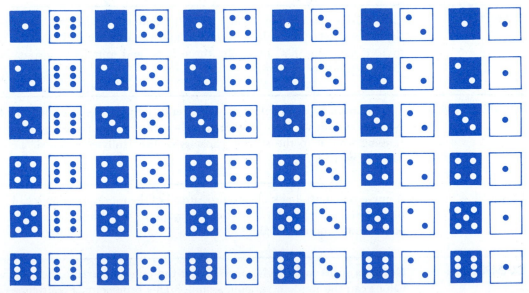

Figure 6.1
Thirty-six possible outcomes when a pair of dice is rolled.

We can then set up the following chart.

Sum on Both Dice	Number of Different Ways in Which Sum Can Be Obtained	Probability
2	1	$\dfrac{1}{36}$
3	2	$\dfrac{2}{36}$
4	3	$\dfrac{3}{36}$
5	4	$\dfrac{4}{36}$
6	5	$\dfrac{5}{36}$
7	6	$\dfrac{6}{36}$
8	5	$\dfrac{5}{36}$
9	4	$\dfrac{4}{36}$
10	3	$\dfrac{3}{36}$
11	2	$\dfrac{2}{36}$
12	1	$\dfrac{1}{36}$

$$Total = \frac{36}{36} = 1$$

In the previous example the random variable of interest to us was the number of dots appearing on both dice together. Often, for mathematical convenience, we use special symbols. We use lowercase letters to represent all possible values of the random variable. We denote the probability of the random variable x by the symbol $p(x)$. Thus, for the preceding example the probability that the random variable assumes the value of 2 is $\dfrac{1}{36}$. Symbolically, $p(2) = \dfrac{1}{36}$. Similarly, $p(7) = \dfrac{6}{36}$.

EXAMPLE 7   An airplane has four engines, each of which operates independently of the other. If the random variable is the number of engines that are functioning properly, the random variable has values 0, 1, 2, 3, 4. For a particular airplane we have the following probability function.

Number of Engines Functioning Properly	Probability
0	$\dfrac{1}{19}$
1	$\dfrac{2}{19}$
2	$\dfrac{4}{19}$
3	$\dfrac{5}{19}$
4	$\dfrac{7}{19}$

$$Total = \frac{19}{19} = 1$$

Notice that in Example 7, as well as in Examples 1 to 6, the sum of all the probabilities is 1. This will be true for any probability function. Also, the probability that the random variable assumes any one particular value is between 0 and 1. Again, this is true in every case. We state this formally as

RULE

a. The sum of all the probabilities of a probability function is always 1, or
$$\sum_{all\ x} p(x) = 1.$$

b. The probability that a random variable assumes any one value in particular is between 0 and 1 inclusive. Zero means that it can never happen and 1 means that it must happen, or $0 \le p(x) \le 1$

COMMENT Some authors distinguish between **probability functions** and **probability distributions.** In this book we will use these terms interchangeably.

In all examples mentioned thus far some of the random variables assumed many values and some took on only two values. The following random variables take on many values:

1. The number of flat tires that Calvin gets when he drives over nails.
2. The sum obtained when rolling a pair of dice. There are 11 possible values, as indicated on page 293.
3. The number of people in an elevator.

Consider the following events:

1. The possible outcome when one coin is flipped (heads or tails).
2. The outcome of either having a cavity or not having a cavity (yes or no).
3. The sex of a newborn child (male or female).
4. The results of an exam (pass or fail only).

The outcomes "head" and "tail" are not the values of random variables because they are not in numerical form. However, if we let "head" equal 0 and "tail" equal 1, the result is a random variable. Remember, a random variable *must* take on numerical values only. We now have a random variable that can take on only two values.

Binomial Variable
Binomial Distribution

    If a random variable has only two possible values, and if the probability of these values remains the same for each trial regardless of what happened on any previous trial (that is, the trials are independent), the variable is called a **binomial variable** and its probability distribution is called a **binomial distribution.** Actually, this is rather a special case of the binomial distribution. Since the binomial distribution is so important in statistics, we will discuss it in great detail in Section 6.5.

EXERCISES FOR SECTION 6.2

1. All students at Lehna University that come to school by car, park their car in the college parking lot. Suppose that the Registrar's office wishes to determine the age of the automobiles used by the students. If x is the random variable representing the age of the automobile, state the possible values that the random variable can attain.
2. For each of the following situations, find the values that the indicated random variables may have.

Situation	Random Variable of Interest
a. A company that employs 227 people	Number of people out of work with the flu
b. A motel that has 26 rooms	Number of rooms rented for the weekend
c. A family that has five pet dogs or cats at home	Number of dogs at home
d. A birthday party that occurs in February	The day of the month that it occurs
e. You plant four seedlings	Number of seedlings that sprout
f. A gas tank of a car that has a capacity of 22 gallons	Number of gallons (rounded off) needed to fill the tank with gas
g. A marriage in an Arab country	Number of wives that a man has

3. Can the following be a probability distribution for a random variable? If your answer is no, explain why.

Random Variable, x	Probability
0	0.31
1	0.08
2	0.22
3	0.17
4	0.23
5	0.02

4. Given the following probability distribution for the random variable x, what is the probability that the random variable has a value of 0?

Random Variable, x	Probability
0	?
1	0.23
2	0.08
3	0.04
4	0.32
5	0.17

5. A family is known to have three children. Let x be the number of girls in the family. Find the probability distribution of x.
6. A coin is tossed twice. Let x be the number of heads that come up in both tosses of the coin. Find the probability distribution of x.
7. A pair of dice is thrown once. Let x be the number of 1's that come up on either or both of the dice. Find the probability distribution of x.
8. A man has a penny, a nickel, a dime, a quarter, a half-dollar, and a dollar piece in his pocket. Two coins are selected at random with the first coin replaced before the second coin is drawn. Let x denote the sum of money obtained using both coins. Find the probability distribution of x.

9. A coin is tossed three times. Let x be the number of heads that come up in all three tosses of the coin. Find the probability distribution of x.

10. *Planting seedlings:* A farmer plants four seedlings. Let x be the number of seedlings that sprout. Find the probability distribution of x. (Assume that a seedling can either sprout or die with equal likelihood.)

11. A bag contains 7 black balls, 4 green balls, 5 yellow balls, and 3 white balls. Three balls are selected at random with each ball replaced before the next ball is drawn. Let x denote the number of yellow balls obtained. Find the probability distribution of x.

12. *Tennis:* Scott, who is a good tennis player, plans to play five games with his friend Christopher who plays tennis just as well as he does. Let x be the number of games that Scott will win. Find the probability distribution of x.

13. Frank knows that two of the eight spark plugs in his car are defective. He selects two spark plugs at random, with the first one replaced before the second is selected. Let x be the number of defective spark plugs obtained. Find the probability distribution of x.

14. Determine if the following formula can be the probability function of some random variable x.

$$p(x) = \frac{6 - x}{15} \qquad \text{for } x = 1, 2, 3, 4, \text{ or } 5$$

Explain your answer.

15. Determine if the following formula can be the probability function of some random variable x.

$$p(x) = \frac{5 - x}{9} \qquad \text{for } x = 0, 1, 2, 3, \text{ or } 4$$

Explain your answer.

16. A discrete random variable x has the following probability distribution.

x	$p(x)$
5	0.40
6	0.05
7	0.18
8	0.29
9	0.08

a. Find $p(x \le 6)$.
b. Find $p(x > 7)$.
c. Find $p(x = 10)$.
d. Find $p(x \le 6 \text{ or } x > 7)$.

17. On the basis of past experience an insurance salesperson knows that the number of long-term health care insurance policies sold per week by the company is a random variable x, with the following probability distribution.

x	p(x)
0	0.145
1	0.243
2	0.252
3	0.106
4	0.098
5	0.088
6	0.039
7	0.029

Suppose that a week is randomly selected.

a. Find $p(x = 3)$.

b. Find $p(x \geq 3)$.

c. Find $p(x \leq 7)$.

d. Find $p(1 \leq x \leq 4)$.

18. Doctors and dentists usually schedule appointments so that patients will not have to wait long before seeing the doctor or dentist. Nevertheless, emergencies do arise and unfortunately patients must wait before being seen. Past records for the Victory Boulevard Medical Center indicate that at any given time (when the center is open), the probability distribution for the number of patients waiting to be seen is as follows.

Number of Patients Waiting, x	Probability p(x)
0	0.03
1	0.05
2	0.08
3	0.15
4	0.23
5	0.24
6	0.22

a. Find $p(x \leq 3)$.

b. Find $p(x > 4)$.

c. Find $p(x < 7)$.

d. Find $p(1 \leq x \leq 6)$.

6.3 THE MEAN OF A RANDOM VARIABLE

Imagine that the traffic department of a city is considering installing traffic signals at the intersection of Main Street and Broadway. The department's statisticians have kept accurate records over the past year on the number of accidents reported per day at this particularly dangerous intersection. They have submitted the report on the number of accidents per day and their respective probabilities shown in Table 6.3.

TABLE 6.3 Report on Accidents at Intersection of Main
Street and Broadway

Number of Accidents x	Probability $p(x)$	Product $x \cdot p(x)$
0	$\dfrac{1}{32}$	$0\left(\dfrac{1}{32}\right) = 0$
1	$\dfrac{1}{32}$	$1\left(\dfrac{1}{32}\right) = \dfrac{1}{32}$
2	$\dfrac{9}{32}$	$2\left(\dfrac{9}{32}\right) = \dfrac{18}{32}$
3	$\dfrac{10}{32}$	$3\left(\dfrac{10}{32}\right) = \dfrac{30}{32}$
4	$\dfrac{8}{32}$	$4\left(\dfrac{8}{32}\right) = \dfrac{32}{32}$
5	$\dfrac{3}{32}$	$5\left(\dfrac{3}{32}\right) = \dfrac{15}{32}$

Let us multiply each of the possible values for the random variable x, which represents the number of accidents given in Table 6.3, by the respective probabilities. The results of these multiplications are shown in the third column of Table 6.3. If we now add the products, we get

$$0 + \left(\frac{1}{32}\right) + \left(\frac{18}{32}\right) + \left(\frac{30}{32}\right) + \left(\frac{32}{32}\right) + \left(\frac{15}{32}\right) = \frac{96}{32} = 3$$

This result is known as the **mean** of the random variable. It tells us that on the average there are three accidents per day at this dangerous intersection.

Recall that in Chapter 3, when we discussed measures of central tendency and measures of variation, we distinguished between sample statistics and population parameters. Thus, we used the symbols $\bar{x}$, s^2, and s as symbols for the sample mean, sample variance, and sample standard deviation, respectively. Also, we used the symbols μ, σ^2, and σ as symbols for the population mean, population variance, and population standard deviation, respectively. In our case the mean number of accidents represents the mean of the entire population of observed data. Therefore, we will use the symbol μ to represent the average number of accidents. Usually, when dealing with probability distributions we work with the entire population. Hence, we use μ as the symbol for mean.

Now consider the following. Suppose we intend to flip a coin four times. What is the average number of heads that we can expect to get? To answer this question we first find the probability distribution. Let x represent the random variable "the number of heads obtained in four flips of the coin." Since the coin is flipped four

Table 6.4 Number of Heads That Can Be Obtained
When a Coin Is Flipped Four Times

Random Variable x	Probability $p(x)$	Product $x \cdot p(x)$
0	$\dfrac{1}{16}$	$0\left(\dfrac{1}{16}\right) = 0$
1	$\dfrac{4}{16}$	$1\left(\dfrac{4}{16}\right) = \dfrac{4}{16}$
2	$\dfrac{6}{16}$	$2\left(\dfrac{6}{16}\right) = \dfrac{12}{16}$
3	$\dfrac{4}{16}$	$3\left(\dfrac{4}{16}\right) = \dfrac{12}{16}$
4	$\dfrac{1}{16}$	$4\left(\dfrac{1}{16}\right) = \dfrac{4}{16}$

times, we may get 0, 1, 2, 3, or 4 heads. Thus, the random variable x can have the values 0, 1, 2, 3, or 4. The probabilities associated with each of these values is indicated in Table 6.4. You should verify that the probabilities given in this table are correct.

We now multiply each possible outcome by its probability, the results of which are shown in the third column of Table 6.4. If we now add these products, the result is again called the mean of the probability distribution. In our case the mean is

$$0 + \left(\frac{4}{16}\right) + \left(\frac{12}{16}\right) + \left(\frac{12}{16}\right) + \left(\frac{4}{16}\right) = \frac{32}{16} = 2$$

This tells us that on the average we can expect to get two heads.

We generalize the results of the previous two examples as follows:

DEFINITION 6.2 Mean	The **mean** of a random variable for a given probability distribution is the number obtained by multiplying all possible values of the random variable having this particular distribution by their respective probabilities and adding these products together.

FORMULA 6.1	The mean of a random variable for a given probability distribution is denoted by the Greek letter μ, read as mu. Thus, $$\mu = \Sigma\, x \cdot p(x)$$ where this summation is taken over all the values that the random variable x can assume and the quantities $p(x)$ are the corresponding probabilities.

Mathematical
Expectation

Expected Value

COMMENT Many books refer to the mean of a random variable for a given probability distribution as its **mathematical expectation** or its **expected value**. We will use the word **mean**. The mean is the mathematical expectation that was discussed in Section 4.6 of Chapter 4.

We illustrate the use of Formula 6.1 with several examples.

EXAMPLE 1

Matthew is a waiter. The following table gives the probabilities that customers will give tips of varying amounts of money.

Amounts of Money (in cents), x	30	35	40	45	50	55	60
Probability, $p(x)$	0.45	0.25	0.12	0.08	0.05	0.03	0.02

Find the mean for this distribution.

Solution

Applying Formula 6.1 we have

$$\mu = \Sigma x \cdot p(x) = 30(0.45) + 35(0.25) + 40(0.12) + 45(0.08)$$
$$+ 50(0.05) + 55(0.03) + 60(0.02)$$
$$= 13.5 + 8.75 + 4.8 + 3.6 + 2.5 + 1.65 + 1.2$$
$$= 36$$

The mean is 36. Thus, Matthew can expect to receive an average tip of 36 cents.

EXAMPLE 2

Rosemary works for the U.S. Census Bureau in Washington. For a particular midwestern town the number of children per family and their respective probabilities is as follows:

Number of Children, x	0	1	2	3	4	5	6
Probability, $p(x)$	0.07	0.17	0.31	0.27	0.11	0.06	0.01

Find the mean for this distribution.

Solution

Applying Formula 6.1 we have

$$\mu = \Sigma x \cdot p(x) = 0(0.07) + 1(0.17) + 2(0.31) + 3(0.27)$$
$$+ 4(0.11) + 5(0.06) + 6(0.01)$$
$$= 0 + 0.17 + 0.62 + 0.81 + 0.44 + 0.30 + 0.06$$
$$= 2.4$$

The mean is 2.4. How can the average number of children per family be 2.4? Should it not be a whole number such as 2 or 3, not 2.4?

COMMENT   It may seem to you that Formula 6.1 is a new formula for calculating the mean. Actually, this is not the case. Recall that in Chapter 3 we defined the mean of a frequency distribution as mean value $= \sum \dfrac{xf}{n}$ or $\sum x \dfrac{f}{n}$. In Chapter 4 we defined the probability of an event as the relative frequency of the event, that is, $p(x) = \dfrac{f}{n}$. If we substitute this value for $\dfrac{f}{n}$ in the preceding formula, we get mean value $= \sum x \cdot p(x)$.

6.4 MEASURING CHANCE VARIATION

Suppose a manufacturer guarantees that a tire will last 20,000 miles under normal driving conditions. If a tire is selected at random and lasts only 12,000 miles, can the difference between what was expected and what actually happened be reasonably attributed to chance, or is there something wrong with the claim?

Similarly, if a coin is flipped 100 times, we would expect the average number of heads to be 50. If a coin was actually flipped 100 times and resulted in only 25 heads, can we conclude that the difference between what was expected and what actually happened is to be attributed to chance, or is it possible that the coin is loaded?

To answer these questions, we need some method of measuring the variations of a random variable that are due to chance. Thus, we will discuss the variance and standard deviation of a probability distribution.

You will recall that in Chapter 3 (page 124) we discussed variation of a set of numbers. We now extend this idea to variation of a probability distribution. We let μ represent the mean, $x - \mu$ represent the difference of any number from the mean, and $(x - \mu)^2$ represent the square of the difference. The difference of a number from the mean is called the **deviation from the mean.** We multiply each of the squared deviations from the mean by their respective probabilities. The sum of these products is called the **variance of a random variable with the given probability distribution.** Formally, we have

Deviation from the Mean

DEFINITION 6.3

Variance

The **variance** of a random variable with a given probability distribution is the number obtained by multiplying each of the squared deviations from the mean by their respective probabilities and by adding these products.

FORMULA 6.2	The variance of a random variable with a given probability distribution is denoted by the Greek letter σ^2 (read as sigma squared). Thus,

$$\sigma^2 = \Sigma(x - \mu)^2 \cdot p(x)$$

where this summation is taken over all the values that random variable x can take on. The quantities $p(x)$ are the corresponding probabilities and $(x - \mu)^2$ is the square of the deviations from the mean.

DEFINITION 6.4 Standard Deviation	The **standard deviation** of a random variable with a given probability distribution is the square root of the variance of the probability distribution. We denote the standard deviation by the symbol σ (sigma). Thus,

$$\sigma = \sqrt{\text{variance}}$$

EXAMPLE 1

A random variable has the following probability distribution:

x	0	1	2	3	4	5
$p(x)$	$\dfrac{7}{24}$	$\dfrac{5}{24}$	$\dfrac{1}{8}$	$\dfrac{1}{8}$	$\dfrac{1}{12}$	$\dfrac{1}{6}$

Find the mean, variance, and standard deviation for this distribution.

Solution

We first find μ by using Formula 6.1 of Section 6.3 and then proceed to use Formula 6.2. We arrange the computations in the form of a chart, as follows:

x	$p(x)$	$x \cdot p(x)$	$x - \mu$	$(x - \mu)^2$	$(x - \mu)^2 \cdot p(x)$
0	$\dfrac{7}{24}$	$0\left(\dfrac{7}{24}\right) = 0$	$0 - 2 = -2$	$(-2)^2 = 4$	$4\left(\dfrac{7}{24}\right) = \dfrac{28}{24}$
1	$\dfrac{5}{24}$	$1\left(\dfrac{5}{24}\right) = \dfrac{5}{24}$	$1 - 2 = -1$	$(-1)^2 = 1$	$1\left(\dfrac{5}{24}\right) = \dfrac{5}{24}$
2	$\dfrac{1}{8}$	$2\left(\dfrac{1}{8}\right) = \dfrac{2}{8}$	$2 - 2 = 0$	$0^2 = 0$	$0\left(\dfrac{1}{8}\right) = 0$
3	$\dfrac{1}{8}$	$3\left(\dfrac{1}{8}\right) = \dfrac{3}{8}$	$3 - 2 = 1$	$1^2 = 1$	$1\left(\dfrac{1}{8}\right) = \dfrac{1}{8}$
4	$\dfrac{1}{12}$	$4\left(\dfrac{1}{12}\right) = \dfrac{4}{12}$	$4 - 2 = 2$	$2^2 = 4$	$4\left(\dfrac{1}{12}\right) = \dfrac{4}{12}$
5	$\dfrac{1}{6}$	$5\left(\dfrac{1}{6}\right) = \dfrac{5}{6}$	$5 - 2 = 3$	$3^2 = 9$	$9\left(\dfrac{1}{6}\right) = \dfrac{9}{6}$

We then have

$$\mu = \Sigma x \cdot p(x)$$

$$= 0\left(\frac{7}{24}\right) + 1\left(\frac{5}{24}\right) + 2\left(\frac{1}{8}\right) + 3\left(\frac{1}{8}\right) + 4\left(\frac{1}{12}\right) + 5\left(\frac{1}{6}\right)$$

$$= 0 + \left(\frac{5}{24}\right) + \left(\frac{2}{8}\right) + \left(\frac{3}{8}\right) + \left(\frac{4}{12}\right) + \left(\frac{5}{6}\right)$$

$$= 0 + \left(\frac{5}{24}\right) + \left(\frac{6}{24}\right) + \left(\frac{9}{24}\right) + \left(\frac{8}{24}\right) + \left(\frac{20}{24}\right)$$

$$= \frac{48}{24} = 2$$

Also

$$\sigma^2 = \Sigma(x - \mu)^2 \cdot p(x)$$

$$= \left(\frac{28}{24}\right) + \left(\frac{5}{24}\right) + 0 + \left(\frac{1}{8}\right) + \left(\frac{4}{12}\right) + \left(\frac{9}{6}\right)$$

$$= \frac{80}{24} \approx 3.33$$

Thus, the mean is 2, the variance is approximately 3.33, and the standard deviation is $\sqrt{3.33}$, or approximately 1.82.

EXAMPLE 2

A dress manufacturer claims that the probability that a customer will buy a particular size dress is as follows:

Size, x	8	10	12	14	16	18
Probability, $p(x)$	0.11	0.21	0.28	0.17	0.13	0.10

Find the mean, variance, and standard deviation for this distribution.

Solution

We first find μ by using Formula 6.2 and then arrange the data in tabular form as follows:

x	$p(x)$	$x \cdot p(x)$	$x - \mu$	$(x - \mu)^2$	$(x - \mu)^2 \cdot p(x)$
8	0.11	0.88	$8 - 12.6 = -4.6$	21.16	$(21.16)(0.11) = 2.33$
10	0.21	2.10	$10 - 12.6 = -2.6$	6.76	$(6.76)(0.21) = 1.42$
12	0.28	3.36	$12 - 12.6 = -0.6$	0.36	$(0.36)(0.28) = 0.10$
14	0.17	2.38	$14 - 12.6 = 1.4$	1.96	$(1.96)(0.17) = 0.33$
16	0.13	2.08	$16 - 12.6 = 3.4$	11.56	$(11.56)(0.13) = 1.50$
18	0.10	1.80	$18 - 12.6 = 5.4$	29.16	$(29.16)(0.10) = 2.92$

$$\mu = \Sigma x \cdot p(x) = 0.88 + 2.10 + 3.36 + 2.38 + 2.08 + 1.80 = 12.6$$

$$\sigma^2 = 2.33 + 1.42 + 0.10 + 0.33 + 1.50 + 2.92 = 8.6$$

Thus the mean is 12.6, the variance is 8.6, and the standard deviation is $\sqrt{8.6}$, or approximately 2.93.

Formula 6.2, like the formula for the variance of a set of numbers (see page 126), requires us first to compute the mean and then to find the square of the deviations from the mean. In many cases we do not wish to do this. For such situations we can use an alternate formula to calculate the variance of a probability distribution.

FORMULA 6.3
The variance of a discrete random variable with a given probability distribution is

$$\sigma^2 = \Sigma x^2 \cdot p(x) - [\Sigma x \cdot p(x)]^2$$

Formula 6.3 may seem strange but it is similar to Formula 3.5 on page 129. Let us see how Formula 6.3 is used.

EXAMPLE 3
Calculate the variance for the probability distribution given in Example 2 by using Formula 6.3.

Solution

We arrange the data as follows:

x	$p(x)$	$x \cdot p(x)$	x^2	$x^2 \cdot p(x)$
8	0.11	0.88	64	64(0.11) = 7.04
10	0.21	2.10	100	100(0.21) = 21.00
12	0.28	3.36	144	144(0.28) = 40.32
14	0.17	2.38	196	196(0.17) = 33.32
16	0.13	2.08	256	256(0.13) = 33.28
18	0.10	1.80	324	324(0.10) = 32.40
		Total = 12.60		Total = 167.36

Using Formula 6.3, we find that the variance is $167.36 - (12.6)^2$

$$\sigma^2 = \Sigma x^2 \cdot p(x) - [\Sigma x \cdot p(x)]^2$$
$$= 167.36 - (12.6)^2$$
$$= 167.36 - 158.76$$
$$= 8.6$$

The variance is 8.6. This is the same result that we got using Formula 6.2.

EXERCISES FOR SECTION 6.4

 1. Brian Borna is in charge of the mailroom of the Boryk Corporation. Each day the company receives several packages via the Express Mail Service of the U.S. Postal system for which he must sign. The following is the probability distribution for the number of daily pieces of mail that comes via Express Mail Service.

Number of Pieces of Mail, x	Probability
0	0.02
1	0.03
2	0.25
3	0.19
4	0.16
5	0.14
6	0.12
7	0.08
8	0.01

Find the mean of this distribution.

 2. Roger Benoit drives a tow truck for the Marcy Collision Corporation. The number of accidents that he responds to per day and the associated probabilities are as follows.

Number of Accidents, x	1	2	3	4	5	6	7	8	9	10	11	12
Probability, p(x)	0.01	0.02	0.01	0.20	0.10	0.18	0.07	0.14	0.12	0.09	0.05	0.01

Find the mean, variance, and standard deviation for this distribution.

 3. Janet Frascilla is a medical laboratory technician. The number of electro-encephalograms (EEGs) that she takes daily and the associated probabilities are as follows.

Number of EEGs, x	Probability
0	0.08
1	0.09
2	0.13
3	0.07
4	0.14
5	0.23
6	0.19
7	0.07

Find the mean, variance, and standard deviation for this distribution.

 4. *Credit card or cash:* Serge Faingold operates the Friendly Travel Agency. The following table gives the probabilities of the daily number of customers who will buy airplane tickets from Serge and pay for them with a credit card.

Number of Customers Paying with a Credit Card, x	Probability
5	0.07
6	0.14
7	0.17
8	0.21
9	0.18
10	0.19
11	0.03
12	0.01

Find the mean, variance, and standard deviation for this distribution.

5. Derek Parilla operates a sporting goods store. The number of hunting permits that he will issue on a weekend and the associated probabilities are as follows.

Number of Hunting Permits, x	Probability, p(x)
0	0.05
1	0.09
2	0.13
3	0.15
4	0.16
5	0.29
6	0.06
7	0.05
8	0.02

Find the mean, variance, and standard deviation for this distribution.

6. Jennifer Robbins operates the Valley Ski Lodge. During the winter skiing season there are numerous skiing accidents. According to her records, the number of skiing accidents per weekend and the corresponding probabilities are as follows.

Number of Accidents, x	Probability, p(x)
0	0.04
1	0.05
2	0.33
3	0.27
4	0.10
5	0.08
6	0.06
7	0.07

Find the mean, variance, and standard deviation for this distribution.

7. Refer to the newspaper clipping shown on the top of the next page. The state officials will be testing for the presence of up to eight possible toxic chemical pollutants. The probability that they will find different types of chemical pollutants is as follows.

STATE OFFICIALS TO INVESTIGATE THE DUMPING OF CHEMICAL WASTES

Atlantic City (Aug. 1): Officials of the New Jersey Environment Protection Bureau announced yesterday that they would begin testing several dumping sites not too far from Atlantic City for the presence of different toxic chemical wastes. These chemical wastes are now beginning to work their way into the water supply of several neighboring communities.

Saturday, August 1, 1991

Number of Toxic Pollutants Found x	Probability $p(x)$
0	0.08
1	0.14
2	0.16
3	0.18
4	0.21
5	0.17
6	0.03
7	0.02
8	0.01

Find the mean, variance, and standard deviation for this distribution.

 8. A die is rolled once. Let x be the number of dots that show. Find the probability distribution of x, and then find σ^2 and σ for this distribution.

 9. A die is rolled twice. Let x be the sum appearing on both dice.
   a. Find the probability distribution of x.
   b. Find μ.
   c. Find σ^2.

10. A die is altered by painting an additional dot on the face that originally had one dot as shown below. Let x be the number of dots that shows when the die is rolled.
   a. Find the probability distribution of x.
   b. Find μ, σ^2, and σ for this distribution.

11. Wilson's Camera Store intends to run a promotion on a particular camcorder. The dealer has analyzed the sales performance over a long period of time when similar promotions were run and has come up with the following probability distribution of the number, x, of camcorders to be sold.

Number of Camcorders Sold, x	Probability, $p(x)$
0	0.02
1	0.06
2	0.19
3	0.33
4	0.23
5	0.13
6	0.03
7	0.01

a. Find the expected number of camcorders to be sold; that is, find μ.

b. Find σ^2 and σ.

 12. Many local telephone companies charge as much as $1.00 per request for directory assistance. Of course, the number of calls for directory assistance is a random variable that varies from time to time. Nevertheless, on the basis of past records it is known that the probability distribution for one local telephone company for this random variable during the morning period (9:00 A.M. to 12:00 Noon) is as follows.

Number of Calls for Directory Assistance, x	Probability $p(x)$
0	0.02
1	0.08
2	0.12
3	0.14
4	0.22
5	0.20
6	0.15
7	0.04
8	0.03

a. Find the expected number of requests for directory assistance, that is, find μ.

b. Find σ^2 and σ.

13. There are eight toll booths available for travelers to use on the Garden Bay Parkway that are always manned by toll collectors. Depending on the volume of traffic, at times it is possible for a commuter to drive up directly to an available toll collector. At other times the commuter must wait in line to pay the toll. On the basis of past records it is estimated that the number of busy toll collectors at noon is as follows.

Number of Busy Toll Collectors, x	Probability, $p(x)$
0	0.04
1	0.05
2	0.09
3	0.17
4	0.22
5	0.14
6	0.12
7	0.11
8	0.06

a. Find the expected number of busy toll collectors, that is, find μ.

b. Find σ^2 and σ.

14. A manufacturer has just introduced a new laundry detergent and would like to promote it by advertising on television, on the radio, in magazines, and by distributing free samples. Market research indicates the following sales to be generated from each advertisement medium and the associated probability.

Advertising Medium	Potential Sales, x	Probability, $p(x)$
Television	$70,000	0.37
Radio	37,000	0.56
Magazines	45,000	0.49
Distributing free samples	50,000	0.42

a. Find the expected sales from each medium.

b. If the manufacturer wishes to use only one of these advertisement media, which one should be selected? Explain your answer.

6.5 THE BINOMIAL DISTRIBUTION

Consider the following probability problem:

EXAMPLE 1

Paula is about to take a five-question, true–false quiz. She is not prepared for the exam and decides to guess the answers without reading the questions.

ANSWER SHEET

Directions: For each question darken the appropriate space.

1. [True]   [False]
2. [True]   [False]
3. [True]   [False]
4. [True]   [False]
5. [True]   [False]

Find the probability that she gets

a. all the answers correct.
b. all the answers wrong.
c. three out of the five answers correct.

Solution

Let us denote a correct answer by the letter "c" and a wrong answer by the letter "w." There are 2 equally likely possible outcomes for question 1, c or w. Similarly, there are 2 equally likely possible outcomes for question 2 regardless of whether the first question was correct or incorrect. There are 2 equally likely possible outcomes for each of the remaining questions 3, 4, and 5. Thus, there are 32 equally likely possible outcomes since

$$2 \times 2 \times 2 \times 2 \times 2 = 32$$

These outcomes we list as follows:

ccccc	cwccc	wcccc	wwccc
cccwc	cwcwc	wccwc	wwcwc
ccccw	cwccw	wcccw	wwccw
cccww	cwcww	wccww	wwcww
ccwcc	cwwcc	wcwcc	wwwcc
ccwcw	cwwcw	wcwcw	wwwcw
ccwwc	cwwwc	wcwwc	wwwwc
ccwww	cwwww	wcwww	wwwww

Once we have listed all possible outcomes, we can construct a chart similar to Table 6.5.

TABLE 6.5 Number of Correct Answers

0	1	2	3	4	5
wwwww	cwwww	ccwww	cccww	wcccc	ccccc
	wcwww	cwcww	ccwcw	cwccc	
	wwcww	cwwcw	ccwwc	ccwcc	
	wwwcw	cwwwc	cwcwc	cccwc	
	wwwwc	wccww	cwccw	ccccw	
		wcwcw	cwwcc		
		wcwwc	wccwc		
		wwcwc	wcccw		
		wwccw	wcwcc		
		wwwcc	wwccc		

Now we can calculate the probability associated with each outcome. We have

$$p(0 \text{ correct}) = \frac{1}{32}$$

$$p(1 \text{ correct}) = \frac{5}{32}$$

$$p(2 \text{ correct}) = \frac{10}{32}$$

$$p(3 \text{ correct}) = \frac{10}{32}$$

$$p(4 \text{ correct}) = \frac{5}{32}$$

$$p(5 \text{ correct}) = \frac{1}{32}$$

We can picture these results in the form of a histogram, as shown in Figure 6.2. The relative frequency histogram for Figure 6.2 is shown in Figure 6.3.

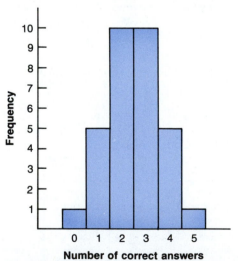

Figure 6.2
Histogram of the number of correct answers obtained by guessing at five true–false questions.

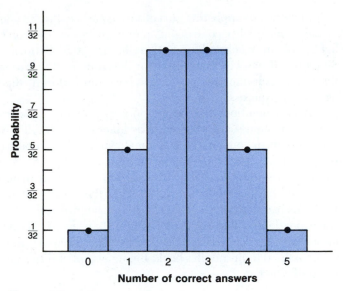

Figure 6.3
Relative frequency histogram for the frequency distribution of Figure 6.2

We can now answer the question raised at the beginning of the problem. We have

a. $p(\text{all correct answers}) = \dfrac{1}{32}$

b. $p(\text{all wrong answers}) = \dfrac{1}{32}$

c. $p(3 \text{ out of } 5 \text{ correct answers}) = \dfrac{10}{32}$

COMMENT   These answers assume that since Paula is not prepared, then $p(\text{correct answer}) = p(\text{wrong answer}) = \dfrac{1}{2}$. On the other hand, these answers would be wrong if Paula were better prepared and had a better probability of, say, 0.8, of getting a correct answer.

Many experiments or probability problems result in outcomes that can be grouped into two categories, success or failure. For example, when a coin is flipped there are two possible outcomes, heads or tails; when a hunter shoots at a target there are two possible outcomes, hit or miss; and when a baseball player is at bat the result is get on base or not get on base.

Statisticians apply this idea of success and failure to wide-ranging problems. For instance, if a quality-control engineer is interested in determining the life of a typical light bulb in a large shipment, each time a bulb burns out he has a "success." Similarly, if we were interested in determining the probability of a family having ten boys, assuming they planned to have ten children, then each time a boy is born we have a "success."

As we mentioned in Section 6.2 (page 295), experiments that have only two possible outcomes are referred to as **binomial probability experiments** or **Bernoulli experiments** in honor of the mathematician Jacob Bernoulli (1654–1705), who studied them in detail. His contributions to probability theory are contained in his book *Ars Conjectandi*, published after his death in 1713. This book also contains a reprint of an earlier treatise of Huygens. In 1657 the great Dutch mathematician Christian Huygens had written the first formal book on probability based on the Pascal–Fermat correspondences discussed in Chapter 1 (page 11). Huygens also introduced the important ideas of mathematical expectation discussed in Chapter 3. All these ideas are contained in Bernoulli's book. Binomial probability experiments are characterized by the following.

Binomial Probability Experiment

Bernoulli Experiment

DEFINITION 6.5   A **binomial probability experiment** is an experiment that satisfies the following properties:

1. There is a fixed number, n, of repeated trials whose outcomes are independent.

2. Each trial results in one of two possible outcomes. We call one outcome a success and denote it by the letter S and the other outcome a failure, denoted by F.

3. The probability of success on a single trial equals p and remains the same from trial to trial. The probability of a failure equals $1 - p = q$. Symbolically,

$$p(\text{success}) = p \quad \text{and} \quad p(\text{failure}) = q = 1 - p$$

4. We are interested in the number of successes obtained in n trials of the experiment.

COMMENT   Although very few real-life situations satisfy all of the above requirements, many of them can be thought of as satisfying, at least approximately, these requirements. Thus, we can apply the binomial distribution to many different problems.

Since all binomial probability experiments are similar in nature and result in either success or failure for each trial of the experiment, we seek a formula for determining the probability of obtaining x successes out of n trials of the experiment, where the probability of success on any one trial is p and the probability of failure is q. To achieve this goal let us consider the following.

A coin is tossed four times. What is the probability of getting exactly one head? We could list all the possible outcomes and count the number of favorable ones.

This is shown below with the favorable outcomes circled.

HHHH   HHTT   THHH   (THTT)
HHHT   HHTH   THHT   THTH
HTHH   (HTTT)   TTHH   TTTT
HTTH   HTHT   (TTTH)   (TTHT)

Thus, the probability of getting exactly one head is $\dfrac{4}{16}$

However, in many cases it is not possible or advisable to list all the possible outcomes. It is for this reason that we consider an alternate approach.

When a coin is tossed there are two possible outcomes, heads or tails. Thus,

$$p(\text{heads}) = \frac{1}{2} \quad \text{and} \quad p(\text{tails}) = \frac{1}{2}$$

Since we are interested in getting only a head, we classify this outcome as a success and write

$$p(\text{head}) = p(\text{success}) = \frac{1}{2} \quad \text{and} \quad p(\text{tail}) = p(\text{failure}) = \frac{1}{2}$$

Therefore,

$$p = \frac{1}{2},\, q = \frac{1}{2} \quad \text{and} \quad p + q = \frac{1}{2} + \frac{1}{2} = 1$$

Each toss is independent of what happened on the preceding toss. We are interested in obtaining one head in four tosses. One possible way in which this can happen, along with the corresponding probabilities, is as follows.

Outcome	head	tail	tail	tail
Success or Failure	Success	Failure	Failure	Failure
Probability	$\frac{1}{2}$	$\frac{1}{2}$	$\frac{1}{2}$	$\frac{1}{2}$
Symbolically	p	q	q	q

Since the probability of success is p and the probability of failure is q, we can summarize this as $p \cdot q^3$. Remember q^3 means $q \cdot q \cdot q$. We would then say that the probability of getting one head is

$$\left(\frac{1}{2}\right) \cdot \left(\frac{1}{2}\right)^3 = \left(\frac{1}{2}\right) \cdot \left(\frac{1}{2}\right) \cdot \left(\frac{1}{2}\right) \cdot \left(\frac{1}{2}\right) = \frac{1}{16}$$

However, we have forgotten one thing. The head may occur on the second, third, or fourth toss.

Different ways in which one head can be obtained when a coin is tossed 4 times	head	tail	tail	tail
	tail	head	tail	tail
	tail	tail	head	tail
	tail	tail	tail	head

There are then four ways in which we can get one head. Thus, the $\dfrac{1}{16}$ that we calculated before can occur in four different ways. Therefore,

$$p(\text{exactly 1 head}) = 4\left(\frac{1}{16}\right) = \frac{4}{16}$$

Notice that this is exactly the same answer we obtained by listing all the possible outcomes.

Similarly, if we were interested in the probability of getting exactly two heads, we could consider one particular way in which this can happen.

Outcome	head	head	tail	tail
Success or Failure	Success	Success	Failure	Failure
Probability	$\dfrac{1}{2}$	$\dfrac{1}{2}$	$\dfrac{1}{2}$	$\dfrac{1}{2}$
Symbolically	p	p	q	q

The probability is thus

$$p^2 \cdot q^2 = \left(\frac{1}{2}\right)^2 \left(\frac{1}{2}\right)^2 = \left(\frac{1}{2}\right)\left(\frac{1}{2}\right)\left(\frac{1}{2}\right)\left(\frac{1}{2}\right) = \frac{1}{16}$$

Again, we must multiply this result by the number of ways that these two heads can occur in the four trials. This is the number of combinations of four things taken two at a time. We can use Formula 4.4 of Chapter 4. We get

$$_4C_2 = \frac{4!}{2!\,(4-2)!} = \frac{4!}{2!\,2!} = 6$$

Thus, the probability of getting two heads in four flips of a coin is

$$6 \left(\frac{1}{16} \right) = \frac{6}{16} = \frac{3}{8}$$

More generally, if we are interested in the probability of getting x successes out of n trials of an experiment, then we consider one way in which this can happen. Here we have assumed that all the x successes occur first and all the failures occur on the remaining $n - x$ trials.

Success or Failure Probability	Success · Success · Success · · · Failure · Failure · Failure
	$\underbrace{p \cdot p \cdot p}_{x \text{ of them}}$ · · · $\underbrace{q \cdot q \cdot q}_{n - x \text{ of them}}$

This gives $p^x q^{n-x}$. We then multiply this result by the number of ways that exactly x successes can occur in n trials.

The number of ways that exactly x successes can occur in a set of n trials is given by

$$_nC_x = \frac{n!}{x!(n - x)!}$$

Here we have replaced r by x in the number of possible combinations formula (Formula 4.4). This leads us to the following **binomial distribution formula.**

FORMULA 6.4　　Binomial Distribution Formula

Binomial Distribution Formula　Consider a binomial experiment that has two possible outcomes, success or failure. Let $p(\text{success}) = p$ and $p(\text{failure}) = q$. If this experiment is performed n times, then the probability of getting x successes out of the n trials is

$$p(x \text{ successes}) = {}_nC_x \, p^x q^{n-x} = \frac{n!}{x!(n - x)!} p^x q^{n-x}$$

We illustrate the use of Formula 6.4 with numerous examples.

EXAMPLE 2　　Admission to Medical School

Ninety percent of the graduates of State University who apply to a particular medical school are admitted. This year six graduates from State University have applied for admission to the medical school. Find the probability that only four of them will be accepted.

Solution

Since six students have applied for admission, $n = 6$. We are interested in the probability that four are accepted, so $x = 4$. Also, 90% of the graduates who apply are admitted, so $p = 0.90$, and 10% of the graduates who apply are not admitted, so $q = 0.10$. Now we apply Formula 6.4. We have

$$\text{Prob}(4 \text{ are accepted}) = \text{Prob}(x = 4)$$

$$= \frac{6!}{4! \, (6 - 4)!} (0.90)^4 (0.10)^{6-4}$$

$$= \frac{6!}{4! \, 2!} (0.9)^4 (0.1)^2$$

$$= \frac{6 \cdot 5 \cdot 4 \cdot 3 \cdot 2 \cdot 1}{4 \cdot 3 \cdot 2 \cdot 1 \cdot 2 \cdot 1} (0.9)(0.9)(0.9)(0.9)(0.1)(0.1)$$

$$= 0.0984$$

Thus, the probability that only four of them will be accepted is 0.0984.

EXAMPLE 3 Consider the accompanying newspaper article. Bill and his friends own five cars, which they park in front of his house. Find the probability that none of them will be stolen this year.

CAR THEFTS ON THE RISE AGAIN

Washington (April 7)—Look out your window. Is your car still in your driveway or in front of your house? If it is there, then you're lucky. According to the latest FBI study released yesterday, there are an average of 2300 vehicles stolen per day in the United States. This puts the chances of your car being stolen at about 1 in 120. According to the survey, about 60% of the vehicles are stolen from private residences, apartments, or streets in residential areas between the hours of 6:00 P.M. and 6:00 A.M..

   Most of the cars are stolen to be stripped for their parts or to be used for joyriding.

Wednesday—April 7, 1989

Reprinted by permission.

Solution

Since Bill and his friends own five cars, $n = 5$. We are interested in the probability that none of the cars will be stolen so that $x = 0$. According to the newspaper article, 1 out of every 120 cars is stolen so that $p = \dfrac{1}{120} = 0.008$ and $q = \dfrac{119}{120} = 0.992$. Applying Formula 6.4 gives

$$\text{Prob(0 stolen cars)} = \text{Prob}(x = 0) = \frac{5!}{0! \; 5!} (0.008)^0 (0.992)^5$$

Remember that $(0.008)^0 = 1$, so that

$$\text{Prob(0 stolen cars)} = 1(0.992)^5 = 0.9606$$

Thus, the probability that none of these cars will be stolen this year is 0.9606.

EXAMPLE 4

Mario is taking a multiple-choice examination that consists of five questions. Each question has four possible answers. Mario guesses at every answer. What is the probability that he passes the exam if he needs *at least* four correct answers to pass?

Solution

In order to pass, Mario needs to get at least four correct answers. Thus, he passes if he gets four answers correct or five answers correct. Each question has four possible answers so that $p(\text{correct answer}) = \dfrac{1}{4}$ and $p(\text{wrong answer}) = \dfrac{3}{4}$. Also, there are five questions, so $n = 5$. Therefore,

$$p(4 \text{ answers correct}) = \frac{5!}{4! \; 1!} \left(\frac{1}{4}\right)^4 \left(\frac{3}{4}\right)$$

$$= \frac{5 \cdot \cancel{4} \cdot \cancel{3} \cdot \cancel{2} \cdot \cancel{1}}{\cancel{4} \cdot \cancel{3} \cdot \cancel{2} \cdot \cancel{1} \cdot \cancel{1}} \left(\frac{1}{4}\right)\left(\frac{1}{4}\right)\left(\frac{1}{4}\right)\left(\frac{1}{4}\right)\left(\frac{3}{4}\right)$$

$$= \frac{15}{1024}$$

and

$$p(5 \text{ answers correct}) = \frac{5!}{5! \, 0!} \left(\frac{1}{4}\right)^5 \left(\frac{3}{4}\right)^0 \qquad \text{(Any number to the 0 power is 1.)}$$

$$= \frac{\cancel{5} \cdot \cancel{4} \cdot \cancel{3} \cdot \cancel{2} \cdot 1}{\cancel{5} \cdot \cancel{4} \cdot \cancel{3} \cdot \cancel{2} \cdot \cancel{1} \cdot 1} \left(\frac{1}{4}\right)\left(\frac{1}{4}\right)\left(\frac{1}{4}\right)\left(\frac{1}{4}\right)\left(\frac{1}{4}\right) \cdot 1$$

$$= \frac{1}{1024}$$

Adding the two probabilities, we get

$$p(\text{at least 4 correct answers}) = p(\text{4 answers correct}) + p(\text{5 answers correct})$$

$$= \frac{15}{1024} + \frac{1}{1024}$$

$$= \frac{16}{1024} = \frac{1}{64}$$

Hence, the probability that Mario passes is $\frac{1}{64}$.

EXAMPLE 5

A shipment of 100 tires from the Apex Tire Corporation is known to contain 20 defective tires. Five tires are selected at random and each tire is replaced before the next tire is selected. What is the probability of getting *at most* 2 defective tires?

Solution

We are interested in the probability of getting *at most* two defective tires. This means zero defective tires, one defective tire, or two defective tires. Thus, the probability of at most two defective tires equals

$$p(\text{0 defective}) + p(\text{1 defective}) + p(\text{2 defective})$$

The probability of a defective tire is $\frac{20}{100}$, or $\frac{1}{5}$. Therefore, the probability of getting a non-defective tire is $\frac{4}{5}$. Now

$$p(\text{0 defective}) = \frac{5!}{0! \, 5!} \left(\frac{1}{5}\right)^0 \left(\frac{4}{5}\right)^5 = \frac{1024}{3125}$$

$$p(\text{1 defective}) = \frac{5!}{1! \, 4!} \left(\frac{1}{5}\right)^1 \left(\frac{4}{5}\right)^4 = \frac{1280}{3125}$$

$$p(\text{2 defective}) = \frac{5!}{2! \, 3!} \left(\frac{1}{5}\right)^2 \left(\frac{4}{5}\right)^3 = \frac{640}{3125}$$

Adding, we get

$$p(\text{at most 2 defectives}) = \frac{1024}{3125} + \frac{1280}{3125} + \frac{640}{3125} = \frac{2944}{3125}$$

Hence, the probability of getting at most two defective tires is $\frac{2944}{3125}$, or approximately 0.94.

EXAMPLE 6

If the conditions are the same as in the previous problem except that now 15 tires are selected, what is the probability of getting *at least* 1 defective tire?

Solution

We could proceed as we did in Example 4. Thus

p(at least 1 defective) $= p$(1 defective) $+ p$(2 defective) $+ \cdots + p$(15 defective)

However, this involves a tremendous amount of computation. Recall (see the Rule on page 294) that the sum of all the values of a probability function must be 1. Thus,

p(0 defective) $+ p$(1 defective) $+ p$(2 defective) $+ \cdots + p$(15 defective) $= 1$

Therefore, if we subtract p(0 defective) from both sides, we have

p(1 defective) $+ p$(2 defective) $+ \cdots + p$(15 defective) $= 1 - p$(0 defective)

Now

$$p(0\,\text{defective}) = \frac{15!}{0!\,15!}\left(\frac{1}{5}\right)^0\left(\frac{4}{5}\right)^{15} = 0.035$$

Consequently, the probability of obtaining at least 1 defective tire is $1 - 0.035$, or 0.965.

COMMENT   Calculating binomial probabilities can sometimes be quite time-consuming. To make the job a little easier, we can use Table III of the Appendix, which gives us the binomial probabilities for different values of n, x, and p. No computations are needed. We only need to know the values of n, x, and p.

EXAMPLE 7

Given a binomial distribution with $n = 11$ and $p = 0.4$, use Table III of the Appendix to find the probability of getting
a. exactly four successes.
b. at most three successes.
c. five or more successes.

Solution

We use Table III of the Appendix with $n = 11$ and $p = 0.4$. We first locate $n = 11$ and then move across the top of the table until we reach the $p = 0.4$ column.
a. To find the probability of exactly four successes, we look for the value given in the table for $x = 4$. It is 0.236. Thus, when $n = 11$ and $p = 0.4$, the probability of exactly four successes is 0.236.
b. To find the probability of getting at most three successes, we look in the table for the values given for $x = 0$, $x = 1$, $x = 2$, and $x = 3$. These probabilities are 0.004, 0.027, 0.089, and 0.177, respectively. We add these (Why?) and get

$0.004 + 0.027 + 0.089 + 0.177 = 0.297$

Thus, the probability of at most three successes is 0.297.

c. To find the probability of five or more successes, we look in the chart for the values given for $x = 5$, $x = 6$, $x = 7$, $x = 8$, $x = 9$, $x = 10$, and $x = 11$. These probabilities are 0.221, 0.147, 0.070, 0.023, 0.005, 0.001, and 0, respectively. We add these and get

$$0.221 + 0.147 + 0.070 + 0.023 + 0.005 + 0.001 + 0 = 0.467$$

Thus, the probability of five or more successes is 0.467.

COMMENT   In the previous example you will notice that there is no value given in Table III when $x = 11$. It is left blank. Whenever there is a blank in the chart, this means that the probability is approximately 0. This is the reason that we used 0 as the probability in our calculations.

When $p = 0.5$, then the binomial distribution is symmetrical and begins to resemble the normal distribution as the value of n gets larger. This is shown in Figure 6.4.

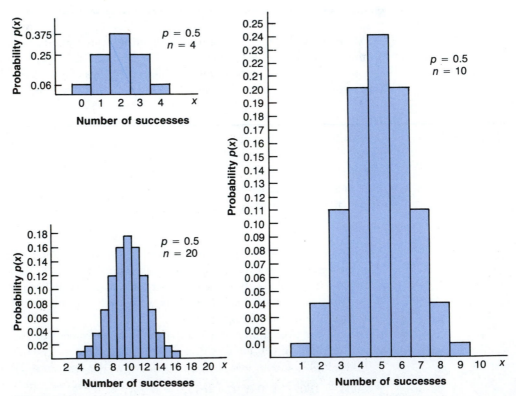

Figure 6.4
Binomial distributions with $p = 0.50$ and different values of n.

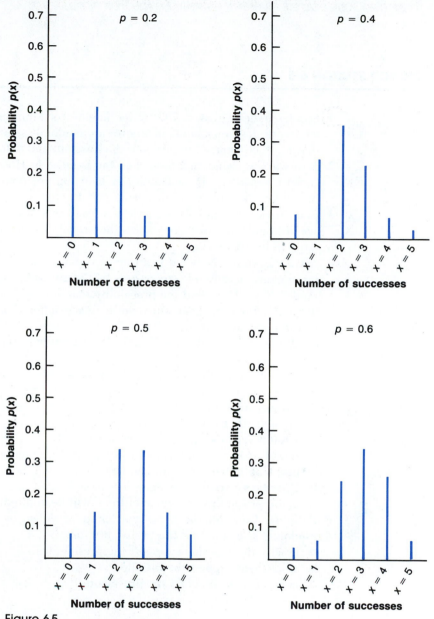

Figure 6.5
Graphs of binomial probabilities for $n = 5$.

As a matter of fact, we can use the values given in Table III and draw the graphs of binomial probabilities. In Figure 6.5 we have drawn several such graphs using different values of p. In each case, however, $n = 5$. What do the graphs look like? We will have more to say about this in the next chapter.

EXERCISES FOR SECTION 6.5

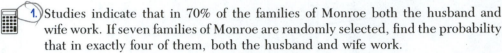

1. Studies indicate that in 70% of the families of Monroe both the husband and wife work. If seven families of Monroe are randomly selected, find the probability that in exactly four of them, both the husband and wife work.

2. A researcher claims that 80% of all the families in Houston have a VCR (video cassette recorder). If six families in Houston are randomly selected, find the probability that all six families have a VCR.

3. Refer back to the newspaper article on page 288. If ten women over 40 years of age are randomly selected, what is the probability that *at most* two of them will develop breast cancer?

4. According to the National Institute of Health,* 32% of all women will suffer a hip fracture because of osteoporosis by the age of 90. If six women aged 90 are randomly selected, find the probability that

   a. exactly three of them will suffer (or have suffered) a hip fracture because of osteoporosis.

   b. at most three of them will suffer (or have suffered) a hip fracture because of osteoporosis.

   c. at least three of them will suffer (or have suffered) a hip fracture because of osteoporosis.

5. According to the Internal Revenue Service (IRS), the chances of your tax return being audited are about 3 in 100 if your income is less than $50,000 and about 8 in 100 if your income is more than $50,000.

   a. If six taxpayers with incomes over $50,000 are randomly selected, what is the probability that no more than half of them will be audited?

   b. If six taxpayers with incomes less than $50,000 are randomly selected, what is the probability that none of them will be audited?

6. *Pollsters:* Consider the newspaper clipping shown below. If seven voters are randomly selected, find the probability that

   a. exactly three of them support the president.

   b. all of them support the president.

   c. none of them supports the president.

*American Demographics, October 1985, p. 20.

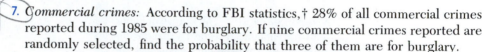

THREE OUT OF FIVE SUPPORT THE PRESIDENT

Washington (Jan. 10)—A survey conducted by Hadly and Kolb indicates that 3 out of 5 Americans support the President's foreign policy and think that it is in the country's best interest. The pollsters sampled 10,000 people across the country to obtain a cross section of voter sentiment.

Thursday—January 10, 1991

7. *Commercial crimes:* According to FBI statistics,† 28% of all commercial crimes reported during 1985 were for burglary. If nine commercial crimes reported are randomly selected, find the probability that three of them are for burglary.

8. *Rape victims:* According to FBI statistics,† only 52% of all rape cases are reported to the police. If seven rape victims are randomly selected, find the probability that at most three of these victims reported the crime to the police.

9. A research organization is interested in determining public opinion on legalizing mercy deaths in the United States. It decides to mail out questionnaires to ten people selected at random. If the probability of any one person answering the questionnaire is $\frac{1}{10}$, find the probability that at least four of the people will fill out the questionnaire and return it.

10. Many people jog to maintain their physical fitness. In a certain community the probability that a man and wife jog together is 0.25. If five men are selected at random, what is the probability that two of them jog with their wives?

11. The probability that a driver making a gas purchase will pay by credit card is $\frac{3}{5}$. If six cars pull up to a gas station to buy gas, what is the probability that four of the drivers will pay for the gas by credit card?

12. On a grocer's shelf there are 40 loaves of bread. Ten of these are known to be stale. Gail purchases 4 loaves of bread from the grocer. Find the probability that
   a. all the loaves of bread purchased are fresh.

†*Source: Criminal Victimization in the United States, 1985, A National Crime Survey Report*; U.S. Department of Justice: Washington, D.C., 1986.

      b. half the number of the loaves of bread purchased are fresh.

      c. none of the loaves of bread purchased is fresh.

13. A drug firm claims that a certain medicine is 93% effective in relieving arthritic pain. If nine people who suffer from arthritis are randomly selected and given this medicine, find the probability that (assuming the claim to be correct):

      a. all of them will be relieved of their arthritic pain.

      b. five of them will be relieved of their arthritic pain.

14. Consider the following newspaper article. If eight people are randomly selected, find the probability that at most two of them are unemployed.

UNEMPLOYMENT RATE UP AGAIN

Newark (March 10)—The Bureau of Labor Statistics announced yesterday that the unemployment rate for our region was up again and now stood at a seasonally adjusted 9.6%. This news, coupled with last week's announcement of the rise in the cost of living, forecasts economic hard times in the near future for our region.

March 10, 1990

If there are many vacant toll booths, what is the probability that cars will use some particular toll booths over others?

15. According to the Mayoca Bridge Authority, 31% of all drivers use the $1.25 exact change lanes when paying the bridge toll. If eight cars pull up to the toll plaza, find the probability that
   a. exactly two of them will use the exact change lane.
   b. at most two of them will use the exact change lane.
   c. at least two of them will use the exact change lane.
16. *False alarms:* 44% of all fire alarms in a certain neighborhood are false alarms. If 12 fire alarms are randomly selected, find the probability that
   a. at least four of them are false alarms.
   b. at most two of them are false alarms.
   c. none of them is a false alarm.

6.6 THE MEAN AND STANDARD DEVIATION OF THE BINOMIAL DISTRIBUTION

Consider the binomial distribution given on page 310, which is also repeated below. Recall that x represents the number of correct answers that Paula obtained on an exam of five questions, where the probability of a correct answer is $\frac{1}{2}$.

x	$p(x)$	$x \cdot p(x)$	x^2	$x^2 \cdot p(x)$
0	$\frac{1}{32}$	0	0	0
1	$\frac{5}{32}$	$\frac{5}{32}$	1	$\frac{5}{32}$
2	$\frac{10}{32}$	$\frac{20}{32}$	4	$\frac{40}{32}$
3	$\frac{10}{32}$	$\frac{30}{32}$	9	$\frac{90}{32}$
4	$\frac{5}{32}$	$\frac{20}{32}$	16	$\frac{80}{32}$
5	$\frac{1}{32}$	$\frac{5}{32}$	25	$\frac{25}{32}$
		$Total = \frac{80}{32} = \frac{5}{2}$		$Total = \frac{240}{32} = \frac{15}{2}$

Let us calculate the mean and variance for this distribution. Using Formula 6.1 of Section 6.4 we get

$$\mu = \Sigma x \cdot p(x) = 0 + \left(\frac{5}{32}\right) + \left(\frac{20}{32}\right) + \left(\frac{30}{32}\right) + \left(\frac{20}{32}\right) + \left(\frac{5}{32}\right)$$

$$= \frac{80}{32} = \frac{5}{2}$$

To calculate the variance we use Formula 6.3 of Section 6.3. From the preceding table we have

$$\sigma^2 = \Sigma x^2 \cdot p(x) - [\Sigma x \cdot p(x)]^2$$

$$= \frac{15}{2} - \left(\frac{5}{2}\right)^2$$

$$= \frac{30}{4} - \frac{25}{4} = \frac{5}{4}$$

Thus, $\mu = \dfrac{5}{2}$, or 2.5, and $\sigma^2 = \dfrac{5}{4}$, or 1.25

Notice that if we multiply the total number of exam questions, which is 5, by the probability of a correct answer, which is $\dfrac{1}{2}$, we get $5\left(\dfrac{1}{2}\right) = 2.5$. This is exactly the same answer we get for the mean by applying Formula 6.1. We might be tempted to conclude that $\mu = np$. This is indeed the case.

Similarly, if we multiply the total number of questions with the probability of a correct answer and with the probability of a wrong answer, we get $5\left(\dfrac{1}{2}\right)\left(\dfrac{1}{2}\right) = 1.25$

Here we might conclude that $\sigma^2 = npq$. Again, this is indeed the case. We can generalize theses ideas in the following.

FORMULA 6.5
Mean of a Binomial Distribution

The **mean of a binomial distribution**, μ, is found by multiplying the total number of trials with the probability of success on each trial. If there are n trials of the experiment and if the probability of success on each trial is p, then

$$\mu = np$$

FORMULA 6.6

The **variance of a binomial distribution** is given by

$$\sigma^2 = npq$$

Variance of a Binomial Distribution

The **standard deviation** σ is the square root of the variance. Thus,

$$\sigma = \sqrt{\text{variance}} = \sqrt{npq}$$

The proofs of Formulas 6.5 and 6.6 will be given as an exercise at the end of this section.

EXAMPLE 1

A die is rolled 600 times. Find the mean and standard deviation of the number of 1's that show.

Solution

This is a binomial distribution. Since the die is rolled 600 times, $n = 600$. Also, there are six possible outcomes so that $p = \dfrac{1}{6}$ and $q = \dfrac{5}{6}$. Thus, using Formulas 6.5 and 6.6, we have

$$\mu = np$$

$$= 600 \left(\frac{1}{6} \right) = 100$$

and

$$\sigma^2 = npq$$

$$= 600 \left(\frac{1}{6} \right)\left(\frac{5}{6} \right) = 83.33$$

Therefore, the mean is 100 and the standard deviation, which is the square root of the variance, is $\sqrt{83.33}$, or approximately 9.13. This tells us that if this experiment were to be repeated many times, we could expect an average of 100 1's per trial with a standard deviation of $\sqrt{83.33}$, or 9.13.

EXAMPLE 2

A large mail-order department store finds that approximately 17% of all purchases are returned for credit. If the store sells 100,000 different items this year, about how many items will be returned? Find the standard deviation.

Solution

This is a binomial distribution. Since 100,000 items were sold, $n = 100{,}000$. Also, the probability that a customer will return the item is 0.17 so that $p = 0.17$. Therefore, the probability that the customer will not return the item is 0.83. Thus, using Formulas 6.5 and 6.6, we have

$$\mu = np$$
$$= (100{,}000)(0.17)$$
$$= 17{,}000$$

and

$$\sigma = \sqrt{npq}$$
$$= \sqrt{(100{,}000)(0.17)(0.83)}$$
$$= \sqrt{14110}$$
$$\approx 118.79$$

Thus, the department store can expect about 17,000 items to be returned with a standard deviation of 118.79.

Formulas 6.5 and 6.6 will be applied in greater detail in Chapter 7.

EXERCISES FOR SECTION 6.6

 1. Consider the newspaper clipping below. If 600 packages of meat are randomly selected, about how many of them can be expected to weigh at least 2 ounces less than its stated weight? Find the standard deviation.

> ### CONSUMER'S AFFAIRS DEPT. TO PROSECUTE APEX
>
> *Marborough (Nov. 8)*—The Consumer's Affairs Department announced this morning that it would begin prosecuting the Apex Supermarket Chain for short weighting the packages of meat it sells. In random samples of packages of meat at all of Apex's stores, it was found that approximately 38% of the packages weighed at least 2 ounces less than what was indicated on the package. The unsuspecting customers conceivably were bilked out of millions of dollars.
>
> November 8, 1990

2. Each year about 2% of the police force of a large city in the South either die, resign, or retire. There are 14,278 police officers currently on the police force. Due to budgetary problems, the mayor announces that 280 police officers will have to be laid off. Neglecting any other considerations, can these cutbacks be accomplished through attrition (that is, by deaths, resignations, or retirements)?

 3. *Air Pollution control:* It is estimated that 18% of the cars registered in the state cannot meet the state's tough new air pollution exhaust system standards. If there are 965,425 cars registered in the state, about how many cars will not pass the state's new air pollution standards? Find the standard deviation.

4. 43% of all prescriptions filled at Hy's Pharmacy are paid for by medicare. If Hy's pharmacy expects to process 375 prescriptions this week, about how many of them can be expected to be paid for by medicare?

5. About two out of every three gas purchases at Pete's Exxon Station are paid for by credit cards. If 495 customers will buy gas at the station one day, about how many of them will pay for the gas with a credit card? Find the standard deviation.

6. Records show that approximately 8% of all refrigerators produced by a company fail before their two-year guarantee expires. If the company plans to manufacture 60,000 refrigerators in a year, about how many of them can the company expect to fail to operate properly before the two-year guarantee expires? Find the standard deviation.

7. 85% of all people who buy a particular camera also buy the carrying case for the camera. If it is estimated that two thousand cameras will be sold this year, about how many carrying cases will be sold? Find the standard deivation.

8. A student has not studied for the upcoming final examination. The exam consists of 75 multiple-choice questions with each question having four possible answers. If the student decides to guess at each question, find the mean and standard deviation of the number of questions answered correctly.

9. According to the Bureau of Vital Statistics, about 53% of all births in the United States are male; that is, the probability that a newborn child will be a male is about 0.53. If Washington Hospital expects 1450 live births this year, what is the expected number of males to be born? Find the standard deviation.

10. According to the U.S. Census Bureau, about 22% of American children under the age of 6 live in households with incomes below the official poverty level. A random sample of 500 children under the age of 6 is taken. What is the mean number of children in the sample who come from households with incomes below the official poverty level? Find the standard deviation.

11. Many colleges nationwide find that not all applicants who are accepted for admission to a college will actually attend that college. Past experience at Bismark College shows that about 85% of all the students accepted will actually attend the college. This year the college has decided to send out 2000 acceptance letters. Assuming that the students make their decisions independently, find the mean and standard deviation of the number of students who will attend the college.

12. Refer back to the previous question. If the college would like to have an entering freshmen class of 1500 students, how many acceptance letters should it send out?

13. It is an accepted practice in the airline industry to overbook flights. One particular airline company has found that approximately 12% of the people who make reservations do not show. If the airline has accepted 240 reservations for a plane that has 210 seats, will the airline have a seat for each passenger who has reserved a seat and who shows up?

*14. Using the formulas for summation, verify that Formulas 6.5 and 6.6 are indeed valid for a binomial distribution.

*6.7 THE POISSON DISTRIBUTION

There are many practical problems where we may be interested in finding the probability that x "successes" will occur over a given interval of time or a region of space. This is especially true when we do not expect many successes to occur over the time interval (which may be of any length, such as a minute, a day, a week, a month, or a year). For example, we may be interested in determining the number of days that school will be closed due to snowstorms, or we may be interested in determining the number of days that a baseball game will be postponed in a given season because of rain. For these and similar problems we can use the Poisson probability function formula.

Before giving this formula, we wish to emphasize that certain underlying assumptions must be satisfied for this formula to be applicable. Among these assumptions are the following:

1. Each "success" occurs independently of the others.
2. The probability of "success" in any interval or region is very small (or that most of the probability is concentrated at the low end of the domain of x).
3. The probability of a success in any one small interval is the same as that for any other small interval of the same size.
4. The number of successes in any interval is independent of the number of successes in any other nonoverlapping interval.

When these assumptions are satisfied we can use the following:

FORMULA 6.7   Poisson Distribution

Poisson Probability
Distribution

The **Poisson probability distribution** representing the number of successes occurring in a given time interval or specified region is given by

$$p(x \text{ successes}) = \frac{e^{-\mu}\mu^x}{x!} \qquad x = 0, 1, 2, 3, \ldots$$

where μ is the average number of successes occurring in the given time interval or specified region. Thus, for the Poisson distribution, the mean is μ, the variance is μ, and the standard deviation is $\sqrt{\mu}$.

COMMENT   The symbol e that appears in the formula is used often in mathematics to represent an irrational number whose value is approximately equal to 2.71828. Table XII in the Appendix gives the values of $e^{-\mu}$ for different values of μ.

Let us illustrate the use of this formula with several examples.

EXAMPLE 1   An animal trainer finds that the number, x, of animal bites per month that her crew experiences follows an approximate Poisson distribution with a mean of 7.5. Find the variance and standard deviation of x, the number of animal bites per month.

Solution

For a Poisson distribution the mean and variance are both equal to μ. Thus, in our case, we have

$$\mu = 7.5 \quad \text{and} \quad \sigma^2 = 7.5$$

The standard deviation is $\sqrt{7.5}$, or approximately 2.7.

EXAMPLE 2   School Closings

Official records in a particular city show that the average number of school closings in a school year due to snowstorms is four. What is the probability that there will be six school closings this year because of snowstorms?

Solution

Based on the given information, $\mu = 4$. We are interested in the probability that $x = 6$. Using the Poisson distribution formula, we get

$$p(x = 6) = \frac{e^{-4}(4)^6}{6!}$$

From Table XII, $e^{-4} = 0.0183156$ so that

$$p(x = 6) = \frac{e^{-4}(4)^6}{6!}$$

$$= \frac{(0.0183156)(4096)}{720} = 0.1042$$

Thus, the probability that there will be six school closings this year because of snowstorms is 0.1042.

EXAMPLE 3   Rainouts

From past experience a baseball club owner knows that about six games, on average, will have to be postponed during the season because of rain. Find the probability that this season

a. three games will have to be postponed because of rain.
b. no games will have to be postponed because of rain.

Solution

Based on the given information, $\mu = 6$

a. Here we are interested in the probability that $x = 3$. We have

$$p(x = 3) = \frac{e^{-6}(6)^3}{3!}$$

From Table XII, $e^{-6} = 0.00247875$ so that

$$p(x = 3) = \frac{(0.00247875)(216)}{6} = 0.0892$$

Thus, the probability that three games will have to be postponed during the season because of rain is 0.0892.

b. Here we are interested in the probability that $x = 0$. We have

$$p(x = 0) = \frac{e^{-6}(6)^0}{0!} \qquad (\text{Remember } 6^0 = 1 \text{ and } 0! = 1)$$

$$= \frac{(0.00247875)1}{1}$$

$$= 0.00247875$$

Thus, the probability that no games will have to be postponed is 0.00247875.

COMMENT The Poisson probability distribution can also be used as an approximation to the binomial probability function. However, such considerations are beyond the scope of this book.

EXERCISES FOR SECTION 6.7

 1. Bill Boyers operates a newspaper stand in a major bus terminal. During the rush hour customers arrive to buy newspapers or magazines at an average of five per minute. Assuming that the arrivals may be represented by a Poisson process, what is the probability that in a particular minute
   a. at most two customers will arrive?
   b. at least two customers will arrive?
   c. exactly two customers will arrive?

 2. The Beck Corporation receives an average of $\mu = 3$ pieces of Express mail each delivery day. Assuming that the arrivals may be represented by a Poisson process, what is the probability that they will get no Express mail on the next two consecutive business days?

3. Assume that the number of industrial accidents per month at a large manufacturing plant in the South follows a Poisson distribution with $\mu = 7$. What is the probability that the number of industrial accidents in a given month will be less than or equal to 4?

4. The maintenance department of the Blue Valley Bus Company reports that the number of bus breakdowns per 8-hour shift follows a Poisson distribution with a mean of 1.8. What is the probability that during a randomly selected 8-hour shift there will be
   a. no bus breakdowns?
   b. between one and four breakdowns?

5. Bob Baily operates the Superior Collision Company. He receives an average of about eight calls for towing per day. What is the probability that on a particular day Bob will receive
   a. at least three calls for towing?
   b. at most three calls for towing?
   c. exactly three calls for towing?
   d. between two and four calls for towing?

6. Dr. Spector is a cardiologist. She receives an average of five emergency calls per day. What is the probability that on a particular day Dr. Spector will receive
   a. at least one emergency call.
   b. at most one emergency call.
   c. exactly one emergency call.

7. On the average, four Canadian quarters are deposited per day into an American pay telephone located in an American amusement park near the Canadian border. What is the probability that exactly four Canadian coins will be deposited into this pay telephone?

8. A certain foreign diplomatic mission receives an average of three bomb threats per day. What is the probability that on a particular day the diplomatic mission will receive.
   a. four or more bomb threats?
   b. no bomb threats?

9. A secretary averages about four typing errors per page typed. What is the probability that she will make
   a. at least two errors on the next page that she types?
   b. no errors on the next page that she types?

10. Suppose that the number of particles emitted from a radioactive source follows a Poisson distribution with an average emission of four particles per minute. What is the probability that at most three particles will be emitted in one minute?

11. The number of wrong numbers received by a telephone switchboard operator during the 12:00–1:00 lunch period follows a Poisson distribution with $\mu = 5$. What is the probability that the operator will receive at most three wrong numbers during the 12:00–1:00 lunch period?

*12. A grocer finds that the demand for a particular item during the month follows a Poisson distribution with $\mu = 6$. The grocer is anxious to have sufficient stock to satisfy customer demand. How large of a stock of the item should the grocer have on hand so as to be able to supply the customer demand with a probability of at least 0.90?

*6.8 HYPERGEOMETRIC DISTRIBUTION

In our discussion of the binomial distribution, we noted that when we select objects without replacement we cannot use the binomial distribution formula since the probability of success from trial to trial is not constant and the trials are not independent.

Before proceeding with our discussion, let us pause for a moment to review some notation. Recall that when we discussed combinations we used the notation $_nC_r$ to represent the number of combinations of n things taken r at a time. As mentioned, another notation that is often used to represent the same idea is $\binom{n}{r}$.

Thus,

$$_5C_3 = \binom{5}{3} = \frac{5!}{3!\,2!}$$

$$_{10}C_8 = \binom{10}{8} = \frac{10!}{8!\,2!}$$

and

$$_nC_r = \binom{n}{r} = \frac{n!}{r!(n-r)!}$$

Suppose 5 cards are randomly drawn without replacement from a 52-card deck. Can we find the probability of obtaining 3 red and 2 black cards? The answer is yes if we proceed as follows. The 3 red cards can be selected from the 26 available red cards in $\binom{26}{3}$ possible ways. Similarly, the 2 black cards can be selected from the 26 available black cards in $\binom{26}{2}$ possible ways. Thus, the total number of ways of selecting 3 red cards and 2 black cards in five draws is $\binom{26}{3} \cdot \binom{26}{2}$. Now we determine how many different ways there are for selecting 5 cards from a 52–card deck. This is simply $\binom{52}{5}$. Using the definition of probability, we have

$$\text{Prob}\begin{pmatrix}3 \text{ red cards and 2 black} \\ \text{cards in 5 draws from a} \\ 52\text{–card deck}\end{pmatrix} = \frac{\binom{26}{3} \cdot \binom{26}{2}}{\binom{52}{5}}$$

$$= \frac{\left(\dfrac{26!}{3!\,23!}\right) \cdot \left(\dfrac{26!}{2!\,24!}\right)}{\left(\dfrac{52!}{5!\,47!}\right)}$$

$$= \frac{(2600)(325)}{2{,}598{,}960} = 0.3251$$

More generally, suppose we are interested in the probability of selecting x successes from k items labeled success and $n - x$ failures from $N - k$ items labeled failure when a random sample of size n is selected from N items. We can then apply the hypergeometric probability function to determine this probability. We have

---

FORMULA 6.8   Hypergeometric Probability Function

Hypergeometric Probability Function

The probability of obtaining x successes when a sample of size n is selected without replacement from N items of which k are labeled success and $N - k$ are labeled failure is given by

$$\frac{\binom{k}{x}\binom{N-k}{n-x}}{\binom{N}{n}} \qquad x = 0, 1, 2, \ldots, n$$

Let us illustrate the use of this formula with several examples.

EXAMPLE 1   A production run of 100 radios is received by the shipping department. It is known that 10 of the radios in the production are defective. The quality control engineer randomly selects 8 of the radios from the production run. Find the probability that 6 of the radios selected are defective.

Solution

We apply the hypergeometric probability function. Here $N = 100$, $n = 8$, $k = 10$, and $N - k = 100 - 10 = 90$. We are interested in the probability that $x = 6$.

Thus,

$$\text{Prob}(x = 6) = \frac{\binom{10}{6}\binom{90}{2}}{\binom{100}{8}}$$

$$\frac{\left(\dfrac{10!}{6!\ 4!}\right) \cdot \left(\dfrac{90!}{2!\ 88!}\right)}{\left(\dfrac{100!}{8!\ 92!}\right)}$$

$$= \frac{(210)(4005)}{186087894300} = 0.0000045$$

Thus, the probability that the quality-control engineer selects six defective radios is 0.0000045.

EXAMPLE 2   A faculty–student committee is to be selected at random from three students and six faculty members. The committee is to consist of five people. Find the probability that the committee will contain

a. no students.
b. one student.
c. two students.
d. three students.

Solution

Let x be the number of students selected to be on the committee.
a. Here $x = 0$, $N = 9$, $n = 5$, and $k = 3$. Thus,

$$\text{Prob}(0 \text{ students}) = \text{Prob}(x = 0) = \frac{\binom{3}{0}\binom{6}{5}}{\binom{9}{5}}$$

$$= \frac{(1)(6)}{126} = \frac{6}{126}$$

b. Here $x = 1$, $N = 9$, $n = 5$, and $k = 3$. Thus,

$$\text{Prob(1 student)} = \text{Prob}(x = 1) = \frac{\binom{3}{1}\binom{6}{4}}{\binom{9}{5}}$$

$$= \frac{(3)(15)}{126} = \frac{45}{126}$$

c. Here $x = 2$, $N = 9$, $n = 5$, and $k = 3$. Thus,

$$\text{Prob(2 students)} = \text{Prob}(x = 2) = \frac{\binom{3}{2}\binom{6}{3}}{\binom{9}{5}}$$

$$= \frac{(3)(20)}{126} = \frac{60}{126}$$

d. Here $x = 3$, $N = 9$, $n = 5$, and $k = 3$. Thus,

$$\text{Prob(3 students)} = \text{Prob}(x = 3) = \frac{\binom{3}{3}\binom{6}{2}}{\binom{9}{5}}$$

$$= \frac{(1)(15)}{126} = \frac{15}{126}$$

EXERCISES FOR SECTION 6.8

1. There are 100 people in an auditorium. It is known that 10 of these people have high blood pressure. If a random sample of 10 of these people is selected without replacement, what is the probability that the sample contains
   a. two people with high blood pressure.
   b. at most two people with high blood pressure.
2. Thirty-nine of the 100 doctors at City Hospital are union members. What is the probability that in a random sample of 7 of these 100 doctors, 3 of them will be union members?
3. Forty-eight of the 69 faculty members at Bates University contribute part of their salary to a tax-deferred annuity plan. What is the probability that in a random sample of 6 of these 69 faculty members, at most 3 of them contribute part of their salary to a tax-deferred annuity plan?

4. The Bayview Payroll Company just created four assistant manager job titles and has decided to fill these jobs by randomly selecting from a possible 11 men and 14 women current employees who qualify. What is the probability that the vacancies will be filled by 2 men and 2 women?

5. A random sample of 3 disks is to be taken from a box of 12 disks. If it is known that 5 of the disks are defective, what is the probability of getting no defective disk in the sample?

6. It is known that 26 of the 50 salespeople who work for the Harcourt Company are married. If a random sample of 9 of these people is taken (without replacement) what is the probability that the sample contains at most 3 salespeople who are married?

7. A committee of four people is to be randomly selected from nine tenants and six homeowners to discuss increased community safety patrols with the police department. What is the probability that the committee will consist of
   a. 0 tenants?
   b. 1 tenant?
   c. 2 tenants?
   d. 3 tenants?
   e. 4 tenants?

8. There are 72 students enrolled in a biology lecture class. It is known that one-fourth of these students are receiving scholarships. If 8 students are randomly selected from this class, what is the probability that none of them is receiving a scholarship?

9. A package of mixed nuts contains 18 walnuts and 9 almonds. Three nuts are randomly selected without replacement. What is the probability that all the nuts selected are almonds?

10. Marx Clothing Store receives a shipment of 100 suits. Unfortunately, 10% of the suits are defective as the buttons were improperly sewn on them. It is decided to randomly select 5 suits from this shipment. Find the probability of obtaining no defective suits.

6.9 USING COMPUTER PACKAGES

The MINITAB statistical package can be used to compute binomial probabilities without having to perform any computations. The MINITAB command **PDF; BINOMIAL** accomplishes this. We illustrate the use of this command with the following example.

EXAMPLE 1   Medical researchers have found that approximately 15% of the population who take a particular antihistamine drug will develop some form of reaction to the drug. If the drug is administered to six people, find the probability that

a. exactly two of these people will develop some form of reaction to the drug.
b. exactly three of these people will develop some form of reaction to the drug.

Solution

In order to get MINITAB to compute these probabilities, we first type the command PDF followed by a semicolon. Then, on the next line, we type the subcommand BINOMIAL followed by the values of n and p. In our case $n = 6$ and $p = 0.15$.

When these commands are applied to our case we obtain the following MINITAB printout:

```
MTB > PDF;
SUBC> BINOMIAL N=6, P=0.15.
    BINOMIAL WITH N =   6   P = 0.150000
        K              P( X = K )
        0                 0.3771
        1                 0.3993
        2                 0.1762
        3                 0.0415
        4                 0.0055
        5                 0.0004
        6                 0.0000
```

From this printout we can easily read our answers. The probability that exactly two of these people ($X = 2$) will develop some form of reaction is 0.1762. Also, the probability that exactly three of these people will develop some form of reaction is 0.0415.

By using the above printout, we can easily answer such questions as "What is the probability that at most five of these people will develop some form of reaction?" we simply add the appropriate probabilities as listed; that is, we add $0.3771 + 0.3993 + 0.1762 + 0.0415 + 0.0055 + 0.0004$.

We can also get MINITAB to give us individual probabilities. Thus, in the previous example, if we were interested only in the probability that exactly three of these people will develop some form of reaction, we would simply type a "3" after the PDF command. We would then have the following MINITAB printout:

```
MTB > PDF 3;
SUBC> BINOMIAL N=6, P=0.15.
        K              P( X = K )
     3.00                 0.0415
```

6.10 SUMMARY

In this chapter we discussed how the ideas of probability can be combined with frequency distributions. Specifically, we introduced the idea of a random variable and its probability distribution. These enable an experimenter to analyze outcomes of experiments and to speak about the probability of different outcomes.

We then discussed the mean and variance of a probability distribution. These allow us to determine the expected number of favorable outcomes of an experiment.

Although we did not emphasize the point, all the events discussed were mutually exclusive. Thus, we were able to add probabilities by the addition rule for probabilities. Also, the events were independent. This allowed us to multiply probabilities by the multiplication rule for probabilities of independent events.

We discussed one particular distribution in detail, the binomial distribution, since it is one of the most widely used distributions in statistics. In addition to the binomial distribution formula itself, which allows us to calculate the probability of getting a specified number of successes in repeated trials of an experiment, formulas for calculating its mean, variance, and standard deviation were given. These formulas were applied to numerous examples. Because of its importance, we will discuss the binomial distribution further in Chapter 7.

We also discussed the Poisson distribution, which can be used only when certain assumptions are satisfied. These were given on page 332.

Finally we presented the hypergeometric probability function, which is used when the sampling is without replacement. Thus, we cannot use the binomial distribution since the probability of success is not the same from trial to trial.

Study Guide

The following is a chapter outline in capsule form. You should now be able to demonstrate your knowledge of the ideas mentioned by giving definitions, descriptions, or specific examples. Page references are given in parentheses.

If an experiment is performed and some quantitative variable denoted by x is measured or observed, the quantitative variable x is called a **random variable** since the values that x may assume in the given experiment depend on chance. A random variable can take on numerical values only. (page 291)

Whenever all the possible values that a random variable may assume can be listed (or counted), the random variable is said to be **discrete**. (page 291)

Random variables that can assume values corresponding to any of the points contained in one or more intervals on a line are called **continuous random variables**. (page 291)

A **probability function** or **probability distribution** is a correspondence that assigns probabilities to the values of a random variable. The sum of all the probabilities of a probability function is always 1. (page 293)

If a random variable has only two possible values, and if the probability of these values remains the same for each trial regardless of what happened on any previous trial, the variable is called a **binomial variable** and its probability distribution is called a **binomial distribution**. (page 295)

The **mean** of a random variable, denoted by μ, for a given probability distribution is the number obtained by multiplying all the possible values of the random variable having this particular distribution by their respective probabilities and adding these products together. The mean is often called the **mathematical expectation** or the **expected value**. (page 300)

The difference of a number from the mean is called the **deviation from the mean**. (page 302)

The **variance** of random variable with a given probability distribution is the number obtained by multiplying each of the squared deviations from the mean by their respective probabilities and adding these products. (page 302)

The **standard deviation** of a random variable with a given probability distribution is the square root of its variance. (page 303)

A **Bernoulli experiment** or **binomial probability experiment** is an experiment that satisfies the following properties:

1. There is a fixed number, n, of repeated trials whose outcomes are independent.
2. Each trial results in one of two possible outcomes. We call one outcome a success and denote it by the letter S and the other outcome a failure denoted by F.
3. The probability of success on a single trial equals p and remains the same from trial to trial. The probability of a failure equals $1 - p = q$. Symbolically,

$$p(\text{success}) = p \quad \text{and} \quad p\,(\text{failure}) = q = 1 - p$$

4. We are interested in the number of successes obtained in n trials of the experiment. (page 314)

The **binomial distribution formula** states the following: Consider a binomial experiment that has two possible outcomes, success or failure. Let $p(\text{success}) = p$ and $p(\text{failure}) = q$. If this experiment is performed n times, then the probability of getting x successes out of the n trials is

$$p(x \text{ successes}) = {}_nC_x p^x q^{n-x} = \frac{n!}{x!\,(n-x)!}\, p^x q^{n-x} \qquad \text{(page 317)}$$

The **mean of a binomial distribution**, μ, is found by multiplying the total number of trials with the probability of success on each trial. If there are n trials of the experiment, and if the probability of success on each trial is p, then $\mu = \boldsymbol{np}$. (page 328)

The **variance of a binomial distribution** is given by $\sigma^2 = npq$. (page 328)

The **standard deviation** is the square root of the variance. Thus,

$$\sigma = \sqrt{\text{variance}} = \sqrt{npq} \qquad \text{(page 328)}$$

The **Poisson probability distribution** representing the number of successes occurring in a given time interval or specified region is given by

$$p(x \text{ successes}) = \frac{e^{-\mu} \mu^x}{x!} \qquad x = 0, 1, 2, 3, \ldots$$

where μ is the average number of successes occurring in the given time interval or specified region. Thus, for the Poisson distribution the mean is μ, the variance is μ, and the standard deviation is $\sqrt{\mu}$. (page 332)

Hypergeometric probability function. The probability of obtaining x successes when a sample of size n is selected without replacement from N items of which k are labeled success and $N - k$ are labeled failure is given by

$$\frac{\binom{k}{x} \binom{N - k}{n - x}}{\binom{N}{n}} \qquad x = 0, 1, 2, \ldots, n \qquad \text{(page 337)}$$

Formulas to Remember

You should be able to identify each symbol in the following formulas, understand the relationships among the symbols expressed in each formula, understand the significance of each formula, and use the formulas in solving problems.

1. Mean of a probability distribution: $\mu = \Sigma x \cdot p(x)$
2. Variance of a probability distribution: $\sigma^2 = \Sigma(x - \mu)^2 \cdot p(x)$
3. Variance of a probability distribution: $\sigma^2 = \Sigma x^2 \cdot p(x) - [\Sigma x \cdot p(x)]^2$
4. Standard deviation of a probability distribution: $\sigma = \sqrt{\text{variance}}$
5. Binomial distribution

$$p(x \text{ successes out of } n \text{ trials}) = \frac{n!}{x! \, (n - x)!} p^x q^{n - x}$$

6. Mean of the binomial distribution: $\mu = np$
7. Variance of the binomial distribution: $\sigma^2 = npq$
8. Poisson distribution: $p(x \text{ successes}) = \dfrac{e^{-\mu} \mu^x}{x!} \qquad x = 0, 1, 2, 3, \ldots$
9. Mean of the Poisson distribution is μ.
10. Variance of a Poisson distribution is μ.
11. $_nC_x = \dbinom{n}{x} = \dfrac{n!}{x! \, (n - x)!}$

12. Hypergeometric distribution:

$$p(x \text{ successes}) = \frac{\binom{k}{x}\binom{N-k}{n-x}}{\binom{N}{n}} \qquad x = 0, 1, 2, \ldots, n$$

Testing Your Understanding of This Chapter's Concepts

1. A die is altered by painting an additional dot on the face that originally had one dot (see the following diagram). A pair of such altered dice is rolled. Let x be the number of dots showing on both dice together.

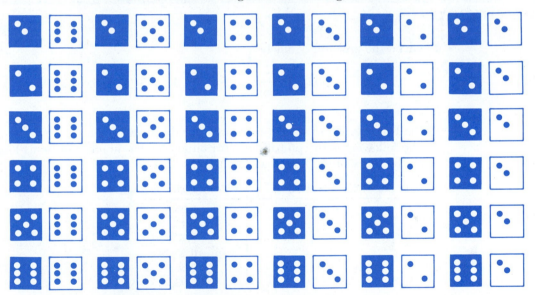

   a. Find the probability distribution of x.
   b. Find μ, σ^2, and σ for this distribution.
2. Consider a random variable x that has the following probability distribution.

x	$p(x)$
0	0.35
1	0.25
2	0.32
3	0.08

Does x have a binomial distribution? Explain your answer.

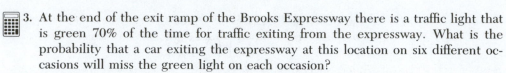

3. At the end of the exit ramp of the Brooks Expressway there is a traffic light that is green 70% of the time for traffic exiting from the expressway. What is the probability that a car exiting the expressway at this location on six different occasions will miss the green light on each occasion?

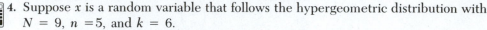

4. Suppose x is a random variable that follows the hypergeometric distribution with $N = 9$, $n = 5$, and $k = 6$.

   a. Find the probability distribution of x.
   b. Find the mean and variance of x.
   c. What is the probability that x will fall within the interval $\mu - 2\sigma$ and $\mu + 2\sigma$?

5. If a binomial experiment is repeated n times, then the number of branches in the tree diagram of the outcomes is 2^n. If a binomial experiment is repeated two times, draw a tree diagram showing that the number of branches is 2^2 or 4.

THINKING CRITICALLY

1. Given the probability function

$$p(x) = \frac{8 - x}{28} \qquad \text{for } x = 1, 2, 3, 4, 5, 6, \text{ or } 7$$

Find the mean and standard deviation for this distribution.

2. The following is the probability distribution for the number of items lost daily on the buses of the Suburban Pupil Transit Lines.

Number of Items Lost Daily, x	Probability, $p(x)$
1	0.05
2	0.43
3	0.17
4	0.25
5	0.06
6	0.03
7	0.01

   a. Find μ and σ.
   b. Find the probability that at least x items lost will be in the interval $\mu \pm 2\sigma$ (*Hint*: Use Chebyshev's Theorem discussed in Chapter 3.)

3. Using the summations given in Chaper 4, show how Formula 6.3 can be derived from Formula 6.2.

4. When we square the binomial expression $(p + q)$, that is, when we evaluate $(p + q)^2$, we get $(p + q)^2 = p^2 + 2pq^2 + q^2$ (Verify this.) When a binomial experiment is performed twice, the probability of two successes in the both trials is p^2, the probability of one success in the two trials is $2pq$, and the probability of no successes in both trials is q^2. Note that these are the first, second, and

third terms, respectively, in the expansion of $(p + q)^2$. Evaluate $(p + q)^3$ and compare your results with a binomial experiment where $n = 3$. Comment. Can you generalize?

5. Consider a binomial probability distribution for which $n = 8$ and $p = 0.5$.
   a. Draw the graph of this probability distribution.
   b. Locate the interval $\mu \pm 2\sigma$ on this graph.

Review Exercises for Chapter 6

1. In the following probability distribution, what is the probability that $x = 4$?

x	3	4	5	6	7	8
$p(x)$	0.13	?	0.36	0.08	0.27	0.13

2. A survey of thousands of sports spectators found that 80% of them believed that atheletes should be barred from participating in any professional sports event if the athlete was treated with any steroids during the month prior to the sports event. If a random sample of nine spectators is taken, what is the probability that
   a. at least five of them believe that athletes should be barred from participating in any professional sports event if they were treated with any steroids during the month prior to the sports event?
   b. at most three of them believe that such athletes should be barred?

3. In analyzing many income tax returns that were flagged by the computer, an Internal Revenue Service agent claims that about 23% of the returns contain arithmetic errors. If 400 flagged tax returns are randomly selected, about how many of them can be expected to contain arithmetic errors? Find the standard deviation.

4. A toy store owner finds that the daily demand for a particular electronic toy follows a Poisson distribution with $\mu = 5$. The store owner wishes to have a sufficient supply of the toy so as to satisfy customer demand. How large a stock should the owner have on hand to be able to satisfy the customer demand with a probability of *at least* 0.90?

5. Find the mean and standard deviation of the following probability distribution:

x	3	4	5	6
$p(x)$	0.3	0.4	0.2	0.1

6. Many large supermarkets use computers that scan the items at the checkout counter. The scanner then prints the correct price of the item on the register. Some stores have such faith in these computers that they make the following claim: "If the computer scanner generates a price that is different from the price marked on the item, we will pay you double the difference." If the computer

scanners are alleged to be 95% accurate, that is, they will correctly scan the price 95% of the time, what is the probability that in eight items scanned, there will be *at most* one item scanned incorrectly?

7. Medicare regulations allow recipients of medicare payments to request a review of a claim when the recipient disagrees with the amount approved. Based upon past experience, it is known that the number of requests for reevaluation of medicare payments received daily by one particular medicare office follows a Poisson distribution with $\mu = 4$. What is the probability that this medicare office will receive at most three requests for review of claims on any given day?

8. A car rental agency finds that approximately 12% of the people who reserve cars cancel their reservations at the last minute or simply do not show. Three hundred twenty people have reserved cars for the coming Memorial Day weekend. About how many people can actually be expected to show up? Find the standard deviation.

9. Refer back to Question 8. The agency manager plans to overbook so that 310 people will actually show up. How many reservations should the manager accept?

10. A random sample of 4 families is taken from a region in which it is known that 15 of the 25 families have members currently serving or have already served in the armed forces. What is the probability that none of the families selected have members currently serving or have already served in the armed forces?

11. *Cancer:* According to medical researchers, about 20% of individuals diagnosed as having one particular form of cancer die within a year. If 12 individuals are diagnosed as having this form of cancer, find the probability that *at most* 3 of them will die within a year.

ONE FORM OF CANCER UNDER CONTROL

Los Angeles (May 6)—Cancer researchers announced yesterday that they had developed a treatment for one form of deadly cancer. Currently, the rate of success with this new treatment is 80%. With further improvement, this rate of success is expected to rise substantially.

May 6, 1988

If 20 people who are suffering from the particular form of cancer mentioned are randomly selected and are subjected to this new treatment, what is the probability that 8 of them will recover? Can we determine such probabilities?

Chapter Test

1. The Brookview Day Care Center consists of five female and three male teachers. The district supervisor decides to visit two teachers' classes. The supervisor writes each teacher's name on a separate piece of paper and places the papers in a hat. Two papers are then selected, with the first one replaced before the second is selected. Let x be the number of females selected. Find the probability distribution of x.

2. The number of commuters x, arriving per minute at a turnstile behaves much like a Poisson random variable whose mean is 6. What is the probability that in a particular minute

   a. at most four commuters will arrive?

   b. at least two commuters will arrive?

3. Doctors estimate that the probability of a pregnant woman having identical twins is approximately 0.004. If ten random births at Beth Israel Medical Center are selected, what is the probability that at least one of the births in the sample was an identical twin?

4. Three cards are selected, one at a time, from an ordinary deck of 52 cards. Let x represent the number of picture (face) cards obtained. Find the probability distribution of x if the selection is made

   a. with replacement.

   b. without replacemnt.

5. A pediatrician finds that about 15% of the infants now injected with the MMR (measles, mumps, and rubella) vaccine, as required by law in many states, develop some form of minor reaction. If the doctor intends to inject 400 infants during the year with this vaccine, about how many can be expected *not* to develop any form of reaction? Find the standard deviation.

6. An automobile manufacturer finds that 12% of all cars manufactured by the company during the year have defective electronic ignition systems. If 12 of the recalled cars are carefully analyzed, what is the probability that more than 5 of them have defective electronic ignition systems?

7. Consider the following probability distribution.

x	1	2	3	4	5	6
$p(x)$	0.001	0.028	0.134	0.312	0.356	0.169

   a. Find μ.

   b. Find σ.

   c. Find the probability that x will fall within the interval $\mu \pm 2\sigma$?

8. A committe of five people is to be selected from among seven male bank managers and eight female bank managers to study the problem of bad real estate loans. What is the probability that the committee will

   a. consist of only women?

   b. consist of only men?

   c. consist of more women than men?

9. Environmentalists estimate that two out of every five fish caught in a certain river are contaminated with cancer causing PCB's. If ten fish caught in this river are randomly selected, what is the probability that at most two of them are contaminated with the cancer causing PCB?

10. It is known that of 45 retired people in an auditorium 11 are members of the American Association of Retired People (AARP). If a sample of 8 retired people

in the auditorium is randomly selected (without replacement), what is the probability that the sample contains at most 3 people who are members of the AARP?

11. Consider the following probability distribution. Find the probability that $x = 12$.

x	5	8	12	17	28	34	41
$p(x)$	0.051	0.155	?	0.241	0.167	0.103	0.045

12. A farmer plants 7 trees. If the probability that any one tree sprouts is $\frac{1}{3}$, find the probability that exactly 4 of the trees will sprout.

   a. $\dfrac{7!}{4!}\left(\dfrac{1}{3}\right)^4\left(\dfrac{2}{3}\right)^3$ b. $\dfrac{7!}{3!}\left(\dfrac{1}{3}\right)^4\left(\dfrac{2}{3}\right)^4$

   c. $\dfrac{7!}{4!\,3!}\left(\dfrac{1}{3}\right)^4\left(\dfrac{2}{3}\right)^3$ d. $\dfrac{7!}{4!\,3!}\left(\dfrac{2}{3}\right)^4\left(\dfrac{1}{3}\right)^3$ e. none of these

13. A car rental agency has reserved 420 cars for this coming weekend's reservations. If the probability of a cancellation is 0.10, about how many cars will the agency rent for the coming weekend?
   a. 42   b. 420   c. 90   d. 378   e. none of these

14. Refer back to the previous question. Find the standard deviation.
   a. 0.3   b. 0.9   c. 6.148   d. 37.8   e. none of these

15. Consider the following probability distribution. Find its mean.

x	0	1	2	3	4
$p(x)$	0.12	0.16	0.39	0.19	0.14

   a. 2.19   b. 2.07   c. 2.50   d. 2.53   e. none of these

16. Refer back to the previous question. Find the variance.
   a. 5.67   b. 1.1769   c. 4.2849   d. 1.3851   e. none of these

Suggested Further Reading

1. Albright, S. Christian. *Statistics for Business and Economics*. New York: Macmillan Publishing Co., 1987.
2. "Computers and Computation." Readings from *Scientific American*, Article 16. San Francisco, CA: Freeman, 1971.
3. Daniel, Wayne W. and James C. Terrell. *Business Statistics*. 4th ed. Boston: Houghton Mifflin Company, 1986.
4. Hamburg, Morris. *Statistical Analysis for Decision Making*, 4th ed. Orlando, FL: Harcourt Brace Jovanovich, Inc., 1987.

5. Hoel, Paul G., Sidney C. Port, and Charles J. Stone. *Introduction to Probability Theory.* Boston: Houghton Mifflin, 1972.

6. Mendenhall, William. *Introduction to Probability and Statistics*, 7th ed. Boston: Duxbury Press, 1987. Chapters 5 and 6.

7. Mosteller, Frederick, R. E. K. Rourke, and G. B. Thomas, Jr. *Probability with Statistical Applications*, 2nd ed. Reading, MA: Addison-Wesley, 1970.

8. Ryan, Barbara F., Brian L. Joiner, and Thomas A. Ryan, Jr. *Minitab Handbook*, 2nd ed. Boston: Duxbury Press, 1985.

7 The Normal Distribution

Is the amount of pollution emitted by cars as they wait for the signal to change at a busy intersection normally distributed? See discussion on p. 376. (© *Bror Karlsson, Fran Heyl Associates*)

In this chapter we introduce what is considered to be the most important continuous probability distribution. There are many reasons why great emphasis is placed on the normal distribution. As indicated in Chapter 2, a number of physical characteristics have probability distributions that are very similar to the bell-shaped normal curve. Hence, that accounts for interest in it. Also, as we will see in subsequent chapters, the normal distribution is used extensively in statistical inference. Thus, in this chapter, we will analyze the normal distribution in great detail.

CHAPTER OBJECTIVES

After studying the material in this chapter, you should be able:

- **To discuss** in detail a probability distribution that is bell-shaped or mound-shaped, called the normal distribution. (Section 7.2)

- **To indicate** how the normal distribution can be used to calculate probabilities. (Section 7.3)

- **To apply** the normal distribution to many different situations. This will be accomplished by converting to *z*-scores or standard scores and using the standard normal distribution. (Section 7.4)

- **To show** how the normal distribution can be used to simplify lengthy computations involving the binomial distribution. (Section 7.5)

- **To point out** how statistical quality-control charts are used in industry. This represents an additional application of the normal distribution. (Section 7.6)

- **To mention** briefly some historical facts about some mathematicians who used the normal distribution. (Throughout Chapter)

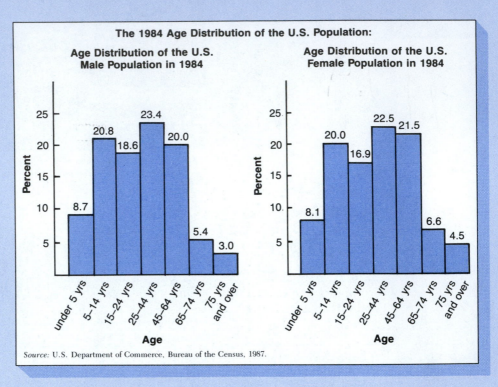

The 1984 Age Distribution of the U.S. Population:

Age Distribution of the U.S. Male Population in 1984

Age Distribution of the U.S. Female Population in 1984

Source: U.S. Department of Commerce, Bureau of the Census, 1987.

The above graph gives the age distribution of the U.S. population for both males and females. Contrary to expectation, the ages for both sexes are not normally distributed. Many people believe that in order to achieve zero population growth, the age distribution should look like the following graph, where we have drawn the bars side by side horizontally for ease in comparison.

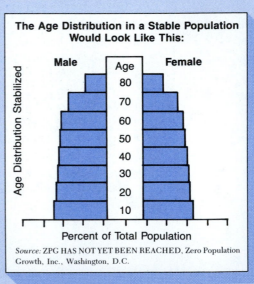

The Age Distribution in a Stable Population Would Look Like This:

Source: ZPG HAS NOT YET BEEN REACHED, Zero Population Growth, Inc., Washington, D.C.

7.1 INTRODUCTION

Until now the random variables that we discussed in detail assumed only the limited values of 0, 1, 2, ... Thus, when a coin is flipped many times the number of heads that comes up is 0, 1, 2, 3, ... Also, the number of defective bulbs in a shipment of 100 bulbs is 0, 1, 2, 3, . . . , 100. Since these random variables can assume only the values 0, 1, 2, . . . , they are called **discrete random variables.**

Discrete Random Variable

As opposed to the preceding examples, consider the following random variables: the length of a page of this book, the height of your pet dog, the temperature at noon on New Year's Day, or the weight of a bag of sugar on your grocer's shelf. Since each of these variables can assume an infinite number of values on a measuring scale, they are called **continuous random variables.** Thus, the weight of a bag of sugar can be 5 pounds, 5.1 pounds, 5.161 pounds, 5.16158 pounds, 5.161581 pounds, and so forth, depending on the accuracy of the scale. Similarly, the temperature at noon on New Year's Day may be 38°, 38.2°, 38.216°, and so forth.

Continuous Random Variables

Among the many different kinds of distributions of random variables that are used by statisticians, the normal distribution is by far the most important. This type of distribution was first discovered by the English mathematician Abraham De Moivre (1667–1754). De Moivre spent many hours with London gamblers. In his *Annuities upon Lives*, which played an important role in the history of actuarial mathematics, and his *Doctrine of Chances*, which is a manual for gamblers, he essentially developed the first treatment of the normal probability curve, which is important in the study of statistics. De Moivre also developed a formula, known as Stirling's formula, that is used for approximating factorials of large numbers.

A rather interesting story is told of De Moivre's death. According to the story, De Moivre was ill and each day he noticed that he slept a quarter of an hour longer than on the preceding day. Using progressions, he computed that he would die in his sleep on the day after he slept 23 hours and 45 minutes. On the day following a sleep of 24 hours De Moivre died.

Many years later the French mathematician Pierre-Simon Laplace (1749–1827) applied the normal distribution to astronomy and other practical problems. The normal distribution was also used extensively by the German mathematician Carl Friedrich Gauss (1777–1855) in his studies of physics and astronomy. Gauss is considered by many as the greatest mathematician of the nineteenth century. At the age of 3 he is alleged to have detected an error in his father's bookkeeping records.

Bell-shaped Distribution

Gaussian Distribution

The normal distribution is sometimes known as the **bell-shaped** or **Gaussian distribution** in honor of Gauss, who studied it extensively.

Although there are other distributions of continuous random variables that are important in statistics, the normal distribution is by far the most important. In this chapter we discuss in detail the nature of the normal distribution, its properties, and its uses.

7.2 THE GENERAL NORMAL CURVE

Refer back to the frequency polygon given on page 49 of Chapter 2. It is reproduced below. Experience has taught us that for many frequency distributions drawn from large populations, the frequency polygons approximate what is known as a **normal** or **bell-shaped curve** as shown in Figure 7.1.

Heights and weights of people, IQ scores, waist sizes, or even life expectancy of cars, to name but a few, are all examples of distributions whose frequency polygons approach a normal curve when the samples taken are from large populations. When the graph of a frequency distribution resembles the bell-shaped curve shown in Figure 7.2 the graph is called a **normal curve** and its frequency distribution is known as a **normal distribution.** The word normal is simply a name for this particular distribution. It does not indicate that this distribution is more typical than any other.

Since the normal distribution has wide-ranging applications, we need a careful description of a normal curve and some of its properties.

As mentioned before, the graph of a normal distribution is a bell-shaped curve. It extends in both directions. Although the curve gets closer and closer to the horizontal axis, it never really crosses it, no matter how far it is extended. The normal distribution is a probability distribution satisfying the following properties:

1. The mean is at the center of the distribution and the curve is symmetrical about the mean. This tells us that we can fold the curve along the dotted line shown in Figure 7.3 and either portion of the curve will correspond with the other portion.

Normal Curve
Normal Distribution

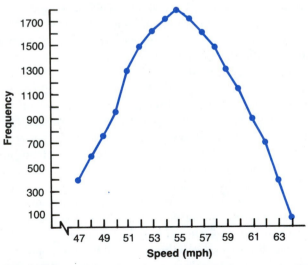

Figure 7.1
Distribution of the speeds of cars as they passed through a radar trap.

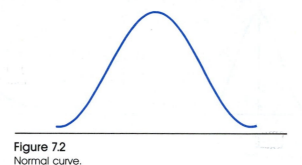

Figure 7.2
Normal curve.

2. Its graph is the bell-shaped curve, most often referred to as the normal curve.
3. The mean equals the median.
4. The scores that make up the normal distribution tend to cluster around the middle, with very few values more than 3 standard deviations away from the mean on either side.

Normal distributions can come in different sizes and shapes. Some are tall and skinny or flat and spread out as shown in Figure 7.4. However, for a given mean and a given standard deviation, there is one and only one normal distribution. The normal distribution is completely specified once we know its mean and standard deviation.

We mentioned in Chapter 2 (see page 29) that the area under a particular rectangle of a histogram gives us the relative frequency and hence the probability of obtaining values within that rectangle. We can generalize this idea to any distribution. We say that *the area under the curve between any two points a and b gives us the probability that the random variable having this particular continuous distribution will assume values between a and b*. This idea is very important since calculating probabilities for the normal distribution will depend on the areas under

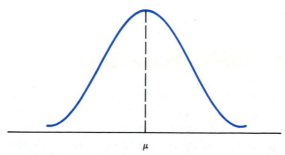

μ

Figure 7.3

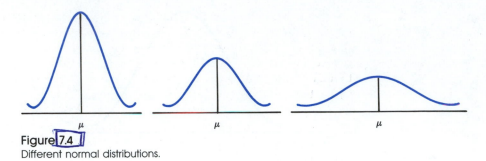

Figure 7.4
Different normal distributions.

the curve. Also, since the sum of the probabilities of a random variable assuming all possible values must be 1 (see page 294), the total area under its probability curve must also be 1.

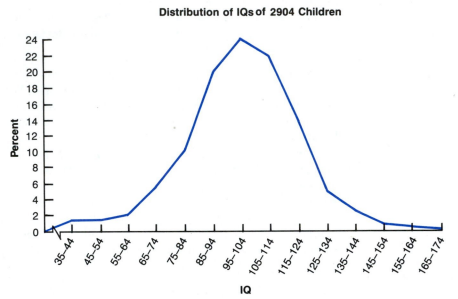

Distribution of IQs of 2904 Children

The graph gives the distribution of IQ scores of 2904 children. Notice the type of frequency distribution that is illustrated. This is approximately a normal distribution.

7.3 THE STANDARD NORMAL CURVE

A normal distribution is completely specified by its mean and standard deviation. Thus, although all normal distributions are basically mound-shaped, different means and different standard deviations will describe different bell-shaped curves. However, it is possible to convert each of these different normal distributions into one standardized form. You may be wondering, why bother? The answer is rather simple.

Since areas under a normal curve are related to probability, we can use special normal distribution tables for calculating probabilities. Such a table is given in the Appendix at the end of this book. However, because the mean and standard deviation can be any values, it would seem that we need an endless number of tables. Fortunately, this is not the case. We only need one standardized table. Thus, the area under the curve between 40 and 60 of a normal distribution with a mean of 50 and a standard deviation of 10 will be the same as the area between 70 and 80 of another normally distributed random variable with mean 75 and standard deviation 5. They are both within 1 standard deviation unit from the mean. It is for this reason that statisticians use a standard normal distribution.

DEFINITION 7.1
Standardized
Normal Distribution

A **standardized normal distribution** is a normal distribution with a mean of 0 and a standard deviation of 1. The curve of a typical standard distribution is shown in Figure 7.5.

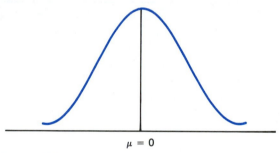

$\mu = 0$

Figure 7.5
Curve of a typical standard distribution.

Table IV in the Appendix gives us the areas of a standard normal distribution between $z = 0$ and $z = 3.90$. We read this table as follows: The first two digits of the z-score are under the column headed by z, the third digit heads the other columns. Thus, to find the area from $z = 0$ to $z = 2.43$, we first look under z to 2.4 and then read across from $z = 2.4$ to the column headed by 0.03. The area is 0.4925, or 49.25%.

Similarly, to find the area from $z = 0$ to $z = 1.69$, we first look under $z = 1.6$ and then read across from $z = 1.6$ to the column headed by 0.09. The area is 0.4545.

EXAMPLE 1

Find the area between $z = 0$ and $z = 1$ in a standard normal curve.

Solution

We first draw a sketch as shown in Figure 7.6. Then using Table IV for $z = 1.00$, we find that the area between $z = 0$ and $z = 1$ is 0.3413. This means

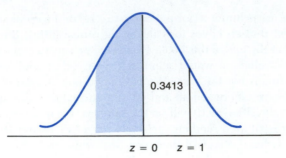

$z = 0$ $z = 1$

Figure 7.6

that the probability of a score with this normal distribution falling between $z = 0$ and $z = 1$ is 0.3413.

EXAMPLE 2   Find the area between $z = -1.15$ and $z = 0$ in a standard normal curve.

Solution

We first draw a sketch as shown in Figure 7.7. Then using Table IV, we look up the area between $z = 0$ and $z = 1.15$. The area is 0.3749, not -0.3749. A negative value of z just indicates that the value is to the left of the mean. The area under the curve (and the resulting probability) is *always* a positive number. Thus, the probability of getting a z-score between 0 and -1.15 is 0.3749.

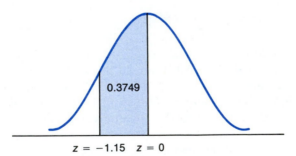

$z = -1.15$ $z = 0$

Figure 7.7

EXAMPLE 3   Find the area between $z = -1.63$ and $z = 2.22$ in a standard normal curve.

Solution

We draw the sketch shown in Figure 7.8. Since Table IV gives the area only from $z = 0$ on, we first look under the normal curve from $z = 0$ to $z = 1.63$. We get 0.4484. Then we look up the area between $z = 0$ and $z = 2.22$. We get 0.4868.

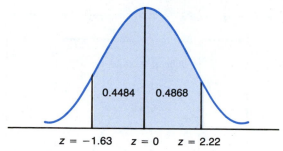

Figure 7.8

Finally, we add these two together and get

$$0.4484 + 0.4868 = 0.9352$$

Thus, the probability that a z-score is between $z = -1.63$ and $z = 2.22$ is 0.9352.

By following a procedure similar to that used in Example 3, you should verify the following.

Properties of Standard Normal Distribution

1. The probability that a z-score falls within 1 standard deviation of the mean on either side, that is, between $z = -1$ and $z = 1$, is approximately 68%.

2. The probability that a z-score falls within 2 standard deviations of the mean, that is, between $z = -2$ and $z = 2$, is approximately 95%.

3. The probability that a z-score falls within 3 standard deviations of the mean is approximately 99.7%.

Thus, approximately 99.7% of z-scores fall within $z = -3$ and $z = 3$ (see Figure 7.9).

It should be noted that the preceding three statements are true because we are dealing with a normal distribution. A more general result, which is true for any set of measurements—population or sample—and regardless of the shape of the frequency distribution is known as **Chebyshev's Theorem**, which we discussed in Chapter 4. We restate it here.

Chebyshev's Theorem

CHEBYSHEV'S THEOREM

Let k be any number equal to or greater than 1. Then the proportion of any distribution that lies within k standard deviations of the mean is at least $1 - \dfrac{1}{k^2}$:

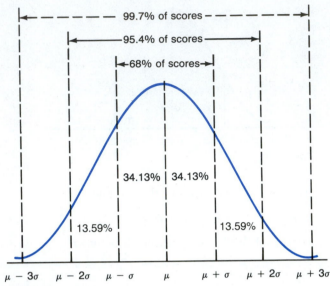

Figure 7.9
The normal distribution.

Although Chebyshev's Theorem is more general in scope, if we know that we have a normal distribution, we use the results given above. Since in this chapter our interest is in the normal distribution, we will not use Chebyshev's Theorem.

In many cases we have to find areas between two given values of z or areas to the right or left of some value of z. Finding these areas is an easy task provided we remember that the area under the entire normal distribution is 1. Thus, since the normal distribution is symmetrical about $z = 0$, we conclude that the area to the right of $z = 0$ and the area to the left of $z = 0$ are both equal to 0.5000.

EXAMPLE 4   Find the area between $z = 0.87$ and $z = 2.57$ in a standard normal distribution (see Figure 7.10).

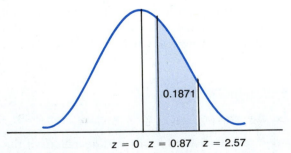

Figure 7.10

Solution

We cannot look this up directly since the chart starts at 0, not at 0.87. However, we can look up the area between $z = 0$ and $z = 2.57$ and get 0.4949, then look up the area between $z = 0$ and $z = 0.87$ and get 0.3078. Next we take the difference between the two and get

$$0.4949 - 0.3078 = 0.1871$$

EXAMPLE 5   Find the probability of getting a z-value less than 0.43 in a standard normal distribution.

Solution

The probability of getting a z-value less than 0.43 really means the area under the curve to the left of $z = 0.43$. This represents the shaded portion of Figure 7.11. We look up the area from $z = 0$ to $z = 0.43$ and get 0.1664 and add this to 0.5000 to get

$$0.5000 + 0.1664 = 0.6664$$

Thus, the probability of getting a z-value less than 0.43 is 0.6664.

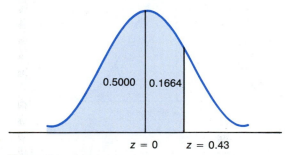

0.5000   0.1664

$z = 0$ $z = 0.43$

Figure 7.11

EXAMPLE 6   Find the probability of getting a z-value in a standard normal distribution that is

a. greater than -2.47
b. greater than 1.82
c. less than -1.53

Solution

a. Using Table IV, we first find the area between $z = 0$ and $z = 2.47$ (see Figure 7.12). We get 0.4932. Then we add this to 0.5000, which is the area to the right of $z = 0$, and get

$$0.4932 + 0.5000 = 0.9932$$

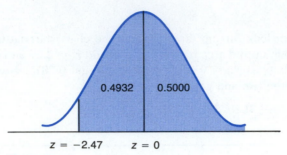

z = −2.47   z = 0

Figure 7.12

b. Here we are interested in finding the area to the right of $z = 1.82$. See Figure 7.13. We find the area from $z = 0$ to $z = 1.82$. It is 0.4656. Since we are interested in the area to the right of $z = 1.82$, we must *subtract* 0.4656 from 0.5000, which represents the *entire* area to the right of $z = 0$. We get

$$0.5000 - 0.4656 = 0.0344$$

Thus, the probability that a z-score is greater than $z = 1.82$ is 0.0344.

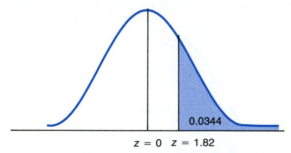

z = 0   z = 1.82

Figure 7.13

c. Here we are interested in the area to the left of $z = -1.53$. See Figure 7.14. We calculate the area between $z = 0$ and $z = -1.53$. It is 0.4370. We subtract

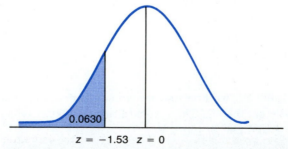

z = −1.53   z = 0

Figure 7.14

this from 0.5000. Our result is

0.5000 − 0.4370 = 0.0630

Thus, the probability that a z-score is less than $z = -1.53$ is 0.0630.

In the preceding examples we interpreted the area under the normal curve as a probability. If we know the probability of an event, we can look at the probability chart and find the z-value that corresponds to this probability.

EXAMPLE 7

If the probability of getting less than a certain z-value is 0.1190, what is the z-value?

Solution

We first draw the sketch shown in Figure 7.15. Since the probability is 0.1190, which is less than 0.5000, we know that the z-value must be to the left of the mean. We subtract 0.1190 from 0.5000 and get

0.5000 − 0.1190 = 0.3810

See Figure 7.15. This means that the area between $z = 0$ and some z-value is 0.3810. Table IV tells us that the z-value is 1.18. However, this is to the left of the mean. Therefore, our answer is $z = -1.18$.

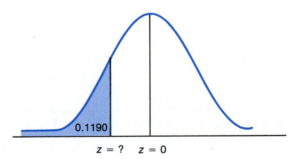

z = ? z = 0

Figure 7.15

EXAMPLE 8

If the probability of getting larger than a certain z-value is 0.0129, what is the z-value?

Solution

In this case we are told that the area to the right of some z-value is 0.0129. This z-value must be on the right side. If it were on the left side, the area would have to be at least 0.5000. Why? See Figure 7.16. Thus, we subtract 0.0129 from 0.5000

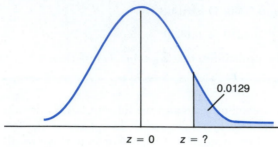

Figure 7.16

and get 0.4871. Then we look up this probability in Table IV and find the z-value that gives this probability. It is $z = 2.23$. If $z = 2.23$, the probability of getting a z-value greater than 2.23 is 0.0129.

The General Normal Distribution

Converting to Standard Scores

If we are given a normal distribution with a mean different from 0 and a standard deviation different from 1, we can **convert** this normal distribution into a standardized normal distribution by converting each of its scores into standard scores. To accomplish this we use Formula 3.8 of Chapter 3, which we will now call Formula 7.1:

FORMULA 7.1

$$z = \frac{x - \mu}{\sigma}$$

Expressing the scores of a normal distribution as standard scores allows us to calculate different probabilities, as the following examples show.

EXAMPLE 9

In a normal distribution $\mu = 25$ and $\sigma = 5$. What is the probability of obtaining a value

a. greater than 30?
b. less than 15?

Solution

a. We use Formula 7.1. We have $\mu = 25$, $x = 30$, and $\sigma = 5$ so that

$$z = \frac{x - \mu}{\sigma} = \frac{30 - 25}{5} = \frac{5}{5} = 1$$

See Figure 7.17. Thus, we are really interested in the area to the right of

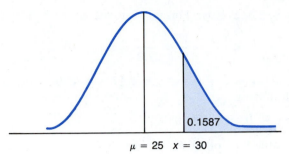

Figure 7.17

$z = 1$ of a standardized normal curve. The area from $z = 0$ to $z = 1$ is 0.3413. The area to the right of $z = 1$ is then

$$0.5000 - 0.3413 = 0.1587$$

Therefore, the probability of obtaining a value greater than 30 is 0.1587.

b. We use Formula 7.1. We have $\mu = 25$, $x = 15$, and $\sigma = 5$ so that

$$z = \frac{x - \mu}{\sigma} = \frac{15 - 25}{5} = \frac{-10}{5} = -2$$

See Figure 7.18. Thus, we are interested in the area to the left of $z = -2$. The area from $z = 0$ to $z = -2$ is 0.4772. Thus, the area to the left of $z = -2$ is

$$0.5000 - 0.4772 = 0.0228$$

The probability of obtaining a value less than 15 is therefore 0.0228.

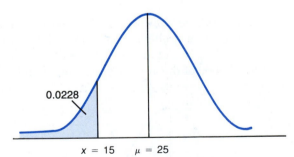

Figure 7.18

EXAMPLE 10

Find the percentile rank of 20 in a normal distribution with $\mu = 15$ and $\sigma = 2.3$.

Solution

The problem is to find the area to the left of 20 in a normal distribution with $\mu =$

15 and $\sigma = 2.3$. We use Formula 7.1 with $\mu = 15$, $x = 20$, and $\sigma = 2.3$ so that

$$z = \frac{x - \mu}{\sigma} = \frac{20 - 15}{2.3} = \frac{5}{2.3} = 2.17$$

The area between $z = 0$ and $z = 2.17$ is 0.4850. See Figure 7.19. Thus, the area to the left of 20 is

$$0.5000 + 0.4850 = 0.9850$$

The percentile rank of 20 is therefore 98.5.

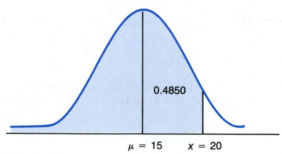

0.4850

$\mu = 15$ $x = 20$

Figure 7.19

EXAMPLE 11

In a certain club heights of members are normally distributed with $\mu = 63$ inches and $\sigma = 2$ inches. If Sam is in the 90th percentile, find his height.

Solution

Since Sam is in the 90th percentile, this means that 90% of the club members are shorter than he is. So, the problem here is to find a z-value that has 90 percent of the area to the left of z. Therefore, we look in the area portion of Table IV to find a z-value that has 0.4000 of the area to its left. See Figure 7.20. We use 0.4000, not

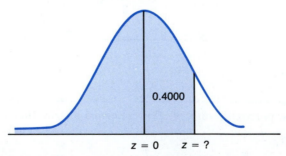

0.4000

$z = 0$ $z = ?$

Figure 7.20

0.9000, since 0.5000 of this is to the left of $z = 0$. The closest entry is 0.3997, which corresponds to $z = 1.28$. Now we convert this score into a raw score by using Formula 3.9, on page 156. We have

$$x = \mu + z\sigma$$
$$= 63 + 1.28(2)$$
$$= 63 + 2.56$$
$$= 65.56$$

Thus, Sam's height is approximately 65.56 inches.

EXAMPLE 12

Use the same information as Example 11 of this section except Bill's percentile rank is 40. Find his height.

Solution

Since Bill's percentile rank is 40, this means that 40% of the club members are shorter than he is. The problem here is to find a z-value that has 40 percent of the area to the left of z. See Figure 7.21. Since we are given the area to the left of z, we must subtract 0.4000 from 0.5000:

$$0.5000 - 0.4000 = 0.1000$$

Then we look in the area portion of Table IV to find a z-value that has 0.1000 as the area between $z = 0$ and that z-value. The closest entry is 0.0987, which corresponds to $z = 0.25$. Since this z-value is to the left of the mean, we have $z = -0.25$. When we convert this score into a raw score, we have

$$x = \mu + z\sigma$$
$$= 63 + (-0.25)2$$
$$= 63 - 0.50 = 62.50$$

Therefore, Bill's height is approximately 62.50 inches.

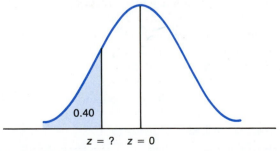

0.40

z = ?   z = 0

Figure 7.21

EXERCISES FOR SECTION 7.3

1. In a standard normal distribution, find the area that lies
   a. between $z = 0$ and $z = 1.47$ *0.4292*
   b. between $z = -0.56$ and $z = 0$
   c. to the right of $z = 1.93$
   d. to the left of $z = -0.78$
   e. to the right of $z = 2.31$
   f. between $z = -1.56$ and $z = 1.69$
   g. between $z = -1.82$ and $z = 2.56$
   h. between $z = -1.66$ and $z = -0.92$ *Subtract*

2. Find the percentage of z-scores in a standard normal distribution that are
   a. above $z = -2.46$ *add*
   b. below $z = 0.86$ *add*
   c. between $z = 1.24$ and $z = 2.09$
   d. above $z = 2.51$ *Subtract*
   e. above $z = 3.94$
   f. between $z = -1.42$ and $z = -0.93$
   g. between $z = -1.38$ and $z = 1.09$ *add the two*
   h. between $z = -2.91$ and $z = -1.82$ *take difference*

3. In a normal distribution find the z-score that cuts off the bottom
   a. 15 percent
   b. 20 percent
   c. 25 percent
   d. 30 percent

4. In a normal distribution, find the z-score(s) that cut(s) off the middle
   a. 20 percent
   b. 30 percent
   c. 40 percent

5. Find z if the area under a standard normal curve
   a. between $z = 0$ and z is 0.4957   *2.63*
   b. to the left of z is 0.9850   *2.17*
   c. to the left of z is 0.0158   *2.15*
   d. to the right of z is 0.0011   *3.05 — or 3.06*
   e. betwen $z = 1.5$ and z is 0.0617
   f. between $z = 2.03$ and z is 0.0163

6. In a normal distribution with $\mu = 80$ and $\sigma = 12$, find the percentage of scores that are
   a. between 82 and 88
   b. greater than 75
   c. less than 69

5.0000
-.4931
.0069

.9850
-.5000
.4850

.5000
-.0011
.4989

$$z = \frac{x - \mu}{\sigma}$$

find the z then find

d. between 71 and 85
e. between 67 and 74

7. In a normal distribution with $\mu = 47$ and $\sigma = 3$ find the percentile rank of
   a. a score of 40
   b. a score of 52
   c. a score of 39

8. In a certain school the bowling scores of the students in a physical education class are approximately normally distributed with a mean of 180 points and a standard deviation of 9 points. Find the score of
   a. Joyce, if she is in the 45th percentile
   b. Marilyn, if she is in the 79th percentile
   c. Frank, if he is in the 95th percentile

9. A normal distribution has mean 70 and unknown standard deviation σ. However, it is known that 5% of the area lies to left of 62. Find σ.

10. A normal distribution has unknown mean, μ, with a standard deviation of $\sigma = 15.3$. However, it is known that the probability that a score is more than 45 is 0.9772. Find μ.

11. A certain normal distribution has unknown mean, μ, and unknown standard deviation, σ. However, it is known that 15.39% of the scores are more than 12.19 and 3.44% of the scores are less than 9.63. Find μ and σ.

12. The following is a standard normal curve with various z-values marked on it. If this curve also represents a normal distribution with $\mu = 24$ and $\sigma = 5$, replace the z-values with raw scores.

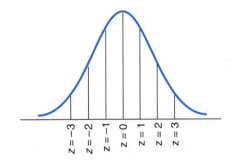

13. Determine the shaded area under the following standard normal curves.

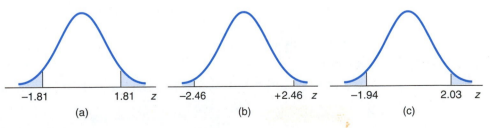

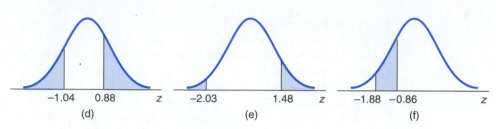

−1.04 0.88	−2.03 1.48	−1.88 −0.86
(d)	(e)	(f)

14. The following is a sketch of a normal curve whose mean μ is 10 and whose standard deviation σ is 2. Find the z-scores corresponding to all the x-values given and also find the areas of the indicated regions.

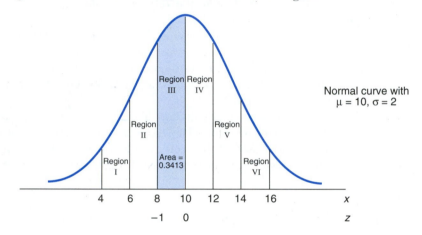

Normal curve with
$\mu = 10$, $\sigma = 2$

7.4 SOME APPLICATIONS OF THE NORMAL DISTRIBUTION

As mentioned earlier, the importance of the normal distribution lies in its wide-ranging applications. In this section we apply the normal distribution to some concrete examples.

EXAMPLE 1

From past experience it has been found that the weight of a newborn infant at a maternity hospital is normally distributed with a mean $7\frac{1}{2}$ pounds (which equals 120 ounces) and a standard deviation of 21 ounces. If a newborn baby is selected at random, what is the probability that the infant weighs less than 4 pounds 15 ounces (which equals 79 ounces)?

Solution

We use Formula 7.1 of Section 7.3. Here $\mu = 120$, $\sigma = 21$, and $x = 79$ so that

$$z = \frac{x - \mu}{\sigma} = \frac{79 - 120}{21} = \frac{-41}{21}, \text{ or } -1.95$$

Thus, we are interested in the area to the left of $z = -1.95$. The area from $z = 0$ to $z = -1.95$ is 0.4744 so that the area to the left of $z = -1.95$ is $0.5000 - 0.4744$, or 0.0256. See Figure 7.22. Therefore, the probability that a randomly selected baby weighs less than 4 pounds 15 ounces is 0.0256.

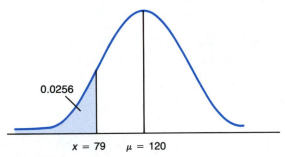

0.0256

$x = 79$ $\mu = 120$

Figure 7.22

EXAMPLE 2

The Flatt Tire Corporation claims that the useful life of its tires is normally distributed with a mean life of 28,000 miles and with a standard deviation of 4000 miles. What percentage of the tires are expected to last more than 35,000 miles?

Solution

Here $\mu = 28,000$, $\sigma = 4000$, and $x = 35,000$. Using Formula 7.1, we get

$$z = \frac{x - \mu}{\sigma} = \frac{35,000 - 28,000}{4000} = 1.75$$

See Figure 7.23. We are interested in the area to the right of $z = 1.75$. The area between $z = 0$ and $z = 1.75$ is 0.4599, so the area to the right of $z = 1.75$ is 0.0401. Thus, approximately 4% of the tires can be expected to last more than 35,000 miles.

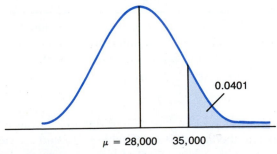

0.0401

$\mu = 28,000$ 35,000

Figure 7.23

EXAMPLE 3

In a recent study it was found that in one town the number of hours that a typical 10-year-old child watches television per week is normally distributed with a mean of 12 hours and a standard deviation of 1.5 hours. If Gary is a typical 10-year-old child in this town, what is the probability that he watches between 9 and 14 hours of television per week?

Solution

We first find the probability that Gary will watch television between 12 and 14 hours per week and add to this the probability that he will watch television between 9 and 12 hours per week. Using Formula 7.1, we have

$$z = \frac{x - \mu}{\sigma} = \frac{14 - 12}{1.5} = 1.33$$

The area between $z = 0$ and $z = 1.33$ is 0.4082. See Figure 7.24. Similarly,

$$z = \frac{x - \mu}{\sigma} = \frac{9 - 12}{1.5} = \frac{-3}{1.5} = -2$$

The area between $z = 0$ and $z = -2$ is 0.4772. See Figure 7.24. Adding these two probabilities, we get $0.4082 + 0.4772 = 0.8854$. Thus, the probability that Gary watches between 9 and 14 hours of television per week is 0.8854.

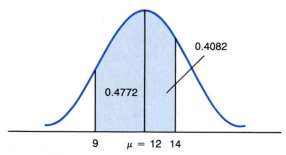

0.4082

0.4772

9      $\mu = 12$ 14

Figure 7.24

EXAMPLE 4

Daisy discovers that the amount of time it takes her to drive to work is normally distributed with a mean of 35 minutes and a standard deviation of 7 minutes. At what time should Daisy leave her home so that she has a 95% chance of arriving at work by 9 A.M.?

Solution

We first find a z-value that has 95% of the area to the left of z. See Figure 7.25. Thus, we look in the area portion of Table IV to find a z-value that has 0.4500 of the area to its left. (Remember 0.5000 of the area is to the left of $z = 0$.) From Table

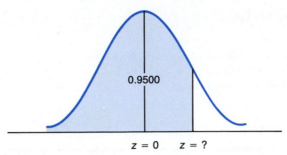

Figure 7.25

IV we find that z is midway between 1.64 and 1.65. We will use 1.645. We then find the raw score corresponding to $\mu = 35$, $\sigma = 7$, and $z = 1.645$. We have

$$x = \mu + z\sigma$$
$$= 35 + (1.645)7$$
$$= 35 + 11.515$$
$$= 46.515$$

So, if Daisy leaves her house 46.515 minutes before 9 A.M., she will arrive on time about 95% of the time. She should leave her home at 8:13 A.M.

EXAMPLE 5

In one study a major television manufacturing corporation found that the life of a typical color television tube is normally distributed with a standard deviation of 1.53 years. If 7% of these tubes last more than 6.9944 years, find the mean life of a television tube.

We are told that 7% of the tubes last more than 6.9944 years. See Figure 7.26. Thus, approximately 43% of the tubes last between μ and 6.9944 years. This means that on a standardized normal distribution the area between $z = 0$ and z is 0.43. Using

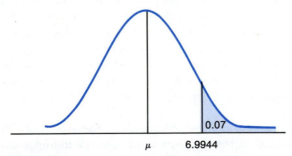

Figure 7.26

Table IV, we find that 0.43 of the area is between $z = 0$ and $z = 1.48$. Then

$$x = \mu + z\sigma$$
$$6.9944 = \mu + (1.48)(1.53)$$
$$6.9944 = \mu + 2.2644$$
$$6.9944 - 2.2644 = \mu$$
$$4.73 = \mu$$

Thus, the mean life of a television tube is 4.73 years.

$z = \dfrac{x - \mu}{\sigma}$

EXERCISES FOR SECTION 7.4

$\mu = 11 \, days$

$\sigma = 2.9 \, days$

$x = 10$

$x = 15$

1. The shelf life of a particular dairy product is approximately normally distributed with a mean of 11 days and a standard deviation of 2.9 days. If one such product is randomly selected, what is the probability that it will last between 10 and 15 days?

2. The amount of time needed to complete a mathematics placement test at Bobryk College is approximately normally distributed with a mean of 45 minutes and a standard deviation of 6 minutes. What is the probability that students will complete the exam within 55 minutes?

3. An auto mechanic finds that the useful life of a spark plug in a car is approximately normally distributed with a mean of 9 months and a standard deviation of 2.3 months. If a spark plug is randomly selected find the probability that it will last less than 6 months.

4. A traffic control engineer of the State Department of Transportation has determined that the waiting time for a commuter in the toll plaza at a certain bridge is approximately normally distributed with a mean of 4.3 minutes and a standard deviation of 1.8 minutes. If a commuter in the toll plaza is randomly selected, what is the probability that the commuter will have to wait
   a. less than 2.0 minutes?
   b. between 2.5 and 5.0 minutes?

5. The college entrance examination scores are approximately normally distributed with a mean of 520 and a standard deviation of 100 for graduates of McDonald High School. Find the probability that a randomly selected student graduate from McDonald High School scored lower than 450 on this exam.

6. The life of a washing machine produced by one major company is known to be normally distributed with a mean life of 5.3 years and a standard deviation of 1.93 years. Find the probability that a randomly selected machine will last more than 6 years.

7. The XYZ Corporation sells photocopying machines. The number of copies that a machine will give before requiring new toner is normally distributed with a mean of 22,000 copies and a standard deviation of 3,500 copies. Find the prob-

ability that a typical machine will give between 15,000 to 25,000 copies before new toner is needed.

8. Mortgage statistics collected by a bank indicate that the number of years that the average new homeowner will occupy the house before moving or selling is normally distributed with a mean of 6.3 years and a standard deviation of 2.31 years. If a homeowner is selected at random, what is the probability that the owner will sell the house after 3 years?

9. Professor Milford has found that the grades on a final exam are normally distributed with a mean of 62 and a standard deviation of 12.
   a. If the passing grade is 52, what percent of the class will fail?
   b. If Professor Milford wants only 80% of the class to pass, what should be the passing grade?
   c. If Professor Milford wants only 8% of the class to get A's, what grade must a student have in order to get an A?

10. A school is administering a scholarship exam. Assume that the length of time required to complete the test is normally distributed with a mean of 50 minutes and a standard deviation of 17 minutes. How much time must be allowed by the proctors if the school wishes to assure enough time for only 85% of the applicants to complete the exam?

11. It is known that IQ scores of workers in a large factory are normally distributed with a mean of 109 and a standard deviation of 11. Furthermore, it is also known that certain factory jobs that demand only a minimum IQ of 102 bore people who have IQ scores over 123. On the basis of IQ what percentage of the company's employees can be used for this particular job?

12. The life of a pressure cooker is normally distributed with a mean of 4.7 years and a standard deviation of 0.9 years. The manufacturer will replace any defective pressure cooker free of charge while under the guarantee. For how many years should the manufacturer guarantee the pressure cookers if not more than 5% of them are to be replaced?

13. In a certain doctor's office the waiting time is normally distributed with a mean of 25 minutes and a standard deviation of 5.3 minutes. Laura has just walked into the doctor's office. Find the probability that she will have to wait at least 18 minutes before seeing the doctor.

14. According to weather bureau records, the number of degree-days (fuel consumption units) for a large northeastern U.S. city during a typical winter season is normally distributed with a mean of 3900 and a standard deviation of 400. Find the probability that next winter will be a mild winter and that as a result at most 4000 degree-days will be recorded.

15. The height of a member of the Jorgan Basketball team is approximately normally distributed with a mean of 74 inches. It is also known that 15% of the members are more than 75 inches. Find the standard deviation.

16. The time it takes Mary to drive to the doctor's office is approximately normally distributed with a mean of 25 minutes and a standard deviation of 6 minutes.

At what time should Mary leave her home so that she has a 95% probability of arriving on time for her 8:00 A.M. appointment?

17. A vending machine is supposed to fill cups with 7 ounces of soda. However, the machine is not functioning properly and can be adjusted according to the vendor's specifications. It is known that the filling process is approximately normally distributed with a standard deviation of 0.3 ounces. At what level should the mean be set so that

  a. only 5% of the filled cups will contain less than 7 ounces?

  b. at most 5% of the filled cups will contain more than 7.5 ounces?

18. Refer back to Exercise 17. The machine has been adjusted so that the amount of soda dispensed is normally distributed with a standard deviation of 0.21 ounces. If 10% of the cups filled with soda by the machine contain more than 7.44 ounces, what is the new mean amount of soda dispensed by the machine?

7.5 THE NORMAL CURVE APPROXIMATION TO THE BINOMIAL DISTRIBUTION

An important application of the normal distribution is the approximation of the binomial distribution. (The binomial distribution, which we discussed in Chapter 6, is often used in sampling without replacement situations if the sample size is small in comparison to the population size.) To see why such an approximation is needed, suppose we wish to determine the probability of getting at least eight heads when a coin is flipped ten times. This is a binomial distribution problem where $n = 10$, $p = \dfrac{1}{2}$, $q = \dfrac{1}{2}$, and $x = 8, 9,$ or 10. We can use Formula 6.4 of Chapter 6 (page 317) to determine these probabilities. We get

$$p(8 \text{ heads}) = \frac{10!}{8!\ 2!} \left(\frac{1}{2}\right)^8 \left(\frac{1}{2}\right)^2 = 0.0439$$

$$p(9 \text{ heads}) = \frac{10!}{9!\ 1!} \left(\frac{1}{2}\right)^9 \left(\frac{1}{2}\right)^1 = 0.0098$$

$$p(10 \text{ heads}) = \frac{10!}{10!\ 0!} \left(\frac{1}{2}\right)^{10} \left(\frac{1}{2}\right)^0 = 0.0010$$

Adding these probabilities gives

$$
\begin{aligned}
p(\text{at least 8 heads}) &= p(8 \text{ heads}) + p(9 \text{ heads}) + p(10 \text{ heads}) \\
&= 0.0439 \quad\; + 0.0098 \quad\; + 0.0010 \\
&= 0.0547
\end{aligned}
$$

Although evaluating the probabilities in this problem is not difficult, the calculations involved are time-consuming. It turns out that the normal distribution can be used as a fairly good approximation to the binomial distribution. To accomplish this, let us first reexamine the previous problem. Since x is a variable that has a binomial distribution, we can use the binomial distribution formula to compute the following probabilities.

Number of Heads, x	Probability, $p(x)$
0	0.0010
1	0.0098
2	0.0439
3	0.1172
4	0.2051
5	0.2461
6	0.2051
7	0.1172
8	0.0439
9	0.0098
10	0.0010

The histogram for this probability distribution is shown in Figure 7.27 where the height of each rectangle represents the probability that x will assume a particular value.

Recall that for a binomial distribution the mean $\mu = np$ and the standard deviation $\sigma = \sqrt{npq}$. These formulas were given in Section 6.6 of Chapter 6. Applying them to our example gives

$$\mu = np = 10\left(\frac{1}{2}\right) = 5$$

and

$$\sigma = \sqrt{npq} = \sqrt{10\left(\frac{1}{2}\right)\left(\frac{1}{2}\right)} = \sqrt{2.5} \approx 1.58$$

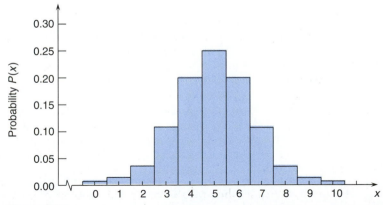

Figure 7.27

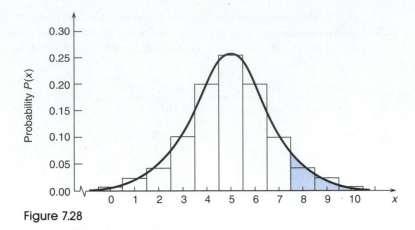

Figure 7.28

If we analyze the histogram given in Figure 7.27, we note that it is bell-shaped. In Figure 7.28 we have superimposed a normal curve whose mean is 5 and standard deviation is 1.58.

In the previous example we were interested in the probability that $x = 8$, $x = 9$, or $x = 10$. This represents the cross-hatched area in Figure 7.28. Note that this region can be approximated by the area under the normal curve between $x = 7.5$ and $x = 10.5$. This represents the shaded portion of the diagram. The 0.5 that we subtracted from 8 and the 0.5 that we added to 10 are referred to as **correction factors for continuity**. Thus, we can approximate our binomial distribution by a normal curve with $\mu = 5$ and $\sigma = 1.58$. We get this approximation by calculating the area under the normal curve between $x = 7.5$ and $x = 10.5$. See Figure 7.29.

Correction Factors for Continuity

When using the normal curve as an approximation for the binomial, we must calculate probabilities using an extra 0.5 either added to or subtracted from the number as a correction factor.

A rigorous mathematical justification as to why we add or subtract 0.5 is beyond the scope of this text. (Suffice it to say that we are representing an integer

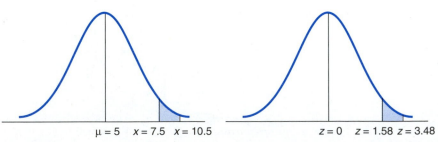

Figure 7.29

value in the binomial distribution by those values of a normal distribution that round off to the integer.)

Returning to our example, we have

$$z = \frac{x - \mu}{\sigma} = \frac{7.5 - 5}{1.58} = 1.58$$

and

$$z = \frac{x - \mu}{\sigma} = \frac{10.5 - 5}{1.58} = 3.48$$

The area between $z = 0$ and $z = 1.58$ is 0.4429 and the area between $z = 0$ and $z = 3.48$ is 0.4997 so that the area between $z = 1.58$ and $z = 3.48$ is $0.4997 - 0.4429 = 0.0568$.

Using the binomial distribution, we find that the probability of getting at least eight heads is 0.0547. Using the normal curve approximation to the binomial distribution, we find the probability of getting at least eight heads is 0.0568. Although the answers differ slightly, the answer we get by using the normal curve approximation is accurate enough for most applied problems. Furthermore, it is considerably easier to calculate.

More generally, if we were interested in the probability of getting 13 heads in 20 tosses of a coin, we can approximate this by calculating the area between $x = 12.5$ and $x = 13.5$.

COMMENT Any time you approximate a binomial probability with the normal distribution, depending on the situation make sure to add or subtract 0.5 from the number.

The normal curve approximation to the binomial distribution is especially helpful when we must calculate the probability of many different values. The following examples illustrate its usefulness.

EXAMPLE 1

Melissa is a nurse at Maternity Hospital. From past experience she determines that the probability that a newborn child is a boy is $\frac{1}{2}$. (In the United States today the probability that a newborn child is a boy is approximately 0.53, not 0.50.) What is the probability that among 100 newborn babies there are at least 60 boys?

Solution

We can determine the probability *exactly* by using the binomial distribution or we can get an *approximation* by using the normal curve approximation. To determine the answer exactly, we say that a newborn child is either a boy or a girl with equal probability. Thus, $p = \dfrac{1}{2}$ and $q = \dfrac{1}{2}$. The probability that there are at least 60 boys means that we must calculate the probability of having 60 boys, 61 boys, . . . , and, finally, the probability of having 100 boys. Using the binomial distribution formula, we have

$$p(\text{at least 60 boys}) = p(60 \text{ boys}) + p(61 \text{ boys}) + \cdots + p(100 \text{ boys})$$

$$= \frac{100!}{60! \, (40)!} \left(\frac{1}{2}\right)^{60} \left(\frac{1}{2}\right)^{40} + \frac{100!}{61! \, 39!} \left(\frac{1}{2}\right)^{61} \left(\frac{1}{2}\right)^{39}$$

$$+ \cdots + \frac{100!}{100! \, 0!} \left(\frac{1}{2}\right)^{100} \left(\frac{1}{2}\right)^{0}$$

Calculating these probabilities requires some lengthy computations. However, the same answer can be closely approximated, and more quickly, by using the normal curve approximation. We first determine the mean and standard deviation.

$$\mu = np = 100 \left(\frac{1}{2}\right) = 50$$

$$\sigma = \sqrt{npq} = \sqrt{100 \left(\frac{1}{2}\right) \left(\frac{1}{2}\right)} = \sqrt{25} = 5$$

Then we find the area to the right of 59.5 as shown in Figure 7.30. We use 59.5 rather than 60.5 in our approximation since we want to include exactly 60 boys in our calculations. If the problem had specified more than 60 boys, we would have used 60.5 rather than 59.5 since more than 60 means do not include 60.

In our case we have

$$z = \frac{x - \mu}{\sigma} = \frac{59.5 - 50}{5} = \frac{9.5}{5} = 1.9$$

From Table IV in the Appendix we find that the area to the right of $z = 1.9$ is $0.5000 - 0.4713$, or 0.0287.

Thus, the probability that among 100 newborn children there are at least 60 boys is 0.0287.

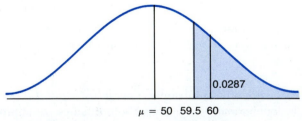

0.0287

$\mu = 50$ 59.5 60

Figure 7.30

EXAMPLE 2

Refer back to Example 1 of this section. Find the probability that there will be between 45 and 60 boys (not including these numbers) among 100 newborn babies at Maternity Hospital.

Solution

The probability is approximated by the area under the normal curve between 45.5 and 59.5. We do not use 44.5 or 60.5 since 45 and 60 are not to be included. To find the area between 45.5 and 50, we have

$$z = \frac{x - \mu}{\sigma} = \frac{45.5 - 50}{5} = -0.90$$

The area is thus 0.3159.

Also, to find the area between 50 and 59.5, we have

$$z = \frac{x - \mu}{\sigma} = \frac{59.5 - 50}{5} = 1.9$$

This area is 0.4713.

Adding these two areas, we get

$$0.3159 + 0.4713 = 0.7872$$

Therefore, the probability that there are between 45 and 60 boys among 100 newborn babies at Maternity Hospital is 0.7872.

EXAMPLE 3

A large television network is considering canceling its weekly 7:30 P.M. comedy show because of a decrease in the show's viewing audience. The network decides to phone 5000 randomly selected viewers and to cancel the show if fewer than 1900 viewers are actually watching the show. What is the probability that the show will be canceled if

a. only 40% of all television viewers actually watch the comedy show?
b. only 39% of all television viewers actually watch the comedy show?

Solution

a. Since a randomly selected television viewer that is phoned either watches the show or does not watch the show, we can consider this as a binomial distribution with $n = 5000$ and $p = 0.40$. We first calculate μ and σ:

$$\mu = 5000(0.40) = 2000$$

$$\sigma = \sqrt{5000(0.40)(0.60)} = \sqrt{1200} \approx 34.64$$

Since the show will be canceled only if fewer than 1900 people watch it, we are interested in the probability of having 0, 1, 2, . . . , 1899 viewers. Using a normal curve approximation, we calculate the area to the left of 1899.5. We have

$$z = \frac{x - \mu}{\sigma} = \frac{1899.5 - 2000}{34.64} = -2.90$$

The area to the left of $z = -2.90$ is

$$0.5000 - 0.4981 = 0.0019$$

See Figure 7.31. Thus, the probability that the show is canceled is 0.0019.

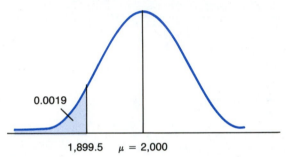

0.0019

1,899.5 $\mu = 2,000$

Figure 7.31

b. In this case the values of μ and σ are different since the value of p is 0.39. We have

$$\mu = 5000(0.39) = 1950$$

$$\sigma = \sqrt{5000(0.39)(0.61)} = \sqrt{1189.5} \approx 34.49$$

Since the show will be canceled if fewer than 1900 people watch it, we calculate the area to the left of 1899.5. We have

$$z = \frac{1899.5 - 1950}{34.49} = -1.46$$

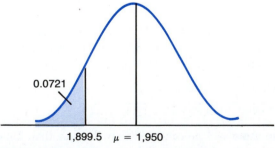

0.0721

1,899.5 $\mu = 1,950$

Figure 7.32

The area to the left of $z = -1.46$ is

$0.5000 - 0.4279 = 0.0721$

See Figure 7.32. Thus, the probability that the show is canceled is 0.0721.

COMMENT The degree of accuracy of the normal curve approximation to the binomial distribution depends on the values of n and p. Figure 7.33 indicates how fast the histogram for a binomial distribution approaches that of a normal distribution as n gets larger. As a rule, the normal curve approximation can be used with fairly accurate results when *both* np and nq are greater than 5.

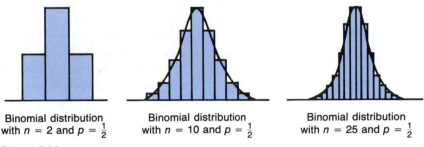

Binomial distribution with $n = 2$ and $p = \frac{1}{2}$

Binomial distribution with $n = 10$ and $p = \frac{1}{2}$

Binomial distribution with $n = 25$ and $p = \frac{1}{2}$

Figure 7.33
Histogram for a binomial distribution approaches that of a normal distribution as n gets larger.

COMMENT The normal curve approximation to the binomial distribution actually depends on a very important theorem in statistics known as the Central Limit Theorem. This theorem will be discussed in a later chapter.

EXERCISES FOR SECTION 7.5

In each of the following exercises use the normal curve approximation to the binomial distribution.

1. It is claimed that 63% of all married men in a particular city have life insurance policies. If 500 married men are randomly selected, find the probability that at most 320 of them have life insurance policies.

2. If 68% of all students attending State University have type O blood, what is the probability that a random sample of 700 students will contain 510 or more students with type O blood?

3. If 55% of all people in Bayerville speak two languages, what is the probability that a random sample of 80 people will contain at most 46 people who speak two languages?

4. Thirty-seven percent of all calculators sold by a leading department store are not fully charged when originally purchased. If 200 calculators were sold during the month of January, find the probability that between 125 to 145 of the calculators *were* fully charged.

5. It is claimed that 1 out of every 5 people in York City is receiving some form of government financial aid. If 5000 people are selected at random, what is the probability that at most 960 of these people are receiving some form of government aid?

6. 80% of all turkeys sold by Farmer McDonald are injected with a certain growth hormone. In a random sample of 400 turkeys find the probability that at most 310 of them will have been injected with this growth hormone.

7. 95% of all calls for assistance to a service station on a cold winter day are for cars that will not start because of a dead battery. If the service station receives 50 calls for assistance, find the probability that 49 of these calls for assistance are for cars that will not start because of a dead battery.

8. The Miller Car Rental Agency finds that 8% of all people making reservations for a car do not actually show up for the car. If the car rental agency has accepted reservations for 90 cars although it has only 81 cars available, what is the probability that it will have a car for each person that has reserved one?

9. Sid is taking a true–false examination consisting of 80 questions. Since he has not studied for the exam, he decides to guess at the answers. If Sid needs at least 50 correct answers to pass, what is the probability that he passes the exam?

10. Statistics for a certain district indicate that 8% of all people who borrow books from the library fail to return them on time. If 450 people borrowed books from the library today, find the probability that 28 of them will not return the books on time.

11. A large department store has found that 12% of all purchases are returned for one reason or another. If 1000 purchases were made, what is the probability that at most 110 of these will be returned for credit?

12. A sociologist claims that 55% of the families in Springfield have two or more children. If a sample of 400 families residing in Springfield is selected, find the probability that
   a. fewer than 240 of them have two or more children.
   b. more than 210 of them have two or more children.

13. *Advertising:* A drug manufacturer claims that 90% of all patients with a particular disease will be cured when treated with a new medicine. Find the probability that in a sample of 100 patients with this particular disease
   a. fewer than 90 of them will be cured.
   b. exactly 90 of them will be cured.

14. *Overbooking:* An airline company finds that 13% of all people who make reservations do not actually show up. If the airline has accepted 150 reservations for a particular flight and if there are 140 available seats, find the probability that the airline will have a seat for each person who has reserved one and shows up.

Government regulations require that airlines that routinely overbook flights offer cash incentives to persuade passengers to reschedule a flight when they are overbooked.

15. The traffic department of Redville has found that the probability that a traffic signal functions in sequence now that all the signals have been staggered is 0.95. If 300 signals are randomly selected, find the probability that at least 285 of them will be functioning in sequence.

16. If 55% of all spectators attending a baseball game buy some form of soft drink, what is the probability that a random sample of 500 spectators will contain at least 275 people who buy some form of soft drink?

17. The first year divorce rate for a certain city is 29%. If 2000 marriages have been recorded this year in the city, what is the probability that fewer than 550 of these marriages will end in divorce within the first year?

18. If 60% of all the adult residents in a city are in favor of new gun control legislation, find the probability that a sample of size 300 will contain fewer than 160 people who favor new gun control legislation.

19. Mario is taking a multiple-choice examination where only one choice of the five possibilities given is the correct answer. There are 60 questions on the exam. Mario decides to guess at the answers. If Mario needs at least 19 correct answers in order to pass, what is the probability that he passes the exam?

20. *Calculators:* According to a recent survey conducted by Swanson and Gelb,*

*Swanson and Gelb, Denver, 1990.

75% of all elementary school teachers believe that calculators should be allowed in the classroom. If 75 teachers are randomly selected, find the probability that fewer than 48 of them believe that calculators should be allowed in the classroom.

7.6 APPLICATION TO STATISTICAL QUALITY CONTROL

In recent years numerous articles and books have been written on how statistical quality controls operate. This is an important branch of applied statistics. What are quality-control charts? To answer this question, we must consider the mass production process.

Industrial experience shows that most production processes can be thought of as normally distributed. So, when a manufacturer adjusts the machines to fill a jar with 10 ounces of coffee, although not all the jars will actually weigh 10 ounces, the weight of a typical jar will be very close to 10 ounces. When too many jars weigh more than 10 ounces, the manufacturer will lose money. When too many jars weigh less than 10 ounces, he or she will lose customers. The manufacturer is therefore interested in maintaining the weight of the jars as close as possible to 10 ounces.

If the production process behaves in the manner just described, the weight of a typical jar of coffee is either acceptable or not acceptable. Thus, it can be thought of as a binomial variable. We can then use the normal approximation.

Quality-Control Charts

Rather than weigh each individual jar of coffee, the manufacturer can use **quality-control charts.** This is a simple graphical method that has been found to be highly useful in the solution to problems of this type.

Figure 7.34 is a typical quality-control chart. The horizontal line represents the time scale. The vertical line has three markings: μ, $\mu + 3\sigma$, and $\mu - 3\sigma$.

The middle line is thought of as the mean of the production process although in reality it is usually the mean weight of past daily samples. The two other lines serve as control limits for the daily production process. These lines have been spaced 3 standard deviation units above and below the mean. From the normal distribution

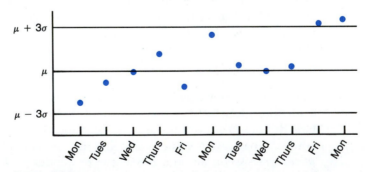

Figure 7.34
Typical quality control chart.

table we find that approximately 99.7% of the area should be between $\mu - 3\sigma$ and $\mu + 3\sigma$. Thus, the probability that the average weight of many jars of coffee falls outside the control bands is only 0.003. This is a relatively small probability. Therefore, if a sample of sufficient size is taken and the average weight is outside the control bands, the manufacturer can then assume that the production process is not operating properly and that immediate adjustment is necessary to avoid losing money or customers. Each of the dots plotted in Figure 7.34 represents the mean of the weights of n jars of coffee determined on different days.

Figure 7.34 indicates that the production process went out of control on a Friday.

7.7 SUMMARY

In this chapter we discussed the difference between a discrete random variable and a continuous random variable. We studied the normal distribution in detail since it is the most important probability distribution of a continuous random variable. Because of their usefulness, normal curve area charts have been constructed. These charts allow us to calculate the area under the standard normal curve. Thus, we can determine the probability that a random variable will fall within a specified range.

Not only can these charts be used to calculate probabilities for variables that are normally distributed, but they can also be used to obtain a fairly good approximation to binomial probabilities. This is especially helpful when we must calculate the probability that a binomial random variable assumes many different values. Numerous applications of these ideas were given.

The normal distribution was then applied to the construction of quality-control charts, which are so important in many industrial processes. Today statistical quality control is an important branch of applied statistics. Without discussing them in detail, we indicated the usefulness of these quality-control charts.

Study Guide

The following is a chapter outline in capsule form. You should now be able to demonstrate your knowledge of the ideas mentioned by giving definitions, descriptions, or specific examples. Page references are given in parentheses.

Discrete random variables can assume only the values 0, 1, 2, ... (page 355)

Continuous random variables can assume an infinite number of values on a measuring scale. (page 355)

When the graph of a frequency distribution resembles the bell-shaped curve, the graph is called a **normal curve** and its frequency distribution is known as a **normal distribution** (also called a bell-shaped or Gaussian distribution). (page 356)

The **normal distribution** has the following properties:

1. The mean is at the center of the distribution and the curve is symmetrical about the mean.

2. Its graph is a bell-shaped curve.
3. The mean equals the median.
4. The scores that make up the normal distribution tend to cluster around the middle with very few values more than 3 standard deviations away form the mean on either side. (page 356)

A **standardized normal distribution** is a normal distribution with a mean of 0 and a standard deviation of 1. (page 359)

For a **standard normal distribution**:

1. The probability that a z-score falls within 1 standard deviation of the mean on either side, that is, between $z = -1$ and $z = 1$, is approximately 68%.
2. The probability that a z-score falls within 2 standard deviations of the mean, that is, between $z = -2$ and $z = 2$, is approximately 95%.
3. The probability that a z-score falls within 3 standard deviations of the mean is approximately 99.7%. (page 361)

Chebyshev's Theorem: Let k be any number equal to or greater than 1. Then the proportion of any distribution that lies within k standard deviations of the mean is at least $1 - \dfrac{1}{k^2}$ (page 361)

If we are given a normal distribution with a mean different from 0 and a standard deviation different from 1, we can **convert** this normal distribution into a standardized normal distribution by converting each of its scores into standard scores. We use the formula $z = \dfrac{x - \mu}{\sigma}$. Thus, to find the probability that a normal random variable falls between the values x_1 and x_2, we first find the z-values corresponding to x_1 and x_2. We get $z_1 = \dfrac{x_1 - \mu}{\sigma}$ and $z_2 = \dfrac{x_2 - \mu}{\sigma}$. Then locate the values of x_1 and x_2 on the x-distribution and the values of z_1 and z_2 on a z-distribution as shown below.

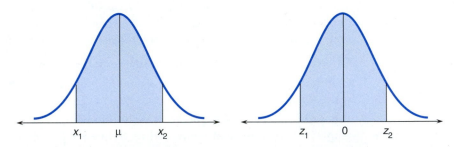

The areas under the standard normal curve between z_1 and z_2 (obtained from Table IV in the Appendix) gives us the probability that x will fall between x_1 and x_2. (page 366)

If we want to determine percentiles when working with a normal random variable, then the *P*th **percentile** of the normal random variable *x* is that value of *x* that has an area of *P*/100 to its left. Thus, the 15th percentile has an area of 0.15 to its left, the 50th percentile has an area of 0.50 to its left, and so on. (page 367)

The **normal distribution can be used as a fairly good approximation to the binomial distribution** provided that *np* and *nq* are *both* greater than 5. First calculate the mean using the formula $\mu = np$ and standard deviation $\sigma = \sqrt{npq}$. Then if the binomial probability to be approximated is of the form "find the probability that $x \le a$ or $x > a$," the continuity correction factor is $(a + 0.5)$ and the approximating standard normal *z*-value is $z = \dfrac{(a + 0.5) - \mu}{\sigma}$ as shown below. If the binomial

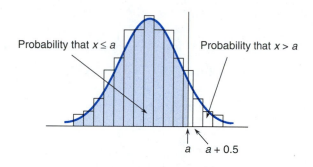

probability to be approximated is of the form "find the probability that $x \ge a$ or $x < a$," the continuity correction factor is $(a - 0.5)$ and the standard normal *z*-value is $z = \dfrac{(a - 0.5) - \mu}{\sigma}$ as shown below.

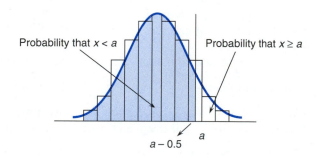

If the binomial probabilities to be approximated is an interval of the form "find the probability that *x* is between *a* and *b* inclusive," treat the ends of the intervals separately. Calculate the two *z*-values by using one of the procedures given above. (page 378)

Quality control charts are often used in industry to determine if the production process is operating smoothly. (page 388)

Formulas to Remember

You should be able to identify each symbol in the following formulas, understand the relationships among the symbols expressed in each formula, understand the significance of each formula, and use the formulas in solving problems.

1. $z = \dfrac{x - \mu}{\sigma}$

2. $x = \mu + z\sigma$

3. Chebyshev's Theorem: At least $1 - \dfrac{1}{k^2}$ of a set of measurements will lie within k standard deviations of the mean: $k = 1, 2, \ldots$.

Testing Your Understanding of This Chapter's Concepts

1. Which standard normal curve has a greater spread, one with a mean of 5 and a standard deviation of 2 or one with a mean of 2 and a standard deviation of 5?
2. Draw a normal curve (making sure to label the first 3 standard deviations on it) that has the following characteristics:
   a. mean = 5 and standard deviation = 2
   b. mean = 2 and standard deviation = 5
   c. mean = 5 and standard deviation = 5
3. A random variable x has a normal distribution with a standard deviation of 6. It is known that the probability that x exceeds 102 is 0.90. Find the mean μ of the probability distribution.
4. Would it be appropriate to use the normal curve approximation to calculate binomial probabilities where $n = 8$ and $p = \dfrac{1}{2}$? Explain your answer.
5. A random variable x has a normal distribution with mean $\mu = 25$ and variance 16. Find the value of x, call it x_0, such that the probability $18 \le x < x_0$ equals 0.8638.

THINKING CRITICALLY

1. It is known that the grades on the Scholastic Aptitude Test (SAT) for a certain school are approximately normally distributed. Furthermore, 12.71% of the stu-

dents scored above 523 and 14.92% of the students scored below 402. What is the mean and standard deviation of the test scores on this exam?

2. Using Chebyshev's Theorem, how many standard deviations on either side of the mean would have to be considered in order for us to include at least 99.7% of the numbers?

3. The life of a new energy-efficient, hot-water tank is normally distributed with a mean of 10 years. If the quality-control engineers require at least 95% of these tanks to have lives exceeding 8.9 years, what is the largest value that σ can have and still keep the engineers satisfied?

4. A student claims that when drawing normal curves the value of the mean, μ, has no effect on the shape of the curve. Do you agree? Explain your answer.

5. A random variable x is normally distributed with mean $\mu = 38$ and standard deviation $\sigma = 3$. How many standard deviations is $x = 42$ away from the mean of x?

Review Exercises for Chapter 7

1. In a standard normal distribution the mean is at
   a. $z = -1$ b. $z = 0$ c. $z = 1$ d. $z = 0.5$ e. none of these

2. Find the total area under the standard normal curve to the left of $z = 1.48$.

3. What is the area under the standard normal curve below $z = -1.63$?

4. A manufacturer claims that "Four out of 5 dentists recommend sugarless gum for their patients who chew gum." Assuming that this claim is true, find the probability that in a randomly selected group of 80 dentists 64 or more will recommend sugarless gum for their patients who chew gum.

5. *Cheating on income taxes:* It is alleged that 15% of the people who file their tax forms with the Internal Revenue Service cheat on the taxes by not reporting all the income earned. If 600 tax returns are randomly selected by auditors, what is the probability that at least 80 of them will be found not to have correctly reported all the income earned?

6. A certain pesticide spray claims to be 95% effective against the fruit fly. If 100 fruit flies are sprayed with this spray, what is the probability that 95 of them will die as a result?

7. Highway patrol officials claim that 12% of all the drivers driving along State Highway 15 on New Year's Day are under the influence of alcohol. If 50 randomly selected drivers are stopped and given an alcohometer test, find the probability that fewer than 4 of them will be under the influence of alcohol.

8. The Hugo Candy Company packages its 12-ounce candy boxes in plastic wrappers. It is known that the lengths of these wrappers (which are produced by a certain machine) are normally distributed. Furthermore, 8.08% of the wrappers have lengths that are more than 15.948 inches and 18.67% of the wrappers have

lengths that are less than 15.2152 inches. What is the mean length of a wrapper and the standard deviation of the wrapper length produced by this machine?

9. The average salary of a major league baseball player was $580,000 in 1987 (*New York Times*, April 12, 1987). Assuming that the 1987 salaries of major league baseball players are normally distributed with a mean of $580,000 and a standard deviation of $150,000, what is the probability that a randomly selected major league baseball player earned
   a. more than $750,000 in 1987?
   b. at least $400,000 in 1987?

10. Surveys indicate that about 30% of all entering freshmen at many colleges nationwide drop out by the end of their first year for a variety of reasons. If Alexville College expects to have an entering class of 1400 students this year, what is the probability that at most 400 of these students will drop out by the end of their first year?

11. It is an accepted practice in many cities to replace *all* light bulbs in a train station even before they burn out. In one city the maintenance department estimates that the life of a typical light bulb currently being used in the train stations is approximately normally distributed with a mean of 1200 hours and a standard deviation of 85 hours. When should *all* the light bulbs of a train station in this city be replaced so that no more than 8% of them will burn out while in use?

12. The County Medical Evaluation Unit is administering a medical exam that will enable successful candidates who pass the exam to practice medicine within the state. Past experience indicates that the time required to complete the exam is approximately normally distributed with a mean of 210 minutes and a standard deviation of 38 minutes. If the Evaluation Unit wishes to ensure enough time so that only 88% of the candidates complete the exam, how much time should they allow?

13. The Argyle Knitting Mill has a daily production that is normally distributed with a mean of 2000 sweaters and a standard deviation of 175 sweaters. As an incentive to increase productivity, management has decided to offer a cash bonus of $75 per day to each worker when the daily production exceeds the 92nd percentile of the distribution. At what level of production will management pay the incentive cash bonus?

14. According to government records (*Statistical Abstract of the United States*, 1987), approximately 23.6% of the country's households are inhabited by one person only. If 500 households are randomly selected, what is the probability that at most 100 of these households are inhabited by one person only?

15. Consider the following newspaper article. If 200 people from the labor force are randomly selected, find the probability that
   a. at least 20 of them are unemployed?
   b. at most 16 of them are unemployed?
   c. exactly 18 of them are unemployed?

UNEMPLOYMENT RATE UP AGAIN

Winchester (March 1)—According to figures released yesterday, the rate of unemployment for our area increased for a third month in a row and now stands at 9.2%. Such a high rate of unemployment for our area can only mean that more jobs must be created by the state.

May 1, 1989

---

16. A manufacturer of computer components claims that the life of one particular computer component is normally distributed with a mean of 2100 hours and a standard deviation of 60 hours. If one such component is randomly selected, find the probability that it will last less than 2000 hours.

Chapter Test

1. In a standard normal distribution what is the closest z-score corresponding to the 22nd percentile?
2. In a standard normal curve find the z-score that cuts off the bottom 14%.
3. In a certain girls school the heights of students are approximately normally distributed such that $\mu = 63$ inches and $\sigma = 2.1$ inches. If Rita is in the 90th percentile, find her height.
4. A psychological introvert-extrovert test yields scores that are normally distributed with a mean of 82 and a standard deviation of 3. People scoring in the top 10% will be classified as extroverts. What score is the cut-off point for an extrovert?
5. An airline company finds that 92% of the people who make reservations actually keep them. If the airline company accepts reservations from 1000 people, what is the probability that 920 people actually show up?
6. *Air pollution:* It is claimed that 30% of all the cars registered in a certain state are polluting the air. Assuming this claim to be true, a random sample is taken of 200 cars registered within the state. Find the probability that the number of cars found to be polluting the air is
   a. less than 50.
   b. more than 65.
   c. at most 55.
   d. exactly 60.
7. A shirt manufacturer claims that the sleeve lengths for men's long-sleeved shirts are normally distributed with a mean of 54 centimers (cm) and a standard deviation of 5 cm. What percentage of the shirts are expected to have sleeve lengths between 49 and 57 cm?

8. Residents along Mulberry Lane claim that two-thirds of the drivers who drive along Mulberry Lane do so at speeds in excess of 35 mph. If 75 cars are randomly observed driving along Mulberry Lane, find the probability that fewer than 49 of them will be driving at speeds in excess of 35 mph.

9. A machine has been set so that it will automatically fill each empty plastic bottle with 1 gallon [128 ounces (oz)] of oil. However, due to the complex nature of the machine, it is observed that the amount of oil in each of the bottles filled by the machine is approximately normally distributed with a standard deviation of 1.88 oz. The manufacturer wishes to recalibrate the machine so that only 5% of the plastic bottles will receive less than 120 oz. At what mean amount should the machine be calibrated (that is, find μ) so as to meet these requirements?

10. Certain computer chips are manufactured by a process that gives them a useful mean life of 6000 hours with a standard deviation of 310 hours. Assume that the useful life of these computer chips is normally distributed. If a random sample of these chips is taken, what percent will have a mean life of between 5700 and 6200 hours?

11. It is known that 38% of all summonses issued by a particular police precinct are for parking violations. If 300 tickets will be issued by this department in January, what is the probability that at least 100 of them will be for parking violations?

12. Jeremiah recently participated in a bowling tournament where the scores were normally distributed with a mean of 250. Jeremiah, who scored 278, was in the 90th percentile. Find the standard deviation.

13. Consider the newspaper article that follows. One researchers claims that despite medical claims to the contrary, two out of seven potential donors are reluctant to donate blood because of their fear of contracting AIDS. Assuming this researcher's claim is true, find the probability that in a random survey of 75 potential donors, at most 20 of them will be reluctant to donate blood because of their fear of contracting AIDS.

BLOOD DONATIONS AT A LOW

December 7—Blood bank officials announced that many volunteers were reluctant to donate blood because of their fear of contracting AIDS. As a result, the blood supply at the city's hospitals is dropping rapidly.

December 7, 1990

14. In an effort to become more cost-efficient, several oil distributors have been experimenting with self-service automatic gas pumping machines that dispense a particular number of gallons of gas after a specified amount of money has been deposited. One such machine has been set so that it will dispense 4 gallons of unleaded high-test gasoline when a $5 bill is inserted. However, the pump is not functioning properly and can be adjusted according to the vendor's specifications. It is known that the pumping and filling process are approximately normally distributed with a standard deviation of 0.33 gallons. At what level should the mean be set so that

    a. only 5% of the time will the pump deliver less than 4 gallons of gas when $5 is deposited?

    b. at most 5% of the time will the pump deliver more than 4.25 gallons of gas when $5 is deposited?

15. Refer back to Exercise 14. The gas pump has been adjusted by mechanics so that the number of gallons of gas dispensed when $5 is deposited is approximately normally distributed with a standard deviation of 0.29 gallons. If 10% of the time more than 4.47 gallons of gas are dispensed, what is the new mean amount of gasoline dispensed by this pump when $5 is deposited?

16. The amount of soup dispensed by a machine is approximately normally distributed with a mean of 7 ounces and a standard deviation of 1.1 ounces. What is the probability that a cup filled by this machine will contain between 5 and 8 ounces?

    a. 0.7842 b. 0.4656 c. 0.3186 d. 0.1470 e. none of these

17. In a certain baseball league, the batting averages are normally distributed with $\mu = 0.285$ and $\sigma = 0.009$. If John McAllister is in the 90th percentile, find his batting average.

    a. 0.297 b. 0.291 c. 0.299 d. 0.294 e. none of these

18. The life of a certain photocopier machine part is normally distributed with a mean of 20,000 hours and a standard deviation of 585 hours. What percentage of these parts will last more than 21,000 hours?

    a. 0.1709 b. 0.4564 c. 0.0436 d. 0.9564 e. none of these

19. A travel agent finds that about 7% of all prepaid reservations are cancelled for one reason or another. If the agent has booked 400 prepaid reservations for the Thanksgiving Day weekend, what is the probability that at most 25 of them will be cancelled?

    a. 0.1879 b. 0.3121 c. 0.2451 d. 0.2776 e. none of these

20. It is claimed that 60% of all full-time employees of the Scott Corporation earn at least $30,000 annually. What is the probability that a survey of 80 randomly selected full-time employees will contain at least 50 employees who earn at least $30,000 annually?

    a. 0.3669 b. 0.1331 c. 0.4090 d. 0.4562 e. none of these

Suggested Further Reading

1. Freund, John E., and Ronald Walpole. *Mathematical Statistics*, 4th ed. Englewood Cliffs, NJ: Prentice-Hall, 1987.
2. Mason, Robert D. *Statistical Techniques in Business and Economics*, 6th ed. Homewood, IL: Irwin, 1986.
3. Mendenhall, W. *Introduction to Probability and Statistics*, 7th ed. Boston: Duxbury, 1987.
4. Wonnacott, Ronald, and Thomas Wonnacott. *Understanding Statistics*, 4th ed. New York: John Wiley, 1985.

8 Sampling

How does a polltaker conduct a random sample? (© Caroline Brown, Fran Heyl Associates)

In this chapter we discuss several sampling techniques and how the results obtained from random samples, for example, the sample mean, $\bar{x}$, can be used to make inferences about the population mean μ. We also consider an important theorem—the Central Limit Theorem—which gives us some very significant information about the sampling distribution of the mean. These ideas will be used extensively in later chapters when we discuss statistical inferences.

CHAPTER OBJECTIVES

After studying the material in this chapter, you should be able:

- **To discuss** what a random sample is and how it is obtained. (Section 8.2)

- **To work with** a table of random digits where each number that appears is obtained by a process that gives every digit an equally likely chance of being selected. (Section 8.2)

- **To understand** stratified sampling, which is a sampling procedure that is used when we want to obtain a sample with a specified number of people from different categories. (Section 8.3)

- **To see** that when repeated samples are taken from a population the frequency distribution of the values of the sample means is called the distribution of the sample means. (Section 8.5)

- **To analyze** the standard error of the mean, which represents the standard deviation of the distribution of sample means. (Section 8.5)

- **To apply** the Central Limit Theorem. This tells us that the distribution of the sample means is basically a normal distribution. We discuss how to use this theorem to make predictions about and calculate probabilities for the sample means. (Sections 8.6 and 8.7)

Gary Settle/NYT Pictures.

On April 6, 1976 both ABC and NBC television networks projected that Morris Udall would win the Democratic primary in Wisconsin. Their predictions were based on samples from selected precincts and did not take certain districts into consideration. When all the rural votes were counted, Jimmy Carter came out on top. Many newspapers were so confident of their predictions that they printed, erroneously, the morning editions of their newspapers with the headline "CARTER UPSET BY UDALL."

CITY TO CALL $5 MILLION OF THE 1995 SERIES 14 BOND

New Dorp (June 30)—City officials announced yesterday that their financial condition had improved sufficiently to enable it to call $5 million of the 1995 series 14 bonds. These bonds, which were originally issued in 1975, have a 12% annual interest rate and are costing the city $7 million in annual interest charges. The bonds to be recalled will be selected in a random manner and will be made public tomorrow. The holders of a called bonds will no longer earn any interest as of tomorrow.

Tuesday, June 30, 1989

The second article specifies that the bonds to be recalled will be selected in a random process. How are such random selections made?

8.1 INTRODUCTION

Suppose we are interested in determining how many people in the United States believe in the fiscal soundness of the Social Security system. Must we ask each person in the country before making any statement concerning the fiscal soundness of the Social Security system? As there are so many people in the United States, it would require an enormous amount of work to interview each person and gather all the data.

Do we actually need the complete population data? Can a properly selected sample give us enough information to make predictions about the entire population? In many cases obtaining the complete population data may be quite costly or even impossible.

Similarly, suppose we are interested in purchasing an electric bulb. Must we use (or test) all electric bulbs produced by a particular company in order to determine the bulb's average life? This is very impractical. Maybe we can estimate the average life of a bulb by testing a sample of only 100 bulbs.

As we noticed earlier in Chapter 1, Definition 1.2, this is exactly what inferential statistics involves. Samples are studied to obtain valuable information about a larger group called the **population.** Any part of the population is called a **sample.** The purpose of sampling is to select the part that truly represents the entire population.

**Population
Sample**

Any sample provides only partial information about the population from which it is selected. Thus, it follows that any statement we make (based on a sample) concerning the population may be subject to error. One way of minimizing this error is to make sure that the sample is randomly selected. How is this done?

In this chapter we discuss how to select a random sample and how to interpret different sample results.

8.2 SELECTING A RANDOM SAMPLE

The purpose of most statistical studies is to make generalizations from samples about the entire population. Yet not all samples lend themselves to such generalizations. Thus, we cannot generalize about the average income of a working person in the United States by sampling only lawyers and doctors. Similarly, we cannot make any generalizations about the Social Security system by sampling only people who are receiving Social Security benefits.

Over the years many incorrect predictions have been made on the basis of nonrandom samples. For example, in 1936 the *Literary Digest* was interested in determining who would win the coming presidential election. It decided to poll the voters by mailing ten million ballots. On the basis of the approximately two million ballots returned, it predicted that Alfred E. Landon would be elected. An October 31 headline read

Landon	1,293,669
Roosevelt	972,897

Final returns in the Digest's poll of ten million voters

Source: *The Literary Digest*, 1936.

Actually, Franklin Roosevelt carried 46 of the 48 states and many of them by a landslide. The ten million people to whom the *Digest* sent ballots were selected from telephone listings and from the list of its own subscribers. The year 1936 was a depression year and many people could not afford telephones or magazine subscriptions. Thus, the *Digest* did not select a random sample of the voters of the United States. The *Literary Digest* soon went out of business. In 1976 Maurice Bryson* argued persuasively that the major problem was the *Digest's* reliance on voluntary response.

Again, in 1948 the polls predicted that Dewey would win the presidential election. One newspaper even printed the morning edition of its newspaper with the headline "DEWEY WINS BY A LANDSLIDE." Of course, Truman won the election and laughed when presented with a copy of the newspaper predicting his defeat.

In both examples the reason for the incorrect prediction is that it was based on information obtained from poor samples. It is for this reason that statisticians insist that samples be randomly selected.

DEFINITION 8.1
Random Sample

A **random sample** of n items is a sample selected from a population in such a way that every different sample of size n from the population has an equal chance of being selected.

DEFINITION 8.2
Random Sampling

Random sampling is the procedure by which a random sample is obtained.

It may seem that the selection of a random sample is an easy task. Unfortunately, this is not the case. You may think that we can get a random sample of voters by opening a telephone book and selecting every tenth name. This will not give a random sample since many voters either do not have phones or else have unlisted numbers. Furthermore, many young voters and most women are not listed. These people, who are members of the voting population, do not have an equal chance of being selected.

---

*M. Bryson, "The *Literary Digest* Poll: Making of a Statistical Myth," *American Statistician*, November 1976.

To illustrate further the nature of random sampling, suppose that the administration of a large southern college with an enrollment of 30,000 students is considering revising its grading system. The administration is interested in replacing its present grading system with a pass–fail system. Since not all students agree with this proposed change, the administration has decided to poll 1000 students. How is this to be done? Polling a thousand students in the school cafeteria or in the student lounge will not result in a random sample since there may be many students who neither eat in the cafeteria nor go to the student lounge.

One way of obtaining a random sample is to write each student's name on a separate piece of paper and then put all the pieces in a large bowl where they can be thoroughly mixed. A paper is then selected from the bowl. This procedure is repeated until 1000 names are obtained. In this manner a random sample of 1000 names can be obtained. Great care must be exercised to make sure that the bowl is thoroughly mixed after a piece of paper is selected. Otherwise the papers on the bottom of the bowl do not have an equal chance of being selected and the sample will not be random.

Are statistics gathered by the U.S. Census Bureau based on random samples?

Although using slips of paper in a bowl will result in a random sample if properly done, this fish bowl method becomes unmanageable as the number of people in the population increases. The job of numbering slips of paper can be completely avoided by using a **table of random numbers** or a **table of random digits.**

What are random numbers? They are the digits 0, 1, 2, . . . , 9 arranged in a random fashion, that is, in such a way that all the digits appear with approximately the same frequency. Table VI in the Appendix is an example of a table of random numbers. How are such tables constructed? Today most tables of random numbers are constructed with the help of electronic computers. Yet a simple spinner such as the one shown in Figure 8.1 will also generate such a table of random numbers. After each spin we use the digit selected by the arrow. The resulting sequence of random digits could then be used to construct random number tables of five digits each, for example, as that shown in Table VI. Although this method will generate random numbers, it is not practical. After a while some numbers will be favored over others as the spinner begins to wear out.

Thus, the best way to select a random sample is to use a table of random digits obtained with the help of an electronic computer. Table VI in the Appendix is such a table. In this table the various digits are scattered at random and arranged in groups of five for greater legibility. Today even some calculators can generate random digits.

Let us now return to our example. As a first step in using this table each student is assigned a number from 00001 to 30,000. To obtain a sample of 1000 students we merely read down column 1 and select the first 1000 students whose numbers are listed. Thus, the following students would be selected:

10,480   22,368   24,130   28,918   09,429 . . .

Notice that we skip the number 42,167 since no student has this number. The same is true for the numbers 37,570, 77,921, 99,562. . . . We disregard any numbers larger than 30,000 that are obtained from this table since no students are associated with these numbers.

To illustrate further the proper use of this table, suppose that a hotel has 150 guests registered for a weekend. The management wishes to select a sample of 15 people at random to rate the quality of its service. It should proceed as follows: Assign each guest a different number from 1 to 150 as 001, 002, . . . , 150. Then

Table of Random Numbers

Table of Random Digits

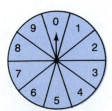

Figure 8.1
A spinner.

select any column in the table of random numbers. Suppose the fourth column is selected. Three-digit numbers are then read off the table by reading down the column. If necessary, the management continues on to another column or another page. Starting on top of column 4, they get

020 853 972 616 166 427 699, . . .

Since the guests have numbers between 001 and 150 only, they ignore any number larger than 150. From column 4 they get

020 079 102 034 081 099 143 073 129

From column 5 they get

078 061 091 133 040 023

Thus, the management should interview the guests whose numbers are 20, 79, 102, 34, 81, 99, 143, 73, 129, 78, 61, 91, 133, 40, and 23.

COMMENT Whenever we speak of a random sample in this text we will assume that it has been selected in the manner just described.

EXERCISES FOR SECTION 8.2

1. A certain math book has 632 pages. To check for typographical errors, the publisher decides to randomly select 25 pages and to check them thoroughly. By using column 1 of Table VI in the Appendix, which pages should be checked?

2. There are 987 inmates in the Pinelawn Correctional Facility, each of whom has been assigned a number from 1 to 987. In recent weeks there has been much unrest among the prisoners. Fearing a possible riot, the warden has decided to meet with a committee of 18 prisoners to be randomly selected from among them to discuss their grievances. By using column 5 of Table VI in the Appendix, which prisoners should be selected?

3. Each member of the security force at one particular major airport carries a badge. The badges are numbered in order from 1 to 869. A random sample of 35 is to be selected for training in new antiterrorist activities and hostage negotiations. Use column 12 of Table VI in the Appendix to find out which security force members should be selected for training.

4. Each welfare recipient in Lawrence has a case number from 1 to 3000. Government officials have decided to verify that a welfare recipient is actually eligible for such aid by selecting a random sample of 20 recipients and thoroughly investigating each claim. If the officials use columns 13 and 14 of Table VI in the Appendix, which case numbers will be selected?

5. Each refrigerator that is manufactured in the United States has a serial number. After a series of complaints about defective thermostats one of the manufacturers has decided to randomly check 40 refrigerators sold in Chicago. According to

company records, refrigerators sold in Chicago during 1990 had serial numbers whose last five digits were between 30,000 and 71,000. If columns 10 and 11 of Table VI in the Appendix are used, which refrigerators will be checked?

6. One large bank in New York City issued 91,469 credit cards during 1990. Bank officials have decided to offer 30 of these credit card holders the opportunity of obtaining a gold card at no annual charge. If column 10 of Table VI in the Appendix is used, which credit card holders will be offered the opportunity of obtaining the gold card?

7. Emergency Room patients often complain that they have to wait long periods of time before receiving medical attention. During the first week of January 1991, 476 patients were seen by doctors in the Emergency Room at Maimomides Medical Center. These patients were assigned numbers 1, 2, 3, . . , 476. Officials of the hospital have decided to interview 18 of these patients to determine how long each had to wait before seeing a doctor. If column 3 of Table VI in the Appendix is used, which patients will be selected to be interviewed?

8. *Bond recalls:* Many municipalities issue bonds to finance a variety of public projects. Most of these bonds have a call provision that allows the issuer to prerefund the bonds before the stated maturity date. In 1980 Clinton issued 10,000 bonds, each numbered in order, 1, 2, 3, . . . , 10,000. In 1992, town officials decide to redeem 15 of these boards ahead of their maturity date by a random selection process. If they use columns 4, 5, and 6 of Table VI in the Appendix, which bonds will they select to be redeemed?

9. Each medical waste container that is produced by the Hughes Waste Disposal Company during a year carries a four-digit serial number from 1 to 9000, in addition to the date of manufacture. During 1990 the company manufactured 5638 of these medical waste containers and numbered them in order, 1, 2, 3, . . . , 5638. Twenty of these containers are to be randomly selected and inspected for their ability to inhibit corrosion. If column 9 of Table VI in the Appendix is used, which containers will be inspected?

10. The Sloan Medical Research Organization has a list of 4769 volunteers, each assigned a number of 1 to 4769, who have agreed to be candidates for a new experimental drug. It is decided that only 20 of these volunteers (randomly selected) will be given the new drug. By using columns 5 and 6 of the Table VI in the Appendix, decide which of the volunteers will be given the experimental drug.

8.3 STRATIFIED SAMPLING

Stratified Samples

Although random sampling, as discussed in Section 8.2, is the most popular way of selecting a sample, there are times when **stratified samples** are preferred. To obtain a stratified sample, we divide the entire population into a number of groups or strata. The purpose of such stratification is to obtain groups of people that are more or less equal in some respect. We select a random sample, as discussed in Section 8.2 from each of these groups or from each stratum. This stratified sampling procedure ensures that no group is missed and improves the precision of our estimates. If we use stratified sampling, then in order to estimate the population mean, we must use *weighted averages* of the strata means, weighted by the population size for that stratum.

Thus, in the example discussed in the beginning of Section 8.2 the administration of the college may first divide the entire student body into four groups: freshmen, sophomores, juniors, and seniors. Then it can select a random sample from each of these groups. The groups are often sampled in proportion to their actual percentages. In this manner the administration can obtain a more accurate poll of student opinion by stratified sampling. However, the cost of obtaining a stratified sample is often higher than that of obtaining a random sample since the administration must spend money to research dividing the student body into four groups.

The method of stratifying samples is especially useful in pre-election polls. Past experience indicates that different subpopulations often demonstrate particularly different voting preferences.

COMMENT Statistical analyses and tests based on data obtained from stratified samples are somewhat different from what we have discussed in this book. We will not analyze such procedures here.

In addition to the random and stratified sampling techniques discussed until now, there are other sampling techniques that can be used. Among these are the following:

Systematic Sampling

Systematic sampling: This is a sampling procedure that is commonly used in business surveys, production processes, and for selections from name files. To use systematic sampling, we first categorize the elements of the population in some way such as alphabetically or numerically. Then we randomly select a starting point. We include in our sample every ith item of the categorized population until we obtain a sample of size n. For example, a systematic sampling procedure might begin with a randomly chosen start (from 0001 to 1200) selected from a pile of invoices. After the first invoice is randomly selected we choose, for example, every 80th invoice thereafter. In this manner we are able to obtain a good cross-sectional representation of the population.

Cluster Sampling

Primary Subgroups
Clusters

Cluster sampling: This is another sampling technique that is often used. With this method, the target population that we wish to analyze is divided into mutually exclusive subgroups called **primary subgroups** or **clusters**, each of which should be representative of the entire population. Then a random sample of these clusters is selected so as to provide estimates of the population values. The objective in using this procedure is to form clusters (or subgroups) that are small images of the target population. By localizing the sample units to relatively few clusters or regions, we can realize substantial cost reduction.

To illustrate the technique of cluster sampling, suppose a newspaper company cannot decide if it should institute home delivery of newspapers in a particular city.

The company plans to conduct a survey to determine the fraction of households in the city that would use the newspaper home delivery service. The sampling procedure that it can use is to choose a city block (cluster) at random and then to survey every household on that block. This cluster sampling procedure is considerably more economical than simple random sampling.

8.4 CHANCE VARIATION AMONG SAMPLES

Imagine that a cigarette manufacturer is interested in knowing the average tar content of a new brand of cigarettes that is about to be sold. The Food and Drug Administration requires such information to be indicated alongside all advertisements that appear in magazines, newspapers, and so on.

The manufacturer decides to send random samples of 100 cigarettes each to 20 different testing laboratories. With the information obtained from these samples, the manufacturer hopes to be able to estimate the mean or average milligram tar content of the cigarette.

Since we will be discussing both samples and populations, let us pause for a moment to indicate the notation that we will use to distinguish between samples and populations. See Table 8.1.

From Chapter 3 we have the following formulas:

FORMULA 8.1 Mean

Sample	Population
$\bar{x} = \dfrac{\Sigma x}{n}$	$\mu = \dfrac{\Sigma x}{N}$

FORMULA 8.2 Standard Deviation

Sample	Population
$s = \sqrt{\dfrac{\Sigma(x - \bar{x})^2}{n - 1}}$	$\sigma = \sqrt{\dfrac{\Sigma(x - \mu)^2}{N}}$

TABLE 8.1 Notation for Sample and Population

Term	Sample	Population
Mean	$\bar{x}$	μ
Standard deviation	s	σ
Number	n	N

Let us now return to our example. Since each sample sent to a laboratory is randomly selected, it is reasonably safe to assume that there will be differences among the means of each sample. The 20 laboratories report the following average milligram content per cigarette:

14.8   16.2   14.8   15.8   15.3   13.9   16.9   15.9   14.3   15.2
14.9   16.2   15.6   15.5   13.4   15.1   15.7   14.8   14.4   15.3

These figures indicate that the sample means vary considerably from sample to sample.

The manufacturer decides to take the average of these 20 sample means and gets

$$\text{Average of 20 sample means} = \frac{\Sigma \bar{x}}{n} = \frac{14.8 + 16.2 + \cdots + 15.3}{20} = \frac{304}{20} = 15.2$$

The manufacturer now uses this overall average of the sample means, 15.2, as an estimate of the true population mean.

How reliable is this estimate? Although we cannot claim for certain that the population mean is 15.2, we can feel reasonably confident that 15.2 is not a bad estimate of the population mean since it is based on 20 × 100, or 2000, observations.

Thus, we can obtain a fairly good estimate of the population mean by calculating the mean of samples. If we let $\mu_{\bar{x}}$, read as mu sub x bar, represent the mean of the samples, then we say that $\mu_{\bar{x}}$ is a good estimate of μ. Generally speaking, if a random sample of size n is taken from a population with mean μ, then the mean of $\bar{x}$ will always equal the mean of the population (regardless of sample size), that is, $\mu_{\bar{x}} = \mu$.

What about the standard deviation? Let us calculate the standard deviation of the sample means. Recall that the formula for the population standard deviation is

$$\sqrt{\frac{\Sigma(x - \mu)^2}{N}}$$

Since μ is unknown, we have to replace it with an estimate. The most obvious replacement is $\mu_{\bar{x}}$. To account for this replacement, we divide by $N - 1$ instead of by N. Thus, the formula for the standard deviation for the sample means is given by Formula 8.3.

FORMULA 8.3

Standard Deviation of the Sample Means

The **standard deviation of the sample means** is given by

$$\sqrt{\frac{\Sigma(\bar{x} - \mu_{\bar{x}})^2}{n - 1}}$$

where n is the number of sample means.

EXAMPLE 1   Calculate the standard deviation of the sample means for the data of the average tar content of the 20 laboratories.

Solution

We arrange the data as follows:

$\bar{x}$	$\bar{x} - \mu_{\bar{x}}$	$(\bar{x} - \mu_{\bar{x}})^2$
14.8	$14.8 - 15.2 = -0.4$	0.16
16.2	$16.2 - 15.2 = 1$	1.00
14.8	$14.8 - 15.2 = -0.4$	0.16
15.8	$15.8 - 15.2 = 0.6$	0.36
15.3	$15.3 - 15.2 = 0.1$	0.01
13.9	$13.9 - 15.2 = -1.3$	1.69
16.9	$16.9 - 15.2 = 1.7$	2.89
15.9	$15.9 - 15.2 = 0.7$	0.49
14.3	$14.3 - 15.2 = -0.9$	0.81
15.2	$15.2 - 15.2 = 0$	0
14.9	$14.9 - 15.2 = -0.3$	0.09
16.2	$16.2 - 15.2 = 1$	1.00
15.6	$15.6 - 15.2 = 0.4$	0.16
15.5	$15.5 - 15.2 = 0.3$	0.09
13.4	$13.4 - 15.2 = -1.8$	3.24
15.1	$15.1 - 15.2 = -0.1$	0.01
15.7	$15.7 - 15.2 = 0.5$	0.25
14.8	$14.8 - 15.2 = -0.4$	0.16
14.4	$14.4 - 15.2 = -0.8$	0.64
_15.3_	$15.3 - 15.2 = 0.1$	_0.01_
304		13.22
$\Sigma \bar{x} = 304$		$\Sigma(\bar{x} - \mu_{\bar{x}})^2 = 13.22$

Using Formula 8.3, we have

$$\text{Standard deviation of sample means} = \sqrt{\frac{\Sigma(\bar{x} - \mu_{\bar{x}})^2}{n - 1}}$$

$$= \sqrt{\frac{13.22}{20 - 1}} = \sqrt{\frac{13.22}{19}}$$

$$= \sqrt{0.696}$$

$$\approx 0.83$$

Thus, the standard deviation of the sample means is approximately 0.83.

In practice, the standard deviation is not calculated by using Formula 8.3 since the computations required are time-consuming. Instead, we can use a shortcut formula given as Formula 8.4. The advantage in using Formula 8.4 is that we do not have to calculate $\mu_{\bar{x}}$ and $\bar{x} - \mu_{\bar{x}}$ and square $\bar{x} - \mu_{\bar{x}}$. We only have to calculate $\Sigma\,\bar{x}$ and $\Sigma\,\bar{x}^2$. These represent the sum of the $\bar{x}$'s and the sum of the squares of the $\bar{x}$'s, respectively. Then we use Formula 8.4.

FORMULA 8.4

Standard Deviation of the Sample Means

The **standard deviation of the sample means** is given by

$$\sqrt{\frac{n(\Sigma\,\bar{x}^2) - (\Sigma\,\bar{x})^2}{n(n-1)}}$$

where n is the number of sample means.

EXAMPLE 2

Using Formula 8.4, find the standard deviation of the sample means for the data of Example 1 in this section.

Solution

We arrange the data as follows:

$\bar{x}$	$\bar{x}^2$
14.8	219.04
16.2	262.44
14.8	219.04
15.8	249.64
15.3	234.09
13.9	193.21
16.9	285.61
15.9	252.81
14.3	204.49
15.2	231.04
14.9	222.01
16.2	262.44
15.6	243.36
15.5	240.25
13.4	179.56
15.1	228.01
15.7	246.49
14.8	219.04
14.4	207.36
15.3	234.09
304	4634.02
$\Sigma\,\bar{x} = 304$	$\Sigma\,\bar{x}^2 = 4634.02$

Using Formula 8.4, we have

$$\text{Standard deviation of sample means} = \sqrt{\frac{n(\Sigma \, \bar{x}^2) - (\Sigma \, \bar{x})^2}{n(n-1)}}$$

$$= \sqrt{\frac{20(4634.02) - (304)^2}{20(19)}}$$

$$= \sqrt{\frac{92680.4 - 92416}{380}}$$

$$= \sqrt{\frac{264.4}{380}} = \sqrt{0.696} \approx 0.83$$

Thus, the standard deviation of the sample means is approximately 0.83. This is the same result we obtained using Formula 8.3.

EXAMPLE 3

A large office building has six elevators, each with a capacity for ten people. The operator of each elevator has determined the average weight of the people in the elevators when operating at full capacity. The results follow.

Elevator	1	2	3	4	5	6
Average Weight (lb.)	125	138	145	137	155	140

Find the overall average of these sample means. Also, find the standard deviation of these sample means by first using Formula 8.3 and then by using Formula 8.4.

Solution

We arrange the data as follows.

$\bar{x}$	$\bar{x} - \mu_{\bar{x}}$	$(\bar{x} - \mu_{\bar{x}})^2$	$\bar{x}^2$
125	$125 - 140 = -15$	225	15,625
138	$138 - 140 = -2$	4	19,044
145	$145 - 140 = 5$	25	21,025
137	$137 - 140 = -3$	9	18,769
155	$155 - 140 = 15$	225	24,025
140	$140 - 140 = 0$	0	19,600
840		488	118,088
$\Sigma \, \bar{x} = 840$		$\Sigma(\bar{x} - \mu_{\bar{x}})^2 = 488$	$\Sigma \, \bar{x}^2 = 118,088$

Then

$$\mu_{\bar{x}} = \frac{\Sigma \, \bar{x}}{n} = \frac{840}{6} = 140$$

Using Formula 8.3, we get

$$\text{Standard deviation of sample means} = \sqrt{\frac{\Sigma(\bar{x} - \mu_{\bar{x}})^2}{n-1}} = \sqrt{\frac{488}{5}} = \sqrt{97.6} \approx 9.88$$

Using Formula 8.4, we get

$$\text{Standard deviation of sample means} = \sqrt{\frac{n(\Sigma\,\bar{x}^2) - (\Sigma\,\bar{x})^2}{n(n-1)}} = \sqrt{\frac{6(118,088) - (840)^2}{6(5)}}$$

$$= \sqrt{\frac{708528 - 705600}{30}} = \sqrt{97.6} \approx 9.88$$

Thus, the mean of the samples is 140 and the standard deviation of the sample means (by Formula 8.3 or Formula 8.4) is approximately 9.88.

8.5 DISTRIBUTION OF SAMPLE MEANS

Let us refer back to the example discussed at the beginning of Section 8.4. The manufacturer decides to draw the histogram for the average cigarette tar content that was obtained from the 20 laboratories. This is shown in Figure 8.2. Notice that the value of $\bar{x}$ is actually a random variable since its value is different from sample to sample. In repeated samples different values of $\bar{x}$ were obtained. Yet they are all close to the 15.2 we obtained as the average of the sample means. Moreover, exactly 70% of the sample means are between 14.37 and 16.03, which represents 1 standard deviation away from the mean in either direction. Also, 90% of the sample means are between 13.54 and 16.86, which represents 2 standard deviations away from the mean in either direction. Thus, Figure 8.2 actually represents the distribution of $\bar{x}$

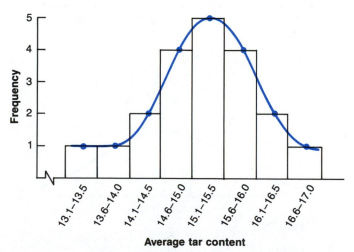

Figure 8.2

Distribution of Sample Means

Sampling Distribution of the Mean

since it tells us how the means of the samples vary from sample to sample. We refer to the distribution of $\bar{x}$ as the **distribution of the sample means** or as **the sampling distribution of the mean.** Although the first terminology is much clearer, the second is more commonly used.

Strictly speaking, Figure 8.2 is not a complete distribution of $\bar{x}$ since it is based on only 20 sample means. To obtain the complete distribution of sample means, we would have to take thousands of samples of 100 cigarettes each. Of course, in practice, we do not take thousands of samples from the same population.

COMMENT   Notice that the sample means form an approximate normal distribution. We will have more to say about this in Section 8.6.

What can we say about this distribution? What is its mean? Its standard deviation? How does this distribution compare with the distribution of *all* the cigarettes? To answer this question, the manufacturer decides to draw the frequency polygon for the tar content of all 2000 cigarettes. This is shown in Figure 8.3.

Let us now compare these two distributions. Notice that both distributions are centered around the same number, 15.2. Thus, it is reasonable to assume that $\mu_{\bar{x}} = \mu$. Also, notice that the distribution of the sample means is not spread out as much as (that is, has a smaller standard deviation than) the distribution of the tar content of all the cigarettes. The reason for this should be obvious. When *all* the cigarettes are considered, several have a very high tar content and several have a very low tar content. These appear on the tail ends of the distribution of Figure 8.3. However, it is unlikely that an entire sample of 100 cigarettes will have a tar content of 18.5. Thus, the distribution of $\bar{x}$ has very little frequency at large distances from the mean.

We use the symbol $\sigma_{\bar{x}}$ to represent the standard deviation of the sampling distribution of the mean. We have the following formula for $\sigma_{\bar{x}}$.

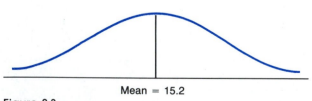

Mean = 15.2

Figure 8.3

FORMULA 8.5
Standard Error of the
Mean

The standard deviation of the sampling distribution of the mean is referred to as the **standard error of the mean.** If random samples of size n are selected from a population whose mean is μ and whose standard deviation is σ, then the theoretical sampling distribution of $\bar{x}$ has mean $\mu_{\bar{x}} = \mu$ and a standard deviation of

$$\sigma_{\bar{x}} = \frac{\sigma}{\sqrt{n}} \cdot \sqrt{\frac{N - n}{N - 1}} \qquad \text{standard error of the mean for finite populations of size } N$$

and

$$\sigma_{\bar{x}} = \frac{\sigma}{\sqrt{n}} \qquad \text{standard error of the mean for infinite populations}$$

Finite Population
Correction Factor

COMMENT   The factor $\sqrt{\dfrac{N - n}{N - 1}}$ in the first formula for $\sigma_{\bar{x}}$ is referred to as the **finite population correction factor.** It is usually ignored; that is, it has very little effect in the calculation of $\sigma_{\bar{x}}$, unless the sample constitutes at least 5% of the population.

COMMENT   It should be obvious from Formula 8.5 that the larger the sample size is, the smaller the variation of the means will be. Thus, as we take larger and larger samples, we can expect the mean of the samples, $\mu_{\bar{x}}$, to be close to the mean of the population, μ. This is illustrated in Figure 8.4.

The standard error of the mean, $\sigma_{\bar{x}}$, plays a very important rule in statistics as will be illustrated in the remainder of this chapter.

To illustrate the concept of distribution of sample means and to see how Formula 8.5 is used, let us consider Mr. and Mrs. Avery, who have five children. The ages of the children are 2, 5, 8, 11, and 14 years. Suppose we first calculate the mean and standard deviation of these ages. We have

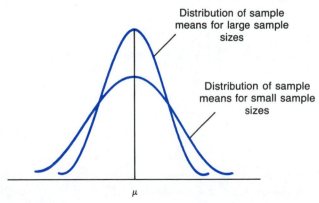

Distribution of sample means for large sample sizes

Distribution of sample means for small sample sizes

μ

Figure 8.4

Age x	$x - \mu$	$(x - \mu)^2$
2	$2 - 8 = -6$	36
5	$5 - 8 = -3$	9
8	$8 - 8 = 0$	0
11	$11 - 8 = 3$	9
14	$14 - 8 = 6$	36
40		90
$\Sigma x = 40$		$\Sigma (x - \mu)^2 = 90$

Then

$$\mu = \frac{\Sigma x}{N} = \frac{40}{5} = 8 \quad \text{and} \quad \sigma = \sqrt{\frac{\Sigma(x - \mu)^2}{N}}$$

$$= \sqrt{\frac{90}{5}}$$

$$= \sqrt{18} \approx 4.2426$$

Thus, the mean age is 8 and the standard deviation is approximately 4.2426 years.

Let us now calculate the mean of each sample of size 2 that can be formed from these ages. We have

If Ages Selected Are	Sample Mean, $\bar{x}$ Is
2 and 5 years	3.5
2 and 8 years	5
2 and 11 years	6.5
2 and 14 years	8
5 and 8 years	6.5
5 and 11 years	8
5 and 14 years	9.5
8 and 11 years	9.5
8 and 14 years	11
11 and 14 years	12.5

There are ten possible samples of size 2 that can be formed from these five ages. What is the average and standard deviation of these sample means? To determine these, we set up the following chart using the sample means as data.

$\bar{x}$	$\bar{x} - \mu_{\bar{x}}$	$(\bar{x} - \mu_{\bar{x}})^2$
3.5	$3.5 - 8 = -4.5$	20.25
5.0	$5.0 - 8 = -3$	9.00
6.5	$6.5 - 8 = -1.5$	2.25
8.0	$8.0 - 8 = 0$	0.00
6.5	$6.5 - 8 = -1.5$	2.25
8.0	$8.0 - 8 = 0$	0.00
9.5	$9.5 - 8 = 1.5$	2.25
9.5	$9.5 - 8 = 1.5$	2.25
11.0	$11.0 - 8 = 3$	9.00
12.5	$12.5 - 8 = 4.5$	20.25
80		67.5
$\Sigma \bar{x} = 80$		$\Sigma(\bar{x} - \mu_{\bar{x}})^2 = 67.5$

Thus, the average of the sample means is $\mu_{\bar{x}} = \dfrac{\Sigma \bar{x}}{\text{No. of samples}} = \dfrac{80}{10} = 8$ and the standard deviation of the sample means is

$$\sigma_{\bar{x}} = \sqrt{\frac{\Sigma(\bar{x} - \mu_{\bar{x}})^2}{\text{No. of samples}}} = \sqrt{\frac{67.5}{10}} = \sqrt{6.75} \approx 2.598.$$

In our case the population size, N, is 5 and the sample size n, is 2 so that the sample size is $\dfrac{2}{5}$ or 40% of the population size. Using Formula 8.5, we get

$$\sigma_{\bar{x}} = \frac{\sigma}{\sqrt{n}} \sqrt{\frac{N - n}{N - 1}}$$

$$= \frac{4.2426}{\sqrt{2}} \sqrt{\frac{5 - 2}{5 - 1}} \qquad \text{(Remember } \sigma = 4.2426, \text{ as previously calculated on page 416.)}$$

$$= \frac{4.2426}{1.414} \sqrt{\frac{3}{4}} \approx \frac{4.2426}{1.414} (0.866)$$

$$\approx 2.598$$

This is exactly the value that we obtained previously. Thus, the average of the sample means is exactly the same as the population mean, that is, $\mu_{\bar{x}} = \mu$, and the standard deviation of the sample means is considerably less than the population standard deviation.

COMMENT In the balance of this chapter we will always assume that our sample size is less than 5% of the population size. Consequently, we will use $\sigma_{\bar{x}} = \dfrac{\sigma}{\sqrt{n}}$ as the standard error of the mean.

Let us summarize our discussion up to this point. Using the distribution of the sample means of the laboratories and the distribution of the tar content of all 2000 cigarettes, or the distribution of the sample means of the ages, we conclude the following:

The mean of the distribution of sample means and the mean of the original population are the same.

The standard deviation of the distribution of the sample means is less than the standard deviation of the original population. The exact relationship is referred to as the standard error of the mean and is found by using Formula 8.5.

The distribution of the sample means is approximately normally distributed.

COMMENT The last statement is so important that it is referred to as the **Central Limit Theorem.** Since much of the work of statistical inference is based on this theorem, we will discuss its importance, as well as its applications in detail, in the following sections.

EXERCISES FOR SECTION 8.5

1. Dr. Raleigh McAllister, a leading obstetrician, reports having delivered 9, 10, 5, 8, 7, and 6 healthy babies during the first six weeks of 1991.
   a. Make a list of all possible samples of size 2 that can be drawn from this list of numbers.
   b. Determine the mean of each of these samples and form a sampling distribution of these sample means.
   c. Find the mean, $\mu_{\bar{x}}$, of this sampling distribution.
   d. Find the standard deviation, $\sigma_{\bar{x}}$, of this sampling distribution.

2. The State Unemployment Insurance Board indicates that there was the following number of new claimants for unemployment insurance benefits during the week of February 1 through 5.

Feb. 1	Feb. 2	Feb. 3	Feb. 4	Feb. 5
54	118	62	79	92

   a. Make a list of all possible samples of size 2 that can be drawn from these numbers.
   b. Determine the mean of each of these samples and form a sampling distribution of the sample means.
   c. Find the mean, $\mu_{\bar{x}}$, of this sampling distribution.
   d. Find the standard deviation, $\sigma_{\bar{x}}$, of this sampling distribution.

3. The Billings Volunteer Ambulance Corps responds to emergency situations and provides medical care as needed. During one week it responded to the following number of medical emergencies.

Mon.	Tues.	Wed.	Thurs.	Fri.	Sat.
45	48	39	52	37	44

a. Make a list of all possible samples of size 3 that can be drawn from these numbers.

b. Determine the mean of each of these samples and form a sampling distribution of the sample means.

c. Find the mean, $\mu_{\bar{x}}$, of this sampling distribution.

d. Find the standard deviation, $\sigma_{\bar{x}}$, of this sampling distribution.

4. The Lincoln Taxi Service keeps accurate records on the average number of trips made to the airport per week by its drivers. During the first six weeks of 1991 the drivers made an average of 14, 28, 11, 13, 9, and 19 trips. Find the mean and standard deviation of the average number of trips that the taxi drivers make to the airport.

5. U.S. Custom agents inspect all incoming luggage at the nation's airports. The number of arrests by custom agents of people attempting to smuggle contraband into the United States on February 12 at five of our international airports was 9, 18, 13, 12, and 23.

a. Determine the mean, μ, and standard deviation, σ, of the population of the number of arrests that these agents made.

b. List all possible samples of size 2 *and* of size 3 that can be selected from these numbers and determine the mean for each of these samples.

c. Find the mean, $\mu_{\bar{x}}$, and standard deviation, $\sigma_{\bar{x}}$, for the sample means in each case.

d. Show that for both samples of size 2 and samples of size 3

$$\sigma_{\bar{x}} = \frac{\sigma}{\sqrt{n}} \sqrt{\frac{N-n}{N-1}}$$

6. Banking officials claim that the average amount of money in savings accounts at all branches of one particular bank is $6428, with a standard deviation of $461., Many samples of size 100 are taken. Find the mean of these samples and the standard error of the mean.

7. Refer back to Exercise 6. What would the sample mean *and* standard error of the mean be if the samples are of

a. size 144 each?   b. size 64 each?

8.6 THE CENTRAL LIMIT THEOREM

One of the most important theorems in probability is the **Central Limit Theorem.** This theorem, first established by De Moivre in 1733 (see the discussion on page 355), was named "The Central Limit Theorem of Probability" by G. Polya in 1920. The theorem may be summarized as follows.

THE CENTRAL LIMIT THEOREM

If large random samples of size n (usually samples of size $n > 30$) are taken from a population with mean μ and standard deviation σ, and if a sample mean $\bar{x}$ is computed for each sample, then the following three facts will be true about the distribution of sample means.

1. The distribution of the sample means will be approximately normally distributed.
2. The mean of the sampling distribution will be equal to the mean of the population. Symbolically,

$$\mu_{\bar{x}} = \mu$$

3. The standard deviation of the sampling distribution will be equal to the standard deviation of the population divided by the square root of the number of items in each sample. Symbolically,

$$\sigma_{\bar{x}} = \frac{\sigma}{\sqrt{n}}$$

COMMENT If the sample size is large enough, the sampling distribution will be normal, even if the original distribution is not. "Large enough" usually means larger than 30 items in the sample.

COMMENT The n referred to in the theorem refers to the size of each sample, not to the number of samples.

Since the Central Limit Theorem is so important, we will discuss its applications in the next section.

8.7 APPLICATIONS OF THE CENTRAL LIMIT THEOREM

In this section we use the Central Limit Theorem to predict the behavior of sample means. To apply the standardized normal distribution discussed in Chapter 7, we have to change Formula 7.1 somewhat. Recall that

$$z = \frac{x - \mu}{\sigma}$$

It can be shown that when dealing with sample means this formula becomes that given as Fomula 8.6.

FORMULA 8.6

$$z = \frac{\bar{x} - \mu}{\sigma/\sqrt{n}}$$

The following examples illustrate how the Central Limit Theorem is applied.

EXAMPLE 1

The average height of all the workers in a hospital is known to be 65 inches with a standard deviation of 2.3 inches. If a sample of 36 people is selected at random, what is the probability that the average height of these 36 people will be between 64 and 65.5 inches?

Solution

We use Formula 8.6. Here $\mu = 65$, $\sigma = 2.3$, and $n = 36$. Thus, $\bar{x} = 64$ corresponds to

$$z = \frac{64 - 65}{2.3/\sqrt{36}} = \frac{-1}{0.3833} = -2.61$$

and $\bar{x} = 65.5$ corresponds to

$$z = \frac{65.5 - 65}{2.3/\sqrt{36}} = \frac{0.5}{0.3833} = 1.30$$

Thus, we are interested in the area of a standard normal distribution between $z = -2.61$ and $z = 1.31$. See Figure 8.5.

From Table IV in the Appendix we find that the area between $z = 0$ and $z = -2.61$ is 0.4955 and that the area between $z = 0$ and $z = 1.30$ is 0.4032. Adding, we get

$$0.4955 + 0.4032 = 0.8987$$

Thus, the probability that the average height of the sample of 36 people is between 64 and 65.5 inches is 0.8987.

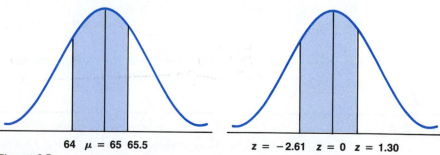

64 $\mu = 65$ 65.5 $z = -2.61$ $z = 0$ $z = 1.30$

Figure 8.5

EXAMPLE 2

(handwritten:) $\mu = \$5000$
$\sigma = \$650$
$N = 36$

The average amount of money that a depositor of the Second National City Bank has in an account is $5000 with a standard deviation of $650. A random sample of 36 accounts is taken. What is the probability that the average amount of money that these 36 depositors have in their accounts is between $4800 and $5300?

Solution

We use Formula 8.6. Here $\mu = 5000$, $\sigma = 650$, and $n = 36$. Thus, $\bar{x} = 4800$ corresponds to

$$z = \frac{4800 - 5000}{650/\sqrt{36}} = \frac{-200}{108.33} = -1.85$$

and $\bar{x} = 5300$ corresponds to

$$z = \frac{5300 - 5000}{650/\sqrt{36}} = \frac{300}{108.33} = 2.77$$

Thus, we are interested in the area between $z = -1.85$ and $z = 2.77$. See Figure 8.6.

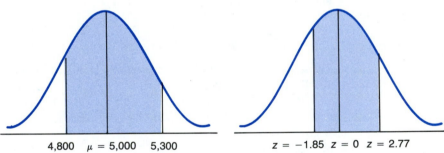

4,800 $\mu = 5,000$ 5,300 $z = -1.85$ $z = 0$ $z = 2.77$

Figure 8.6

From Table IV in the Appendix we find that the area between $z = 0$ and $z = -1.85$ is 0.4678 and that the area between $z = 0$ and $z = 2.77$ is 0.4972. Adding these two, we get

$$0.4678 + 0.4972 = 0.9650$$

Thus, the probability is 0.9650 that the average amount of money these depositors have in their accounts is between $4800 and $5300.

EXAMPLE 3

The average purchase by a customer in a large novelty store is $4.00 with a standard deviation of $0.85. If 49 customers are selected at random, what is the probability that their average purchases will be less than $3.70?

(handwritten:) $\mu = \$4.00$
$\sigma = \$0.85$
$N = 49$
$z = \dfrac{3.70 - 4.00}{0.85/\sqrt{49}}$ $\dfrac{-.30}{.1214} = -2.47$

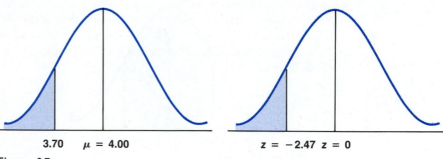

3.70 $\mu = 4.00$ $z = -2.47 \quad z = 0$

Figure 8.7

Solution

We use Formula 8.6. Here $\mu = 4.00$, $\sigma = 0.85$, and $n = 49$. Thus, $\bar{x} = 3.70$ corresponds to

$$z = \frac{3.70 - 4.00}{0.85/\sqrt{49}} = \frac{-0.30}{0.1214} = -2.47$$

Therefore, we are interested in the area to the left of $z = -2.47$ (Fig. 8.7).

From Table IV in the Appendix we find that the area from $z = 0$ to $z = -2.47$ is 0.4932. Thus, the area to the left of $z = -2.47$ is $0.5000 - 0.4932$, or 0.0068. Therefore, the probability that the average purchase will be less than \$3.70 is 0.0068.

EXAMPLE 4

The Smith Trucking Company claims that the average weight of its delivery trucks when fully loaded is 6000 pounds with a standard deviation of 120 pounds. 36 trucks are selected at random and their weights recorded. Within what limits will the average weights of 90% of the 36 trucks lie?

Solution

Here $\mu = 6000$ and $\sigma = 120$. We are looking for two values within which the weights of 90% of the 36 trucks will lie. From Table IV we find that the area between $z = 0$ and $z = 1.65$ is approximately 0.45. Similarly, the area between $z = 0$ and $z = -1.65$ is approximately 0.45. See Figure 8.8. Using Formula 8.6, we have

$$z = \frac{\bar{x} - \mu}{\sigma/\sqrt{n}}$$

If $z = 1.65$, then

$$1.65 = \frac{\bar{x} - 6000}{120/\sqrt{36}}$$

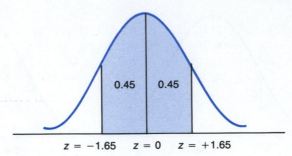

$$z = -1.65 \qquad z = 0 \qquad z = +1.65$$

Figure 8.8

$$= \frac{\bar{x} - 6000}{20}$$

$$1.65(20) = \bar{x} - 6000$$

$$33 + 6000 = \bar{x}$$

$$6033 = \bar{x}$$

If $z = -1.65$, then

$$-1.65 = \frac{\bar{x} - 6000}{120/\sqrt{36}}$$

$$= \frac{\bar{x} - 6000}{20}$$

$$-1.65(20) = \bar{x} - 6000$$

$$6000 - 33 = \bar{x}$$

$$5967 = \bar{x}$$

Thus, 90% of the trucks will weigh between 5967 and 6033 pounds.

EXERCISES FOR SECTION 8.7 *Review*

Find z then %

1. Many automobile insurance companies charge premiums for auto coverage depending on the number of miles driven annually by the principal driver. Moreover, the Tri-Cal Insurance Company claims that it insures automobiles whose principal operators drive an average of 10,200 miles a year with a standard deviation of 860 miles. A random survey of 49 of this company's automobile policies is taken. What is the probability that the average number of miles driven annually by the principal operator is between 9,800 and 10,500 miles?

2. Global Airways claims that the average number of people who pay for in-flight movies when the plane is fully loaded is 23 with a standard deviation of 2.9. A sample of 36 fully loaded planes is taken. Find the probability that fewer than 22 people paid for in-flight movies.

3. Insurance claims indicate that the average charge by an obstetrician in a certain large northeastern city during 1991 for a Caesarean section delivery was $3900 with a standard deviation of $525. A random survey of 36 obstetricians from this city is taken. What is the probability that the average charge by these doctors will be more than $4100?

4. The mean breaking strength of a certain metal alloy is 3000 pounds with a standard deviation of 640 pounds. A random sample of 64 pieces of this metal alloy is taken. What is the probability that these pieces will have a mean breaking strength that is more than 3100 pounds?

5. An auditor for a utilities company claims that the average amount owed on delinquent accounts is $86.40 with a standard deviation of $7.95. What is the probability that a random sample of 81 delinquent accounts will have an average amount owed of $88.00 or more?

6. The average age at which a woman has her first child in a certain city is 26 years with a standard deviation of 5.6 years. A random sample of 49 women is selected. What is the probability that the average age at which these women had their first child is between 24 and 25 years?

7. The Delrio Baking Company claims that the average weight of all its chocolate brownie cupcakes is 5 ounces with a standard deviation of 0.19 ounces. The local consumer's complaint bureau selects a sample of 35 cupcakes and records their weight. Within what limits should the weights of 95% of the cupcakes lie?

8. A certain farm cooperative claims that the average weight of its packaged bags of potatoes is 5 pounds or 80 ounces with a standard deviation of 4 ounces. To check on the accuracy of the stated weight, a local consumer's complaint bureau official selects a random sample of 35 bags of potatoes. Within what limits should 95% of these bags lie?

9.

NEW HOUSING STARTS DOWN DRAMATICALLY

January 25—The economic recession that is currently gripping our nation is affecting the building industry particularly hard. In our region only 17 new housing starts were reported this past month. In a city where the average age of a house is 43.3 years (with a standard deviation of 8.2 years) such devastating facts can wreak havoc on our already staggering economy.

January 25, 1991

$$z = \frac{\bar{x} - m}{\sigma / \sqrt{n}}$$

Consider the preceding newspaper article. If 100 homes in this city are randomly selected, what is the probability that the average age of these houses will be between 42 and 45 years old?

10. A survey conducted by Giardino and Johnson found that the average amount of money spent by an individual in Morganville on automobile repairs is $921 per year with a standard deviation of $102. If a survey of 49 randomly selected individuals in Morganville is taken, what is the probability that they spend an average of at most $900 on automobile repairs?

11. Banking officials claim that the average amount of money that a person keeps on deposit in a checking account is $438 with a standard deviation of $87. If 100 checking accounts are randomly selected, find the probability that the average amount of money on deposit in these accounts is between $400 and $450.

8.8 USING COMPUTER PACKAGES

The **MINITAB** statistical software can be used to generate, for example, a sampling distribution of x based on 25 samples of size $n = 4$ measurements, each drawn from a normal population with $\mu = 10$ and $\sigma = 3$. We first have to use RANDOM and AVERAGE to generate each sample of size $n = 4$ measurements and to compute the sample mean. To accomplish this, we generate one sample of size $n = 4$ as follows:

Sample 1

```
MTB  > RANDOM 4 OBS, INTO C1;
SUBC > NORMAL MU = 10, SIGMA = 3.
```

The following appears in C1: 8.9321 11.6106 10.1234 9.7862

```
MTB > AVERAGE THE OBSERVATIONS IN C1
      MEAN = 10.113
```

To generate a second sample, we repeat the RANDOM and AVERAGE statements again and set the results into C2. We repeatedly use the same procedure until we obtain the 25 samples of size $n = 4$ measurements that we desire. This can be very time-consuming. (We could use the MINITAB command

```
MTB  > RANDOM 4 OBS, INTO C1 - C25;
SUBC > NORMAL MU = 10, SIGMA = 3.
```

to accomplish the same result.) After the 25 sample means are obtained we enter these results into MINITAB to obtain some numerical descriptive measures as was done in Chapter 3. For our example we enter the 25 sample means into MINITAB as follows:

```
MTB  > SET THE FOLLOWING DATA INTO C26
DATA >   10.113    10.101    9.996    10.002    10.123
DATA >    9.468     9.823    9.982    10.176    10.082
            .         .         .         .         .
            .         .         .         .         .
            .         .         .         .         .

DATA >  END
```

Then we have MINITAB obtain numerical descriptive measures for the data. We get

```
MTB > DESCRIBE C26
```

The computer output for our example will be

	N	MEAN	MEDIAN	TRMEAN	ST DEV	SEMEAN
C26	25	10.016	10.183	10.213	1.386	0.0812

	MIN	MAX	Q1	Q3
C26	7.938	10.816	8.096	11.547

```
MTB > STOP
```

COMMENT The mean of the sample means is 10.016. This is fairly close to the theoretical value of $\mu = 10$. Similarly, the standard deviation of the sample means is 1.386, which is close to the theoretical value of $\dfrac{\sigma}{\sqrt{n}} = \dfrac{3}{\sqrt{4}} = 1.5$. These experimental values would be closer to their respective theoretical values had we taken many more samples of size n = 4 rather than only the 25 samples that we generated.

8.9 SUMMARY

In this chapter we discussed the nature of random sampling and how to go about selecting a random sample. The most convenient way of selecting a random sample is to use a table of random digits.

In some situations, as we pointed out, stratified samples are preferred.

When repeated random samples are taken from the same population, different sample means are obtained. The average of these sample means can be used as an estimate of the population mean. Of course, these sample means form a distribution. If the samples are large enough, the Central Limit Theorem tells us that they will be normally distributed. Furthermore, the mean of the sampling distribution is the same as the population mean. The standard deviation of the sampling distribution is less than the population standard deviation. The Central Limit Theorem led to many useful applications.

Study Guide

The following is a chapter outline in capsule form. You should now be able to demonstrate your knowledge of the ideas mentioned by giving definitions, descriptions, or specific examples. Page references are given in parentheses.

Samples are studied to obtain valuable information about a larger group called the **population**. Any part of the population is called a **sample**. (page 401)

A **random sample** of n items is a sample selected from a population in such a way that every different sample of size n from the population has an equal chance of being selected. (page 402)

Random sampling is the procedure by which a random sample is obtained. (page 402)

A **table of random numbers** or a **table of random digits** is a chart that lists the digits 0, 1, 2, ..., 9 arranged in such a way that all the digits appear with approximately the same frequency. (page 404)

To obtain a **stratified sample**, we first divide the entire population into a number of groups or strata. Then we select a random sample from each of these groups or from each stratum. (page 406)

To use **systematic sampling**, we first categorize the elements of the population in some way, such as alphabetically or numerically. Then we randomly select a starting point. We include in our sample every ith term of the categorized population until we obtain a sample of size n. (page 407)

When using **cluster sampling**, we first divide the target population to be analyzed into mutually exclusive subgroups called **primary subgroups** or **clusters**, each of which should be representative of the entire population. Then a random sample of these clusters is selected so as to provide estimates of the population values. (page 407)

A fairly good estimate of the population mean can be obtained by calculating the mean of samples from the population; that is, $\mu_{\bar{x}} = \mu$. (page 409)

The **standard deviation of the sample means** is given by

$$\sqrt{\frac{\Sigma(\bar{x} - \mu_{\bar{x}})^2}{n - 1}} \quad \text{or} \quad \sqrt{\frac{n(\Sigma \bar{x}) - (\Sigma \bar{x})^2}{n(n - 1)}}$$

n always represents the number of sample means. (pages 409 and 411)

The **distribution of the sample means**, $\bar{x}$, shows us how the means vary from sample to sample. It is also referred to as the **sampling distribution of the mean**. (page 413)

The standard deviation of the sampling distribution of the mean is referred to as the **standard error of the mean** and is given by

$$\sigma_{\bar{x}} = \frac{\sigma}{\sqrt{n}} \cdot \sqrt{\frac{N - n}{N - 1}} \qquad \text{for finite populations of size } N$$

and

$$\sigma_{\bar{x}} = \frac{\sigma}{\sqrt{n}} \qquad \text{for infinite populations} \qquad \text{(page 414)}$$

The factor $\sqrt{\dfrac{N-n}{N-1}}$ used in calculating the standard error of the mean is referred to as the **finite population correction factor**. (page 415)

If large random samples of size n (usually samples of size $n > 30$) are taken from a population with mean μ and standard deviation σ, and if a sample mean $\bar{x}$ is computed for each sample, then the **Central Limit Theorem** says that the following three facts will be true about the distribution of sample means.

1. The distribution of the sample means will be approximately normally distributed.
2. The mean of the sampling distribution will be equal to the mean of the population; that is, $\mu_{\bar{x}} = \mu$.
3. The standard deviation of the sampling distribution will be equal to the standard deviation of the population divided by the square root of the number of items in each sample; that is, $\sigma_{\bar{x}} = \dfrac{\sigma}{\sqrt{n}}$. (page 420)

Formulas to Remember

You should be able to identify each symbol in the following formulas, understand the relationships among the symbols expressed in each formula, understand the significance of each formula, and use the formulas in solving problems.

1. Population mean: $\mu = \dfrac{\Sigma x}{N}$

2. Sample mean: $\bar{x} = \dfrac{\Sigma x}{n}$

3. Population standard deviation: $\sigma = \sqrt{\dfrac{\Sigma(x - \mu)^2}{N}}$

4. Standard deviation of sample means: $\sqrt{\dfrac{\Sigma(\bar{x} - \mu_{\bar{x}})^2}{n - 1}}$

5. Computational formula for standard deviation of sample means:

   $$\sqrt{\dfrac{n(\Sigma \bar{x}^2) - (\Sigma \bar{x})^2}{n(n - 1)}}$$

6. Standard error of the mean when sample size is less than 5% of the population size: $\sigma_{\bar{x}} = \dfrac{\sigma}{\sqrt{n}}$. The finite population correction factor $\sqrt{\dfrac{N-n}{N-1}}$ adjusts the standard error to describe most accurately the amount of variation. It should be applied whenever the sample size is 5% or more of the population size.

7. When dealing with sample means (Central Limit Theorem): $z = \dfrac{\bar{x} - \mu}{\sigma/\sqrt{n}}$

Testing Your Understanding of This Chapter's Concepts

1. What relationship exists, if any, between the population standard deviation, the standard error of the mean, and the sample size?

2. A student claims that if a sample is selected from an infinite population, then we can assume that the sample is less than 5% of the population, and so we can ignore the finite population correction factor. Do you agree? Explain your answer.

3. A supermarket chain has 70 stores in the five boroughs of New York City as follows.

Borough	Number of stores
Brooklyn	17
Staten Island	12
Queens	22
Bronx	8
Manhattan	11

   a. Determine the mean, μ, and standard deviation, σ, of the population of the number of stores that this supermarket chain has.

   b. List all possible samples of size 2 *and* size 3 that can be selected from these numbers and determine the mean of each of these samples.

   c. Find the mean, $\mu_{\bar{x}}$, and the standard deviation, $\sigma_{\bar{x}}$, for the sample means in each case.

   d. Show that for both the samples of size 2 and size 3

   $$\sigma_{\bar{x}} = \frac{\sigma}{\sqrt{n}} \cdot \sqrt{\frac{N-n}{N-1}}$$

   e. Draw the histogram for these sample means (similar to what was done in Figure 8.2).

THINKING CRITICALLY

1. The lengths of fish caught in a certain lake over the years are approximately normally distributed with a mean of 14 inches and a standard deviation of 2.63 inches.

   a. What is the probability that the mean length of 36 fish caught in this lake by Joe will be more than 16 inches long?

   b. What is the probability that the mean length of 36 fish caught from this lake by Joe and his friends will be greater than 16 inches?

2. Explain why it is true that a sample mean from a large sample size gives us a better estimate of the population mean than a sample mean from a smaller sample size.

3. Explain the difference between σ_x and $\sigma_{\bar{x}}$. (*Hint:* Use sketches of normal distributions to illustrate your answers.)

Review Exercises for Chapter 8

1. There are 58 math majors at Brooks University. The Mathematics Department wishes to sample 20 of these majors to determine whether a course in discrete mathematics should be offered in the coming semester. Assume that each of these students has a number 1, 2, 3, ..., 58. If it is decided to use column 7 of Table VI in the Appendix to obtain this information, which students will be surveyed?

2. Studies show that the average amount of time that a 7-year-old child watches television per week is 17.4 hours with a standard deviation of 4.96 hours. If a sample of fifty 7-year-old children is taken, what is the probability that these youngsters watch television more than an average of 19 hours per week?

3. A spokesperson for the Neeko Photocopy Machine Corp. claims that the P-21 gear in the finishing unit of the photocopying machines that it services will provide an average of 275,000 copies before replacement is needed. The standard deviation is 14,275 copies. If a sample of 100 such gears is randomly selected, find the probability that the average number of copies provided by these gears before replacement is needed will be at least 270,000 copies.

4. A certain soda filling machine fills each bottle of soda with an average of 28 ounces of soda with a standard deviation of 1.69 ounces. If a sample of 50 bottles of soda filled by this machine is randomly selected, within what limits will the average amount of soda contained in 90% of these bottles lie?

5. The average 36-month home equity loan at Wesley National Bank is for $8900 with a standard deviation of $987. If a sample of 75 such home equity loans is randomly selected, what is the probability that the average amount of such a loan will be for at least $8700?

6. When a person defaults on an auto loan, the lending bank will take title of the car and then attempt to recover its money by selling the car at an auction. Past experience indicates that the average price of a repossessed car at The Galaxy Auto Auctions is $3200 with a standard deviation of $862. If a sample of 30 such auctioned cars is randomly selected, what is the probability that the average price of these cars is at most $3500?

7. Students at McDougald High School had an average score of 529 on the mathematics part of the Scholastic Aptitude Test (SAT). The standard deviation was 27. If a sample of 35 student scores at this school is randomly selected, above what score will the average of 90% of these student scores lie?

8. In 1991 there were 13,491 MD license plates issued by the Motor Vehicle Bureau of the state. Many newspapers in the state claim that some doctors do not drive at all but merely register the car for their friends to enable them to park freely in the city. It is decided to take a random survey of 25 doctors to determine the extent of this practice. If columns 2 through 5 of Table VI in the appendix are used, which doctors will be surveyed?

9. Each of five police stations has been testing 100 tires of a particular brand to determine the durability of the tires. The average life of the tires at each of these

stations was 49,000, 38,000, 31,000, 33,000, and 29,000 miles. Find the mean and standard deviation of these samples.

10. Insurance industry officials report that the average amount of life insurance in effect per covered family in Goshen is $68,500 with a standard deviation of $8750. A random survey of 64 covered families in Goshen is selected. What is the probability that the average amount of insurance in effect for these families is at least $70,000?

Chapter Test

1. An official of the U.S. Social Security Administration claims that the average monthly Social Security payment that a retired worker receives is $423 with a standard deviation of $78. To verify the accuracy of this claim, a random sample of 49 retired workers receiving Social Security payments is selected. Within what limits will the mean of the sample lie with a probability of 0.90?

2. Coastway Airlines claims that the average weight of luggage checked in by each passenger is 36 pounds with a standard deviation of 3.61 pounds. If a random sample of 40 pieces of checked-in luggage is taken, what is the probability that the average weight of these pieces of luggage will be at least 38 pounds?

3. The length of time between car arrivals at a drive up window of a fast food restaurant on a major state highway and the length of time required by the attendant to fill the order are two random variables that are important to management in helping decide how many drive-up windows to install and how many attendants to hire for efficient operations. Studies of one restaurant show that the interarrival time (the time between the arrivals of two consecutive cars) has a mean of 4.87 minutes and a standard deviation of 2.69 minutes. A random sample of the interarrival times of 64 cars is selected. What is the probability that the average interarrival time will exceed 5 minutes?

4. Government records indicate that the average selling price of a two-family house in Harris was $170,000 with a standard deviation of $12,000. If a random sample of 40 recently sold homes in Harris is selected, what is the probability that the average selling price of these houses will be at least $165,000?

5. U.S. Department of Energy officials report that the average annual expenditure for home heating oil (for a two-family home) in a particular northeastern city during 1990 was $2175. The standard deviation was $182. Suppose 30 two-family homes are randomly selected. What is the probability that the average annual expenditure for home heating oil for these homes will be between $2200 and $2300?

6. The number of children in each of the five families that were winners in a recent New York State Lottery is as follows.

Family	Number of Children
A	4
B	3
C	6
D	5
E	2

    a. Make a list of all possible samples of size 2 that can be drawn from these numbers.

    b. Determine the mean of each of these samples and form a sampling distribution of these sampling means.

    c. Find the mean, $\mu_{\bar{x}}$, of this sampling distribution.

    d. Find the standard deviation of this sampling distribution.

  7. Each car that is manufactured in the United States has a serial number. After numerous accidents involving a sticking accelerator rod, one of the auto manufacturers has decided to randomly check 35 cars sold on the East Coast. According to company records, cars sold on the East Coast during 1990 had serial numbers whose last five digits were between 30,000 and 71,000. If columns 6 and 7 of Table VI in the Appendix are used, which cars will be checked?

  8. Banking industry records indicate that the average weekly income for all part-time tellers in a large metropolitan area is $188.72 with a standard deviation is $12.41. A random sample of 49 part-time tellers is taken. What is the probability that the average weekly income of these tellers will be less than $190?

  9. A tire manufacturer claims that the best quality tire sold by the company will last an average of 35,000 miles before wearing out. The standard deviation is 5600 miles. To check on the accuracy of this claim, a consumer's group randomly selects numerous samples each consisting of 100 such tires manufactured by this company. Within what limits should the average mileage of 90% of these samples lie?

 10. A population is known to have a mean of 168 and a standard deviation of 53. Many samples of size 64 are taken. Find the mean of these samples and the standard error of the mean.

 11. There are five female pediatricians at the Rao Clinic. The number of years that each has been with the clinic is 5, 10, 4, 8, and 13.

    a. Determine the mean, μ, and standard deviation, σ, of the population of the number of years that these five female pediatricians have been at the clinic.

    b. List all possible samples of size 2 *and* size 3 that can be selected from these numbers, and determine the mean for each of these samples.

    c. Find the mean, $\mu_{\bar{x}}$, and standard deviation, $\sigma_{\bar{x}}$, for the ten sample means that are obtained in each case.

    d. Show that for both the samples of size 2 and size 3

$$\sigma_{\bar{x}} = \frac{\sigma}{N}\sqrt{\frac{n-N}{n-1}}$$

12. The average strength of a certain plastic material (on a special scale) is 124 with a standard deviation of 9.3. What is the probability that the average strength of 36 randomly selected pieces of this plastic material will be greater than 127?
    a. 0.2560   **b.** 0.5256   **c.** 0.4744   **d.** 0.0256   **e.** none of these

13. Five police stations have each been testing 100 tires of a particular brand. The average life of the tires at each of these stations was 41,000, 36,000, 25,000, 31,000 and 29,000 miles. Find the overall average of these sample means.
    a. 33,700   **b.** 34,200   **c.** 32,400   **d.** 40,000   **e.** none of these

14. In the previous exercise find the standard deviation of these samples.
    a. 6228.9646   **b.** 78.9238   **c.** 4169.38   **d.** 3326.814   **e.** none of these

15. The average life of a transistor produced by the Apex Company is 200 hours with a standard deviation of 11.2 hours. If a random sample of 49 transistors produced by this company is selected, what is the probability that the average life will be between 196 and 203 hours?
    a. 0.4637   **b.** 0.9637   **c.** 0.0363   **d.** 0.0726   **e.** none of these

16. Refer back to the previous question. If a random survey of 49 transistors is selected, within what limits should the life of 95% of these transistors lie?
    a. between 196.864 and 203.136 hours   **b.** between 198.21 and 201.39 hours   **c.** between 197.426 and 201.738 hours   **d.** between 196.213 and 202.897 hours   **e.** none of these

17. The average weight of a student at Bork College is 125 pounds with a standard deviation of 9 pounds. Within what limits should the average weights of 95% of 50 randomly selected students of this college lie?
    a. between 138.39 and 149.72 pounds   **b.** between 129.79 and 142.21 pounds   **c.** between 125.59 and 126.41 pounds   **d.** between 122.51 and 127.49 pounds   **e.** none of these

Suggested Further Reading

1. Lapin, L. *Statistics for Modern Business Decisions*, 4th ed. Orlando, FL: Harcourt, Brace, Jovanovich, 1987.
2. Mansfield, Edwin. *Statistics for Business and Economics*, 3rd ed. New York: W.W. Norton, 1987.
3. Mason, Robert D. *Statistical Techniques in Business and Economics*, 6th ed. Homewood, IL: Irwin, 1986.
4. Mendenhall, W. L. *Introduction to Probability and Statistics*, 7th ed. Boston, MA: Duxbury Press, 1987.
5. Mendenhall, W. L., L. Ott, and R. L. Scheaffer. *Elementary Survey Sampling*. Belmont, CA: Wadsworth, 1968.
6. Raj, Des. *The Design of Sample Surveys*. New York: McGraw-Hill, 1972.
7. Ryan, Barbara F., Brian L. Joiner, and Thomas A. Ryan Jr. *Minitab Handbook*, 2nd ed. Boston: Prindle, Weber and Schmidt, 1985.
8. Slonim, M. *Sampling*. New York: Simon & Schuster, 1971.
9. "For Better Polls." *Business Week*, June 25, 1949, p. 24.

9

Estimation

Using data from some sample runners, we can often obtain confidence intervals for results about all runners. See problem 1 in exercises for Section 9.6. (© *Chromosohm/Joseph Sonn, Fran Heyl Associates*)

In this chapter we indicate how sample statistics can be used to provide estimates for the mean of a population, the proportion (percentage) of a population that has a particular attribute, and the standard deviation of the population. Since such estimates are necessarily subject to sampling error, we can provide information about the accuracy of such estimates by setting up confidence intervals and by selecting a sample of the appropriate size.

CHAPTER OBJECTIVES

After studying the material in this chapter, you should be able:

- **To discuss** how sample data can often be used to estimate certain unknown quantities. This use of samples is called statistical estimation. (Section 9.1)

- **To point out** that population parameters are statistical descriptions of the population. (Section 9.2)

- **To understand** that sample data can be used to obtain both point and interval estimates. Point estimates give us a single number, whereas interval estimates set up an interval within which the parameter is expected to lie. (Section 9.2)

- **To see** that degrees of confidence give us the probability that the interval will actually contain the quantity that we are trying to estimate. (Section 9.3)

- **To apply** the Central Limit Theorem to set up confidence intervals for the mean and standard deviation. (Section 9.4)

- **To use** the Student's *t*-distribution to set up confidence intervals when the sample size is small. (Sections 9.4 and 9.5)

- **To analyze** how we determine the correct size of a sample for a given allowable error. (Section 9.6)

- **To indicate** how the Central Limit Theorem is also used to set up confidence intervals for population proportions. (Section 9.7)

LIFE EXPECTATION BY SEX FOR SOME COUNTRIES AVERAGE LIFETIME (in years)

Country	Male	Female
North America		
Canada	68.7	75.1
United States	64.7	75.2
Mexico	59.4	63.4
Puerto Rico	68.9	75.2
Europe		
Belgium	67.7	73.5
Denmark	70.7	75.9
England and Wales	68.9	75.1
France	68.5	76.1
Italy	67.9	73.4
Norway	71.1	76.8
Poland	66.8	73.8
Sweden	72.0	77.4
Switzerland	69.2	75.0
U.S.S.R.	65.0	74.0
Yugoslavia	65.3	70.1
Asia		
India	41.9	40.6
Israel	70.1	72.8
Japan	70.5	75.9
Jordan	52.6	52.0
Korea	59.7	64.1
Africa		
Egypt	51.6	53.8
Nigeria	37.2	36.7

Source: Information Please Yearbook.

Consider the information contained in the preceding table. How do we calculate the average life expectancy for both males and females? Does life expectancy depend on the sex of the person and on the country in which the person lives? Such information has important implications for demographers.

Very often we use currently available information to make predictions about the future. Great care must be exercised in making such predictions as there are many factors that can offset the reliability of such estimates. The same is true for the population forecasts presented in the article that follows.

U.S. POPULATION—NEW FORECAST

The U.S. population will rise from 238.2 million now to 267.5 million by the year 2000—the result of a growth rate slowed from 1.1 percent annually in the '70s to 0.8 for the rest of the century.

That's the latest projection of the National Planning Association, which also predicts that 84 percent of the population expansion will occur in the West and South. Arizona, Nevada, and Florida will have the fastest growth rates, and California, Texas, and Florida will gain the most people.

	1985	2000		1985	2000
California	25.8 mil.	30.4 mil.	Oklahoma	3.2 mil.	3.7 mil.
New York	17.6 mil.	17.5 mil.	Connecticut	3.2 mil.	3.3 mil.
Texas	15.9 mil.	20.0 mil.	Iowa	2.9 mil.	3.0 mil.
Pennsylvania	11.9 mil.	12.1 mil.	Oregon	2.8 mil.	3.4 mil.
Florida	11.6 mil.	15.6 mil.	Mississippi	2.6 mil.	2.9 mil.
Illinois	11.6 mil.	11.9 mil.	Arkansas	2.4 mil.	2.8 mil.
Ohio	10.9 mil.	11.2 mil.	Kansas	2.4 mil.	2.6 mil.
Michigan	9.3 mil.	9.8 mil	West Virginia	1.9 mil.	1.8 mil.
New Jersey	7.5 mil.	7.8 mil.	Utah	1.6 mil.	2.2 mil.
North Carolina	6.3 mil.	7.6 mil.	Nebraska	1.6 mil.	1.6 mil.
Georgia	5.9 mil.	6.9 mil.	New Mexico	1.4 mil.	1.7 mil.
Massachusetts	5.9 mil.	6.3 mil.	Maine	1.2 mil.	1.3 mil.
Virginia	5.6 mil.	6.4 mil.	New Hampshire	1.0 mil.	1.3 mil.
Indiana	5.6 mil.	5.9 mil.	Idaho	1.0 mil.	1.2 mil.
Missouri	5.0 mil.	5.2 mil.	Hawaii	1.0 mil.	1.1 mil
Tennessee	4.8 mil.	5.5 mil.	Rhode Island	997,000	1.2 mil.
Wisconsin	4.8 mil.	5.1 mil.	Nevada	944,000	1.3 mil.
Washington	4.5 mil.	5.3 mil.	Montana	801,000	807,000
Louisiana	4.4 mil.	4.7 mil.	South Dakota	692,000	713,000
Maryland	4.3 mil.	4.7 mil.	North Dakota	666,000	672,000
Minnesota	4.2 mil.	4.4 mil.	Delaware	615,000	657,000
Alabama	4.1 mil.	4.6 mil.	Dist. of Columbia	597,000	535,000
Kentucky	3.7 mil.	4.1 mil	Vermont	541,000	625,000
South Carolina	3.4 mil.	4.0 mil.	Wyoming	510,000	616,000
Arizona	3.2 mil.	4.7 mil.	Alaska	460,000	583,000
Colorado	3.2 mil.	4.1 mil.	**U.S. Total**	**238.2 mil.**	**267.5 mil**

June 17, 1985

Source: *U.S. News & World Report*, June 17, 1985.

9.1 INTRODUCTION

We have mentioned on several occasions that statistical inference is the process by which statisticians make predictions about a population on the basis of samples. Much information can be gained from a sample. As we mentioned in Chapter 8, the average of the sample means can be used as an estimate of the population mean. Also, we can obtain an estimate of the population standard deviation on the basis of samples. Thus, one use of sample data is to *estimate* certain unknown quantities of the population. This use of samples is referred to as **statistical estimation.**

Statistical Estimation

On the other hand, sample data can also be used either to accept or reject specific claims about populations. To illustrate this use of sample data, suppose a manufacturer claims that the average milligram tar content per cigarette of a particular brand is 15 with a standard deviation of 0.5. Repeated samples are taken to test this claim. If these samples show that the average tar content per cigarette is 22, then the manufacturer's claim is incorrect. If these samples show that the average tar content is within "predictable limits," then the claim cannot be rejected. Thus, sample data can also be used either to accept or reject specific claims about populations. This use of samples is referred to as **hypothesis testing.**

Hypothesis Testing

Statistical inference can be divided into two main categories: problems of estimation and tests of hypotheses. In this chapter we discuss statistical estimation. In Chapter 10 we will analyze the nature of hypothesis testing.

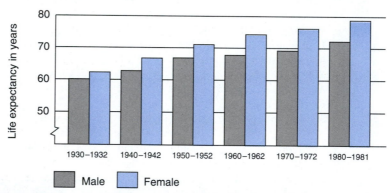

Figure 9.1
The graph gives us an estimate of the life expectancy of men and women. Such information is used by insurance companies in determining the premium to charge for life insurance.

Courtesy of the Metropolitan Life Insurance Company and the Population Reference Bureau (Washington).

9.2 POINT AND INTERVAL ESTIMATES

Population Parameters

In most statistical problems we do not know certain population values such as the mean and the standard deviation. Somehow we want to use the information obtained from samples to estimate their values. These values, which really are statistical descriptions of the population, are often referred to as **population parameters.**

Suppose we are interested in determining the average life of an electric refrigerator under normal operating conditions. A sample of 50 refrigerators is taken and their lives are recorded as shown below:

6.9	7.6	5.7	3.6	7.7	6.6	7.2	7.3	10.6	5.9
8.2	5.7	7.6	8.7	7.9	8.8	7.0	8.1	7.3	6.8
5.7	11.1	8.5	8.9	7.6	5.6	9.0	9.2	6.8	8.3
6.1	9.7	9.8	7.4	6.8	7.3	8.3	9.9	7.5	7.8
7.7	7.4	9.1	7.3	5.5	8.1	6.7	8.8	7.6	5.3

Point Estimate

The average life of these refrigerators is 7.6 years. Since this is the only information available to us, we would logically say that the mean life of *all* similar refrigerators is 7.6 years. This estimate of 7.6 years for the population mean is called a **point estimate** since this estimate is a single number. Of course, this estimate may be a poor estimate, but it is the best we can get under the circumstances.

Our confidence in this estimate would be improved considerably if the sample size were larger. Thus, we would have much greater confidence in an estimate that is based on 5000 refrigerators or 50,000 refrigerators than in one that is based only on 50 refrigerators.

One major disadvantage with a point estimate is that the estimate does not indicate the extent of the possible error. Furthermore, a point estimate does not specify how confident we can be that the estimate is close in value to the parameter that it is estimating. Yet point estimates are often used to estimate population parameters.

Interval Estimation

Another type of estimate that is often used by statisticians, which overcomes the disadvantages mentioned in the previous paragraph, is **interval estimation.** In this method we first find a point estimate. Then we use this estimate to construct an interval within which we can be reasonably sure that the true parameter will lie. Thus, in our example a statistician may say that the mean life of the refrigerators will be between 7.2 and 8.0 years with a 95% degree of confidence. An interval such as this is called a **confidence interval.** The lower and upper boundaries, 7.2 and 8.0, respectively, of the interval are called **confidence limits.** The probability that the procedure used will give a correct interval is called the **degree of confidence.**

Confidence Interval
Confidence Limits
Degree of Confidence

Generally speaking, as we increase the degree of certainty, namely, the degree of confidence, the confidence interval will become wider. Thus, if the length of an interval is very small (with a specific degree of confidence), then a fairly accurate estimate has been obtained.

When estimating the parameters of a population, statisticians use both point and interval estimates. In the next few sections we will indicate how point and interval estimates are obtained.

9.3 ESTIMATING THE POPULATION MEAN ON THE BASIS OF A LARGE SAMPLE

In Chapter 8 we indicated that the average of the sample means can be used as an estimate of the population mean, μ. Moreover, the larger the sample size is, the better the estimate will be. Yet, as we pointed out in Section 9.2, there are some disadvantages with using point estimates.

The Central Limit Theorem (see page 420) says that the sample means will be normally distributed if the sample sizes are large enough. Generally speaking, statisticians say that a sample size is considered large if it is greater than 30. We can use the Central Limit Theorem to help us construct confidence intervals. This is done as follows.

Since the sample means are approximately normally distributed, we can expect 95% of the $\bar{x}$'s to fall between

$$\mu - 1.96\sigma_{\bar{x}} \quad \text{and} \quad \mu + 1.96\sigma_{\bar{x}}$$

(since from a normal distribution chart we note that 0.95 probability implies that $z = 1.96$ or -1.96)

Recall (Formula 8.5, page 414) that

$$\sigma_{\bar{x}} = \frac{\sigma}{\sqrt{n}}$$

Thus, 95% of the $\bar{x}$'s are expected to fall between

$$\mu - 1.96\frac{\sigma}{\sqrt{n}} \quad \text{and} \quad \mu + 1.96\frac{\sigma}{\sqrt{n}}$$

This is shown in Figure 9.2. If all possible samples of size n are selected, and the interval $\bar{x} \pm 1.96\frac{\sigma}{\sqrt{n}}$ is established for each sample, then 95% of all such intervals

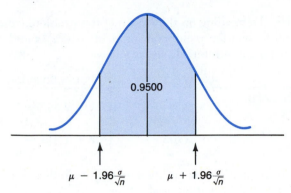

Figure 9.2

are expected to contain μ. Thus, a 95% confidence interval for μ is $\bar{x} \pm 1.96 \dfrac{\sigma}{\sqrt{n}}$.

In order to determine the interval estimate of μ, we must first know the value of the population standard deviation, σ. Although this value is usually unknown since the sample size is large, we can use the sample standard deviation as an approximation for σ. We then have the following confidence interval for μ.

FORMULA 9.1
95% Confidence Interval for μ

Let $\bar{x}$ be a sample mean and s be the sample standard deviation. Then an interval is called a **95% confidence interval for μ** if the lower boundary of the confidence interval is

$$\bar{x} - 1.96 \frac{s}{\sqrt{n}}$$

and if the upper boundary of the confidence interval is

$$\bar{x} + 1.96 \frac{s}{\sqrt{n}}$$

COMMENT   Formula 9.1 tells us how to find a 95% confidence interval for μ. This means that in the long run 95% of such intervals will contain μ. We can be 95% confident that μ lies within the specified interval. We must still realize that 5% of the time the population mean will fall outside this interval. This is true because the sample means are normally distributed and 5% of the values of a random variable in a normal distribution will fall farther away than 2 standard deviations from the mean (see page 361).

COMMENT   Depending on the nature of the problem, some statisticians will often prefer a 99% confidence interval for μ or a 90% confidence interval for μ. The boundaries for these intervals are as follows.

	Lower Boundary	Upper Boundary
99% Confidence Interval	$\bar{x} - 2.58 \dfrac{s}{\sqrt{n}}$	$\bar{x} + 2.58 \dfrac{s}{\sqrt{n}}$
90% Confidence Interval	$\bar{x} - 1.645 \dfrac{s}{\sqrt{n}}$	$\bar{x} + 1.645 \dfrac{s}{\sqrt{n}}$

Thus, as we reduce the size of the interval, we reduce our confidence that the true mean will fall within that interval.

The following examples illustrate how we establish confidence intervals.

EXAMPLE 1

A coffee vending machine fills 100 cups of coffee before it has to be refilled. On Monday the mean number of ounces in a filled cup of coffee was 7.5. The population standard deviation is known to be 0.25 ounces. Find 95% and 99% confidence intervals for the mean number of ounces of coffee dispensed by this machine.

Solution

We use Formula 9.1. Here $n = 100$, $\sigma = 0.25$, and $\bar{x} = 7.5$. To construct a 95% confidence interval for μ, we have

Lower Boundary	Upper Boundary
$= \bar{x} - 1.96 \dfrac{\sigma}{\sqrt{n}}$	$= \bar{x} + 1.96 \dfrac{\sigma}{\sqrt{n}}$
$= 7.5 - 1.96 \left(\dfrac{0.25}{\sqrt{100}} \right)$	$= 7.5 + 1.96 \left(\dfrac{0.25}{\sqrt{100}} \right)$
$= 7.5 - 0.05$	$= 7.5 + 0.05$
$= 7.45$	$= 7.55$

Thus, we conclude that the population mean will lie between 7.45 and 7.55 ounces with a confidence of 0.95.

To construct a 99% confidence interval for μ, we have

Lower Boundary	Upper Boundary
$= \bar{x} - 2.58 \dfrac{\sigma}{\sqrt{n}}$	$= \bar{x} + 2.58 \dfrac{\sigma}{\sqrt{n}}$
$= 7.5 - 2.58 \left(\dfrac{0.25}{\sqrt{100}} \right)$	$= 7.5 + 2.58 \left(\dfrac{0.25}{\sqrt{100}} \right)$
$= 7.5 - 0.06$	$= 7.5 + 0.06$
$= 7.44$	$= 7.56$

Thus, we conclude with a 99% confidence that the population mean will lie between 7.44 and 7.56 ounces. In this example we did not have to use s as an estimate of σ since we were told that the population standard deviation was known to be 0.25.

EXAMPLE 2

A sample survey of 81 movie theaters showed that the average length of the main feature film was 90 minutes with a standard deviation of 20 minutes. Find a

a. 90% confidence interval for the mean of the population.
b. 95% confidence interval for the mean of the population.

Solution

We use Formula 9.1. Here $n = 81$, $s = 20$, and $\bar{x} = 90$.

a. To construct a 90% confidence interval for μ, we have

Lower Boundary	Upper Boundary
$= \bar{x} - 1.645 \dfrac{s}{\sqrt{n}}$	$= \bar{x} + 1.645 \dfrac{s}{\sqrt{n}}$
$= 90 - 1.645 \left(\dfrac{20}{\sqrt{81}} \right)$	$= 90 + 1.645 \left(\dfrac{20}{\sqrt{81}} \right)$
$= 90 - 3.66$	$= 90 + 3.66$
$= 86.34$	$= 93.66$

Thus, a 90% confidence interval for μ is 86.34 to 93.66 minutes.

b. To construct a 95% confidence interval for μ, we have

Lower Boundary	Upper Boundary
$= \bar{x} - 1.96 \dfrac{s}{\sqrt{n}}$	$= \bar{x} + 1.96 \dfrac{s}{\sqrt{n}}$
$= 90 - 1.96 \left(\dfrac{20}{\sqrt{81}} \right)$	$= 90 + 1.96 \left(\dfrac{20}{\sqrt{81}} \right)$
$= 90 - 4.36$	$= 90 + 4.36$
$= 85.64$	$= 94.36$

Thus, a 95% confidence interval for μ is 85.64 to 94.36 minutes.

Notice that as we increase the size of the confidence interval, our confidence that this interval contains μ also increases.

EXAMPLE 3

The management of the Night-All Corporation recently conducted a survey of 196 of its employees to determine the average number of hours that each employee sleeps at night. The company statistician submitted the following information to the management:

$$\Sigma\, x = 1479.8 \qquad \text{and} \qquad \Sigma(x - \bar{x})^2 = 1755$$

where x is the number of hours slept by each employee. Find a 95% confidence interval estimate for the average number of hours that each employee sleeps at night.

Solution

In order to use Formula 9.1, we must first calculate $\bar{x}$ and s. Using the given information, we have

$$\bar{x} = \frac{\Sigma\, x}{n}$$

$$= \frac{1479.8}{196}$$

$$= 7.55$$

and

$$s = \sqrt{\frac{\Sigma(x - \bar{x})^2}{n - 1}}$$

$$= \sqrt{\frac{1755}{195}}$$

$$= \sqrt{9} = 3$$

[handwritten:]
$$X - 1.96 \frac{\sigma}{\sqrt{N}}$$
$$10 - 1.96\, (1.23)$$
$$10 - 0.11 = \frac{22.36}{9.89}$$
$$10 \pm .11 \approx 10.11$$

Now we can use Formula 9.1, with $\bar{x} = 7.55$, $s = 3$, and $n = 196$. To find the 95% confidence for μ, we have

Lower Boundary	Upper Boundary
$= \bar{x} - 1.96\,\dfrac{s}{\sqrt{n}}$	$= \bar{x} + 1.96\,\dfrac{s}{\sqrt{n}}$
$= 7.55 - 1.96\left(\dfrac{3}{\sqrt{196}}\right)$	$= 7.55 + 1.96\left(\dfrac{3}{\sqrt{196}}\right)$
$= 7.55 - 0.42$	$= 7.55 + 0.42$
$= 7.13$	$= 7.97$

Thus, the management can conclude with a 95% confidence that the average number of hours that an employee sleeps at night is between 7.13 and 7.97 hours.

EXERCISES FOR SECTION 9.3

1. A sample of 49 banks in the metropolitan area found that the average charge for a bank money order or teller's check was $2.75 for amounts up to $1000. Furthermore, the standard deviation was $0.25. Find a 90% confidence interval for the average charge for a bank money order or teller's check.

2.

MORE BANK FAILURES PREDICTED

Washington (Feb. 15)—The Federal Deposit Insurance Corp (FDIC) disclosed yesterday that hundreds of banks were in financial trouble and that quite a few of them would have to be closed or merged with other banks. One of the primary causes cited for the current problems facing the banking industry is bad real estate loans in earlier years. At many banks today it is very difficult, if not impossible, to obtain a home mortgage loan.

February 15, 1991

Consider the preceding newspaper article. Most banks now charge points to obtain a home mortgage loan or require a larger down payment to be made by the buyer. A survey of 64 randomly selected banks in New York disclosed that these banks now require average down payments of 35% of the purchase price to be paid by the buyer. The standard deviation is 5%. Find a 99% confidence interval for the percentage of the purchase price to be paid by the buyer that is now required as a down payment by banks.

3. Many school districts nationwide will not allow a youngster to enroll in school unless the child has been inoculated with the MMR (measles, mumps, rubella) vaccine. In a recent study involving 500 children inoculated with this vaccine, it was found that some children developed a mild reaction an average of 10 days after inoculation. The standard deviation was 1.23 days. Find a 95% confidence interval for the average time for reaction by a child inoculated with this vaccine.

4. In the wake of the recent threats of terrorism, one airport insurance agent claims that many travelers are increasing the amount of flight insurance that they buy. A survey of 100 randomly selected travelers buying flight insurance indicated that these travelers had purchased an average of $30,000 additional flight insurance. The standard deviation was $7,000. Find a 90% confidence interval for the average amount of additional flight insurance purchased.

5. Car rental agencies in large cities often allow a person to rent a car at one location and return the car at another location. However, the car rental agencies often

charge a "drop-off fee" for this privilege. A survey of 36 car rental agencies in one city found that they charged an average of $15 for this privilege. The standard deviation was $2.98. Find a 95% confidence interval for the average "drop-off fee" charged by car rental agencies.

6. Tourism dollars are often the lifeline for many underdeveloped or Third World countries. To stimulate tourism, such countries advertise extensively. A survey of 64 randomly selected tourists returning from one of these countries found that they spent an average of $627 in these countries. The standard deviation was $83. Find a 99% confidence interval for the average amount of money spent by a tourist traveling to one of these countries.

7. A consumer's group sampled 49 different stores in the city and found that the average price of a particular model and brand of washing machine was $522.85 with a standard deviation of $29.55. Find a 95% confidence interval for the average price of this particular washing machine.

8. Oil companies try to keep substantial inventories of gasoline on hand to avoid any possible shortages. One oil company claims that it has an inventory sufficient to last an average of 16 days at 64 of its storage facilities located throughout the southwestern part of the United States. The standard deviation is 2.13 days. Find a 99% confidence interval for the average amount of gasoline in storage.

*9.

> ## HOSPITAL COSTS EXPECTED TO RISE AGAIN
>
> *January 17*—Citing rising costs for labor and supplies, a spokesperson for the voluntary league of hospitals said yesterday that hospital costs will rise an average of 6.2% effective March 1, 1991. This represents the third consecutive year in which hospital costs have risen.
>
> January 17, 1991

Consider the preceding article. The average charge at 36 of these hospitals for a routine (noncaesarean section) maternity delivery and a subsequent 3-day stay for mother and child is $1375 with a standard deviation of $178. These rates are in effect before any raises. Find an 85% confidence interval for the average charge for a routine maternity delivery.

10. 75 people paid lawyers an average of $425 to write a will and a right-to-die agreement. The standard deviation was $51. Find a 90% confidence interval for the average charge by a lawyer for writing a will and a right-to-die agreement.

11. A random sample of n measurements was taken from a population with unknown

mean, μ, and standard deviation, σ. The following data were obtained: $\Sigma\,x =$ 850, $\Sigma\,x^2 = 18{,}400$, and $n = 40$.

a. Find a 95% confidence interval for μ.
b. Find a 99% confidence interval for μ.

9.4 ESTIMATING THE POPULATION MEAN ON THE BASIS OF A SMALL SAMPLE

In Section 9.3 we indicated how to determine confidence intervals for μ when the sample size is larger than 30. Unfortunately, this is not always the case. Suppose a sample of 16 bulbs is randomly selected from a large shipment and has a mean life of 100 hours with a standard deviation of 5 hours. Using only the methods of Section 9.3, we cannot determine confidence intervals for the mean life of a bulb since the sample size is less than 30.

Student's t-distribution

Fortunately, in such situations we can base confidence intervals for μ on a distribution that is in many respects similar to the normal distribution. This is the **Student's t-distribution.** This distribution was first studied by William S. Gosset, who was a statistician for Guinness, an Irish brewing company. Gosset was the first to develop methods for interpreting information obtained from small samples. Yet his company did not allow any of its employees to publish anything. So, Gosset secretly published his findings in 1907 under the name "Student." To this day this distribution is referred to as the Student's t-distribution.

Number of Degrees of Freedom

Figure 9.3 indicates the relationship between the normal distribution and the t-distribution. Notice that the t-distribution is also symmetrical about zero, which is its mean. However, the shape of the t-distribution depends on a parameter called the **number of degrees of freedom.** In our case the number of degrees of freedom, abbreviated as d.f., is equal to the sample size minus 1. If the population sampled

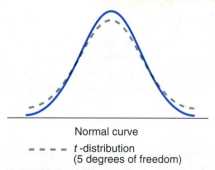

Normal curve

- - - - t-distribution
(5 degrees of freedom)

Figure 9.3
Relationship between the normal distribution and the t-distribution

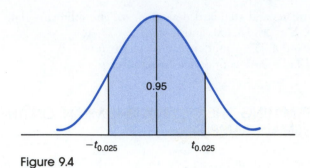

Figure 9.4

is normally distributed, then $\dfrac{\bar{x} - \mu}{s/\sqrt{n}}$ has a t-distribution. This standardized t-distribution is symmetrical, bell-shaped, and has zero as its mean.

Table VII in the Appendix indicates the value of t for different degrees of freedom. Thus, the 1.96 of Formula 9.1 of Section 9.3 has to be replaced by the $t_{0.025}$ value as listed in this table, depending on the number of degrees of freedom. When using the $t_{0.025}$ value of Table VII, 95% of the area under the curve of the t-distribution will fall between $-t_{0.025}$ and $t_{0.025}$, as shown in Figure 9.4.

We then have the following formula:

FORMULA 9.2   Let $\bar{x}$ be a sample mean and let s be the sample standard deviation. We have the following 95% small-sample confidence interval for μ:

$$\text{Lower boundary} = \bar{x} - t_{0.025}\,\frac{s}{\sqrt{n}}$$

$$\text{Upper boundary} = \bar{x} + t_{0.025}\,\frac{s}{\sqrt{n}}$$

COMMENT   In addition to the $t_{0.025}$ values, Table VII in the Appendix also lists many other values of t. Thus the $t_{0.005}$ values are used when we want a 99% confidence interval for μ, the $t_{0.050}$ values are used when we want a 90% confidence interval for μ, the $t_{0.10}$ values are used when we want an 80% confidence interval for μ, and so on.

EXAMPLE 1   A survey of 16 taxi drivers found that the average tip they receive is 40 cents with a standard deviation of 8 cents. Find a 95% confidence interval for estimating the average amount of money that a taxi driver receives as a tip.

$n = 16$
$\bar{x} = 40\text{¢}$
$s = 8\text{¢}$

Solution

We use Formula 9.2. Here we have $n = 16$, $\bar{x} = 40$, and $s = 8$. We first find the number of degrees of freedom, which is $n - 1$, or $16 - 1 = 15$. Then we find the appropriate value of t from Table VII. The $t_{0.025}$ value with 15 degrees of freedom is 2.131. Finally, we establish the confidence interval. We have

Lower Boundary	Upper Boundary
$= \bar{x} - t_{0.025} \dfrac{s}{\sqrt{n}}$	$= \bar{x} + t_{0.025} \dfrac{s}{\sqrt{n}}$
$= 40 - 2.131 \left(\dfrac{8}{\sqrt{16}} \right)$	$= 40 + 2.131 \left(\dfrac{8}{\sqrt{16}} \right)$
$= 40 - 4.26$	$= 40 + 4.26$
$= 35.74$	$= 44.26$

Thus, a 95% confidence interval for the average amount of money that a taxi driver will receive as a tip is 35.74 to 44.26 cents.

EXAMPLE 2

A survey of the hospital records of 25 randomly selected patients suffering from a particular disease indicated that the average length of stay in the hospital is 10 days. The standard deviation is estimated to be 2.1 days. Find a 99% confidence interval for estimating the mean length of stay in the hospital.

Solution

We use Formula 9.2. Here $n = 25$, $\bar{x} = 10$, and $s = 2.1$. We first find the number of degrees of freedom, which is $n - 1$, or $25 - 1 = 24$. Then we find the appropriate value of t from Table VII in the Appendix. The $t_{0.005}$ value with 24 degrees of freedom is 2.797. Finally, we establish the confidence interval. We have

Lower Boundary	Upper Boundary
$= \bar{x} - t_{0.005} \dfrac{s}{\sqrt{n}}$	$= \bar{x} + t_{0.005} \dfrac{s}{\sqrt{n}}$
$= 10 - 2.797 \left(\dfrac{2.1}{\sqrt{25}} \right)$	$= 10 + 2.797 \left(\dfrac{2.1}{\sqrt{25}} \right)$
$= 10 - 1.17$	$= 10 + 1.17$
$= 8.83$	$= 11.17$

Thus, a 99% confidence interval for the average length of stay in the hospital is 8.83 to 11.17 days.

EXAMPLE 3

A music teacher asks six randomly selected students how many hours a week each practices playing the electric guitar. The teacher receives the following answers: 10, 12, 8, 9, 16, 5. Find a 90% confidence interval for the average length of time that a student practices playing the electric guitar.

Solution

In order to use Formula 9.2, we must first calculate the sample mean and the sample standard deviation. We have

$$\bar{x} = \frac{\Sigma x}{n} = \frac{10 + 12 + 8 + 9 + 16 + 5}{6} = \frac{60}{6} = 10$$

To calculate the sample standard deviation, we arrange the data as follows.

x	$x - \bar{x}$	$(x - \bar{x})^2$
10	$10 - 10 = 0$	0
12	$12 - 10 = 2$	4
8	$8 - 10 = -2$	4
9	$9 - 10 = -1$	1
16	$16 - 10 = 6$	36
5	$5 - 10 = -5$	25
	Total =	70

$$s = \sqrt{\frac{\Sigma(x - \bar{x})^2}{n - 1}} = \sqrt{\frac{70}{6 - 1}} = \sqrt{14} \approx 3.74$$

Now we find the appropriate value of t from Table VII. The $t_{0.05}$ value with $n - 1$, or $6 - 1 = 5$, degrees of freedom is 2.015. Finally, we establish the confidence interval. We have

Lower Boundary	Upper Boundary
$= \bar{x} - t_{0.05}\dfrac{s}{\sqrt{n}}$	$= \bar{x} + t_{0.05}\dfrac{s}{\sqrt{n}}$
$= 10 - 2.015\left(\dfrac{3.74}{\sqrt{6}}\right)$	$= 10 + 2.015\left(\dfrac{3.74}{\sqrt{6}}\right)$
$= 10 - 3.08$	$= 10 + 3.08$
$= 6.92$	$= 13.08$

Thus, a 90% confidence interval for the average length of time that a student practices playing the electric guitar is 6.92 to 13.08 hours.

COMMENT We wish to emphasize a point made earlier. For small sample inferences to be valid it is assumed that the sample is obtained from some normally distributed population. Often this may not be true. In this case we have to use other statistical procedures.

EXERCISES FOR SECTION 9.4

1. It always pays to shop around before purchasing a used car. Nine customers purchased the same model car (equipped with the same options and accumulated mileage) from nine different used-car dealers and paid an average of $7200 for the car. The standard deviation was $812. Construct a 99% confidence interval for the average used-car price of this particular model car.

A rather common scene these days in many supermarkets is the appearance of consumer comparison shopping and the use of magazine or newspaper coupons.

2. Consider the newspaper article on the top of the next page.

   A survey of 12 randomly selected city dwellers found that they were wasting an average of 61 gallons of water per day. The standard deviation was 8.3 gallons per day. Construct a 90% confidence interval for the average amount of water wasted daily by these people.

3. The Trans-Union Insurance Company is revising its rate schedules that it charges for jewelry insurance. A staff statistician wishes to estimate the average payment for claims resulting from lost or stolen jewelry. The statistician randomly selects 15 claim payments for lost or stolen jewelry that were settled during 1990. The average claim settlement was for $2371 with a standard deviation of $278. Construct a 95% confidence interval estimate of the mean payment for claims resulting from lost or stolen jewelry.

CITY STILL IN WATER CRISIS

Los Angeles (Feb. 12)—Despite the rains that fell over the weekend, the city is still in the grip of a severe water drought and emergency. The reservoirs are well below their normal levels for this time of the year.

Many city residents are still wasting a lot of water each day in a variety of ways, including letting the water run needlessly, leaky faucets and pipes, unnecessary flushes of the toilet, etc.

The mayor is pleading with the citizens to conserve water.

February 14, 1991

4. Engineers at the Maxwell Products Company have designed a new metal bearing rod that is alleged to be extremely durable. To estimate the mean useful length of life, engineers test nine of these rods and find that the useful length of life (in hours) of each is as follows:

25,000	25,200	25,600
25,280	25,400	25,014
25,300	25,250	25,350

Previous experience with similar such rods indicates that the useful length of life of individual rods of this particular type is normally distributed. Construct a 90% confidence interval for the mean useful length of life of all rods of this new type.

5. A zoologist is interested in determining the average life span of a certain species of elephant bred in captivity. The zoologist collects data on 15 elephants of this type from 15 different randomly selected zoos throughout America. The average life spans of these elephants is found to be 29.9 years with a standard deviation of 7.4 years. The zoologist knows that previous experience with similar elephants indicates that the life spans of elephants are normally distributed. Construct a 95% confidence interval for the true average life span of an elephant from this species.

6. Medical records of nine hospitals in a large city indicate that the hospitals treated an average of 28 patients per week for AIDS-related sicknesses. The standard deviation was 2.82. Construct a 90% confidence interval for the mean number of AIDS-related sicknesses treated per week. State any assumptions that are required for the interval estimation procedure to be valid.

7. The kidneys of a human being normally remove toxic wastes from the blood. When a person's kidneys deteriorate or malfunction, the person must be treated

by dialysis, which is a machine that performs the toxic removal function normally performed by healthy kidneys. However, it is known that dialysis can result in retention of phosphorus in the blood. This condition must be monitored carefully. The phosphorus levels of eight patients on dialysis were measured and the following results were obtained (in milligrams of phosphorus per deciliter of blood):

5.8   5.2   4.9   5.1   5.3   6.1   5.9   6.5

Assuming that the phosphorus levels of patients on dialysis is normally distributed, construct a 90% confidence interval for the average level of phosphorus in the blood of patients on dialysis.

8. A researcher is performing a biofeedback experiment on 15 randomly selected subjects. The researcher attaches a special instrument to the forehead of each subject in an attempt to measure muscle tension. The mean and standard deviation for these subjects turns out to be 6.1 and 1.3 microvolts, respectively. Assuming that the average muscle tension is normally distributed, construct a 90% confidence interval for the average muscle tension for all patients.

9. A consumer's group tested 12 randomly selected cars (of a particular make and model) for fuel consumption. On a combination of highway and city driving, the mean and standard deviation was found to be 26.8 mi/gal (miles per gallon) and 2.31 mi/gal, respectively. Construct a 95% confidence interval for the average fuel consumption (in mi/gal) for all such cars.

10. A chemist has developed a new synthetic fiber. To measure the fiber's strength, experiments are performed on ten strands of this fiber to determine the average number of pounds that can be supported by strands of this fiber before breaking. The following breaking-load results (in pounds) were obtained:

3.3   2.2   3.1   2.8   2.6   2.9   3.0   2.4   2.7   3.2

Construct a 95% confidence interval for the mean breaking strength of a strand of this new fiber.

The exhaust fumes created by the many vehicles using the toll booths make it very unhealthy for the toll collectors

11. Refer to the article shown below. Health officials tested the air at seven tunnels to determine the pollution content. The officials found an average of 18 ppm (parts per million) of a certain pollutant in the air at these tunnels. The standard deviation was 3.2 ppm. Find a 95% confidence interval for the average ppm content of this pollutant in the air at these tunnels.

TOLL COLLECTORS PROTEST UNHEALTHY WORKING CONDITIONS

New Dorp (May 17)—Toll collectors staged a 3-hour protest yesterday against the unhealthy working conditions occurring at different times of the day at the toll plazas of the city's seven tunnels. A spokesperson for the toll collectors indicated that the level of sulfur oxide pollutants in the air at the plazas was far in excess of the recommended safety level. Between the hours of 10 A.M and 1 P.M., motorists were able to use the tunnel without paying any toll. In an effort to resolve the dispute, Bill Sigowsky has been appointed as mediator.

May 17, 1990

9.5 THE ESTIMATION OF THE STANDARD DEVIATION

Until now we have been discussing how to obtain point and interval estimates of the population mean μ. In this section we discuss point and interval estimates for the population standard deviation.

Suppose we took a sample of 100 cigarettes of a particular brand and determined that the sample mean tar content was 15.2 with a standard deviation of 1.1. If this procedure were to be repeated many times, each time we would obtain different estimates for the population mean and standard deviation. Nevertheless, the mean of these estimates will approach the true population mean as the sample size gets larger. When this happens we say that the average of the sample means is an **unbiased estimator** of the population mean.

Recall that on page 408 two formulas were given for the variance and standard deviation. Suppose we were to calculate the variance of each sample by using the formula

$$\frac{\Sigma(x - \bar{x})^2}{n}$$

Unbiased Estimator

Then we cannot use the average of these values as an unbiased estimate of the population variance. Specifically, the average of such estimates would most likely always be too small, no matter how many samples are included. We can compensate for this by dividing by $n - 1$ instead of n. When this is done, it can be shown that

$$\frac{\Sigma(x - \bar{x})^2}{n - 1}$$

is an unbiased estimate of σ^2. This means that as more and more samples are included, the average of the variances of these samples will approach the true population variance. Thus, a point estimate of σ^2 is

$$\frac{\Sigma(x - \bar{x})^2}{n - 1}$$

We therefore say that $\bar{x}$ is an unbiased estimator of the population mean μ and that s^2 is an unbiased estimator of σ^2. However, this does not mean that s is an unbiased estimator of σ. Nevertheless, when n is large, this bias is small. Thus, we can use s as an estimator of σ.

When dealing with large sample sizes, mathematicians have developed the following 95% confidence interval for σ.

FORMULA 9.3

Let s be the sample standard deviation and let n be the number in the sample. We have the following 95% confidence interval for σ:

$$\text{Lower boundary} = \frac{s}{1 + \dfrac{1.96}{\sqrt{2n}}} \qquad \text{Upper boundary} = \frac{s}{1 - \dfrac{1.96}{\sqrt{2n}}}$$

EXAMPLE 1

A sample of 50 cigarettes was taken. The sample mean tar content was 15.2 with a standard deviation of 1.1. Find a 95% confidence interval for the population standard deviation.

Solution

We use Formula 9.3. Here $n = 50$ and $s = 1.1$. The lower boundary of the confidence interval is

$$\frac{s}{1 + \dfrac{1.96}{\sqrt{2n}}} = \frac{1.1}{1 + \dfrac{1.96}{\sqrt{2(50)}}}$$

$$= \frac{1.1}{1 + 0.196}$$

$$= 0.92$$

The upper boundary of the confidence interval is

$$\frac{s}{1 - \dfrac{1.96}{\sqrt{2n}}} = \frac{1.1}{1 - \dfrac{1.96}{\sqrt{2(50)}}}$$

$$= \frac{1.1}{1 - 0.196}$$

$$= 1.37$$

Thus, a 95% confidence interval for σ is 0.92 to 1.37.

9.6 DETERMINING THE SAMPLE SIZE

Until now we have been discussing how sample data can be used to estimate various parameters of a population. Selecting a sample usually involves an expenditure of money. The larger the sample size, the greater the cost. Therefore, before selecting a sample we must determine how large the sample should be.

Generally speaking, the size of the sample is determined by the desired degree of accuracy. Most problems specify the maximum allowable error. Yet we must realize that no matter what the size of a sample is, any estimate may exceed the maximum allowable error. To be more specific, suppose we are interested in estimating the mean life of a calculator on the basis of a sample. Suppose also that we want our estimate to be within 0.75 years of the true value of the mean. In this case the **maximum allowable error**, denoted as e, is 0.75 years. How large a sample must be taken? Of course, the larger the sample size, the smaller the chance that our estimate will not be within 0.75 years of the value. If we want to be 95% confident that our estimate will be within 0.75 years of the true value, then the sample size must satisfy

Maximum Allowable Error

$$1.96\sigma_{\bar{x}} = 0.75$$

More generally, if the maximum allowable error is e, then we must have

$$e = 1.96\sigma_{\bar{x}}$$

Since

$$\sigma_{\bar{x}} = \frac{\sigma}{\sqrt{n}}$$

we get

$$e = 1.96\,\frac{\sigma}{\sqrt{n}}$$

Solving this equation for n gives Formula 9.4.

FORMULA 9.4	Let σ be the population standard deviation, e the maximum allowable error, and n the size of the sample that is to be taken from a large population. Then a sample of size

$$n = \left(\frac{1.96\sigma}{e}\right)^2$$

will result in an estimate of μ, which is less than the maximum allowable error 95% of the time.

EXAMPLE 1

Suppose we wish to estimate the average life of a calculator to within 0.75 years of the true value. Past experience indicates that the standard deviation is 2.6 years. How large a sample must be selected if we want our answer to be within 0.75 years 95% of the time?

Solution

We use Formula 9.4. Here $\sigma = 2.6$ and $e = 0.75$ so that

$$n = \left(\frac{1.96\sigma}{e}\right)^2$$

$$= \left(\frac{1.96(2.6)}{0.75}\right)^2 = (6.79)^2 = 46.10$$

Thus, the sample should consist of 47 calculators. (*Note:* In the determination of sample size any decimal is always rounded to the next highest number.)

EXAMPLE 2

The management of a large company in California desires to estimate the average working experience, measured in years, of its 2000 hourly workers. How large a sample should be taken in order to be 95% confident that the sample mean does not differ from the population mean by more than $\frac{1}{2}$ year? (Past experience indicates that the standard deviation is 2.6 years.)

Solution

We use Formula 9.4. Here $\sigma = 2.6$ and $e = 0.50$. Then

$$n = \left(\frac{1.96\sigma}{e}\right)^2$$

$$= \left(\frac{1.96(2.6)}{0.50}\right)^2$$

$$= (10.19)^2$$

$$= 103.84$$

Thus, the company officials should select 104 hourly workers to determine the average working experience of an hourly worker.

EXERCISES FOR SECTION 9.6

1. A sample of 125 runners in a marathon found that the average time required to run the first lap was 18.64 minutes with a standard deviation of 2.68 minutes. Find a 95% confidence interval for the population standard deviation of *all* the runners.

2. A sample of 36 second graders showed that a second grader spends an average of 8.2 hours per week playing with friends. The standard deviation was 1.23 hours. Find a 95% confidence interval for the population standard deviation.

3. Due to the increase in crime in the Brighton neighborhood, a citizen's community patrol was formed. 75 randomly selected residents of the Brighton neighborhood were asked to indicate how much each would be willing to contribute to cover the expenses of the patrol. The average was computed and found to be $84 with a standard deviation of $8.23. Find a 99% confidence interval for the population standard deviation.

4. Fifty mothers who recently returned to the work force were asked to indicate how much they were paying weekly for baby-sitting services. The standard deviation of the cost was computed and found to be $8.41. Find a 90% confidence interval for the population standard deviation.

For each of Exercises 5–10, assume that the maximum allowable error must not be exceeded with a 95% degree of confidence.

5. *Cost of a college textbook*: A book distributor wishes to estimate the cost of a remedial college-level math textbook. How large of a sample of competing books should be taken if the estimate is to be within 85¢ of the true value? (Assume that σ = $1.49)

6. In an attempt to increase revenue, a health foods manufacturer is planning an extensive promotional campaign. However, the health foods manufacturer wishes first to determine the average amount of money people are willing to spend annually on health foods. How large of a sample should be taken if the manufacturer wishes the estimate to be within $3 of the true value? (Assume that previous similar studies found that σ = $5.93)

7. A sociologist wishes to estimate the average amount of time per week that parents in Dover spend helping their children with their homework. How large of a sample size should the sociologist select if the sociologist wishes that the estimate be within 0.85 hours of the true value. (Assume that previous similar studies found that σ = 3.98)

8. The Merv Chemical Company is interested in determining the average amount of money women spend annually on facial cosmetics. How large of a sample should be taken if the company wishes that the estimate be within $1.86 of the true value? (Assume that $\sigma = \$4.76$)

9. The Commissioner of the State Motor Vehicle Bureau wishes to estimate the average time needed by a mechanic to determine whether a car passes the state's motor vehicle emission requirements. How large of a sample should be selected if the commissioner wishes that the estimate be within 5 minutes of the true value? (Assume that $\sigma = 8.29$ minutes)

10. An economist wishes to estimate the average amount of money spent by a family on medical insurance. How large of a sample should be selected if it is desired that the estimate be within $15 of the true value? (Assume that $\sigma = \$29.95$)

9.7 THE ESTIMATION OF PROPORTIONS

So far we have discussed the estimation of the population mean and the population standard deviation. Very often statistical problems arise for which the data are available in **proportion** or **count form** rather than in measurement form. For example, suppose a doctor has developed a new technique for predicting the sex of an unborn child. The doctor tests the new method on 1000 pregnant women and correctly predicts the sex of 900 of the children. Is the new technique reliable? The doctor has correctly predicted the sex of 900 unborn children. Thus, the **sample proportion** is $\dfrac{900}{1000}$, or 0.90. What is the **true proportion** of unborn children whose sex the doctor can correctly predict?

Proportion Form
Count Form

Sample Proportion

True Proportion

Since we will be discussing both sample proportions and true population proportions, we will use the following notation:

p true population proportion
$\hat{p}$ sample proportion   ($\hat{p}$ is pronounced p hat)

In our case the doctor estimates the true population proportion, p, to be 0.90. This estimate is based on the sample proportion $\hat{p}$. How reliable is the estimate of the true population proportion? Suppose the doctor now tests the technique on 1000 different pregnant women. For what proportion of these 1000 unborn children will the doctor be able to correctly predict the sex?

If the true population proportion is p, repeated sample proportions will be normally distributed. The mean of the sample distribution of proportions will be p. The standard deviation of these sample proportions will equal

$$\sqrt{\frac{p(1 - p)}{n}}$$

Thus, if the doctor's technique is reliable, he or she should find that the sample proportion will be normally distributed with a mean of 0.9. The standard deviation will be

$$\sqrt{\frac{(0.9)(1 - 0.9)}{1000}} = \sqrt{0.00009}, \quad \text{or } 0.0095$$

We can summarize these results as follows.

FORMULA 9.5

Suppose we have a large population, a proportion of which has some particular characteristic. We select random samples of size n and determine the proportion in each sample with this characteristic. Then the sample proportions will be approximately normally distributed with mean p and standard deviation

$$\sigma_{\hat{p}} = \sqrt{\frac{p(1 - p)}{n}}$$

Since the sample proportions are approximately normally distributed, we can apply the standardized normal charts as we did for the sample means. The following examples illustrate how this is done.

EXAMPLE 1

From past experience it is known that 70% of all airplane tickets sold by Global Airways are round-trip tickets. A random sample of 100 passengers is taken. What is the probability that at least 75% of these passengers have round-trip tickets?

Solution

We use Formula 9.5. Here $n = 100$, $p = 0.70$, and the sample proportion $\hat{p}$ is 0.75. Since the sample proportions are normally distributed, we are interested in the area to the right of $x = 0.75$ in a normal distribution whose mean is 0.70. See Figure

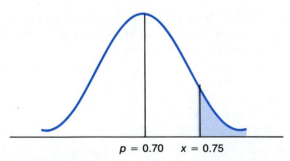

$p = 0.70 \quad x = 0.75$

Figure 9.5

9.5. We first calculate the standard deviation of the sample proportions, denoted as $\sigma_{\hat{p}}$.

$$\sigma_{\hat{p}} = \sqrt{\frac{p(1 - p)}{n}}$$

$$= \sqrt{\frac{0.70(1 - 0.70)}{100}}$$

$$= \sqrt{0.0021}$$

$$= 0.046$$

Then

$$z = \frac{\hat{p} - p}{\sigma_{\hat{p}}}$$

$$= \frac{0.75 - 0.70}{0.046} \qquad \text{($\hat{p}$ replaces $\bar{x}$ and p replaces μ in}$$
$$\text{Formula 8.6 on page 421).}$$

$$= 1.09$$

From Table IV the area between $z = 0$ and $z = 1.09$ is 0.3621. Therefore, the area to the right of $z = 1.09$ is $0.5000 - 0.3621$, or 0.1379. Thus, the probability that at least 75% of the passengers have round-trip tickets is 0.1379.

EXAMPLE 2

54% of all nurses in the day shift at Downtown Hospital have type O blood. 36 nurses from the day shift are selected at random. What is the probability that between 51% and 58% of these members have type O blood?

Solution

We use Formula 9.5. Here $p = 0.54$ and $n = 36$. We first calculate $\sigma_{\hat{p}}$:

$$\sigma_{\hat{p}} = \sqrt{\frac{p(1 - p)}{n}}$$

$$= \sqrt{\frac{0.54(1 - 0.54)}{36}}$$

$$= \sqrt{0.0069}$$

$$= 0.083$$

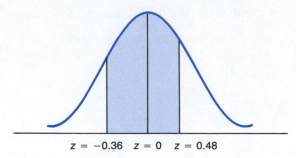

$$z = -0.36 \quad z = 0 \quad z = 0.48$$

Figure 9.6

Now we can use Formula 9.5 with $p = 0.54$, $n = 36$, and $\sigma_{\hat{p}} = 0.083$:

$$\hat{p} = 0.51 \text{ corresponds to} \quad z = \frac{0.51 - 0.54}{0.083} = \frac{-0.03}{0.083} = -0.36$$

$$\hat{p} = 0.58 \text{ corresponds to} \quad z = \frac{0.58 - 0.54}{0.083} = \frac{0.04}{0.083} = 0.48$$

Thus, we are interested in the area between $z = -0.36$ and $z = 0.48$. See Figure 9.6.

From Table IV we find that the area between $z = 0$ and $z = -0.36$ is 0.1406, and the area between $z = 0$ and $z = 0.48$ is 0.1844. Adding, we get

$$0.1406 + 0.1844 = 0.3250$$

Thus, the probability that between 51% and 58% of these nurses have type O blood is 0.3250.

In the same way that the sample mean is used to estimate the population mean we can use the sample proportions, $\hat{p}$, as a point estimate of the population proportion, p. However, this estimate does not indicate the probability of its accuracy. Thus, we set up interval estimates for the true population proportion, p. We have Formula 9.6.

FORMULA 9.6   Let $\hat{p}$ be a sample proportion and let p be the true population proportion. Then we have the following 95% confidence interval for p (assuming we have a large sample):

$$\text{Lower boundary} = \hat{p} - 1.96 \sqrt{\frac{\hat{p}(1 - \hat{p})}{n}}$$

$$\text{Upper boundary} = \hat{p} + 1.96 \sqrt{\frac{\hat{p}(1 - \hat{p})}{n}}$$

EXAMPLE 3

A union member reported that 80 out of 120 workers interviewed supported some form of work stoppage to further its demands for a shorter workweek. Find a 95% confidence estimate of the true proportion of workers supporting the union's stand on a work stoppage.

Solution

We use Formula 9.6. Here $\hat{p} = \dfrac{80}{120}$, or 0.67, and $n = 120$. To construct a 95% confidence interval for p, we have

Lower Boundary	Upper Boundary
$= \hat{p} - 1.96 \sqrt{\dfrac{\hat{p}(1 - \hat{p})}{n}}$	$= \hat{p} + 1.96 \sqrt{\dfrac{\hat{p}(1 - \hat{p})}{n}}$
$= 0.67 - 1.96 \sqrt{\dfrac{0.67(1 - 0.67)}{120}}$	$= 0.67 + 1.96 \sqrt{\dfrac{0.67(1 - 0.67)}{120}}$
$= 0.67 - 1.96 \sqrt{0.0018}$	$= 0.67 + 1.96 \sqrt{0.0018}$
$= 0.67 - 1.96(0.043)$	$= 0.67 + 1.96(0.043)$
$= 0.67 - 0.084$	$= 0.67 + 0.084$
$= 0.586$	$= 0.754$

Thus, the union official concluded with 95% confidence that the true proportion of workers supporting the union claim is between 0.586 and 0.754. This is the 95% confidence interval.

EXAMPLE 4

A new public telephone has been installed in an airport baggage-claim area. A quarter was lost 45 of the first 300 times that it was used. Construct a 95% confidence interval for the true proportion of times that a user will lose a quarter.

Solution

We use Formula 9.6. Here $\hat{p} = \dfrac{45}{300}$, or 0.15, and $n = 300$. To construct a 95% confidence interval for p, we have

Lower Boundary	Upper Boundary
$= \hat{p} - 1.96 \sqrt{\dfrac{\hat{p}(1 - \hat{p})}{n}}$	$= \hat{p} + 1.96 \sqrt{\dfrac{\hat{p}(1 - \hat{p})}{n}}$
$= 0.15 - 1.96 \sqrt{\dfrac{0.15(1 - 0.15)}{300}}$	$= 0.15 + 1.96 \sqrt{\dfrac{0.15(1 - 0.15)}{300}}$
$= 0.15 - 1.96 \sqrt{0.0004}$	$= 0.15 + 1.96 \sqrt{0.0004}$
$= 0.15 - 1.96(0.02)$	$= 0.15 + 1.96(0.02)$
$= 0.15 - 0.039$	$= 0.15 + 0.039$
$= 0.111$	$= 0.189$

Thus, a 95% confidence interval for the true proportion of times that a user will lose a quarter is 0.111 to 0.189.

EXERCISES FOR SECTION 9.7

1. Find the 95% confidence interval for the proportion of defective items in a large shipment of parts when
   a. $n = 1000$ and $\hat{p} = 0.3$ b. $n = 5000$ and $\hat{p} = 0.08$

2. 369 of the 450 drivers interviewed said that they would never use retreaded tires. Find a 95% confidence interval for the true proportion of all drivers who would never use retreaded tires.

3. A public opinion research firm finds that 312 of the 688 families sampled prefer the new brand of orange juice over the one they are currently using. Find a 95% confidence interval for the true proportion of all families who prefer the new brand of orange juice.

4. Government safety officials claim that only about 40% of the drivers on the thruway use their seat belts. A random survey of 200 drivers is taken. What is the probability that more than 45% will be found to be using their seat belts?

5. Police department speed enforcement agents claim that 80% of the drivers on the highways simply do not obey the state's 55-mph (miles per hour) speed limit. If a random survey of 60 cars traveling on the highway is selected, find the probability that fewer than 75% of drivers simply do not obey the state's 55-mph speed limit.

6. A random sample of 390 students taking math courses at a college was conducted. It was found that 260 of them believe that math finals should be optional. Find a 90% confidence interval for the true proportion of math students who believe that math finals should be optional.

7. Administration officials claim that 45% of the student body at Victoria College are part-time students who work during the day. The student government claims that this figure is grossly underestimated. A random sample of 700 part-time students is taken. What is the probability that more than 330 of them work during the day?

8. A bicycle manufacturer claims that 96% of all bikes on the road in a certain town are equipped with nighttime safety reflectors. A random sample of 85 bikes is selected. What is the probability that fewer than 93% of these bikes are not equipped with the reflectors?

9. 175 of the 220 workers interviewed in a large northeastern city said that they would never use public transportation to get to work but would rather drive. Find a 90% confidence interval for the true proportion of the workers who would not use public transportation to get to work.

10. A Wall Street brokerage firm, which employs approximately 50,000 people in all its branch locations nationwide, claims that 46% of its top-paying jobs are held by women. An affirmative action group is challenging this claim. A random sample of 100 of the company's top-paying jobs is taken. What is the probability that the proportion of women having these top-paying jobs is less than 45%?

11. Lincoln Hospital claims that only 4% of the patients treated in the Emergency Room of the hospital are dissatisfied with the services provided. To test this claim, the hospital randomly selects 500 patients and asks them to rate the services provided. What is the probability that the proportion of patients dissatisfied with the service provided is between 2% and 7%?

12. A random survey of 500 pregnant women in New York City conducted by Epstein and Rogers* indicated that 145 of them preferred a female obstetrician to a male obstetrician. Find a 95% confidence interval for the true proportion of all pregnant women living in New York City who prefer a female obstetrician.

13. Consider the newspaper article that follows. If a random survey of 525 families across the state is taken, what is the probability that more than 102 of them will be found to be one-person households?

23% OF AMERICAN HOUSEHOLDS ARE 1-PERSON

Washington (Oct. 21)—According to the latest figures released by the U.S. Census Bureau, about 23% of all households in America today are one-person households. This startling result has important implications for our concept of "family."

Monday, October 21, 1991

9.8 USING COMPUTER PACKAGES

We can use a MINITAB program to determine confidence intervals quite easily. To illustrate, let us use a MINITAB program to determine a 90% confidence interval for the data on the average length of time that a student practices playing the electric guitar given in Example 3 of Section 9.4 (page 450). We have the following:

```
MTB  > SET THE FOLLOWING DATA INTO C1
DATA > 10  12    8    9   16    5
DATA > END
```

*Epstein and Rogers, New York City, 1990.

After the data are entered we instruct the computer to determine a 90% confidence interval. We have the following desired results:

```
MTB > TINTERVAL WITH 90 PERCENT CONFIDENCE FOR DATA IN C1

        N      MEAN    STDEV    SEMEAN          90.0 PERCENT C.I.
C1      6     10.00     3.74     1.53          (  6.92,    13.08)
MTB > STOP
```

COMMENT In a similar manner, we can obtain 95% confidence intervals for μ. Minor changes in the code of the program allows us to obtain any desired confidence intervals for μ.

9.9 SUMMARY

In this chapter we indicated how sample data can be used to estimate the population mean and the population standard deviation. Both point and interval estimates were discussed. In most cases sample data are used to construct confidence intervals within which a given parameter with a specified probability is likely to lie. Sample data can also be used to make estimates about the true population proportion and to construct confidence intervals for the population proportion.

All estimates considered in this chapter were unbiased estimates. This means that the average of these estimates will approach the true population parameter that they are trying to estimate as more and more samples are included. For this reason we divide $\Sigma(x - \bar{x})^2$ by $n - 1$, not by n, in determining the sample variance.

We also indicated how to determine the size of a sample to be used in gathering data. Depending on the maximum allowable error, Formula 9.4 determines the sample size with a 95% degree of confidence.

Study Guide

The following is a chapter outline in capsule form. You should now be able to demonstrate your knowledge of the ideas mentioned by giving definitions, descriptions, or specific examples. Page references are given in parentheses.

Sample data can be used to estimate unknown quantities of the population. This use of sample data is referred to as **statistical estimation.** (page 438)

Sample data can also be used either to accept or reject specific claims about populations. This use of sample data is referred to as **hypothesis testing.** (page 438)

Population values, such as the mean and the standard deviation, are statistical descriptions of the population and are referred to as **population parameters.** (page 439)

An estimate of a population parameter that is a single number is called a **point estimate.** (page 439)

Interval estimation involves constructing an interval, called a **confidence interval,** within which we can be reasonably sure that the true parameter will lie. (page 439)

The lower and upper bounds of a confidence interval are called **confidence limits.** (page 439)

The probability that the procedure used will give a correct interval is called the **degree of confidence.** (page 439)

A **95% confidence interval for** μ means that if all possible samples of size n are selected, and the interval $x \pm 1.96 \dfrac{s}{\sqrt{n}}$ is established for each sample, then 95% of all such intervals are expected to contain μ. Similar statements can be made for other confidence intervals. (page 441)

The **Student's** t**-distribution,** which in many respects is similar to the normal distribution, is used when we deal with information obtained from small samples. (page 447)

The shape of the t-distribution depends on a parameter called the **number of degrees of freedom,** abbreviated as d.f., which equals the sample size minus 1. (page 447)

When estimating population parameters such as the mean we can take repeated samples. If the mean of these estimates approaches the true population mean as the sample size gets larger, then the average of the sample means is an **unbiased estimator** of the population mean. Thus, $\bar{x}$ is an unbiased estimator of the population mean μ and s^2 is an unbiased estimator of σ^2. (page 454)

The size for any sample is determined by the desired degree of accuracy. This is usually done by specifying the **maximum allowable error,** denoted as e. (page 456)

Often statistical data are available in **proportion** or **count form.** In such cases we use the **sample proportion** $\hat{p}$ to estimate the **true population** proportion, p. If the true population proportion is p, repeated sample proportions will be normally distributed. The mean of the sample distribution of proportions will be p. The standard deviation of these sample proportions will equal $\sqrt{\dfrac{p(1 - p)}{n}}$. (page 459)

Formulas to Remember

You should be able to identify each symbol in the following formulas, understand the relationships among the symbols expressed in each formula, understand the significance of each formula, and use the formulas in solving problems.

1.

Size of Sample	Parameter	Degree of Confidence	Lower Boundary	Upper Boundary
Large	Mean	90%	$\bar{x} - 1.645 \dfrac{s}{\sqrt{n}}$	$\bar{x} + 1.645 \dfrac{s}{\sqrt{n}}$
Large	Mean	95%	$\bar{x} - 1.96 \dfrac{s}{\sqrt{n}}$	$\bar{x} + 1.96 \dfrac{s}{\sqrt{n}}$
Large	Mean	99%	$\bar{x} - 2.58 \dfrac{s}{\sqrt{n}}$	$\bar{x} + 2.58 \dfrac{s}{\sqrt{n}}$
Small	Mean	95%	$\bar{x} - t_{0.025} \dfrac{s}{\sqrt{n}}$	$\bar{x} + t_{0.025} \dfrac{s}{\sqrt{n}}$
Large	Standard deviation	95%	$\dfrac{s}{1 + \dfrac{1.96}{\sqrt{2n}}}$	$\dfrac{s}{1 - \dfrac{1.96}{\sqrt{2n}}}$
Large	Proportion	95%	$\hat{p} - 1.96 \sqrt{\dfrac{\hat{p}(1 - \hat{p})}{n}}$	$\hat{p} + 1.96 \sqrt{\dfrac{\hat{p}(1 - \hat{p})}{n}}$

2. Size of sample where maximum allowable error may not be exceeded with a 95% degree of confidence: $n = \left(\dfrac{1.96\sigma}{e}\right)^2$

3. Mean of sampling proportions: p

4. Standard deviation of sampling proportion: $\sqrt{\dfrac{p(1 - p)}{n}}$

Testing Your Understanding of This Chapter's Concepts

1. For a given confidence level, if we wish to double the accuracy of an estimate, we must quadruple the sample size. Do you agree with this statement? Explain your answer.

2. Which of the following best describes the meaning of a 90% confidence interval for the population mean?
   a. The population mean will fall within the specified interval with a probability of 0.90.
   b. The probability is 0.90 that the mean is a correct point estimate.
   c. If many samples of the same size were to be taken and 90% confidence intervals constructed in the same manner, we would expect the population mean to be within the specified intervals 90% of the time.
   d. The sample mean is a good estimate of the population mean 90% of the time.

3. A consumer advocate called seven different car service operators in the metropolitan area to determine the charge for car service from one particular place to the airport. The prices quoted were as follows:

$18.00 $15.00 $22.00 $19.50 $17.00 $21.50 $20.00

Construct a 90% confidence interval for the average charge of car service from the particular place to the airport.

4. *Do consumers use newspaper or magazine coupons?* A recent Nielsen poll found that about 60% of those people surveyed used newspaper or magazine coupons. A random survey of 450 shoppers is taken. What is the probability that the proportion of shoppers who use newspaper or magazine coupons is at least 65%?

5. A psychologist performed a pattern-recognition experiment on 15 volunteers to determine the average time needed to recognize patterns that have been camouflaged. The average time needed to recognize the patterns was 5.28 minutes with a standard deviation of 1.06 minutes. Estimate the true average time needed to recognize the patterns with a 90% confidence interval. State any assumptions that are required for the interval estimation procedure to be valid.

THINKING CRITICALLY

1. A candidate for the U.S. House of Representatives conducted a random survey of 1500 registered voters and found that 650 favored his candidacy over the other candidates.
   a. Construct a 90% interval for the true proportion of the registered voters that favor his candidacy for the U.S. House of Representatives.
   b. How large a sample must be taken to be sure with a confidence of 95% that his error is no larger than ± 0.001?

2. A random sample of 800 executives found that 80% of the executives carry beepers with them at all times. What is the 99.7% confidence interval for the population proportion of executives who carry beepers with them at all times?

3. A random sample of size 300 is selected from a population with unknown mean, μ, and unknown standard deviation, σ. The following values are known: $\Sigma x = 3800$ and $\Sigma x^2 = 140,000$. Find a 95% confidence interval for μ.

4. Under what conditions can a sample standard deviation, s, be used in place of σ in the formula $n = \left(\dfrac{1.96\sigma}{e}\right)^2$ when σ is unknown? Explain your answer.

5. The critical t-values as given in Table VII in the Appendix can be approximated by $t = \sqrt{df \cdot (e^{y^2/df} - 1)}$, where $df = n - 1$, $e = 2.718$, $y = z\left(\dfrac{8\,df + 3}{8\,df + 1}\right)$,

and z is the critical z-score. Using this approximation, find the critical t-score for a 95% confidence interval corresponding to $n = 9$. Compare your answer with that given in Table VII. Some computer programs approximate t-values by using this procedure.

Review Exercises for Chapter 9

1. An airline reservations agent randomly selects 190 tickets for an overseas flight and finds that payment for 164 of them was by credit card. Find a 95% confidence interval for the true proportion of airline tickets for which payment is made by credit card.

2. A consumer opinion research group claims that 32% of the population prefer the new brand of facial tissue that has just been introduced on the market. To test this claim, a random sample of 520 consumers is selected. What is the probability that the proportion of the consumers who prefer the new brand of facial tissue is between 30% and 35%?

3. 45 podiatrists were asked to indicate their charge for the surgical removal of an ingrown toe nail. The standard deviation of their charges was computed and found to be $17. Find a 99% confidence interval for the population standard deviation.

4. Global Airlines claims that 96% of its planes arrive on time. A random sample of 71 Global plane arrivals is taken. What is the probability that fewer than 92% of them arrive on time?

5. A teacher wishes to estimate the average amount of time spent by a student per week doing homework. How large a sample should the teacher select if the teacher wishes that the estimate be within 0.74 hours of the true value? (Assume that $\sigma = 1.93$ hours.)

6. Nine members of a health club were randomly selected and asked to indicate the number of pounds that each lost over the past month since joining the aerobics program. The results are 10, 4, 11, 3, 7, 8, 6, 9, and 5 pounds. Find a 95% confidence interval for the average weight loss of a member since joining the aerobics program.

7. A random sample of 425 computer programmers in Washington disclosed that 295 owned computers equipped with a 40-megabyte hard disk drive. Find a 95% confidence interval for the true proportion of computer programmers in Washington who own a computer equipped with a 40-megabyte hard disk drive.

8. Professor Gonzales has been keeping records on how long it takes a college student to complete a particular chemistry experiment. For a group of 100 students the average time needed was 29 minutes with a standard deviation of 4.3 minutes. Find a 95% confidence interval for the mean time needed by a college student to complete the particular chemistry experiment.

9. Mountainview Ski Lodge would like to determine the proportion of its guests that would be interested in forming a ski club. It randomly samples 425 of its guests and finds that 285 of them would be interested in forming a ski club. Construct a 95% confidence interval for the true proportion of its guests who would be willing to form a ski club.

10.

UNIVERSITY TO CUT BACK ON COMPUTER ACCESSIBILITY

Staten Island (Mar. 1)—Faced with a mounting budget deficit, university officials announced yesterday that they were considering closing the student computer operations on weekends to save thousands of dollars.

Friday, March 1, 1991

Refer to the newspaper article above. The director of the computer center has available the following statistics on how many of the university's 40 personal computers were actually in use on numerous Saturday mornings at 10:00 A.M.

Weekend	Number of personal computers in use
Saturday January 5	21
Saturday January 12	17
Saturday January 19	32
Saturday January 26	19
Saturday February 2	18
Saturday February 9	23
Saturday February 16	25
Saturday February 23	29

Construct a 95% confidence interval for the mean number of the university's computers in use on Saturday mornings at 10:00 A.M.

Chapter Test

1. Babylon officials claim that 70% of the town's families own more than one car. A sample of 270 of the town's families is randomly selected. What is the probability that fewer than 165 of them own more than one car?

2. A consumer's research organization has been hired by a man to determine the true proportion of residents in Columbus that own a VCR. If a random survey of 600 residents of Columbus disclosed that 429 of them owned a VCR, find a 95% confidence interval for estimating the true proportion, p, of residents of Columbus that own a VCR.

3. A market research analyst wishes to determine the true average price of a gallon of unleaded premium gas. She samples ten randomly selected gas stations and discovers that for these stations $\bar{x}$ = \$1.24 and s = 19 cents. Find a 95% confidence interval for μ, the average price of a gallon of unleaded premium gallon of gas.

4. A state environmental cleanup task force is interested in determining the true average amount of pollution in one of the state's rivers. The task force obtains 20 samples of water from the northernmost part of one particular river. These samples are analyzed by an exhaustive chemical process and rated on a special scale. It is found that $\bar{x}$ = 54.3 with s = 4.76. Find a 90% confidence interval for μ, the average amount of pollution in this particular river.

5. The environmental cleanup task force mentioned in the last example has determined that one source of the water pollution is leakage from chemical waste containers that are buried in the ground. From past experience it is known that 60% of all containers containing chemical wastes that are buried in the ground will eventually leak. If a sample of 49 containers containing chemical wastes is randomly selected and the containers are subjected to extensive pressure, what is the probability that between 50% and 56% of these will eventually leak? See the following newspaper article.

RESIDENTS TO RETURN TO LOVE CANAL

Niagara Falls (May 2)—As a result of a massive cleanup campaign, officials from the New York State's Environmental Commission announced yesterday that residents would soon be allowed to return to the Love Canal area. Chemical wastes from the now defunct Hooker Chemical Company had worked their way into the Canal region and created a health hazard in the vicinity of the canal. The massive cleanup campaign cost millions of dollars.

May 2, 1990

6. A city official is interested in determining the average age of all runners who participate in the city's annual 15-mile marathon. A random sample of 100 runners showed a sample mean of $\bar{x}$ = 26 years with s = 7.83 years. Find a 95% confidence interval for the average age of a runner in the city's annual 15-mile marathon.

7. The operators of an express delivery service in New York City that specializes in delivering packages for Wall Street brokerage firms wishes to estimate the

average weight of a package to the nearest 0.6 of an ounce. Previous surveys have shown that $\sigma = 0.92$ ounces. How large a sample should the operators select?

8. Government regulatory authorities recently approved the marketing of a new antihistamine drug. However, it is known that approximately 5% of the population will suffer adverse side effects from the drug. If a random sample of 75 patients who are using the drug is selected, what is the probability that more than 9 of these people will experience some adverse side effects from the drug?

9. A random survey of 800 senior citizens in Detroit found that 475 of them believed that fear of being a victim of a crime occupied much of their thoughts. What is the probability that the proportion of senior citizens in Detroit whose thoughts are occupied with the fear of being the victim of a crime is greater than 55%?

10. Medical records of ten hospitals in a large city indicate that the hospitals treated an average of 35 patients a week suffering from AIDS. The standard deviation was 2.98. Find a 90% confidence interval for the mean number of patients who are suffering from AIDS and are treated weekly by these hospitals.

11. A random sample of 600 pregnant women in Lawrence found that 390 of them had medical coverage for nursery care of newborn children. What is the probability that the number of pregnant women who have this type of medical coverage is less than 59%?

12. A computer manufacturer wishes to estimate the average life of a particular type of computer chip to within 0.21 years of the true value of the mean. If σ is known to be 1.69 years, how large a sample should be selected?

13. A random survey of 325 American adults found that 236 of them consume alcoholic beverages. Find a 90% confidence interval for the true proportion of American adults who consume alcoholic beverages.

14. The average cost of a pack of cigarettes at 45 convenience stores on January 1 was $1.25 with a standard deviation of 9 cents. Find a 95% confidence interval for the average cost of a pack of cigarettes.
    a. between $1.23 and $1.27   b. between $1.22 and $1.28
    c. between $1.21 and $1.29   d. between $1.20 and $1.30   e. none of these

15. A manufacturer wishes to determine the list price for a typewriter. A survey of the list prices of typewriters sold by 8 other competing companies showed an average list price of $195 with a standard deviation of $18. Find a 95% confidence interval for the average list price for a typewriter.
    a. between $179.95 and $210.05   b. between $182.95 and $207.05   c. between $181.76 and $209.24   d. between $180.95 and $209.05   e. none of these

16. Seven police officers of the Brag Police department issued an average of 38 parking tickets during the month of January. The standard deviation was 4. Find a 90% confidence interval for the average number of parking tickets issued by the Brag police department during the month of January.
    a. between 33.06 and 42.94 tickets   b. between 34.06 and 39.94 tickets
    c. between 35.06 and 40.94 tickets   d. between 36.98 and 39.21 tickets
    e. none of these

17. A sample of 60 computer majors at a college indicated that a computer major spends an average of 17.6 hours per week writing programs. The standard deviation was 1.23 hours. Find a 95% confidence interval for the population standard deviation.

    a. between 16.29 and 18.91 hours   **b.** between 17.46 and 17.88 hours   **c.** between 16.86 and 18.24 hours   **d.** between 17.29 and 17.91 hours   **e.** none of these

18. Forty obstetricians were asked to indicate their charge for a normal delivery (no complications). The standard deviation of their charges was computed and found to be $85. Find a 99% confidence interval for the population standard deviation.

    a. between $66.84 and $117.89   **b.** between $65.97 and $119.46   **c.** between $67.21 and $118.78   **d.** between $68.97 and $122.47   **e.** none of these

19. A concerned citizen's group is interested in estimating the average response time by the city's police department to a call for help. How large a sample (with a 95% degree of confidence) should the group select, if the group wishes the estimate to be within 1.2 minutes of the true value. (Assume $\sigma = 2.3$)

    a. 14   **b.** 15   **c.** 16   **d.** 17   **e.** none of these

Suggested Further Reading

1. Freund, John E., and Ronald Walpole. *Mathematical Statistics*, 4th ed. Englewood Cliffs, NJ: Prentice-Hall, 1987.

2. *Gallup Opinion Index*. Princeton, NJ: American Institute of Public Opinion, January 1976.

3. Hamburg, Morris. *Statistical Analysis For Decision-Making*. 4th ed. Orlando, FL: Harcourt, Brace, Jovanovich, 1987.

4. Neter, J., W. Wasserman, and G. A. Whitmore. *Applied Statistics*. Boston: Allyn & Bacon, 1978.

5. Snedecor, George, and William Cochran. *Statistical Methods*, 7th ed. Ames, IA: Iowa State University Press, 1980.

6. Zuwaylif, Fadil H. *Applied Business Statistics*. 2nd ed. Reading, MA: Addison-Wesley, 1984.

10 Hypothesis Testing

Many of our country's lakes and rivers are polluted because garbage and chemical wastes are dumped into them. How do we determine if there is any significant difference between the average amount of pollution found in these different lakes and rivers. Exercise 2 on p. 504.

In the previous chapter we indicated how sample statistics can be used to help us set up confidence intervals for means, standard deviations, and proportions.

In this chapter we indicate how statistics can be used to make decisions about certain assumptions called hypotheses. Statistical inferences of this type are called hypotheses tests. We will analyze procedures for testing hypotheses regarding means, differences between means, and proportions. The appropriate procedure will vary and depend on whether we have a small or large sample size.

CHAPTER OBJECTIVES	After studying the material in this chapter, you should be able:

After studying the material in this chapter, you should be able:

- **To analyze** how sample data can be used to reject or accept a claim about some aspect of a probability distribution. The claim to be tested is called the null hypothesis. (Section 10.2)

- **To see** that a test statistic is a number that we compute to determine when to reject the null hypothesis. (Section 10.2)

- **To determine** when critical rejection regions tell us to reject a null hypothesis. This occurs when the test statistic value falls within this region. How these regions are set up depends on the specifications given within the problem. (Section 10.2)

- **To discuss** the two errors that can be made when we use sample data to accept or reject a null hypothesis. We may incorrectly reject a true hypothesis or we may incorrectly accept a false hypothesis. In both cases, an error is made. (Section 10.3)

- **To indicate** what a level of significance is. (Section 10.3)

- **To distinguish** between tests concerning means, differences between means, and proportions. Thus, we discuss the test statistics used when we wish to use sample data to determine whether observed differences between means and proportions are significant. (Sections 10.4 to 10.8)

- **To apply** the hypothesis-testing procedures to wide-ranging problems. (Throughout chapter)

475

Statistics in the News

MANY AMERICANS SUPPORT HIGHER TAXES

Washington (January 7)—The results of a nationwide survey by Administration officials indicate that many Americans believe that fighting crime and drugs is foremost in their minds. About 75% of those surveyed said that they would be willing to have their sales tax increased to fund any improvement in crime fighting and to make our streets safer.

January 7, 1991

NEW CHOLESTEROL TEST DEVELOPED

Washington (May 20)—Medical researchers announced that they have developed a new machine that can determine an individual's blood serum cholesterol level within 3 minutes. The new procedure allows the doctor or lab technician to inform the patient of the results almost immediately as opposed to the old procedure which required that the blood sample be sent to a medical laboratory. The new machine will be available for nationwide use within a short period of time.

May 20, 1990

Often we read about various claims that have been made. How do we verify the accuracy of such claims? Consider the first article. If we randomly sample 100 Americans, would we expect 75 of them to support this claim? If only 70 of them said that they would be willing to have their sales tax increased to fund any nationwide crime fighting task force, would you say that this is significantly less than the claimed 75%?

Now consider the second article. How do we determine whether the new procedure is accurate? When is any discrepancy between results obtained from the new procedure and the old procedure significant?

10.1 INTRODUCTION

[handwritten: 10.1 - 10.3 - short answer fill in blank]

Many television commercials contain unusual performance claims. For example, consider the following TV commercials:

1. Four out of five dentists recommend Brand X sugarless gum for their patients who chew gum.
2. A particular brand of tire will last an average of 40,000 miles before replacement is needed.
3. A certain detergent produces the cleanest wash.
4. Brand X paper towels are stronger and more absorbent.

    How much confidence can one have in such claims? Can they be verified statistically? Fortunately, in many cases the answer is yes. Samples are taken and claims are tested. We can then make a decision on whether to accept or reject a claim on the basis of sample information. This process is called **hypothesis testing.** Perhaps this is one of the most important uses of samples.

Hypothesis Testing

    As we indicated in Chapter 9, hypothesis testing is an important branch of statistical inference. Sample data provide us with estimates of population parameters. These estimates are in turn used in arriving at a decision either to accept or reject an hypothesis. By an **hypothesis** we mean an assumption about one or more of the population parameters that will either be accepted or rejected on the basis of the information obtained from a sample. In this chapter we discuss methods for determining whether to accept or reject any hypothesis on the basis of sample data.

Hypothesis

10.2 TESTING AGAINST AN ALTERNATIVE HYPOTHESIS

Suppose several players are in a gambling casino rolling a die. A bystander notices that in the first 120 rolls of the die, a 6 showed only eight times. Is this reasonable or is the die loaded? The management claims that this unusual occurrence is to be attributed purely to chance and that the die is an honest die. The bystander claims otherwise.

    If the die is an honest die, then Formula 6.5 (see page 328) for a binomial distribution tells us that the average number of 6's occurring in 120 rolls of the die is 20, as

$$\mu = np$$

$$= 120 \left(\frac{1}{6}\right)$$

$$= 20$$

If the die is loaded, then $\mu \neq 20$. (The symbol $\neq$ means "is not equal to.") Since the die is either an honest die or a loaded die, we must choose between the hypothesis

$\mu = 20$ and the hypothesis $\mu \neq 20$. Thus, sample data will be used either to accept or reject the hypothesis $\mu = 20$. Such an hypothesis is called a **null hypothesis** and is denoted by H_0. Any hypothesis that differs from the null hypothesis is called an **alternative hypothesis** and is denoted as $H_1, H_2, \ldots$, and so on. In our example

$$\text{Null hypothesis, } H_0: \quad \mu = 20$$

$$\text{Alternative hypothesis, } H_1: \quad \mu \neq 20$$

Notice that by formulating the alternative hypothesis as $\mu \neq 20$, we are indicating that we wish to perform a **two-sided,** or **two-tailed test.** This means that if the die is not honest, then it may be loaded in favor of obtaining 6's more often than is expected or less often than is expected.

In our example the bystander strongly suspects that the die is loaded against obtaining 6's. Thus, his alternative hypothesis would be $\mu < 20$. (The symbol $<$ stands for "is less than.") Similarly, if the bystander suspected that the die was loaded in favor of obtaining 6's more often than is expected, the alternative hypothesis would be $\mu > 20$. (The symbol $>$ stands for "is greater than.") In each of these cases the null hypothesis remains the same, $H_0: \mu = 20$. Such alternative hypotheses indicate that we wish to perform a **one-sided,** or **one-tailed test.**

COMMENT   It should be noted that the decision on an alternative hypothesis should be made *before* the results of the sample are known.

A decision as to whether to accept or reject the null hypothesis will be made on the basis of sample data. How is such a decision made? We must realize that even if we know for sure that the die is honest, it is very unlikely that we would get exactly twenty 6's in the 120 rolls of the die. Moreover, if we were to roll the die 120 times on many different occasions, we would find that the number of 6's appearing is around 20, sometimes more and sometimes less. It is therefore obvious that we

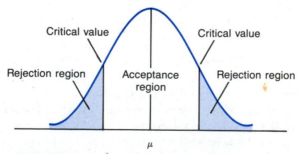

Figure 10.1
A two-tailed rejection region.

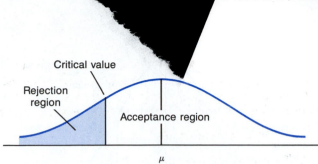

Figure 10.2
A one-tailed rejection region.

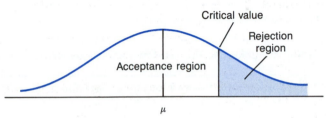

Figure 10.3
A one-tailed rejection region.

Acceptance Region

must set up some interval that we call the **acceptance region.** If the number of 6's appearing in 120 rolls of the die is within this acceptance region, then we will accept the null hypothesis. If the number of 6's obtained is outside this region, then we will reject the null hypothesis that the die is an honest die. These possibilities are shown in Figures 10.1, 10.2, and 10.3. The value that separates the rejection region from the acceptance region is called the **critical value.**

Critical Value

The type of symbol in the alternative hypothesis tells us what type of rejection region to use as shown in the following table:

If the Symbol in the Alternative Hypothesis Is	$<$	$\neq$	$>$
Then the Rejection Region Consists of	one region on the left side.	two regions, one on each side.	one region on the right side.

Suppose we decide to accept the null hypothesis if the number of 6's obtained is between 15 and 25. Since in our case only eight 6's were obtained, we would reject the null hypothesis. In this case the acceptance region is 15 to 25. The two-tailed rejection region corresponds to the two tails, less than 15 and more than 25, as shown in Figure 10.4. When we reject a null hypothesis, we are claiming that the value of the population parameter, that is, the average number of 6's, is some value other than the one specified in the null hypothesis. Also, when the sample data indicate that we should reject a null hypothesis, we say that the observed difference is **significant.**

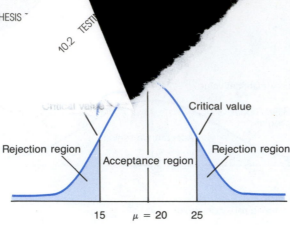

Figure 10.4
A two-tailed rejection region.

Our discussions in the last few paragraphs lead us to the following definitions:

DEFINITION 10.1 Null Hypothesis	The **null hypothesis**, denoted by H_0, is the statistical hypothesis being tested.

DEFINITION 10.2 Alternative Hypothesis	The **alternative hypothesis** denoted by H_1, H_2, . . . , is the hypothesis that will be accepted when the null hypothesis is rejected.

DEFINITION 10.3 One-tailed Test One-sided Test	A **one-sided**, or **one-tailed test** is a statistical test that has the rejection region located in the left tail or the right tail of the distribution.

DEFINITION 10.4 Two-sided Test Two-tailed Test	A **two-sided**, or **two-tailed test** is a statistical test that has the rejection region located in both tails of the distribution.

Let us summarize the steps that need to be followed in hypothesis testing:

1. State the null hypothesis, which indicates the value of the population parameter to be tested.
2. State the alternative hypothesis, which indicates the belief that the population parameter has a value other than the one specified in the null hypothesis.
3. Set up rejection and acceptance regions for the null hypothesis. The **region of rejection** is called the **critical region.** The remaining region is called the **acceptance region.**
4. Compute the value of the test statistic. A **test statistic** is a calculated number that is used to decide whether to reject or accept the null hypothesis. The formula for computing the value of the test statistic depends on the parameter we are testing.

Region of Rejection
Critical Region
Acceptance Region
Test Statistic

5. Reject the null hypothesis if the test statistic value falls within the rejection region, that is, the critical region. Otherwise, do not reject the null hypothesis.
6. State the conclusion for the particular problem.

In this chapter we discuss various tests that enable us to decide whether to reject or accept a null hypothesis. Such tests are called **statistical tests of hypotheses** or **statistical tests of significance.**

10.3 TWO TYPES OF ERRORS

Since any decision either to accept or reject a null hypothesis is to be made on the basis of information obtained from sample data, there is a chance that we will make an error. There are two possible errors that we can make. We may reject a null hypothesis when we really should accept it. Thus, returning to the die problem of Section 10.2, we may reject the claim that the die is honest even though it actually is honest. Alternately, we may accept a null hypothesis when we should reject it. Thus, we may say that the die is honest when it really is a loaded die.

These two errors are referred to as a **type-I** and a **type-II error,** respectively. In either case we have made a wrong decision. We define these formally as follows:

> **DEFINITION 10.5**
> Type-I Error
>
> **A type-I error** is made when a true null hypothesis is rejected; that is, we reject a null hypothesis when we should accept it.

> **DEFINITION 10.6**
> Type-II Error
>
> **A type-II error** is made when a false null hypothesis is accepted; that is, we accept a null hypothesis when we should reject it.

In the following box we indicate how these two errors are made:

		And We Claim That	
		H_0 Is True	H_0 Is False
If	H_0 Is True	Correct decision (no error)	Type-I error
	H_0 Is False	Type-II error	Correct decision (no error)

When deciding whether to accept or reject a null hypothesis, we always wish to minimize the probability of making a type-I error or a type-II error. Unfortunately, the relationship between the probabilities of the two types of errors is of such a nature that if we reduce the probability of making one type of error, we usually increase the probability of making the other type. In most applied problems one type of error is more serious than the other. In such situations careful attention is given to the more serious error.

How much risk should a statistician take in rejecting a true hypothesis, that is, in making a type-I error? Generally speaking, statisticians use the limits of 0.05 and 0.01. Each of these limits is called a **level of significance** or **significance level.** We have the following definition:

> **DEFINITION 10.7**
> Significance Level
>
> The **significance level** of a test is the probability that the test statistic falls within the rejection region when the null hypothesis is true.

5% Level of Significance

1% Level of Significance

The **0.05 level of significance** is used when the statistician wishes that the risk of rejecting a true null hypothesis not exceed 0.05. The **0.01 level of significance** is used when the statistician wishes that the risk of rejecting a true null hypothesis not exceed 0.01.

In this book we will usually assume that we wish to correctly accept the null hypothesis 95% of the time and to incorrectly reject it only 5% of the time. Thus, the maximum probability of a type-I error that we are willing to accept, that is, the significance level, will be 0.05. *The probability of making a type-I error when H_0 is true is denoted by the Greek letter α (pronounced alpha)*. Therefore, the probability of making a correct decision is $1 - \alpha$.

As we indicated on page 479, when dealing with one-tailed tests, the critical region lies to the left or to the right of the mean. This is shown in Figure 10.5. When dealing with a two-tailed test, one-half of the critical region is to the left of the mean and one-half is to the right. The probability of making a type-I error is evenly divided between these two tails, as shown in Figure 10.6.

COMMENT   If the test statistic falls within the acceptance region, we do not reject the null hypothesis. When a null hypothesis is not rejected, this does not mean that what the null hypothesis claims is guaranteed to be true. It simply means that on the basis of the information obtained from the sample data there is not enough evidence to reject the null hypothesis.

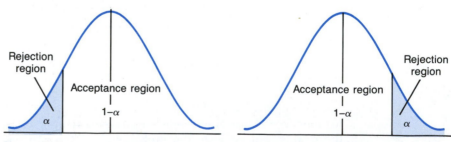

Figure 10.5

Figure 10.6

10.4   TESTS CONCERNING MEANS FOR LARGE SAMPLES

In this section we discuss methods for determining whether we should accept or reject a null hypothesis about the mean of a population. We will illustrate the procedure with several examples.

Suppose a manufacturer claims that each family-size bag of pretzels sold weighs 12 ounces, on the average, with a standard deviation of 0.8 ounces. A consumer's group decides to test this claim by accurately weighing 49 randomly selected bags of pretzels. If the mean weight of the sample is considerably different from the population mean, the manufacturer's claim will definitely be rejected. Thus, if the mean weight is 30 ounces or 5 ounces, the manufacturer's claim will be rejected. Only when the sample mean is close to the claimed population mean do we need statistical procedures to determine when to reject or accept a null hypothesis.

Let us assume that the sample mean of the 49 randomly selected bags of pretzels is 11.8 ounces. Since the sample mean, 11.8, is not the same as the population mean, 12, we wish to test the manufacturer's claim at the 5% level of significance.

The population parameter being tested in this case is the mean weight, μ, and the questioned value is 12 ounces. Thus,

Null hypothesis, H_0: $\mu = 12$

Alternative hypothesis, H_1: $\mu \neq 12$

The alternative hypothesis of not equal suggests a two-tailed rejection region. Therefore, the α of 0.05 is split equally between the two tails, as shown in Figure 10.7.

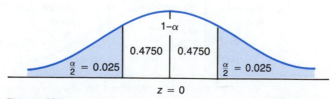

Figure 10.7

We now look in Table IV of the Appendix to determine which z-value has 0.4750 of the area between $z = 0$ and this z-value. From Table IV we find that the z-value is 1.96. We label this on the diagram in Figure 10.7 and get the acceptance-rejection diagram shown in Figure 10.8.

Since the Central Limit Theorem tells us that the sample means are normally distributed, we use

$$z = \frac{\bar{x} - \mu}{\dfrac{\sigma}{\sqrt{n}}}$$

as the test statistic and reject the null hypothesis if the value of the test statistic falls in the rejection region. In using this test statistic, $\bar{x}$ is the sample mean and μ is the population mean as claimed in the null hypothesis. In our case we have

$$z = \frac{\bar{x} - \mu}{\dfrac{\sigma}{\sqrt{n}}}$$

$$= \frac{11.8 - 12}{\dfrac{0.8}{\sqrt{49}}}$$

$$= \frac{-0.2}{\dfrac{0.8}{7}}$$

$$= -1.75$$

Since this calculated value of z falls within the acceptance region, our decision is that we cannot reject H_0. The difference between the sample mean and the assumed

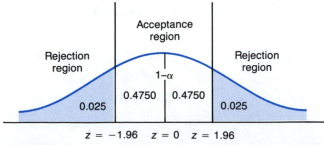

Figure 10.8

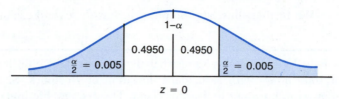

Figure 10.9

value of the population mean may be due purely to chance. We say that the difference is *not statistically significant.*

If the level of significance had been 0.01, we would split the 0.01 into two equal tails as shown in Figure 10.9. From Table IV we find that the z-value that has 0.4950 of the area between $z = 0$ and this z-value is 2.58. Thus, we reject the null hypothesis if the test statistic falls in the critical region shown in Figure 10.10.
Since we obtained a z-value of -1.75, which is in the acceptance region, we do not reject the null hypothesis.

Let us summarize the testing procedure outlined in the previous paragraphs for a two-tailed rejection region.

1. First convert the sample mean into standard units using

$$z = \frac{\bar{x} - \mu}{\dfrac{\sigma}{\sqrt{n}}}$$

2. Then reject the null hypothesis about the population mean if z is less than -1.96 or greater than 1.96 when using a 5% level of significance.

3. If we are using a 1% level of significance, reject the null hypothesis if z is less than -2.58 or greater than 2.58.

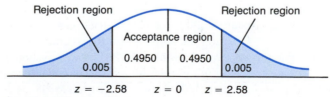

Figure 10.10

We further illustrate the procedure with several examples.

EXAMPLE 1   A light bulb company claims that the 60-watt light bulb it sells has an average life of 1000 hours with a standard deviation of 75 hours. Sixty-four new bulbs were allowed to burn out to test this claim. The average lifetime of these bulbs was found to be 975 hours. Does this indicate that the average life of a bulb is not 1000 hours? (Use a 5% level of significance.)

Solution

In this case the population parameter being tested is μ, the average life of a bulb, and the value questioned is 1000. Since we are testing whether the average life of a bulb is or is not 1000 hours, we have

$$H_0: \quad \mu = 1000$$

$$H_1: \quad \mu \neq 1000$$

[handwritten: $\bar{x} = 975$, $\mu = 1000$, $\sigma = 75$ hrs, $n = 64$]

We are given that $\bar{x} = 975$, $\sigma = 75$, $\mu = 1000$, and $n = 64$. We first calculate the value of the test statistic, z. We have

$$z = \frac{\bar{x} - \mu}{\dfrac{\sigma}{\sqrt{n}}}$$

$$= \frac{975 - 1000}{\dfrac{75}{\sqrt{64}}}$$

$$= -2.67 \quad \text{*[handwritten: would be in the rejection region]*}$$

[handwritten diagram: curve showing RR, Acceptance Region, RR with $z = -1.96$, $z = 0$, $1.96 = z$]

We use the two-tailed rejection region shown in Figure 10.11. The value of $z = -2.67$ falls in the rejection region. Thus, we reject the null hypothesis that the average life of a bulb is 1000 hours. In this case, the test statistic is sufficiently extreme so that we can reject H_0. The difference is statistically significant.

Perhaps you are wondering why we used $\mu \neq 1000$ as the alternative hypothesis and not $\mu < 1000$. After all, who cares if the average life of a bulb is more than 1000 hours. The answer is that the manufacturer cares. When a manufacturer claims that the average life of a bulb is 1000 hours, the manufacturer is concerned when bulbs last more or less than 1000 hours. If the mean life is less than 1000 hours, the manufacturer will lose business and consumer confidence. If the mean life is more than 1000 hours, the company will lose money.

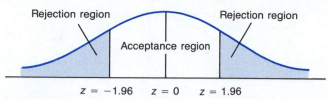

Figure 10.11

EXAMPLE 2

(handwritten: 5000)
(handwritten: $\bar{X} = 4800$)
(handwritten: $\sigma = 460$)
(handwritten: $n = 36$)
(handwritten: $\mu = 4800 \; 5000$)

A bank teller at the Eastern Savings Bank claims that the average amount of money on deposit in a savings account at this bank is $4800 with a standard deviation of $460. A random sample of 36 accounts is taken to test this claim. The average of these accounts is found to be $5000. Does this sample indicate that the average amount of money on deposit is not $4800? (Use a 5% level of significance.)

Solution

In this case the population parameter being tested is μ, the average amount of money on deposit in a savings account. The value questioned is $4800. Since we are testing whether the average amount of money on deposit is $4800 or not, we have

$$H_0: \quad \mu = 4800$$
$$H_1: \quad \mu \neq 4800$$

We are given that $\bar{x} = 5000$, $\mu = 4800$, $\sigma = 460$, and $n = 36$ so that

(handwritten left margin: 4800 5000)
(handwritten: $\dfrac{5000 - 4800}{\dfrac{460}{\sqrt{36}}}$)
(handwritten: $\dfrac{200}{76.7} = 2.61$)

$$z = \frac{\bar{x} - \mu}{\dfrac{\sigma}{\sqrt{n}}}$$

$$= \frac{5000 - 4800}{\dfrac{460}{\sqrt{39}}}$$

$$= 2.61 \quad \text{(handwritten: would be in the rejection region)}$$

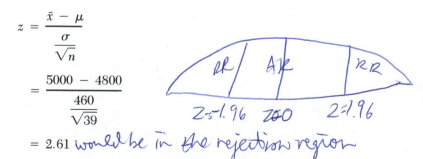

(handwritten in figure: RR / AR / RR, $Z = -1.96$, $Z = 0$, $Z = 1.96$)

We use the same two-tailed rejection region as shown in Figure 10.11. The value of $z = 2.61$ falls in the rejection region. Thus, we reject the null hypothesis that the average amount of money on deposit is $4800.

EXAMPLE 3

The average score of all sixth graders in a certain school district on the 1–2–3 math aptitude exam is 75 with a standard deviation of 8.1. A random sample of 100 students in one school was taken. The mean score of these 100 students was 71. Does this indicate that the students of this school are significantly less skilled in their math-

(handwritten: $n = 100$)
(handwritten: $\bar{X} = 71$)
(handwritten: $\mu = 75$)
(handwritten: $\sigma = 8.1$)

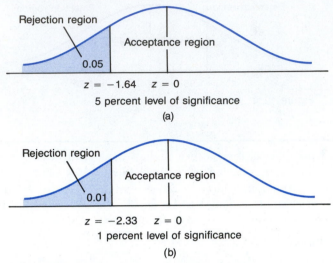

Figure 10.12
One-tailed tests for μ less than some given value.

ematical abilities than the average student in the district? (Use a 5% level of significance.)

Solution

In this case the population parameter being tested is μ, the mean score on the math aptitude exam. The value questioned is $\mu = 75$. We want to determine if the students of this particular school are significantly less skilled in their mathematical abilities. Thus, it is reasonable to set up a one-sided, or a one-tailed, test with the alternative hypothesis being that the population mean for this school is less than 75. We have

$$H_0: \quad \mu = 75$$

$$H_1: \quad \mu < 75$$

When dealing with one-tailed (left-side) alternative hypotheses, we have the rejection regions illustrated in Figure 10.12. These values, like those in the two-tailed tests, are obtained from Table IV in the Appendix. You should verify these results.

Now we can calculate the test statistic, z. Here $\bar{x} = 71$, $\mu = 75$, $\sigma = 8.1$, and $n = 100$. We have

$$z = \frac{\bar{x} - \mu}{\dfrac{\sigma}{\sqrt{n}}}$$

$$= \frac{71 - 75}{\frac{8.1}{\sqrt{100}}}$$

$$= -4.94$$

Since the z-value of -4.94 is in the rejection region, we conclude that the students of this school are significantly less skilled in their mathematical abilities than the average student in the district.

EXAMPLE 4

The We-Haul Trucking Corp. claims that the average hourly salary of its mechanics is \$9.25 with a standard deviation of \$1.55. A random sample of 81 mechanics showed that the average hourly salary of these mechanics was only \$8.95. Does this indicate that the average hourly salary of a mechanic is significantly less than \$9.25? (Use a 1% level of significance.)

Solution

In this case the population parameter being tested is μ, the mean hourly salary. The questioned value is \$9.25. We wish to know if the hourly salary is less than \$9.25. Thus, we use a one-tailed test. We have

$$H_0: \quad \mu = 9.25$$

$$H_1: \quad \mu < 9.25$$

Here we are given that $\bar{x} = 8.95$, $\mu = 9.25$, $\sigma = 1.55$, and $n = 81$ so that

$$z = \frac{\bar{x} - \mu}{\frac{\sigma}{\sqrt{n}}}$$

$$= \frac{8.95 - 9.25}{\frac{1.55}{\sqrt{81}}}$$

$$= -1.74$$

We use the same one-tailed rejection region as shown in Figure 10.12. The value of $z = -1.74$ falls in the acceptance region. Thus, the sample data do not provide us with sufficient justification to reject the null hypothesis. We cannot conclude that the average hourly salary of a mechanic is significantly less than \$9.25.

EXAMPLE 5

A trash company claims that the average weight of any of its fully loaded garbage trucks is 11,000 pounds with a standard deviation of 800 pounds. A highway department inspector decides to check on this claim. She randomly checks 36 trucks and finds that the average weight of these trucks is 12,500 pounds. Does this indicate that the average weight of a garbage truck is more than 11,000 pounds? (Use a 5% level of significance.)

Solution

In this case the population parameter being tested is μ, the average weight of a garbage truck. The value questioned is 11,000 pounds. The sample data suggest that the mean weight is really more than 11,000 pounds. Thus, the alternative hypothesis will be that the population mean is more than 11,000 pounds. We have

$$H_0: \quad \mu = 11,000$$

$$H_1: \quad \mu > 11,000$$

When dealing with a one-tailed (right-side) alternative hypothesis we have the rejection regions shown in Figure 10.13.

Here we are given that $\bar{x} = 12,500$, $\mu = 11,000$, $\sigma = 800$, and $n = 36$ so that

$$z = \frac{\bar{x} - \mu}{\dfrac{\sigma}{\sqrt{n}}}$$

$$= \frac{12,500 - 11,000}{\dfrac{800}{\sqrt{36}}}$$

$$= 11.25$$

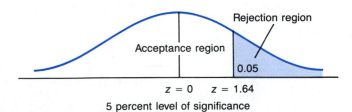

5 percent level of significance

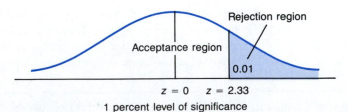

1 percent level of significance

Figure 10.13
One-tailed tests for μ greater than some given value.

Since the z-value of 11.25 falls within the rejection region, we conclude that the average weight of a garbage truck is not 11,000 pounds. We reject the null hypothesis.

EXAMPLE 6

An insurance company advertises that it takes 21 days on the average to process an auto accident claim. The standard deviation is 8 days. To check on the truth of this advertisement, a group of investigators randomly selects 35 people who recently filed claims. They find that it took the company an average of 24 days to process these claims. Does this indicate that it takes the insurance company more than 21 days on the average to process a claim? (Use a 1% level of significance.)

Solution

In this case the population parameter being tested is μ, the average number of days needed to process a claim. The questioned value is $\mu = 21$. The sample data suggest that $\mu > 21$. Thus, we will use a one-tailed test. We have

$$H_0: \quad \mu = 21$$

$$H_1: \quad \mu > 21$$

Here we are given that $\bar{x} = 24$, $\mu = 21$, $\sigma = 8$, and $n = 35$ so that

$$z = \frac{\bar{x} - \mu}{\dfrac{\sigma}{\sqrt{n}}}$$

$$= \frac{24 - 21}{\dfrac{8}{\sqrt{35}}} = 2.22$$

Since the z-value of 2.22 falls within the acceptance region, we cannot reject the null hypothesis. Thus, we cannot conclude that it takes the insurance company more than 21 days to process a claim.

COMMENT   We wish to emphasize again that the large-sample case tests that we have been discussing work only if the sample size is sufficiently large. This means that n should be at least 30. This is necessary so that the distribution of $\bar{x}$ be approximately normal.

COMMENT   Since σ is unknown in many practical applications, we often have no choice but to use the sample standard deviation, s, as an approximation for σ. Again, s provides a good approximation to σ if the sample size is sufficiently large ($n \geq 30$).

We summarize our discussion as follows:

Large-Sample Test of Hypothesis About a Population Mean

	One-Tailed Test	Two-Tailed Test	One-Tailed Test
Null Hypothesis	$H_0: \mu = \mu_0$	$H_0: \mu = \mu_0$	$H_0: \mu = \mu_0$
Alternative Hypothesis	$H_1: \mu < \mu_0$	$H_1: \mu \neq \mu_0$	$H_1: \mu > \mu_0$
Test Statistic	$z = \dfrac{\bar{x} - \mu_0}{\dfrac{\sigma}{\sqrt{n}}}$ $= \dfrac{\bar{x} - \mu_0}{\dfrac{s}{\sqrt{n}}}$	$z = \dfrac{\bar{x} - \mu_0}{\dfrac{\sigma}{\sqrt{n}}}$ $= \dfrac{\bar{x} - \mu_0}{\dfrac{s}{\sqrt{n}}}$	$z = \dfrac{\bar{x} - \mu_0}{\dfrac{\sigma}{\sqrt{n}}}$ $= \dfrac{\bar{x} - \mu_0}{\dfrac{s}{\sqrt{n}}}$
Rejection Region	$z < -z_\alpha$	$z < -z_{\alpha/2}$ or $z > z_{\alpha/2}$	$z > z_\alpha$

In this chart, z_α is the z-value such that prob $(z > z_\alpha) = \alpha$.

$z_{\alpha/2}$ is the z-value such that prob $(z > z_{\alpha/2}) = \dfrac{\alpha}{2}$.

μ_0 represents a particular value for μ as specified in the null hypothesis that we are testing.

CURE FOUND FOR DEADLY VIRUS

Washington:(Jan 27)—A treatment has been found for severe cases of a respiratory virus that infects more than 800,000 U.S. infants a year.

The Food and Drug Administration has approved the drug ribavirin for acute cases of respiratory syncytial virus, or RSV. Administered with an inhalation hood in a hospital for up to seven days, ribavirin stops RSV from reproducing.

Taken orally for 10 days, ribavirin also is the first effective therapy for Lassa fever, a deadly virus found primarily in Africa, say doctors from the Centers for Disease Control.

January 27, 1986

When a new drug is developed, how do we determine whether the average number of people cured by the drug is significant? Is one drug better than another in terms of the average number of people cured?

EXERCISES FOR SECTION 10.4

1. An official in charge of ambulance maintenance suspects that the average lifetime of 33,000 miles (mi) claimed by the manufacturer for a particular type of tire is considerably too high. To check on the accuracy of the claim, the official puts 49 of these tires on the ambulances. These tires last for an average of 30,295 mi with a standard deviation of 1471 mi. Using a 5% level of significance, should we reject the manufacturer's claim?

2. According to a drug manufacturer, the average systolic blood pressure of a person (at rest) while using a particular drug is 140 with a standard deviation of 12.5. A survey of 45 patients using the drug showed that the average systolic blood pressure of these patients was 134. Using a 1% level of significance, should we reject the manufacturer's claim?

3. The average hourly salary of a statistical typist in a certain city is $14 with a standard deviation of $1.69. The Ajax Typing Service, which employs 55 typists, pays its typists an average hourly rate of $11.88. Can the Ajax Typing Service be accused of paying lower than the average hourly rate? (Use a 5% level of significance.)

4. State officials claim that the average number of arrests per week for drunk driving on the thruway is 49 with a standard deviation of 14.8. If a random survey of 40 weeks found an average of 52 arrests per week for drunk driving, should we reject the state official's claims? (Use a 5% level of significance.)

5. Consider the newspaper article that follows.

> ### STATE TO INVESTIGATE ABUSE OF SICK LEAVE PRIVILEGE
>
> *New Haven (Aug. 2)*—The governor's office announced yesterday the establishment of a commission to investigate abuses of the sick leave policy privilege afforded to highway patrol officers. Under this privilege an officer is allowed an unlimited number of sick days per year.
>
> The officials claim that this privilege is being abused by some officers.
>
> August 2, 1990

The union claims that the average number of sick days taken by an officer is 9.4 days per year with a standard deviation of 4.3 days. A survey of 45 highway patrol officers found that these officers were absent an average of 11.9 days per year. Using a 1% level of significance, should we reject the union's claim?

6. There are 484 faculty members at Merck College. Administration officials claim that the average salary of a faculty member is $39,104 with a standard deviation of $3,440. Union representatives claim that his figure is inaccurate. A survey of 42 faculty members finds that the average salary of these faculty members is $36,082. Using a 5% level of significance, should we reject the administration official's claim?

7. A baseball manager claims that the average speed of a fast ball thrown by a particular pitcher is 92 miles per hour (mph) with a standard deviation of 10.83 mph. 50 fast balls thrown by this pitcher are randomly selected. If the average speed of these pitches is found to be 90 mph, should we reject the baseball manager's claim? (Use a 10% level of significance.)

8. One stockbroker reports that the average commission for a transaction is $68 with a standard deviation of $13.12. Fifty of the stockbroker's transactions are randomly selected. If the average commission is found to be $71, should we reject the stockbroker's claim? (Use a 5% level of significance.)

9. City officials claim that the typical fireman retires after an average of 24.2 years on the job. The standard deviation is 3.9 years. A survey of 45 retired firemen indicated that these firemen retired after 28.3 years. Should we reject the city officials' claim? (Use a 1% level of significance.)

10. The game warden of a large park claims that the average length of a trout fish in the lake in the park is 19 inches with a standard deviation of 4 inches. The average length of 58 trout fish caught in the lake by several fishermen is found to be only 17 inches. Does this indicate that the average length is less than 19 inches? (Use a 5% level of significance.)

11. The manufacturer of a certain foreign car sold in the United States claims that it will average 35 miles per gallon of gasoline. To test this claim, a consumer's group randomly selects 40 of these cars and drives them under normal driving conditions. These cars average 28 miles to the gallon with a standard deviation of 9.3 miles. Should we reject the manufacturer's claim? (Use a 5% level of significance.)

10.5 TESTS CONCERNING MEANS FOR SMALL SAMPLES

In the last section we indicated how sample data can be used to reject or accept a null hypothesis about the mean. The sizes of all the samples discussed were large enough to justify the use of the normal distribution. When the sample size is small, we must use the t-distribution discussed in the last chapter instead of the normal distribution. The following examples illustrate how the t-distribution is used in hypothesis testing.

EXAMPLE 1

A manufacturer claims that each can of mixed nuts sold contains an average of 10 cashew nuts. A sample of 15 cans of these mixed nuts has an average of 8 cashew nuts with a standard deviation of 3. Does this indicate that we should reject the manufacturer's claim? (Use a 5% level of significance.)

Solution

Since the sample size is only 15, the test statistic becomes

$$t = \frac{\bar{x} - \mu}{\frac{s}{\sqrt{n}}}$$

instead of z. Depending upon the number of degrees of freedom, we have the acceptance-rejection regions shown in Figures 10.14 and 10.15. In each case the value of t is obtained from Table VII in the Appendix. It depends on the number of degrees of freedom, which is $n - 1$.

Let us now return to our problem. The population parameter being tested is μ, the average number of cashew nuts in a can of mixed nuts. The questioned value is 10. We have

$$H_0: \quad \mu = 10$$

$$H_1: \quad \mu \neq 10$$

We will use a two-tailed rejection region as shown in Figure 10.14. Here we are given that $\bar{x} = 8$, $\mu = 10$, $s = 3$, and $n = 15$ so that

$$t = \frac{\bar{x} - \mu}{\frac{s}{\sqrt{n}}} = \frac{8 - 10}{\frac{3}{\sqrt{15}}} = -2.58$$

From Table VII we find that the $t_{0.025}$ value for $15 - 1$, or 14 degrees of freedom is 2.145, which means that we reject the null hypothesis if the test statistic is less than -2.145 or greater than 2.145. Since $t = -2.58$ falls within the critical region,

Figure 10.14
Two-tailed small-sample rejection region (5% level of significance).

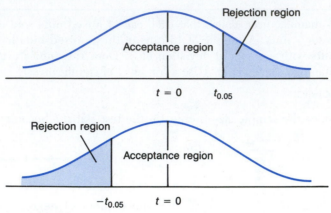

Figure 10.15
One-tailed small-sample rejection region (5% level of significance).

we reject the manufacturer's claim that each can of mixed nuts contains an average of 10 cashew nuts.

EXAMPLE 2

A new weight-reducing pill is being sold in a midwestern city. The manufacturer claims that any overweight person who takes this pill as directed will lose 15 pounds within a month. To test this claim, a doctor gives this pill to six overweight people and finds that they lose an average of only 12 pounds with a standard deviation of 4 pounds. Can we reject the manufacturer's claim? (Use a 5% level of significance.)

Solution

In this case the population parameter being tested is μ, the average number of pounds lost when using this pill. The questioned value is $\mu = 15$. We have

$$H_0: \quad \mu = 15$$

$$H_1: \quad \mu < 15$$

Here we are given that $\bar{x} = 12$, $\mu = 15$, $s = 4$, and $n = 6$ so that

$$t = \frac{\bar{x} - \mu}{\dfrac{s}{\sqrt{n}}} = \frac{12 - 15}{\dfrac{4}{\sqrt{6}}} = -1.84$$

From Table VII we find that the $t_{0.05}$ value for $6 - 1$, or 5 degrees of freedom is 2.015, which means that we reject the null hypothesis if the test statistic is less than -2.015. Since $t = -1.84$ does not fall within the rejection region, we cannot reject the manufacturer's claim that the average number of pounds lost when using this pill is 15.

COMMENT   Again, we wish to emphasize a point made earlier: When testing hypotheses involving small sample sizes, we assume that the relative frequency distribution of the population from which the sample is to be selected is approximately normal.

We summarize the procedure that should be followed when testing hypotheses about a population mean and when we have a small sample size.

Small-Sample Test of Hypothesis About a Population Mean

	One-tailed Test	Two-tailed Test	One-tailed Test
Null Hypothesis	H_0: $\mu = \mu_0$	H_0: $\mu = \mu_0$	H_0: $\mu = \mu_0$
Alternative Hypothesis	H_1: $\mu < \mu_0$	H_1: $\mu \neq \mu_0$	H_1: $\mu > \mu_0$
Test Statistic	$t = \dfrac{\bar{x} - \mu_0}{\dfrac{s}{\sqrt{n}}}$	$t = \dfrac{\bar{x} - \mu_0}{\dfrac{s}{\sqrt{n}}}$	$t = \dfrac{\bar{x} - \mu_0}{\dfrac{s}{\sqrt{n}}}$
Rejection Region	$t < -t_\alpha$	$t < -t_{\alpha/2}$ or $t > t_{\alpha/2}$	$t > t_\alpha$

In this chart t_α is the t-value such that prob $(t > t_\alpha) = \alpha$.

$t_{\alpha/2}$ is the t-value such that prob $(t > t_{\alpha/2}) = \dfrac{\alpha}{2}$.

The distribution of t has $n - 1$ degrees of freedom.

EXERCISES FOR SECTION 10.5

1. Welfare department officials of a certain city claim that the average number of cases of child abuse handled daily by its various agencies is 7.62. A newspaper reporter decides to test this claim. The reporter randomly selects 10 days and determines that the average number of child-abuse claims handled on these days was 5.32 with a standard deviation of 2.06. Using a 5% level of significance, should we reject the welfare department officials' claim?

2. A large oil company claims that the average cost of a gallon of regular unleaded gasoline to a motorist is $1.43. To test this claim, 18 service stations are randomly selected. It is found that the average price of a gallon of regular unleaded gasoline at these stations is $1.52, with a standard deviation of 4 cents. Should we reject the oil company's claim? (Use a 5% level of significance.)

3. *Is service getting better?* A commuter estimates that in the last 40 days his train has been late an average of 25 minutes (min). Due to the installation of a new

power substation, railroad officials claim that service has improved. In the last 15 working days the commuter finds that his train has been late an average of only 18 min with a standard deviation of 4.8 min. Does this indicate an improvement of service? (Use a 1% level of significance.)

4. Industry representatives claim that the average price of an electronic computer game is $20.95. Several consumer groups claim that the average price is much higher. Ten electronic computer games are randomly selected. The average price of these games is found to be $22.42 with a standard deviation of $2.88. Should we reject the industry representative's claim? (Use a 1% level of significance.)

5. Bank officials claim that the average home equity loan is $14,000 with a standard deviation of $1600. To check on this claim, a new loan officer took a sample of 16 home equity loans and found that the average loan was $17,800. Does this indicate that the average home equity loan is more than $14,000? (Use a 1% level of significance.)

6. In an effort to attract new industry to the region a mayor claims that the average age of a worker in the region is 28 years with a standard deviation of 6.72 years. A prospective company is interested in determining whether the mayor's claim is accurate. A random sample of 15 workers in the region reveals an average age of 30 years. Is there sufficient evidence to conclude that the average age is not 28 years? (Use a 5% level of significance.)

7. A coffee vending machine is set so as to dispense 6.8 ounces (oz) of coffee per cup. If the machine is tested 11 times, yielding a mean cupful of 7.1 oz of coffee with a standard deviation of 0.98 oz, does this evidence at the 5% level of significance suggest that the machine is overfilling cups? Explain your answer.

8. A sociologist finds that for the entire population of Clarksville the average number of years of education is 12.71 years with a standard deviation of 2.86 years. In one part of the city a random sample of 12 people indicated that the average number of years of education for these people was 13.17 years. Using a 5% level of significance, test the null hypothesis that the average number of years of education for this part of the city is the same as the average for the entire city.

9. A manager of the Laurel Heights Processing Company cafeteria believes that, based on past experience, approximately 375 lunches are needed daily. The standard deviation is 10.62 lunches. If a random survey of 16 days gives a mean of 345 lunches, does this sample evidence agree with the manager's claim or is it significantly different? (Use a 5% level of significance.)

10. A certain antihistamine/decongestant medicine for babies bears a label indicating that it contains 25 milligrams (mg) of pseudoephedrine HCL per 1 milliliter (mL). The Food and Drug Administration randomly selects eighteen 1-mL samples and finds that the mean pseudoephedrine HCL content for these samples is 29 mg with a standard deviation of 4 mg. Using a 1% level of significance, test the null hypothesis that the mean content for the sample is the same as that indicated on the label. (Assume that the sample standard deviation can be used for σ.)

10.6 TESTS CONCERNING DIFFERENCES BETWEEN MEANS FOR LARGE SAMPLES

There are many instances in which we must decide whether the observed difference between two sample means is due purely to chance or whether the population means from which these samples were selected are really different. For example, suppose a teacher gave an IQ test to 50 girls and 50 boys and obtained the following test scores:

	Boys	Girls
Mean	78	81
Standard Deviation	7	9

Is the observed difference between the scores significant? Are the girls smarter?

In problems of this sort the null hypothesis is that there is no difference between the means. Since we will be discussing more than one sample, we use the following notation. Let $\bar{x}_1$, s_1, and n_1 be the mean, standard deviation, and sample size, respectively, of one of the samples, and let $\bar{x}_2$, s_2, and n_2 be the mean, standard deviation, and sample size, respectively, of the second sample. Decisions as to

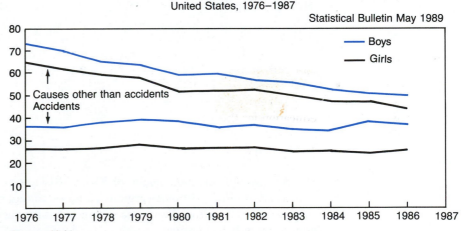

Trend in Death rates Among Children Aged 1–4, by Sex
United States, 1976–1987

Statistical Bulletin May 1989

Figure 10.16
This clipping indicates the trend in death rates, by sex, among children aged 1–4 years. Would you say that the average number of deaths as a result of accidents is significantly higher for boys than for girls? What about child deaths not resulting from accidents? Is the observed difference significant?

(Table courtesy of The Metropolitan Life Insurance Company.)
Source of basic data: Reports of the Division of Vital Statistics, National Center for Health Statistics.

whether to reject or accept the null hypotheses are then based on the test statistic z, where

$$z = \frac{\bar{x}_1 - \bar{x}_2}{\sqrt{\dfrac{s_1^2}{n_1} + \dfrac{s_2^2}{n_2}}}$$

(assuming the samples are independent and both samples are large).

Depending on whether the alternative hypothesis is $\mu_1 \neq \mu_2$, $\mu_1 < \mu_2$, or $\mu_1 > \mu_2$, we have a two-sided test or a one-sided test as indicated in Section 10.4. The following examples illustrate how this test statistic is used.

EXAMPLE 1

Consider the example discussed at the beginning of this section. Is the observed difference between the two IQ scores significant? (Use a 5% level of significance.)

Solution

Let $\bar{x}_1$, s_1, and n_1 represent the boys' mean score, standard deviation, and sample size, and let $\bar{x}_2$, s_2, and n_2 be the corresponding girls' scores. Then the problem is whether the observed difference between the sample means is significant. Thus,

$$H_0: \quad \mu_1 = \mu_2$$

$$H_1: \quad \mu_1 \neq \mu_2$$

Here we are given that $\bar{x}_1 = 78$, $s_1 = 7$, $n_1 = 50$, $\bar{x}_2 = 81$, $s_2 = 9$, and $n_2 = 50$ so that

$$z = \frac{\bar{x}_1 - \bar{x}_2}{\sqrt{\dfrac{s_1^2}{n_1} + \dfrac{s_2^2}{n_2}}} = \frac{78 - 81}{\sqrt{\dfrac{7^2}{50} + \dfrac{9^2}{50}}} = \frac{-3}{\sqrt{\dfrac{49}{50} + \dfrac{81}{50}}}$$

$$= \frac{-3}{\sqrt{2.60}} = -1.86$$

We use the two-tail rejection region of Figure 10.11 (see page 487). The value of $z = -1.86$ falls in the acceptance region. Thus, we cannot conclude that the sample data support the claim that there is a significant difference between the boys' IQ scores and the girls' IQ scores.

EXAMPLE 2

An executive who has two secretaries, Jean and Mark, is interested in knowing whether there is any significant difference in their typing abilities. Mark typed a 40-page report and made an average of 4.6 errors per page. The standard deviation was 0.6. Jean typed a 30-page report and made an average of 2.3 errors per page. The standard deviation was 0.8. Is there any significant difference between their performances? (Use a 5% level of significance.)

Solution

Let $\bar{x}_1$, s_1, and n_1 represent Mark's scores and let $\bar{x}_2$, s_2, and n_2 represent Jean's scores. Then the problem is whether or not the observed difference between the sample means is significant. Thus,

$$H_0: \quad \mu_1 = \mu_2$$

$$H_1: \quad \mu_1 \neq \mu_2$$

Here we are given that $\bar{x}_1 = 4.6$, $s_1 = 0.6$, $n_1 = 40$, $\bar{x}_2 = 2.3$, $s_2 = 0.8$, and $n_2 = 30$ so that

$$
\begin{aligned}
z &= \frac{\bar{x}_1 - \bar{x}_2}{\sqrt{\dfrac{s_1^2}{n_1} + \dfrac{s_2^2}{n_2}}} \\
&= \frac{4.6 - 2.3}{\sqrt{\dfrac{(0.6)^2}{40} + \dfrac{(0.8)^2}{30}}} \\
&= \frac{2.3}{\sqrt{0.009 + 0.0213}} \\
&= \frac{2.3}{\sqrt{0.0303}} = \frac{2.3}{0.174} \\
&= 13.22
\end{aligned}
$$

We use the two-tail rejection region of Figure 10.11 (see page 487). The value of $z = 13.22$ falls in the rejection region. Thus, there is a significant difference between the typing abilities of the two secretaries.

EXAMPLE 3

There are many advertisements on television about toothpastes. One such advertisement claims that children who use Smile toothpaste have fewer cavities than children who use any other brand. To test this claim, a consumer's group selected 100 children and divided them into two groups of 50 each. The children of group I were told to brush daily with only Smile toothpaste. The children of group II were told to brush daily with Vanish toothpaste. The experiment lasted one year. The following number of cavities were identified:

Smile: $\bar{x}_1 = 2.31$ $s_1 = 0.6$

Vanish: $\bar{x}_2 = 2.68$ $s_2 = 0.4$

Is Smile significantly more effective than Vanish in preventing cavities? (Use a 5% level of significance.)

Solution

In this case the question is whether Smile is better than Vanish. This means that people who use Smile toothpaste will have fewer cavities than those who use Vanish. Thus,

$$H_0: \quad \mu_1 = \mu_2$$

$$H_1: \quad \mu_1 < \mu_2$$

Here we are given that $\bar{x}_1 = 2.31$, $s_1 = 0.6$, $n_1 = 50$, $\bar{x}_2 = 2.68$, $s_2 = 0.4$, and $n_2 = 50$ so that

$$z = \frac{\bar{x}_1 - \bar{x}_2}{\sqrt{\dfrac{s_1^2}{n_1} + \dfrac{s_2^2}{n_2}}}$$

$$= \frac{2.31 - 2.68}{\sqrt{\dfrac{(0.6)^2}{50} + \dfrac{(0.4)^2}{50}}} = \frac{-0.37}{\sqrt{0.0104}} = \frac{-0.37}{0.102}$$

$$= -3.63$$

We use the one-tailed rejection region of Figure 10.13 (see page 490). The value of $z = -3.63$ falls in the rejection region. Thus, we reject the null hypothesis. The sample data would seem to support the manufacturer's claim. Actually, further studies are needed before making any definite decision about the effectiveness of Smile in preventing cavities.

EXAMPLE 4

The local chapter of a Women's Liberation group claims that a female college graduate earns less than a male college graduate. A survey of 40 men and 30 women indicated the following results.

	Average Starting Salary	Standard Deviation
Women	$29,000	$600
Men	$29,700	$900

Do these figures support the claim that women earn less? (Use a 1% level of significance.)

Solution

Let $\bar{x}_1$, s_1, and n_1 represent the women's scores, and let $\bar{x}_2$, s_2, and n_2 represent the men's scores. The problem is whether or not the observed difference between the sample means is significant. Thus,

$$H_0: \quad \mu_1 = \mu_2$$

$$H_1: \quad \mu_1 < \mu_2$$

Here we are given that $\bar{x}_1 = 29{,}000$, $s_1 = 600$, $n_1 = 30$, $\bar{x}_2 = 29{,}700$, $s_2 = 900$, and $n_2 = 40$ so that

$$z = \frac{\bar{x}_1 - \bar{x}_2}{\sqrt{\dfrac{s_1^2}{n_1} + \dfrac{s_2^2}{n_2}}}$$

$$= \frac{29{,}000 - 29{,}700}{\sqrt{\dfrac{(600)^2}{30} + \dfrac{(900)^2}{40}}} = \frac{-700}{\sqrt{12{,}000 + 20{,}250}}$$

$$= \frac{-700}{\sqrt{32{,}250}} = \frac{-700}{179.58}$$

$$= -3.9$$

We use the one-tailed rejection region of Figure 10.12 (see page 488). The value of $z = -3.9$ falls in the rejection region. Thus, we reject the null hypothesis. There is a significant difference between the starting salary of men and women.

We summarize the procedure to be used in the following chart:

Large-Sample Test of Hypothesis About the Difference Between Two Population Means			
	One-tailed Test	Two-tailed Test	One-tailed Test
Null Hypothesis	$H_0: \ (\mu_1 - \mu_2) = A$	$H_0: \ (\mu_1 - \mu_2) = A$	$H_0: \ (\mu_1 - \mu_2) = A$
Alternative Hypothesis	$H_1: \ (\mu_1 - \mu_2) < A$	$H_1: \ (\mu_1 - \mu_2) \neq A$	$H_1: \ (\mu_1 - \mu_2) > A$
Test Statistic	$z = \dfrac{(\bar{x}_1 - \bar{x}_2) - A}{\sigma_{(\bar{x}_1 - \bar{x}_2)}}$	$z = \dfrac{(\bar{x}_1 - \bar{x}_2) - A}{\sigma_{(\bar{x}_1 - \bar{x}_2)}}$	$z = \dfrac{(\bar{x}_1 - \bar{x}_2) - A}{\sigma_{(\bar{x}_1 - \bar{x}_2)}}$
	$\approx \dfrac{(\bar{x}_1 - \bar{x}_2) - A}{\sqrt{\dfrac{s_1^2}{n_1} + \dfrac{s_2^2}{n_2}}}$	$\approx \dfrac{(\bar{x}_1 - \bar{x}_2) - A}{\sqrt{\dfrac{s_1^2}{n_1} + \dfrac{s_2^2}{n_2}}}$	$\approx \dfrac{(\bar{x}_1 - \bar{x}_2) - A}{\sqrt{\dfrac{s_1^2}{n_1} + \dfrac{s_2^2}{n_2}}}$
Rejection Region	$z < -z_\alpha$	$z < -z_{\alpha/2}$ or $z > z_{\alpha/2}$	$z > z_\alpha$

In the preceding chart, A is the numerical value for $(\mu_1 - \mu_2)$ as specified in the null hypothesis. In many applied problems we are interested in testing that there is no difference between the population means. Of course, in such a situation $A = 0$.

COMMENT For these tests to work we assume that the sample sizes, n_1 and n_2, are sufficiently large and that the two random samples are selected independently from a large population.

EXERCISES FOR SECTION 10.6

 1. *Bus breakdowns*: A large bus company is considering purchasing new buses for its fleet from one of two manufacturers. The company has been road testing 80 buses manufactured by Company A and 60 buses manufactured by Company B over a period of months and has gathered the following information on the number of breakdowns per week.

	Average Number of Breakdowns per Week	Standard Deviation
Buses Manufactured by Company A	17	5
Buses Manufactured by Company B	15	4

Is there any significant difference between the average number of breakdowns per week for the buses manufactured by each company? (Use a 1% level of significance.)

 2. On a certain scale the average pollution content in a state's 38 lakes was found to be 222 with a standard deviation of 11. On the same scale the average pollution content in a neighboring state's 47 lakes was found to be 213 with a standard deviation of 15. Is there any significant difference between the average pollution content of both state's lakes? (Use a 5% level of significance.)

 3. Forty-five traffic signals that were painted with one brand of paint lasted an average of 3.4 years before repainting was necessary. The standard deviation was 0.69 years. Sixty-two traffic signals painted with a second brand of paint lasted an average of 2.98 years before repainting was necessary. The standard deviation was 0.97 years. Is there any significant difference between the average lasting time of these two brands of paint? (Use a 5% level of significance.)

 4. Sixty-four tires manufacured by one process were subjected to an endurance test. The tires lasted an average of 22,000 miles (mi) before replacement was needed. The standard deviation was 480 mi. Fifty-two tires manufactured by a second process and subjected to the same endurance test lasted an average of 20,000 mi before replacement was needed. The standard deviation was 600 mi.

Is there any significant difference between the average life of a tire manufactured by these two processes? (Use a 5% level of significance.)

5. *Cheating the poor*: A consumer's group claims that many gas stations charge higher prices for a gallon of gas in poorer neighborhoods than in middle-class neighborhoods. To investigate this claim, the group purchased one gallon of unleaded premium gas from each of 41 gas stations located in poorer neighborhoods and one gallon of unleaded premium gas from each of 53 gas stations located in middle-class neighborhoods. The following results were obtained.

	Poorer Neighborhoods	Middle-Class Neighborhoods
Average Price	1.38	1.31
Sample Standard Deviation	0.04	0.09
Sample Size	41	53

Is there any significant difference between the average price of a gallon of gas in these neighborhoods? (Use a 1% level of significance.)

Oil companies have been accused by many groups of charging different prices for the same grade of gasoline depending on the area within the city in which it is sold. Is this true or are the prices determined by competition?

6. *Class size*: The Mathematics Department at Baker College claims that the average class size of its 36 remedial math courses is larger than the average class size of the 31 remedial English courses offered at the college. The administration of the college has obtained the following statistics.

	Math Remedial Courses	English Remedial Courses
Average Class Size	41	38
Standard Deviation	8	7
Number of Classes	36	31

Is there any significant difference between the average class size of the remedial math classes and the remedial English classes? (Use a 1% level of significance.)

7. Forty-five students who were taught multiplication by the Chisanbop method needed an average of 27 minutes (min) to complete a test. The standard deviation was 6.9 min. Sixty-four students who were taught multiplication the traditional way needed an average of 29 min to complete the test. The standard deviation was 8.5 min. Is there any significant difference between the average time needed to complete the test using both methods? (Use a 5% level of significance.)

8. Sixty fences painted with Brand X rust inhibitor lasted an average of 8 years before repainting was necessary. The standard deviation was 2.96 years. Seventy-five fences painted with Brand Y rust inhibitor lasted an average of 9 years before repainting was necessary. The standard deviation was 3.32 years. Is there any significant difference in the durability of the two brands of rust inhibitor? (Use a 5% level of significance.)

9. The State Motor Vehicle Department conducted tests on 200 randomly selected drivers in the state to determine the average time needed to react to a particular driving hazard. It divided the 200 drivers into two groups of 100 drivers each. Group A needed an average of 9 seconds (sec) to react to the hazard. The standard deviation was 0.09 sec. Group B needed 11.5 sec to react to the hazard. The standard deviation was 0.03 sec. Is there any significant difference in the average reaction time of these two groups of drivers? (Use a 5% level of significance.)

10. Mr. Blank, the company payroll supervisor, claims that the 80 workers in the 13th floor take a longer coffee break than the 50 workers on the 8th floor. To support his claim, he has gathered the following information.

	Average Time Spent on Coffee Break (in minutes)	Standard Deviation
Workers on 13th Floor	23	5
Workers on 8th floor	19	8

Is Mr. Blank's claim justified? (Use a 5% level of significance.)

11. Various studies* have shown that the percentage of body fat that a person has

---

*Bandyopadhyav, B., and H. Chattopadhyay "Body Fat in Urban and Rural Male College Students of Eastern India" in *American Journal of Physical Anthropology*, January 1981, Vol. 54, pp. 119–122.

can be a good indicator of the person's overall health. One medical researcher conducted a survey of young American male adults living on the East Coast and young American males living on the West Coast. The following results were obtained.

	East Coast Males	West Coast Males
Average Percentage of Body Fat	172	161
Standard Deviation	11.98	14.62
Number Involved in Survey	72	88

Using a 5% level of significance, test the null hypothesis that the average percentage of body fat is not significantly different for East Coast males than for West Coast males.

10.7 TESTS CONCERNING DIFFERENCES BETWEEN MEANS FOR SMALL SAMPLES

In the last section we indicated how we test the difference between sample means. In all the examples the sample sizes were large enough ($n \geq 30$) to justify our use of the normal distribution. If this is not the case, we must use the t-distribution. We assume that the population from which the samples are selected have approximately normal probability distributions and that the random samples are selected independently. We then have the following small-sample-size hypothesis testing procedures.

Small-Sample Test of Hypothesis About the Difference Between Means

	One-Tailed Test	Two-Tailed Test	One-Tailed Test
Null Hypothesis	H_0: $(\mu_1 - \mu_2) = A$	H_0: $(\mu_1 - \mu_2) = A$	H_0: $(\mu_1 - \mu_2) = A$
Alternative Hypothesis	H_1: $(\mu_1 - \mu_2) < A$	H_1: $(\mu_1 - \mu_2) \neq A$	H_1: $(\mu_1 - \mu_2) > A$
Test Statistic	$t = \dfrac{(\bar{x}_1 - \bar{x}_2) - A}{s_p \sqrt{\dfrac{1}{n_1} + \dfrac{1}{n_2}}}$	$t = \dfrac{(\bar{x}_1 - \bar{x}_2) - A}{s_p \sqrt{\dfrac{1}{n_1} + \dfrac{1}{n_2}}}$	$t = \dfrac{(\bar{x}_1 - \bar{x}_2) - A}{s_p \sqrt{\dfrac{1}{n_1} + \dfrac{1}{n_2}}}$

$$\text{where } s_p = \sqrt{\frac{(n_1 - 1)s_1^2 + (n_2 - 1)s_2^2}{n_1 + n_2 - 2}}$$

	One-Tailed Test	Two-Tailed Test	One-Tailed Test
Rejection Region	$t < -t_\alpha$	$t < -t_{\alpha/2}$ or $t > t_{\alpha/2}$	$t > t_\alpha$

COMMENT   When using the tests outlined in the preceding chart, it is assumed that the variances of the two populations are equal.

COMMENT   When using the tests outlined in the preceding chart, the number of degrees of freedom for the t-distribution is $n_1 + n_2 - 2$.

We illustrate the procedure with an example.

EXAMPLE 1

A chemist at a paint factory claims to have developed a new oil-based paint that will dry very quickly. The manufacturer is interested in comparing this new paint with his currently best-selling paint. In order to accomplish this, he paints each of five different walls with a gallon of his best-selling paint and with a gallon of the new fast-drying paint. The number of minutes needed for each of these paints to dry thoroughly is shown as follows.

Number of Minutes Needed to Dry

Current Best-Selling Paint	New Fast-Drying Paint
48	42
46	43
44	45
46	43
43	44

Using a 5% level of significance, is the new paint significantly more effective in its drying time than the old paint?

Solution

Using the data in the table, we first compute the sample means and the sample standard deviation. We have

Current Best-Selling Paint	New Fast-Drying Paint
$n_1 = 5$	$n_2 = 5$
$\bar{x}_1 = 45.4$	$\bar{x}_2 = 43.4$
$s_1 = 1.949$	$s_2 = 1.14$

In this case the null hypothesis is $\mu_1 = \mu_2$ and the alternative hypothesis is $\mu_1 < \mu_2$, where μ_1 is the average drying time of the current best-selling paint and μ_2 is the average drying time of the new fast-drying paint.

Based on past experience with other paints, the manufacturer knows that the drying time of paint is approximately normally distributed and that the variances for different paints are about the same. Since the samples were randomly and independently selected, we compute the test statistic. We have

$$t = \frac{\bar{x}_1 - \bar{x}_2}{s_p\sqrt{\dfrac{1}{n_1} + \dfrac{1}{n_2}}}$$

and

$$s_p = \sqrt{\frac{(n_1 - 1)s_1^2 + (n_2 - 1)s_2^2}{n_1 + n_2 - 2}}$$

so that

$$s_p = \sqrt{\frac{(5 - 1)(1.949)^2 + (5 - 1)(1.14)^2}{5 + 5 - 2}} \approx 1.597$$

Therefore,

$$t = \frac{45.4 - 43.4}{1.597\sqrt{\frac{1}{5} + \frac{1}{5}}} \approx 1.98$$

The number of degrees of freedom is $n_1 + n_2 - 2 = 5 + 5 - 2$ or 8. From Table VII of the Appendix, the $t_{0.05}$ value with 8 degrees of freedom is 1.86.

The test statistic value of $t = 1.98$ falls in the critical rejection region. Hence, we reject the null hypothesis and conclude that based on the data, the newly developed fast-drying paint does indeed dry faster than the current best-selling paint.

COMMENT   If we know for sure that the samples are from nonnormal populations, then we cannot justifiably use the small-sample t test for $(\mu_1 - \mu_2)$. In this situation it is advisable to use the nonparametric Wilcoxon rank-sum test to be discussed in Chapter 14.

Until now we have assumed that the variances of both sampled populations are equal. If this is not the case, then the chart given on page 507 has to be modified somewhat. We have the following:

Small-Sample Test of Hypotheses About Differences Between Means When Variances of Both Sampled Populations Are Not Equal

Test statistic

$$t = \frac{\bar{x}_1 - \bar{x}_2 - A}{\sqrt{\frac{s_1^2}{n_1} + \frac{s_2^2}{n_2}}}$$

Degrees of Freedom

$$df = \frac{\left(\frac{s_1^2}{n_1} + \frac{s_2^2}{n_2}\right)^2}{\left[\frac{(s_1^2/n_1)^2}{n_1 - 1} + \frac{(s_2^2/n_2)^2}{n_2 - 1}\right]}$$

COMMENT   When using the above formula, *df* may not be an integer. In that case round *down* to the nearest integer.

EXERCISES FOR SECTION 10.7

 1. The management of a supermarket wants to determine whether its new extensive television advertisement campaign has increased its mean daily income. The average daily business income for 10 days before the new advertisement campaign appeared was $84,583, with a standard deviation of $8,143. The average income for 12 days after the new advertisement campaign appeared was $88,586 with a standard deviation of $9,671. Using a 5% level of significance, test the null hypothesis that the advertising campaign has increased the average income.

 2. An industrial engineer has been hired to determine whether there is a significant difference between the number of defective items produced by the day shift and the number produced by the night shift of the company. The engineer has gathered the following data for three weeks: (*Note*: There is no night shift on Friday.)

	Day Shift	Night Shift
Average Number of Defective Items	21.3	25.6
Standard Deviation	2.61	3.04
Number of Days Surveyed	15	13

At the 5% level of significance, do the data provide sufficient evidence to indicate a difference in the mean number of defective items produced by both shifts?

 3. Using gene-splicing techniques, a biochemist has developed two new antiviral drugs. The drugs are administered to humans afflicted with the same type of disease. The average number of hours elapsed before the drug shows any effect is as follows.

	Drug A	Drug B
Average Number of Hours	6.2	7.1
Standard Deviation	1.96	2.01
Number of Individuals Involved in Survey	14	12

Using a 1% level of significance, do the data provide sufficient evidence to indicate a difference in the average number of hours elapsed before either drug shows any effect?

 4. Eleven randomly selected male business majors from one college received job offers with an average starting salary of $24,050. The standard deviation was

$1220. Twelve randomly selected female business majors of the same college were offered jobs with an average starting annual salary of $23,880. The standard deviation was $1410. Is it true that the average annual starting salary for a female business major of this college is significantly less than the average starting salary for a male business major of this college? (Use a 5% level of significance.)

 5. The 17 members of the production department in the Mathematics/Science Division of the Arlow Publishing Company have been with the company for an average of 15 years with a standard deviation of 2.2 years. The 13 members of the Production Department in the Humanities Division of the Arlow Publishing Company have been with the company for an average of 11 years with a standard deviation of 3.2 years. Is it true that the average number of years that the members of the Production Department in the Humanities Division have been with the company is significantly less than the average number of years that the members of the Production Department in the Mathematics/Science Division have been with the company? (Use a 5% level of significance.)

 6. Environmentalists sampled 14 different sites along one polluted lake and found an average coliform bacteria count of 18.3 per unit with a standard deviation of 2.26. They then sampled 15 different sites along a different polluted lake and found an average coliform bacteria count of 24.6 per unit with a standard deviation of 3.88. Is it true that the average number of coliform bacteria count for one lake is significantly lower that the average number of coliform bacteria count for the other lake? (Use a 5% level of significance.)

 7. *How often do you dine out?* In an effort to determine how often New Yorkers dine out, two independent studies were undertaken. One study of 12 families

found that they dined out an average of eight times per month with a standard deviation of 1.22. A second survey of 14 families found that these families dined out an average of six times per month with a standard deviation of 0.86. Is it true that the two surveys do not differ significantly in their determination of the average number of times per month that a New Yorker dines out? (Use a 5% level of significance.)

 8. Ten patients in the Emergency Room at St. Vincent's Hospital waited an average of 41 minutes before receiving medical attention. The standard deviation was 5.63 minutes (min). Seven patients in the Emergency Room at Richmond Memorial Hospital waited an average of 28 min before receiving medical attention. The standard deviation was 4.69 min. Using a 1% level of significance, is there any significant difference between the average amount of time that one must wait in the Emergency Rooms at both hospitals before receiving medical attention?

9. An official for the Department of Consumer's Affairs accurately weighed the contents of 13 one-pound (lb) cans of name-brand coffee and found that they had an average weight of 0.997 lb with a standard deviation of 0.23 lb. The official then accurately weighed the contents of ten randomly selected cans of the store-brand of coffee and found that these had an average weight of 1.013 lb with a standard deviation of 0.12 lb. Is it true that the pound cans of store-brand coffee have an average weight that is significantly greater than the average weight of the other name-brand cans of coffee? (Use a 5% level of significance.)

10. A large Wall Street brokerage house uses high-volume photocopy machines manufactured by either Company A or Company B. However, to cut costs, it would like to use machines manufactured by Company A or B exclusively. The following annual operating cost sample data are available.

Photocopy Machine Manufactured by

Company A			Company B			
$14,368	$12,986	$13,276	$8,947	$9,286	$14,361	$13,726
10,861	11,583	13,992	9,227	12,247	14,883	8,962
12,002	11,983		12,413	13,247	12,777	9,972
			11,654	10,943	10,492	12,592

a. Compute the mean and standard deviation for each sample.
b. Management has decided to use only the photomachines manufactured by Company B if it is found to be significantly cheaper. Using a 5% level of significance, test the null hypothesis that the machines manufactured by Company B are at least as expensive to operate as those manufactured by Company A.
c. What action should management take? Explain your answer.

10.8 TESTS CONCERNING PROPORTIONS

Suppose a congressman claims that 60% of the voters in his district are in favor of lowering the drinking age to 16 years. If a random sample of 400 voters showed that 221 of them favored the proposal, can we reject the congressman's claim? Questions of this type occur quite often and are usually answered on the basis of observed proportions. We assume that we can use the binomial distribution and that the probability of success is the same from trial to trial. We can therefore apply Formulas 6.5 and 6.6 (see page 328), which give us the mean and standard deviation of a binomial distribution. Thus,

$$\text{Mean: } \quad \mu = np$$

$$\text{Standard deviation: } \quad \sigma = \sqrt{np(1 - p)}$$

The null hypothesis in such tests assumes that the observed proportion, $\hat{p}$, is the same as the population proportion, p. Depending on the situation, we have the following alternative hypotheses:

Null hypothesis: H_0: $\hat{p} = p$

Alternative
  hypothesis: H_1: $\hat{p} \neq p$ [which means a two-tailed test]

              H_2: $\hat{p} > p$ [which means a one-tailed (right-side) test]

              H_3: $\hat{p} < p$ [which means a one-tailed (left-side) test].

The test statistic is z, where

$$z = \frac{\hat{p} - p}{\sqrt{\dfrac{p(1 - p)}{n}}}$$

We are assuming that the sample size is large. We reject the null hypothesis if the test statistic falls in the critical, or rejection region.

The following examples illustrate how we test proportions.

EXAMPLE 1

The Dean of Students at a college claims that only 12% of the students commute to school by bike. To test this claim, a students' group takes a sample of 80 students. They find that 14 of these students commute by bike. Is the Dean's claim acceptable? (Use a 5% level of significance.)

Solution

In this case the population parameter being tested is p, the true proportion of students who commute by bike. The questioned value is 0.12. Since we are testing whether or not the true proportion is 0.12, we have

$$H_0: \quad p = 0.12$$

$$H_1: \quad p \neq 0.12$$

We are told that 14 of the 80 sampled students commute to school by bike so that $\hat{p} = \dfrac{14}{80} = 0.175$. Thus,

$$z = \frac{\hat{p} - p}{\sqrt{\dfrac{p(1 - p)}{n}}} = \frac{0.175 - 0.12}{\sqrt{\dfrac{(0.12)(1 - 0.12)}{80}}}$$

$$= \frac{0.055}{\sqrt{0.00132}} = \frac{0.055}{0.036}$$

$$= 1.53$$

We use the two-tailed rejection region of Figure 10.11 (see page 487). The value of $z = 1.53$ falls in the acceptance region. Thus we cannot reject the null hypothesis and the Dean's claim that the true proportion of students who commute to school by bike is 12%.

EXAMPLE 2

In a recent press conference a senator claimed that 55% of the American people supported the president's foreign policy. To test this claim, a newspaper editor selected a random sample of 1000 people and 490 of them said that they supported the president. Is the senator's claim justified? (Use a 1% level of significance.)

Solution

In this case the population parameter being tested is p, the true proportion of Americans who support the president. The questioned value is 0.55. Since we are testing whether or not the true proportion is 0.55, we have

$$H_0: \quad p = 0.55$$

$$H_1: \quad p \neq 0.55$$

We are told that 490 of the 1000 people interviewed supported the president so that

$$\hat{p} = \frac{490}{1000} = 0.49$$

Thus,

$$z = \frac{\hat{p} - p}{\sqrt{\dfrac{p(1 - p)}{n}}} = \frac{0.49 - 0.55}{\sqrt{\dfrac{(0.55)(1 - 0.55)}{1000}}}$$

$$= \frac{-0.06}{\sqrt{0.0002475}} = \frac{-0.06}{0.016}$$

$$= -3.75$$

Since the level of significance is 1%, we use the two-tailed rejection of Figure 10.10 (see page 485). The value of $z = -3.75$ falls in the rejection region. Thus, we reject the null hypothesis and the senator's claim that 55% of the American people support the president's foreign policy.

EXAMPLE 3

A latest government survey indicates that 22% of the people in Camelot are illegally receiving some form of public assistance. The mayor of Camelot believes that the figures are exaggerated. To test this claim, she carefully examines 75 cases and finds that 11 of these people are illegally receiving aid. Does this sample support the government's claim? (Use a 5% level of significance.)

Solution

In this case the population parameter being tested is p, the true proportion of people illegally receiving financial aid. The questioned value is 0.22. Since we are testing whether the true proportion is 0.22 or lower, we have

$$H_0: \quad p = 0.22$$

$$H_1: \quad p < 0.22$$

We are told that 11 of the 75 cases examined are illegally receiving aid so that

$$\hat{p} = \frac{11}{75} = 0.15$$

Thus,

$$z = \frac{\hat{p} - p}{\sqrt{\dfrac{p(1 - p)}{n}}} = \frac{0.15 - 0.22}{\sqrt{\dfrac{(0.22)(1 - 0.22)}{75}}}$$

$$= -1.46$$

We use the one-tail rejection region of Figure 10.12 (see page 488). The value of $z = -1.46$ falls within the acceptance region. Thus, we cannot reject the null hypothesis that 22% of the people in Camelot are illegally receiving financial aid.

EXAMPLE 4   Government officials claim that approximately 29% of the residents of a state are opposed to building a nuclear plant to generate electricity. Local conservation groups claim that the true percentage is much higher. To test the government's claim an independent testing group selects a random sample of 81 state residents and finds that 38 of the people are opposed to the nuclear plant. Can we conclude that the government's claim is inaccurate? (Use a 5% level of significance.)

Solution

In this case the population parameter being tested is p, the true proportion of state residents who are opposed to building the nuclear plant. The questioned value is 0.29. Since we are testing whether the true proportion is 0.29 or higher, we have

$$H_0: \quad p = 0.29$$

$$H_1: \quad p > 0.29$$

We are told that 38 of the 81 residents are opposed to the nuclear plant so that

$$\hat{p} = \frac{38}{81} = 0.47$$

Thus,

$$z = \frac{\hat{p} - p}{\sqrt{\dfrac{p(1 - p)}{n}}} = \frac{0.47 - 0.29}{\sqrt{\dfrac{(0.29)(1 - 0.29)}{81}}}$$

$$= \frac{0.18}{\sqrt{0.002542}} = \frac{0.18}{0.0504} = 3.57$$

We use the one-tail rejection region of Figure 10.13 (see page 490). The value of $z = 3.57$ falls in the rejection region. Thus, we reject the null hypothesis that the true proportion of state residents opposed to the nuclear plant is 0.29.

EXERCISES FOR SECTION 10.8

1. A cosmetics manufacturer claims that 68% of all people who use deodorants in Deer Banks prefer the spray-can type as opposed to the roll-on type. To test this claim, 83 people who use deodorants are randomly selected, and it is found that 45 of them prefer the spray-can type of deodorant as opposed to roll-on type. Can we accept the cosmetics manufacturer's claim? (Use a 5% level of significance.)

2.

> ## 70% OF ALL PEOPLE IN FAVOR OF GUN CONTROL LEGISLATION
>
> *Washington (Dec. 26)*—A nationwide survey conducted by Walinsky and Rogers reveals that 70% of the people interviewed are in favor of tighter gun control legislation. This is a direct outgrowth of the rising incidence of crime in which guns are used that is currently sweeping our country.
>
> December 26, 1990

Consider this newspaper article. If 65 people are randomly selected, and if 40 of them are in favor of tighter gun control legislation, can we accept the claim that 70% of all people are in favor of gun control legislation. (Use a 5% level of significance.)

3. An environmentalist claims that only 40% of all empty cans of soda in a particular state are recycled. Industry representatives claim that the percentage is higher. In a random sample of 90 cans of soda sold, it is found that only 44 of the empty cans were returned for recycling. Can we accept the environmentalist's claim? (Use a 4% level of significance.)

4. The latest statistics released by the U.S. Census Bureau indicate that 23% of all American households are one-person households. A survey of 68 households, selected at random, finds that only 31 of them are one-person households. Can we accept the Census Bureau's claim? (Use a 5% level of significance.)

5. *Affirmative action*: A university claims that 38% of all its faculty members are from minority groups or are women. In a random sample of 84 faculty members it is found that 33 of them are from minority groups or are women. Can we accept the university's claim? (Use a 4% level of significance.)

6. A political scientist claims that 44% of the population believe that our government can deal effectively with our pressing economic problems. A poll of 90 randomly selected people conducted by Johnson and Sayers found that only 35 of them believe that our government can deal effectively with our pressing economic problems. Can we accept the political scientist's claim? (Use a 1% level of significance.)

7. A dermatologist believes that 90% of all people with acne who use a particular medicated soap will have their symptoms disappear. To test this claim, 48 people with acne are tested with this medicated soap and 42 of them have their symptoms disappear. Can we accept the dermatologists claim? (Use a 5% level of significance.)

8. A State Motor Vehicle Bureau official states that 22% of all cars on the state highways are equipped with defective air-pollution control devices. To test this

claim, an independent group randomly selects 400 cars and checks their air-pollution control devices. Ninety-five of the cars are not properly equipped with an air-pollution control device. Can we conclude that the Motor Vehicle Bureau's claim is inaccurate? (Use a 5% level of significance.)

9. The manager of The Marcy Housing Development believes that 20% of all families in the housing complex have a dog or a cat. In a random sample of 40 of the families living in the housing complex, it is found that 11 of these families have a dog or a cat. Can we accept the housing manager's claim? (Use a 5% level of significance.)

10. The ABC Driving School claims that 65% of all its students pass the road test on their first try. In a random sample of 39 students from this school it is found that 28 of the students passed their road test the first time. At a 5% level of significance, should we reject the driving school's claim?

11. A travel agency that specializes in booking cruises to the Caribbean Islands claims that approximately 8% of all people who book reservations cancel these reservations for one reason or another. During the first three months of 1990 there were 4823 reservations and 380 cancellations of these reservations. Should we accept the travel agency's claim? (Use a 5% level of significance.)

10.9 USING COMPUTER PACKAGES

We can use MINITAB to perform hypothesis tests involving a population mean when the sample size is large ($n \geq 30$) and also when the sample size is small ($n \leq 30$). We illustrate the procedure with the following examples.

EXAMPLE 1

Industry representatives claim that the average price of an introductory college-level math textbook is $34.95. Numerous student groups claim that the price is considerably higher. A survey of 36 such books disclosed the following prices:

$32.95	$33.49	$36.45	$34.28	$33.68	$34.95	$32.78	$36.21	$33.89
35.79	31.87	33.51	34.76	35.10	32.69	32.95	33.29	33.78
34.78	35.65	35.29	36.19	31.79	37.25	34.56	35.49	34.95
34.78	35.87	36.85	33.17	35.67	34.99	37.10	36.08	33.09

Using a 5% level of significance, should we reject the industry representative's claim?

Solution

We must determine if H_1: $\mu = 34.95$ or H: $\mu > 34.95$ with $\alpha = 0.05$. In order to use MINITAB for hypothesis testing, we must know the value of the population standard deviation, σ. Unfortunately, this information is not supplied. However, since the sample size is sufficiently large ($n = 36$), we can use MINITAB's DESCRIBE command discussed in Chapter 3 to accomplish this. First we enter the data into the computer. Then we use the MINITAB command **ZTEST** followed by the null hypothesis (MU = 34.95), the estimated value of σ (SIGMA = 1.461), and the storage location of the sample data (C1). When these are executed the machine responds SUBC. We must now inform the machine as to what the alternate hypothesis is. If we want to perform a left-tailed test, we type ALTERNATIVE = -1. If we want to perform a right-tailed test we type ALTERNATIVE = $+1$. For a two-tailed test we type nothing. MINITAB is already programmed for a two-tailed test. When applied to our example we have the following MINITAB printout:

```
MTB > SET THE FOLLOWING DATA INTO C1
DATA> 32.95   33.49   36.45   34.28   33.68   34.95   32.78   36.21   33.89
DATA> 35.79   31.87   33.51   34.76   35.10   32.69   32.95   33.29   33.78
DATA> 34.78   35.65   35.29   36.19   31.79   37.25   34.56   35.49   34.95
DATA> 34.78   35.87   36.85   33.17   35.67   34.99   37.10   36.08   33.09
DATA> END

MTB > DESCRIBE C1
                 N       MEAN     MEDIAN    TRMEAN    STDEV    SEMEAN
C1              36      34.610    34.780    34.624    1.461    0.243

               MIN       MAX        Q1        Q3
C1           31.790    37.250    33.340    35.760

MTB > ZTEST, MU = 34.95, SIGMA > 1.461, C1;
SUBC> ALTERNATIVE = +1.

TEST OF MU = 34.950 VS MU G.T. 34.950
THE ASSUMED SIGMA = 1.46

                 N       MEAN     STDEV    SE MEAN        Z    P VALUE
C1              36      34.610    1.461      0.243    -1.40       0.92

MTB > STOP
```

In this printout we have some very valuable information. It informs us that MINITAB is testing the null hypothesis of $\mu = 34.950$ versus the alternative hypothesis that μ is G.T. (greater than) 34.950. MINITAB then informs us of the assumed value of σ, the column number that it worked with, the sample size, sample mean, sample standard deviation, and standard error of the mean. The z-value of -1.40 represents the value of the test statistic obtained by computing

$$z = \frac{\bar{x} - \mu}{\dfrac{s}{\sqrt{n}}} = \frac{34.610 - 34.95}{\dfrac{1.461}{\sqrt{36}}} = -1.40$$

The *P*-value is really the most important thing in hypothesis testing. We have

---

DEFINITION 10.8
P-value

The **P-value** for a hypothesis test is the smallest significance level at which the null hypothesis can be rejected based on the sample data. If the *P*-value is less than the significance level, α, then we reject H_0; otherwise we do not reject H_0.

In our case the *P*-value is 0.92. Since this value is more than 0.05, we do not reject H_0. We would then say that based on the sample data, we cannot reject the industry representative's claim.

EXAMPLE 2

A potato chip manufacturer packs 32-ounce bags of potato chips. The manufacturer wants the bags to contain, on the average, 32 ounces of potato chips. A quality control engineer selects 13 bags of potato chips and records their weights as follows:

32.5, 32.1, 31.9, 31.8, 32.3, 32.2, 31.8, 31.9, 32.3, 32.4, 32.2, 32.1, 32.3

At the 5% level of significance, can we accept the manufacturer's claim that the bags contain more than 32 ounces of potato chips?

Solution

TTEST

We must determine if H_0: $\mu = 32$ or H_1: $\mu > 32$ with $\alpha = 0.05$. First we enter the data into the computer. Then we use the MINITAB command **TTEST** followed by the null hypothesis (MU = 32.0) and the storage location of the data (C1). When these are executed the machine responds SUBC>. Since we are performing a right-tailed test, we type ALTERNATIVE = +1. When applied to our example we have the following MINITAB printout:

```
MTB > SET THE FOLLOWING DATA IN C1
DATA> 32.5, 32.1, 31.9, 31.8, 32.3, 32.2, 31.8, 31.9, 32.3, 32.4, 32.2, 32.1, 32.3
DATA> END

MTB > TTEST, MU = 32.0, C1;
SUBC> ALTERNATIVE = +1.

TEST OF MU = 32.0000 VS MU G.T. 32.0000

          N      MEAN     STDEV    SE MEAN      T    P VALUE
C1       13    32.1385    0.2293    0.0636    2.18    0.025

MTB > STOP
```

Again, the *P*-value is really the most important thing. In this case the *P*-value of 0.025 is less than $\alpha = 0.05$, so we reject the null hypothesis. We conclude that the data indicate that the average 32-ounce bag of potato chips contains more than 32 ounces of potato chips.

10.10 SUMMARY

In this chapter we discussed hypothesis testing, which is a very important branch of statistical inference. Hypothesis testing is the process by which a decision is made either to reject or accept a null hypothesis about one of the parameters of the distribution. The decision to accept or reject a null hypothesis is based on information obtained from sample data. Since any decision is subject to error, we discussed type-I and type-II errors.

We reject a null hypothesis when the test statistic falls in the critical, or rejection region. The critical region is determined by two things:

1. whether we wish to perform a one-tailed or two-tailed test
2. the level of significance

If a null hypothesis is not rejected, we cannot say that the sample data prove that what the null hypothesis says is necessarily true. It merely does not reject it. Some statisticians prefer to say that in this situation they *reserve judgement* rather than accept the null hypothesis.

Study Guide

The following is a chapter outline in capsule form. You should now be able to demonstrate your knowledge of the ideas mentioned by giving definitions, descriptions, or specific examples. Page references are given in parentheses.

Hypothesis testing is the procedure whereby we make a decision to accept or reject a claim on the basis of sample information. (page 477)

By an **hypothesis** we mean an assumption about one or more of the population parameters that will be either accepted or rejected on the basis of sample information. (page 477)

When testing any hypothesis, if the sample value falls within an **acceptance region,** then we will not reject the null hypothesis. (page 479)

The value that separates the rejection region from the acceptance region is called the **critical value.** (page 479)

The **null hypothesis,** denoted by H_0, is the statistical hypothesis being tested. (page 480)

The **alternative hypothesis,** denoted by H_1, H_2, . . . is the hypothesis that will be accepted when the null hypothesis is rejected. (page 480)

A **one-sided,** or **one-tailed test** is a statistical test that has the rejection region located in the left tail or the right tail of the distribution. (page 480)

A **two-sided** or **two-tailed test** is a statistical test that has the rejection region located in both tails of the distribution. (page 480)

The **region of rejection** is called the **critical region.** The remaining region is called the **acceptance region.** (page 480)

A **test statistic** is a calculated number that is used to decide whether to reject or accept the null hypothesis. (page 480)

Tests that enable us to decide whether to reject or accept a null hypothesis are called **statistical tests of hypotheses** or **statistical tests of significance.** (page 481)

A **type-I error** is made when a true null hypothesis is rejected; that is, we reject it when we should accept it. The probability of making a type-I error is denoted by α. (page 481)

A **type-II error** is made when a false null hypothesis is accepted; that is, we accept a null hypothesis when we should reject it. (page 481)

The **significance level** of a test is the probability that the test statistic falls within the rejection region when the null hypothesis is true. (page 482)

The **5% level of significance** is used when the statistician wishes that the risk of rejecting a true null hypothesis not exceed 5%. (page 482)

The **1% level of significance** is used when the statistician wishes that the risk of rejecting a true null hypothesis not exceed 1%. (page 482)

The MINITAB command **ZTEST** followed by the null hypothesis, the estimated value of μ, and the storage location of the sample data allows us to perform a hypothesis test involving a population mean when the sample size is large $n \geq 30$. If the sample size is small, we use the MINITAB command **TTEST** to accomplish the same thing. (page 519)

The **P-value** for a hypothesis test is the smallest significance level at which the null hypothesis can be rejected on the basis of sample data. If the P-value is less than the significance level, α, then we reject H_0; otherwise we do not reject H_0. (page 520)

Formulas to Remember

We discussed methods for testing means, proportions, and differences between means. All the tests studied are summarized in Table 10.1. You should be able to identify each symbol in the formulas, understand the relationships among the symbols expressed in each formula, understand the significance of each formula, and use the formulas in solving problems.

Testing Your Understanding of This Chapter's Concepts

1. In any test of hypotheses, is it ever possible to make a type-I error *and also* a type-II error at the same time? Explain your answer.
2. In one particular experiment, a researcher tested the null hypothesis $\mu \leq 45$

TABLE 10.1 Various Tests for Accepting or Rejecting a Null Hypothesis

Population Parameter Tested	Sample Size	Type of Test	Significance	Test Statistic	Reject Null Hypothesis if
Mean	Large	Two-tailed	0.05	$z = \dfrac{\bar{x} - \mu}{\sigma/\sqrt{n}}$	$z < -1.96$ or $z > 1.96$
Mean	Large	Two-tailed	0.01		$z < -2.58$ or $z > 2.58$
Mean	Large	One-tailed	0.05		$z < -1.645$ or $z > 1.645$
Mean	Large	One-tailed	0.01		$z < -2.33$ or $z > 2.33$
Mean	Small	Two-tailed	0.05	$t = \dfrac{\bar{x} - \mu}{s/\sqrt{n}}$	$t < -t_{0.025}$ or $t > t_{0.025}$
Mean	Small	One-tailed	0.05		$t < -t_{0.05}$ or $t > t_{0.05}$
Difference between sample means	Large	Two-tailed	0.05 or 0.01	$z = \dfrac{\bar{x}_1 - \bar{x}_2}{\sqrt{\dfrac{s_1^2}{n_1} + \dfrac{s_2^2}{n_2}}}$	Same as above
Difference between sample means	Large	One-tailed	0.05 or 0.01		
Difference between sample means	Small	Two-tailed	0.05 or 0.01	$t = \dfrac{\bar{x}_1 - \bar{x}_2}{s_p\sqrt{\dfrac{1}{n_1} + \dfrac{1}{n_2}}}$ where $s_p = \sqrt{\dfrac{(n_1 - 1)s_1^2 + (n_2 - 1)s_2^2}{n_1 + n_2 - 2}}$ $df = n_1 + n_2 - 2$	Same as above
Difference between sample means	Small	One-tailed	0.05 or 0.01		
Proportion	Large	Two-tailed	0.05 or 0.01	$z = \dfrac{\hat{p} - p}{\sqrt{\dfrac{p(1 - p)}{n}}}$	Same as above
Proportion	Large	One-tailed	0.05 or 0.01		

against the alternative hypothesis $\mu > 45$. Assuming that the sample mean x has a value of 40,

a. could the null hypothesis ever be rejected? Explain your answer.

b. if we accept the null hypothesis, what type of error could we be making? Explain your answer.

3. The formula given on page 507 for the pooled standard deviation is

$$s_p = \sqrt{\frac{(n_1 - 1)s_1^2 + (n_2 - 1)s_2^2}{n_1 + n_2 - 2}}$$

Show that if the sample sizes n_1 and n_2 are equal, then s_p^2 is just the mean of s_1^2 and s_2^2.

4. If the level of significance of a hypothesis test is chosen to be $\alpha = 0$, what is the probability of a type-I or a type-II error? Explain your answer.

THINKING CRITICALLY

1. Due to technological advances, customers who have checking accounts at almost any bank can withdraw funds from their accounts either by going personally to their bank and dealing with a human teller or by using money machines that are available nationwide at numerous locations. The First Maritime Bank is interested in estimating the difference in costs per transaction between a human teller and a cash machine teller. It has gathered the following sample data from its numerous branches.

	Withdrawal Using	
	Human Teller	*Cash Machine*
Average cost per transaction	15 cents	11 cents
Standard Deviation	2.1 cents	1.98 cents
Number involved in survey	12 branch offices	14 cash machine sites

Set up a 90% confidence interval for the mean cost savings per transaction by using the cash machines.

2. To qualify for federal funding, many programs stipulate that the recipients must be economically disadvantaged and have an annual family income below a specified level. The following are the family incomes of ten families residing in the Jonathan Williams Houses and 12 families residing in the Martin Luther King Houses.

Income of Families Residing in

Jonathan Williams Houses		Martin Luther King Houses	
$17,363	$11,807	$17,382	$14,862
$14,802	$16,123	$18,043	$17,098
$17,291	$15,492	$14,196	$16,051
$16,080	$17,032	$11,769	$13,954
$15,923	$14,994	$12,983	$16,083
		$13,888	$13,995

a. Compute the mean and standard deviation for each sample.
b. Construct a 95% confidence interval estimate of the difference in average family income using the families residing in the Williams or King houses.
c. At the 5% level of significance, should we reject or accept the null hypothesis of identical means?

Review Exercises for Chapter 10

1. Forty-five identical cars were driven an average of 21.4 miles (mi) on one gallon of Brand A gasoline. The standard deviation was 4.9 mi. These cars were then driven an average of 19.2 mi on one gallon of Brand B gasoline. The standard deviation was 6.9 mi. Is there a significant difference between the average number of miles driven with a gallon of each of these brands of gasoline? (Use a 5% level of significance.)

2. A large company claims that the average yearly salary of its workers is $18,000. The union believes that the average salary is much less. In a random sample of 200 employees the union finds that the average salary is $16,400 with a standard deviation of $1600. Should we accept the company's claim? (Use a 5% level of significance.)

3. The We-Move-It Company claims that a typical homeowner lives in the house an average of 7 years before selling it. The standard deviation is 3.02 years. To check this claim, a real estate firm interviews 64 homeowners. The results show that these homeowners sold their houses after an average of 7.9 years. Should we reject the moving company's claim? (Use a 5% level of significance.)

4. It is estimated that 25% of all students at Merrick College work full-time. In a random survey of 90 college students at this college, it is found that 34 students work full-time. Should we reject the estimate that 25% of all students at Merrick College work full-time? (Use a 5% level of significance.)

5. A sample of 180 families with four children living in Nova County indicated that each family spends an average of $67.88 per week for food. The standard deviation was $8.49. A similar survey of 160 families in Vega County indicated that each

family spends an average of $63.98 per week for food. The standard deviation was $9.88. Is the difference between the average weekly expenditure for food in these two counties significant? (Use a 1% level of significance.)

6.

FALSE FIRE ALARMS ON THE RISE

Brighton (Dec. 12)—Fire department officials disclosed yesterday that the number of false alarms called in to the department has reached epidemic proportions. Last month 58% of all fire alarms received were false alarms. Cited one official, "While our men are out on a false alarm, a true alarm may come in. Our equipment cannot be two places at one time."

December 12, 1991

According to fire department officials in a certain city, 58% of all fire alarms are false. To check this, a city official examines the department records. If 81 of 132 fire alarms turn out to be false, should we reject the fire department claim? (Use a 5% level of significance.)

7. A farmer claims that by injecting turkeys with a certain chemical the weight of a turkey will increase considerably. To test this claim, another farmer divided 200 turkeys into two groups of 100 each. Group A received the chemical while group B did not. The following results were obtained over a 6-month period.

	Group A	Group B
Average Improvement (lb)	7.1	6.3
Sample Standard Deviation (lb)	3.8	3.2

Is the claim of the first farmer justified? (Use a 5% level of significance.)

8. Fifty families who previously owned a Brand X washing machine indicated that their machines lasted an average of 6.8 years. The standard deviation was 0.98 years. Forty-nine families who previously owned a Brand Y washing machine indicated that their machines lasted an average of 7.9 years. The standard deviation was 1.3 years. Is the difference between the average life of these two brands of washing machines significant? (Use a 5% level of significance.)

9. A tax inspector claims that 11% of the merchants in the state charge the wrong sales tax. The Merchant's Association challenges this claim. In a random sample of 88 merchants it is found that 11 of the merchants charged the wrong sales tax. Should we reject the tax inspector's claim? (Use a 1% level of significance.)

10. A commuter estimates that over the past two years her train was late, on the average, 19 minutes (min) each day with a standard deviation of 5 min. In the last 20 working days, however, her train was late an average of only 12 min. Does this indicate an improvement of service? (Use a 5% level of significance.)

11. An English teacher claims that 70% of all students can pass the school proficiency exam. In a random sample of 72 seniors it is found that 58 seniors passed the exam. Should we reject the teacher's claim? (Use a 1% level of significance.)

12. The average hourly salary of a bank teller in one city is $6.85 with a standard deviation of $0.56. One large bank, which employs 85 tellers, pays an hourly rate of $6.70. Can this bank be accused of paying below the average hourly rate? (Use a 5% level of significance.)

13. A sociologist believes that women from suburban areas marry at an earlier age than women from urban areas. Do the following sample results support this claim? (Use a 5% level of significance.)

	Urban Women	Suburban Women
Average Age at which Women First Married	21.3	19.1
Standard Deviation	2.6	1.9
Number in survey	38	45

14. According to the Red Cross, 45% of all the people in a southern city have type O blood. A random survey of 80 people found that 38 of them have type O blood. Should we reject the Red Cross claim? (Use a 1% level of significance.)

15. Court officials claim that any criminal case on the court's calendar is disposed of in an average of 60 days with a standard deviation of 6 days. The local civil liberties organization claims that the average is considerably higher. An average of 66 days was required to dispose of 43 criminal cases. Should we reject the court officials' claim? (Use a 5% level of significance.)

16. The Chamber of Commerce claims that 10% of all businesses are polluting the atmosphere. Environmentalists claim that the percentage is higher. A random survey of 98 businesses shows that 18 of them are polluting the atmosphere. Should we reject the claim made by the Chamber of Commerce? (Use a 1% level of significance.)

17. Forty typists in the east wing of an office building type an average of 38 business letters per hour with a standard deviation of 7.9 letters. Fifty typists in the west wing type an average of 33 comparable business letters per hour with a standard deviation of 4.8 letters. Is the difference between the average number of letters typed by the workers in both wings significant? (Use a 5% level of significance.)

Chapter Test

1. *Religion:* A sociologist claims that 40% of the people in Woodbern attend religious services each week on a regular basis. If a random survey of 93 people in Woodbern found that only 33 of them attended religious services each week on a regular basis, can we reject the sociologist's claim? (Use a 5% level of significance.)

2. In a certain factory a particular machine must be adjusted whenever the daily production contains more than 7% defective items. The machine usually produces thousands of items a day. On June 2 the quality-control engineer sampled 85 items produced and found 10 of them defective. The manager then ordered a halt in production claiming that the machine needed adjustments. Is the decision an incorrect one? (Use a 5% level of significance.)

3. Sixty-five students, who were taught by one teaching technique, took a test. The average grade was 119 with a standard deviation of 14. Forty-eight other students who were taught by a different teaching technique also took the same test. Their average grade was 124 with a standard deviation of 19. Is there any significant difference between the two teaching techniques? (Use a 4% level of significance.)

4. *Dieting:* Two groups of overweight people were put on special diets. The people in group A, which had 47 people in it, were given a special diet. The people in group B were given a different diet. Group B had 34 people in it. Each person ate only the permissible foods as specified in the person's diet. The average weight loss in group A was 22 pounds (lb) with a standard deviation of 2.8 lb. The average weight loss in group B was 27 lb with a standard deviation of 7.8 lb. Is there any significant difference between the average weight loss resulting from these two diets? (Use a 5% level of significance.)

5. If we toss a coin eight times and obtain 1 heads and 7 tails, can we conclude that the coin is loaded? Explain your answer.

6. A market analyst claims that a homemaker will pay an average of $20 for an AM/FM clock radio. To test this claim, 40 homemakers are randomly selected and the amount of money that each is willing to pay for an AM/FM clock radio is recorded. It is found that the average is $15 with a standard deviation of $2.17. Which of the following is the most appropriate null hypothesis, H_0?
   a. $p = \$15$ b. $\mu = \$20$ c. $\bar{x} = \$15$ d. $\mu = \$15$
   e. none of these

7. A manufacturer of plumbing supplies claims that the new PVC plumbing pipes currently being marketed will last an average of 14 years before replacement is needed. The standard deviation is 1.29 years. If a group of 20 PVC pipes lasted only an average of 10.7 years with a standard deviation of 1.83 years, should we reject the manufacturer's claim? (Use a 1% level of significance.)

8. A telephone company has coated 40 telephone poles with a special antirot coating produced by Company A and 36 other telephone poles with a special antirot coating produced by Company B. The following data are available on how long each of these coatings protected the poles before replacement was needed.

	Poles Coated with Chemical Produced by Company	
	A	B
Average Lifetime (yr)	9	10
Standard Deviation (yr)	1.2	0.91

Is the difference between the average lifetime of a telephone pole coated with these chemicals significant? (Use a 1% level of significance.)

9. An industry representative claims that 68% of all American households had a personal computer in 1990. To test this claim, an independent group randomly selects 123 households. It is found that 77 of these households have a personal computer. Should we reject the industry representative's claim? (Use a 5% level of significance.)

10. In an effort to decrease our dependence on imported oil most new homes that are built today are heated by domestically available gas heat. In an advertisement to induce homeowners to convert from oil heat to gas heat, one northeast utility company claims that 65% of all homes in the city are heated by gas. An oil company official disputes this claim and contends that the percentage is considerably lower. A random sample of 311 homes found that 164 of them are heated by gas. Should we reject the claim that 65% of all homes in the city are heated by gas? (Use a 1% level of significance.)

11. Data gathered from 60 purchasers of a certain type of personal computer suggest that the time it takes a purchaser to fully connect the computer to the monitor and printer is a random variable having a mean of 75 minutes (min) and a standard deviation of 5.8 min. A group of 40 different potential purchasers of the same computer is randomly selected and shown a film on how to connect the computer to the monitor and printer easily. After viewing the film, each purchaser is asked to fully connect the computer. It is found that they now average 66 minutes. Using a 1% level of significance, does viewing the film reduce the time required to connect the computer?

12. Banking industry officials in a large city claim that the average charge for a "bounced" check is $7.00. To test this claim, a random sample is taken of the charge imposed by 15 banks for a bounced check. It is found that the average charge is $8.25 with a standard deviation of $0.96. Using a 1% level of significance, should we reject the banking industry officials' claim?

13. A research botanist is interested in determining whether a special grow lite affects plant growth. Two groups of seven plants each are tested. One group is exposed to the special grow lite constantly. Nothing special is done to the second group. All the plants are placed in a controlled environment. The heights (in centimeters) of the plants after several months are as follows.

Heights of Plants Exposed to Special Grow Lite	Heights of Plants Not Exposed to Grow Lite
13	12
15	14
20	16
18	17
17	15
16	18
19	17

Using a 5% level of significance, does the special grow lite have a significant effect on the plant's growth?

14. *Cheating the poor:* A consumer's group claims that many large supermarkets charge higher prices for a quart of milk in poorer neighborhoods. To investigate this claim, one quart of milk is purchased from supermarkets located in each of these neighborhoods. The prices paid are as follows.

Poorer Neighborhoods	Middle-Class Neighborhoods
65¢	59¢
63¢	61¢
69¢	63¢
61¢	58¢
64¢	60¢
60¢	62¢
59¢	61¢
62¢	64¢
	58¢

Is there any significant difference between the average price of a quart of milk in these neighborhoods? (Use a 5% level of significance.)

15. The average hourly salary of a gas station attendant in a certain city is $4.75 with a standard deviation of 79 cents. One large gas station chain, which employs 89 attendants, pays its attendants an average hourly rate of $4.50. Can this gas station chain be accused of paying lower than the average hourly rate? (Use a 5% level of significance.)

   a. Yes   **b.** No   **c.** Not enough information given   **d.** None of these

16. From past experience, a statistics teacher has found that the average grade on the midterm exam is 81 with a standard deviation of 5.2. This year, a class of 49 students had an average of 75. Does this indicate that we should reject the teacher's claim that the average midterm exam grade is 81? (Use a 5% level of significance.)

   a. Yes   **b.** No   **c.** Not enough information given   **d.** None of these

17. Research indicates that the average reaction time by a rat to certain stimulus is 1.47 seconds. Seven rats have been injected with a drug. Their average reaction time to the stimulus after being injected with the drug is found to be 1.69 seconds

with a standard deviation of 0.93 seconds. Using a 1% level of significance, does this drug significantly affect the average reaction time to the stimulus?

    a. Yes   **b.** No   **c.** Not enough information given   **d.** None of these

18. The average length of a trout in a certain hatchery is 17.6 inches with a standard deviation of 1.1 inches. A fisherman caught 9 fish whose average length was 16.4 inches. At the 1% level of significance, can we reject the claim that the average length of a trout is 17.6 inches?

    a. Yes   **b.** No   **c.** Not enough information given   **d.** None of these

19. Two loading machines are being compared. Machine I needed an average of 25 minutes to process 100 packages. The standard deviation was 3.7 minutes. Machine II needed an average of 31 minutes to process 50 packages. The standard deviation was 9 minutes. At the 5% level of significance, is the difference between the mean processing time of the two machines significant?

    a. Yes   **b.** No   **c.** Not enough information given   **d.** None of these

Suggested Further Reading

1. Freund, John E., and Ronald E. Walpole. *Mathematical Statistics*, 4th ed. Englewood Cliffs, NJ: Prentice-Hall, 1987.
2. Iman, Ronald L., and W. J. Conover. *Modern Business Statistics*. New York: Wiley, 1983.
3. Mendenhall, William. *Introduction to Probability and Statistics*, 7th ed. MA: Duxbury, 1987.
4. Olson, Charles L., and Mario J. Picconi, *Statistics for Business Decision Making*. Glenview, IL: Scott Foresman, 1983.
5. Ott, Lyman. *An Introduction to Statistical Methods and Data Analysis*, 2nd ed. Boston: Prindle, Weber and Schmidt, 1984.
6. Romano, A. *Applied Statistics for Science and Industry*. Boston: Allyn & Bacon, 1977.

11 Linear Correlation and Regression

Is the number of points scored by a basketball player directly related to the number of hours practiced? See discussion on p. 593. (© *Bill Whittaker, Photo Researchers, Inc.*)

Often we are given two or more variables and wish to know whether they are related. If they are, can we find some sort of equation for predicting the value of one variable when the values of the other are known? How reliable are such predictions? In this chapter we analyze procedures that can be used to determine when a relationship exists between variables and also to find an equation expressing such a relationship. Thus, we will study correlation and regression.

CHAPTER OBJECTIVES

After studying the material in this chapter, you should be able:

- **To briefly discuss** some of the pioneers in the field of correlation and regression. (Section 11.1)
- **To decide** whether two variables are related. We draw two lines, one horizontal and one vertical, and then place dots in various places corresponding to the given data. (Section 11.2)
- **To discuss** linear correlation, which tells us whether there is a relationship that will cause an increase (or decrease) in the value of one variable when the other is increased (or decreased). (Section 11.2)
- **To measure** the strength of the linear relationship that exists between two variables. This is the coefficient of linear correlation. (Section 11.3)
- **To determine** the reliability of r. (Section 11.4)
- **To calculate** a regression line that allows us to predict the value of one of the variables if the value of the other variable is known. (Section 11.5)
- **To analyze** the method of least squares that we use to determine the estimated regression line used in prediction. (Section 11.6)
- **To indicate** how the standard error of the estimate tells us how well the least-squares prediction equation describes the relationship between two variables. (Section 11.7)
- **To set up** confidence intervals for our regression estimates. (Sections 11.8 and 11.9)
- **To introduce** the concept of multiple regression where one variable depends on many other variables. (Section 11.10)

Statistics in the News

TURNPIKE USE DOWN

Trenton (March 26)—In the wake of the recent hike in tolls on the New Jersey Turnpike, the number of vehicles using the roadway has dropped sharply. Many truckers have switched to alternate parallel roads to avoid paying the 100% rate increase.

Officials were unable to predict yesterday how the decrease in vehicular use of the turnpike resulting from the rate increase would affect the overall revenue picture.

March 26, 1991

MORE WOES FOR BANKS

Washington (March 15)—Government officials were wrestling with the difficult problem of how to bail out the increasing number of banks experiencing financial difficulties. The FDIC has already closed many financially insolvent banks this year, with the prediction that more closings were sure to come. Officials estimate that more than 150 billion dollars is needed to shore up the ailing banking industry.

March 15, 1991

Consider the preceding newspaper articles. The first implies that overall turnpike revenue is related to the number of vehicles using the toll road. Can we predict the decrease in the number of vehicles that will use the turnpike as a result of a toll increase?

Now consider the second article. Can we accurately predict the number of banks that will be closed or the amount of money that is needed to bail out insolvent banks? How reliable are such predictions?

11.1 INTRODUCTION

Up to this point we have been discussing the many statistical procedures for analyzing a single variable. However, when dealing with the problems of applied statistics in education, psychology, sociology, and so on, we may be interested in determining whether a relationship exists between two or more variables. For example, if college officials have just administered a vocational aptitude test to 1000 entering freshmen, they may be interested in knowing whether there is any relationship between the math aptitude scores and the business aptitude scores. Do students who score well on the math section of the aptitude exam also do well on the business part? On the other hand, is it true that a student who scored poorly on the math part will necessarily score poorly on the business part? Similarly, the college officials may be interested in determining if there is a relationship between high school averages and college performance.

Questions of this nature frequently arise when we have many variables and are interested in determining relationships between these sets of scores. In this chapter we learn how to compute a number that measures the relationship between two sets of scores. This number is called the **correlation coefficient.** The English mathematician Karl Pearson (1857–1936) studied it in great detail and to some extent so did another English mathematician, Sir Francis Galton (1822–1911).

Sir Francis Galton, a cousin of Charles Darwin, undertook a detailed study of human characteristics. He was interested in determining whether a relationship exists between the heights of fathers and the heights of their sons. Do tall parents have tall children? Do intelligent parents or successful parents have intelligent or successful children? In *Natural Inheritance* Galton introduced the idea referred to today as correlation. This mathematical idea allows us to measure the closeness of the relationship between two variables. Galton found that a very close relationship exists between the heights of fathers and the heights of their sons. On the question of whether intelligent parents have intelligent children, it has been found that the correlation is 0.55. As we will see, this means that it is not necessarily true that

Correlation
Coefficient

LIKE PARENTS, LIKE CHILDREN

Jan. 5—Children whose parents are hooked on drugs and alcohol are more likely to fall victim to these substances than other kids. *U.S. News & World Report* says: "Studies show 65 percent of those youths dependent on drugs or alcohol are from homes where at least one parent is also hooked."

Saturday, January 5, 1991

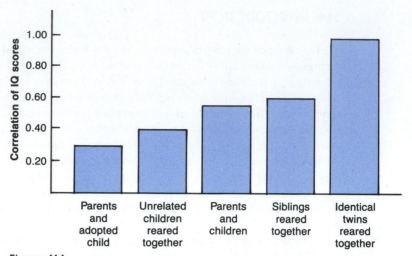

Figure 11.1
Correlations between the intelligence of parents and that of their children.

intelligent parents have intelligent children. In many cases children will score higher or lower than their parents. Figure 11.1 shows how the correlations for IQ range from 0.28 between parents and adopted children to 0.97 for identical twins reared together.

The precise mathematical measure of correlation as we use it today was actually formulated by Karl Pearson.

If there is a high correlation between two variables, we may be interested in representing this correspondence by some form of an equation. So we will discuss the **method of least squares.** The statistical method of least squares was developed by Adrien-Marie Legendre (1752–1833). Although Legendre is best known for his work in geometry, he also did important work in statistics. He developed the method of least squares. This method is used when we want to find the regression equation.

Finally, we will discuss how this equation can be used to make predictions and we will examine the reliability of these predictions.

11.2 SCATTER DIAGRAMS

To help us understand what is meant by a correlation coefficient, let us consider a guidance counselor who has just received the scores of an aptitude test administered to ten students. See Table 11.1.

The counselor may be interested in determining if there is any correlation among these sets of scores. For example, the counselor may wish to know whether a student who scores well on the math aptitude part of the exam will also score well on the

TABLE 11.1 Different Aptitude Scores Received by Ten Students

Student	Math Aptitude	Business Aptitude	Language Aptitude	Music Aptitude
A	52	48	26	22
B	49	49	53	23
C	26	27	48	57
D	28	24	31	54
E	63	59	67	13
F	44	40	75	20
G	70	72	31	9
H	32	31	22	50
I	49	50	11	17
J	51	49	19	24

Scatter Diagram

business aptitude part. She can analyze the situation pictorially by means of a **scatter diagram.**

To make a scatter diagram, we draw two lines, one vertical and one horizontal. On the horizontal line we indicate the math scores, and on the vertical line we indicate the business scores. Although we can put the math scores on the vertical line, we have purposely labeled the math scores on the horizontal line. This is done because we wish to predict the scores on the business aptitude part on the basis of the math scores. After both axes, that is, both lines, are labeled, we use one dot to represent each person's score. The dot is placed directly above the person's math score and directly to the right of the business score. Thus, the dot for Student A's score is placed directly above the 52 score on the math axis and to the right of 48 on the business axis. Similarly, the dot for Student B's score is placed directly above the 49 score on the math axis and to the right of 49 on the business axis. The same procedure is used to locate all the dots of Figure 11.2.

Linear Correlation

Notice that these dots form an approximate straight line. When this happens we say that there is a **linear correlation** between the two variables. Notice also that the higher the math score is, the higher the business score will be. The line moves in a direction that is from lower left to upper right. When this happens, we say that

Positive Correlation

there is a **positive correlation** between the math scores and the business scores. This means that a student with a higher math score will tend to have a higher business score.

Now let us draw the scatter diagram for the business aptitude scores and the music aptitude scores. It is shown in Figure 11.4. In this case we notice that the higher the business score, the lower the music score. Again, the dots arrange themselves in the form of a line, but this time the line moves in a direction that is from

Negative Correlation

upper left to lower right. When this happens we say that there is a **negative correlation** between the business scores and the music scores. This means that a student with a high business score will tend to have a low music score.

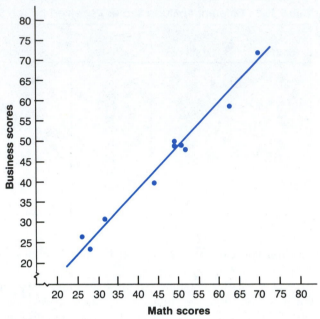

Figure 11.2
Scatter diagram for the math and business scores.

EMPLOYMENT STATUS OF PERSONS 16 TO 24 YEARS OF AGE AND NOT ENROLLED IN SCHOOL OCTOBER 1990			
Educational Attainment	Labor Force Participation Rate	Unemployment Rate	Percent Employed
College graduates	95.2%	5.9%	89.7%
1 to 3 years of college	89.4	8.8	81.5
High school, no college	84.3	12.5	73.7
High school dropouts	67.5	25.3	50.4
TOTAL	**81.8%**	**14.0%**	**70.2%**

Figure 11.3
The "Labor Force Participation Rate" is the percent of the total civilian, noninstitutional population group with the indicated educational characteristic who were employed or seeking employment. The "Unemployment Rate" is the percent of those participating who were not employed. We have all heard the concept that if you want a good job, get a good education. This clipping provides documentation that there is a positive correlation between the education that one has (educational attainment) and the chances of holding or securing a job (labor force participation rate). Furthermore, the clipping indicates that there is a negative correlation between educational attainment and unemployment.
Source: Current Population Survey conducted by the U.S. Bureau of the Census.

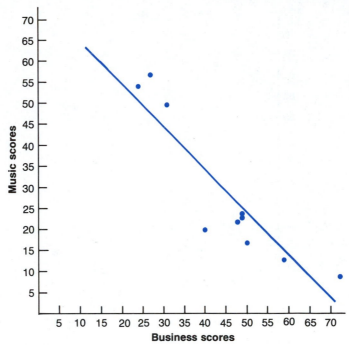

Figure 11.4
Scatter diagram for the business and music scores.

Now let us draw the scatter diagram for the language scores and music scores. It is shown in Figure 11.5. In this case the dots do not form a straight line. When this happens we say that there is little or no correlation between the language scores and the music scores.

COMMENT   Although we will concern ourselves with only linear, that is, a straight line, correlation, the dots may suggest different types of curves. These are studied in detail by statisticians. Several examples of such scatter diagrams are given in Figure 11.6. In this text we will analyze only linear correlation.

11.3 THE COEFFICIENT OF CORRELATION

Coefficient of Linear Correlation

Once we have determined that there is a linear correlation between two variables, we may wish to determine the strength of the linear relationship. Karl Pearson developed a **coefficient of linear correlation,** which measures the strength of a relationship between two variables. The value of the coefficient of linear correlation is calculated by means of Formula 11.1.

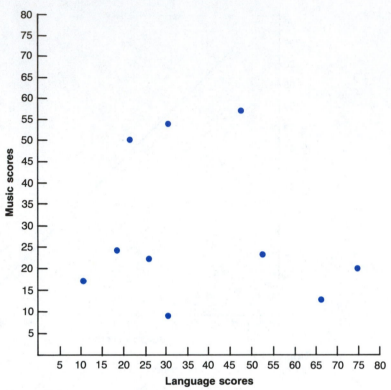

Figure 11.5
Scatter diagram for the language and music scores.

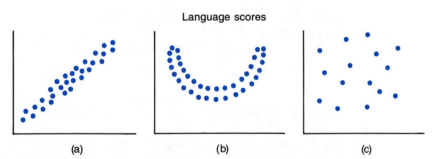

Figure 11.6
Scatter diagrams that suggest (a) a linear relationship, (b) a curvilinear relationship, and (c) no relationship.

| FORMULA 11.1 | The coefficient of linear correlation is given by |

$$r = \frac{n(\Sigma x\,y) - (\Sigma\,x)(\Sigma\,y)}{\sqrt{n(\Sigma\,x^2) - (\Sigma\,x)^2}\,\sqrt{n(\Sigma\,y^2) - (\Sigma\,y)^2}}$$

where

x = label for one of the variables

y = label for the other variable

n = number of pairs of scores

When using Formula 11.1 the coefficient of correlation will always have a value between -1 and $+1$. A value of $+1$ means perfect positive correlation and corresponds to the situation where all the dots lie exactly on a straight line. A value of -1 means perfect negative correlation and again corresponds to the situation where all the points lie exactly on a straight line. Correlation is considered high when it is close to $+1$ or -1 and low when it is close to 0. If the coefficient of linear correlation is zero, we say that there is no linear correlation. These possibilities are indicated in Figures 11.7 and 11.8.

Although Formula 11.1 looks complicated, it is rather easy to use. The only new symbol that appears is $\Sigma\,xy$. This value is found by multiplying the corresponding

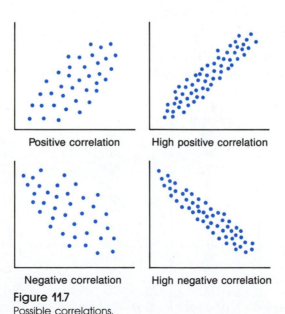

Positive correlation   High positive correlation

Negative correlation   High negative correlation

Figure 11.7
Possible correlations.

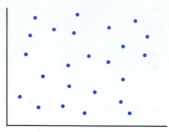

No correlation

Figure 11.8
No correlation.

values of x and y and then adding all the products. The following examples illustrate the procedure.

EXAMPLE 1

Find the correlation coefficient for the data of Figure 11.2 (see page 538). It is repeated in the following chart:

Math Score	52	49	26	28	63	44	70	32	49	51
Business Score	48	49	27	24	59	40	72	31	50	49

Solution

We first let x represent the math score and y represent the business score. Then we arrange the data in tabular form as follows and apply Formula 11.1.

x (Math)	y (Business)	x^2	y^2	xy
52	48	2704	2304	2496
49	49	2401	2401	2401
26	27	676	729	702
28	24	784	576	672
63	59	3969	3481	3717
44	40	1936	1600	1760
70	72	4900	5184	5040
32	31	1024	961	992
49	50	2401	2500	2450
51	49	2601	2401	2499
464	449	23,396	22,137	22,729

$$\Sigma x = 464 \qquad \Sigma y = 449 \qquad \Sigma x^2 = 23,396 \qquad \Sigma y^2 = 22,137 \qquad \Sigma xy = 22,729$$

Then

$$r = \frac{n(\Sigma\ xy) - (\Sigma\ x)(\Sigma\ y)}{\sqrt{n(\Sigma\ x^2) - (\Sigma\ x)^2}\ \sqrt{n(\Sigma\ y^2) - (\Sigma\ y)^2}}$$

$$= \frac{10(22,729) - (464)(449)}{\sqrt{10(23,396) - (464)^2}\ \sqrt{10(22,137) - (449)^2}}$$

$$= \frac{18,954}{\sqrt{18,664}\ \sqrt{19,769}}$$

$$= \frac{18,954}{(136.62)(140.60)}$$

$$= \frac{18,954}{19,208.77} = 0.9867$$

Thus, the coefficient of correlation is 0.9867. Since this value is close to $+1$, we say that there is a high degree of positive correlation. Figure 11.2 also indicated the same result.

EXAMPLE 2

Find the coefficient of correlation for the data of Figure 11.5 (see page 540). It is repeated in the following chart.

Language Score	26	53	48	31	67	75	31	22	11	19
Music Score	22	23	57	54	13	20	9	50	17	24

Solution

We first let x represent the language score and y represent the music score. Then we arrange the data in tabular form as follows and apply Formula 11.1.

x (Language)	y (Music)	x^2	y^2	xy
26	22	676	484	572
53	23	2809	529	1219
48	57	2304	3249	2736
31	54	961	2916	1674
67	13	4489	169	871
75	20	5625	400	1500
31	9	961	81	279
22	50	484	2500	1100
11	17	121	289	187
19	24	361	576	456
383	289	18,791	11,193	10,594

$\Sigma\ x = 383$ $\Sigma\ y = 289$ $\Sigma\ x^2 = 18,791$ $y^2 = 11,193$ $\Sigma\ xy = 10,594$

Now we apply Formula 11.1. We have

$$r = \frac{n(\Sigma\ xy) - (\Sigma\ x)(\Sigma\ y)}{\sqrt{n(\Sigma\ x^2) - (\Sigma\ x)^2}\ \sqrt{n(\Sigma\ y^2) - (\Sigma\ y)^2}}$$

$$= \frac{10(10,594) - (383)(289)}{\sqrt{10(18,791) - (383)^2}\ \sqrt{10(11,193) - (289)^2}}$$

$$= \frac{-4747}{\sqrt{41,221}\ \sqrt{28,409}} = \frac{-4747}{(203.03)(168.55)}$$

$$= \frac{-4747}{34,220.71} = -0.1387$$

Thus, the coefficient of correlation is -0.1387. Since this value is close to zero, there is little correlation. Figure 11.5 indicated the same result.

How will the correlation between x and y be affected if x is coded by adding the same number to (or subtracting the same number from) each score? How is y affected?

Fortunately, it turns out that the correlation coefficient is unaffected by adding or subtracting a number to either x or y or both. Thus, x can be coded in one way—perhaps by adding or subtracting a number—and y can be coded in another way—say, by multiplying by a number. In either case the value of the correlation coefficient is unaffected. Of greater importance is the fact that if we code before calculating the value of r, we do not have to uncode our results.

Let us code the data of Example 2 of this section and see how coding simplifies the computations involved.

EXAMPLE 3

By coding the data, find the coefficient of correlation for the data of Example 2 of this section.

Solution

We will code the data by subtracting 38 from each x-value and 29 from each y-value. Our new distribution of test scores then becomes

Language Score	-12	15	10	-7	29	37	-7	-16	-27	-19
Music Score	-7	-6	28	25	-16	-9	20	21	-12	-5

Now we calculate r from the coded data. We have the following.

x (Language)	y (Music)	x^2	y^2	xy
-12	-7	144	49	84
15	-6	225	36	-90
10	28	100	784	280
-7	25	49	625	-175
29	-16	841	256	-464
37	-9	1369	81	-333
-7	-20	49	400	140
-16	21	256	441	-336
-27	-12	729	144	324
-19	-5	361	25	95
3	-1	4123	2841	-475
$\Sigma x = 3$	$\Sigma y = -1$	$\Sigma x^2 = 4123$	$\Sigma y^2 = 2841$	$\Sigma xy = -475$

Then

$$r = \frac{n(\Sigma\, xy) - (\Sigma\, x)(\Sigma\, y)}{\sqrt{n(\Sigma\, x^2) - (\Sigma\, x)^2}\ \sqrt{n(\Sigma\, y^2) - (\Sigma\, y)^2}}$$

$$= \frac{10(-475) - (3)(-1)}{\sqrt{10(4123) - (3)^2}\ \sqrt{10(2841) - (-1)^2}}$$

$$= \frac{-4750 + 3}{\sqrt{41{,}230 - 9}\ \sqrt{28{,}410 - 1}}$$

$$= \frac{-4747}{\sqrt{41{,}221}\ \sqrt{28{,}409}} = \frac{-4747}{(203.03)(168.55)} = -0.1387$$

Notice that the value of r obtained by coding and the value of r obtained by working with the original uncoded data is exactly the same. Coding simplifies computations if the values with which we code are chosen carefully.

EXERCISES FOR SECTION 11.3

1. For each of the following, indicate whether you would expect a positive correlation, a negative correlation, or zero correlation.
   a. The age of an individual and the incidence of heart attacks.
   b. The hand with which a person writes and the IQ of the individual.
   c. The length of a book and the price of the book.
   d. The quantity of acid rain that falls and the quantity of fish that a lake has.
   e. The attendance at a movie theater and the price of admission to the theater.
   f. The height of an individual and the incidence of cancer.

g. The birthday of a man and the birthday of his wife.

h. The number of credits that a student takes per semester and the tuition charge at the college.

i. Annual income and educational level.

2. An experiment was conducted to determine if there is any correlation between height and shoe size. Five people were randomly selected and their heights (in inches) and shoe size were recorded in the following chart.

Height, x	65	67	67	71	72
Shoe size, y	$7\frac{1}{2}$	$8\frac{1}{2}$	8	10	11

Draw a scatter diagram for the data and then compute the coefficient of correlation.

3. A large pharmaceutical company wishes to force its employees to work overtime. The union claims that the more hours an employee works, the greater is the risk of an accident because of fatigue. To support its claim, the union has gathered the following statistics on the average number of hours worked per week by an employee and the average number of accidents per week.

Number of Hours Worked, x	35	37	39	41	43	45	47	49
Number of Accidents, y	3.2	3.1	3.6	4.4	7.7	10.1	6.9	8.8

a. Draw a scatter diagram for the data and then compute the coefficient of correlation.

b. Does the union claim seem to be justified?

4. The president of the Equity Corporation is interested in knowing if there is any correlation between the number of health insurance policies sold and the number of life insurance policies sold by the company's sales force. Over a 5-week period the following number of policies were sold.

Number of Health Insurance Policies Sold, x	26	34	25	18	24
Number of Life Insurance Policies Sold, y	20	18	29	16	21

Draw a scatter diagram for the data and then compute the coefficient of correlation.

5. An appliance store dealer believes that the weekly sales of air conditioners are dependent on the average outside temperature during the week. In support of this claim the dealer randomly selects 7 weeks and records the average outdoor temperature and the sales of air conditioners during the week. This information is listed in the following chart.

Average Outside Temperature, x	Number of Air Conditioners Sold, y
75	14
83	17
67	12
89	19
95	22
79	16
81	16

a. Draw a scatter diagram for the data and compute the coefficient of correlation.

b. Do these data support the dealer's claim?

6. *Age and blood pressure:* Does a person's systolic blood pressure rise as he or she ages? To check on this claim, the following data have been collected.

Age (in years), x	Blood Pressure, y
38	126
45	129
48	132
54	136
59	141
63	143
68	148
72	153

a. Draw a scatter diagram for the data and compute the coefficient of correlation.

b. Do the data support the claim that age and systolic blood pressure are related?

7. An English teacher believes that an individual's vocabulary and spelling scores are related. To support this claim, the following data have been collected.

Vocabulary, x	Spelling, y
77	78
94	91
65	75
84	80
69	72
60	73
74	76
70	83
80	84

a. Draw a scatter diagram for the data and compute the coefficient of correlation.

b. Is the English teacher's claim justified?

8. A physical therapist measured the chest girth and the weight of ten randomly selected patients. The results are as follows.

Chest Girth (in inches), x	Weight (in pounds), y
37.6	182
34.5	163
35.3	167
30.8	131
31.3	137
34.1	160
33.2	151
37.1	175
32.6	143
31.9	139

a. Draw a scatter diagram for the preceding data and compute the coefficient of correlation.

b. Is weight related to chest girth?

9. Refer back to Figure 11.3 on page 538. The following is a list of the number of years of schooling beyond high school and the salaries of eight randomly selected people in Belleville.

Number of Years of Schooling Beyond High School, x	Salary (in thousands of dollars), y
2	25
3	27
5	30
1	24
0	22
7	38
6	33
4	29

a. Draw a scatter diagram for the data and compute the coefficient of correlation.

b. Is salary related to the number of years of schooling?

10. A high school guidance counselor is analyzing the term grades of six randomly selected students in algebra and trigonometry in an attempt to determine whether performance in algebra and performance in trigonometry are related. The following are their scores.

Student	Algebra Score, x	Trigonometry Score, y
A	86	80
B	79	71
C	68	85
D	95	90
E	72	76
F	75	77

Draw a scatter diagram for the data, and then compute the coefficient of correlation.

11. A study was conducted to investigate the relationship between the weight and diastolic blood pressure of an American male between 40 to 50 years of age. The following results were obtained.

Weight (in pounds), x	Diastolic Blood Pressure, y
182	90
169	81
171	84
193	92
185	88
152	79
141	78
199	90
220	97

a. Compute the coefficient of correlation for the preceding data.

b. Is weight related to diastolic blood pressure? Explain your answer.

12. The following data represent the number of years in practice and the annual income of ten randomly selected tax lawyers in a West Coast City.

Number of Years in Practice, x	Annual Income (in thousands of dollars), y
7	56
9	63
12	79
8	60
6	49
4	45
9	51
11	68
10	53
14	85

a. Compute the coefficient of correlation for the preceding data.

b. Is annual income related to the number of years of practice? Explain your answer.

11.4 THE RELIABILITY OF *r*

Although the coefficient of correlation is usually the first number that is calculated when we are given several sets of scores, great care must be used in interpreting the results. It can undoubtedly be said that among all the statistical measures discussed in this book the correlation coefficient is the one that is most misused. One reason for this misuse is the assumption that because the two variables are related, a change in one will result in a change in the other.

Many people have applied a positive correlation coefficient to prove a cause-and-effect relationship that may not even exist. To illustrate the point, it has been shown that there is a high positive correlation between teacher's salaries and the use of drugs on campus. Does this mean that reducing the teachers' salaries would reduce the use of drugs on campus or does it simply mean that the students at wealthier schools (which pay higher salaries) are more apt to use drugs?

Frequently, two variables may appear to have a high positive correlation even though they are not directly associated with each other. There may be a third variable that is highly correlated to these two variables.

There is another important consideration that is often overlooked. When r is computed on the basis of sample data, we may get a strong correlation, positive or negative, which is due purely to chance, not to some relationship that exists between x and y. For example, if x represents the amount of snowfall and y represents the number of hours that Joe studied on five consecutive days, we may have the following results:

Amount of Snow (in inches), x	1	4	2	6	3
Number of Hours Studied, y	2	6	3	4	4

The value of r in this case is 0.63. Can we conclude that if it snows more, then Joe studies more?

Fortunately, a chart has been constructed that allows us to interpret the value of the correlation coefficient correctly. This is Table V in the Appendix. This chart allows us to determine the significance of a particular value of r. We use this table as follows:

1. First compute the value of r using Formula 11.1.
2. Then look in the chart for the appropriate r-value corresponding to some given n, where n is the number of pairs of scores.
3. The value of r is *not* statistically significant if it is between $-r_{0.025}$ and $r_{0.025}$ for a particular value of n.

Level of Significance

The subscript, that is, the little numbers, attached to r is called the **level of significance.** If we use this chart and the $r_{0.025}$ values of the chart as our guideline, we will be correct in saying that when there is no significant statistical correlation between x and y and we will not reject H_0 95% of the time.

Table V also gives us the values of $r_{0.005}$. We use these chart values when we want to be correct 99% of the time. In this book we use the $r_{0.025}$ values only. Table V is constructed so that r can be expected to fall between $-r_{0.025}$ and $+r_{0.025}$ approximately 95% of the time and between $-r_{0.005}$ and $+r_{0.005}$ approximately 99% of the time when the true correlation between x and y is zero.

Returning to our example, we have $n = 5$ and $r = 0.63$. From Table V we have $r_{0.025} = 0.878$. Thus, r will *not* be statistically significant if it is between -0.878 and $+0.878$. Since $r = 0.63$ is between -0.878 and $+0.878$, we conclude that the correlation may be due purely to chance. We cannot say that if it snows more, then Joe will necessarily study more.

Similarly, in Example 1 of Section 11.3 (see page 542) we found that $r = 0.9867$. There were ten scores, so $n = 10$. The chart values tell us that if r is between -0.632 and $+0.632$, there is *no* significant statistical correlation. Since the value of r that we obtained is greater than $+0.632$, we conclude that there is a *definite* positive correlation between x and y. Thus, we are justified in claiming that there is a relationship between the math aptitude scores and the business aptitude scores.

EXERCISES FOR SECTION 11.4

1. Refer back to Exercise 2, page 546. Determine if r is significant.
2. Refer back to Exercise 3, page 546. Determine if r is significant.
3. Refer back to Exercise 4, page 546. Determine if r is significant.
4. Refer back to Exercise 5, page 546. Determine if r is significant.
5. Refer back to Exercise 6, page 547. Determine if r is significant.
6. Refer back to Exercise 7, page 547. Determine if r is significant.
7. Refer back to Exercise 8, page 547. Determine if r is significant.
8. Refer back to Exercise 9, page 548. Determine if r is significant.
9. Refer back to Exercise 10, page 548. Determine if r is significant.
10. Refer back to Exercise 11, page 549. Determine if r is significant.
11. Refer back to Exercise 12, page 549. Determine if r is significant.

11.5 LINEAR REGRESSION

Let us return to the example discussed in Section 11.2. In that example a guidance counselor was interested in determining whether a relationship existed between the different aptitudes tested. Once a relationship, in the form of an equation, can be found between two variables the counselor can use this relationship to *predict* the value of one of the variables if the value of the other variable is known. Thus, the counselor may wish to predict how well a student will do on the business portion of the test if she knows the student's score on the math part.

Also, the counselor might be analyzing whether any correlation exists between high school averages and college grade-point averages. Her intention would be to try to find some relationship that will *predict* a college student's academic success from a knowledge of the high school average alone.

COMMENT It should be noted that the correlation coefficient merely determines whether two variables are related, but it does not specify how. Thus, the correlation coefficient cannot be used to solve prediction problems.

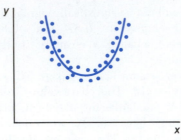

Figure 11.9
Curvilinear regression.

When given a prediction problem, we first locate all the dots on a scatter diagram as we did in Section 11.2. Then we try to fit a straight line to the data in such a way that it best represents the relationship between the two variables. Such a fitted line is called an **estimated regression line.** Once we have such a line we try to find an equation that will determine this line. We can then use this equation to predict the value of y corresponding to a given value of x.

Estimated Regression Line

Fitting a line to a set of numbers is by no means an easy task. Nevertheless, methods have been designed to handle such prediction problems. These methods are known as **regression methods.** In this book we discuss **linear regression** only. This means that we will try to fit a straight line to a set of numbers.

Regression Methods
Linear Regression

COMMENT   Occasionally, the dots are so scattered that a straight line cannot be fitted to the set of numbers. The statistician may then try to fit a **curve** to the set of numbers. This is shown in Figure 11.9. We would then have **curvilinear regression.**

Curvilinear Regression

The following examples illustrate how a knowledge of the regression line enables us to predict the value of y for any given value of x. It is standard notation to call the variable to be predicted the **dependent variable** and to denote it by y. The known variable is called the **independent variable** and is denoted by x.

Dependent Variable
Independent Variable

EXAMPLE 1

A guidance counselor notices that there is a strong positive correlation between the math scores and business scores. Based on a random sample of many students, she draws the scatter diagram shown in Figure 11.10. To this scatter diagram we have drawn a straight line that best represents the relationship between the two variables. This line enables us to predict the value of y for any given value of x. For example, if a student scored 35 on the math portion of the exam, then this line predicts that the student will score about 59 on the business part of the exam. This may be seen by first finding 35 on the horizontal axis, x, then moving straight up until we hit the estimated regression line. Finally, we move directly to the left to see where we cross the vertical axis, y. This is indicated by the dotted line of Figure 11.11. Similarly,

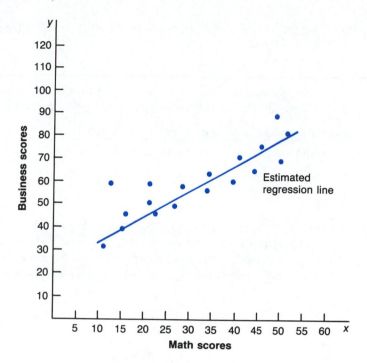

Figure 11.10

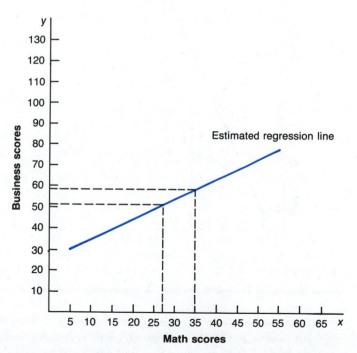

Figure 11.11

we can predict that a student whose math score is 27 will score 51 on the business portion of the exam.

EXAMPLE 2

A certain organization claims to be able to predict a person's height if given the person's weight. It has collected the following data for ten people.

Weight (in pounds), x	140	135	146	160	142	157	138	164	159	150
Height (in inches), y	63	61	68	72	66	65	64	73	70	69

If a person weighs 155 pounds, what is his predicted height?

Solution

We first draw the scatter diagram as shown in Figure 11.12. Then we draw a straight line that best represents the relationship between the two variables. This line now enables us to predict a person's height if we know the person's weight. The estimated regression line predicts that a person who weighs 155 pounds will be about 70 inches tall.

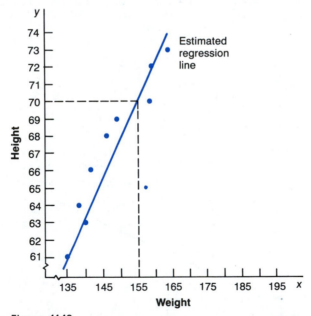

Figure 11.12

How does one draw an estimated regression line? Since there is no set procedure, different people are likely to draw different regression lines. So, although we may speak of finding a straight line that best fits the data, how is one to know when the best fit has been achieved? There are, in fact, several reasonable ways in which best fit can be defined. For this reason statisticians use a mathematical method for determining an equation that best describes the linear relationship between two variables. The method is known as the **least-squares method** and is discussed in detail in the next section.

11.6 THE METHOD OF LEAST SQUARES

Deviation

Least-Squares
Method

Whenever we draw an estimated regression line, not all points will lie on the regression line. Some will be above it and some will be below. The difference between any point and the corresponding point on the regression line is called the (vertical) **deviation** from the line. It represents the difference in value between what we predicted and what actually happened. See Figure 11.13. The **least-squares method** determines the estimated regression line in such a way that the sum of the squares of these vertical deviations is as small as possible. Although a background in calculus is needed to understand how we obtain the formula for the least-square regression line, it is very easy to use the formula.

FORMULA 11.2   Regression Equation

Estimated
Regression Line

The equation of the **estimated regression line** is

$$\hat{y} = b_0 + b_1 x$$

where

$$b_1 = \frac{n(\sum xy) - (\sum x)(\sum y)}{n(\sum x^2) - (\sum x)^2} \qquad b_0 = \frac{1}{n}(\sum y - b_1 \cdot \sum x)$$

and n is the number of pairs of scores.

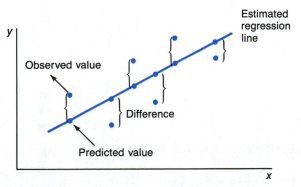

Figure 11.13
Difference between y-values and the estimated regression line.

Regression Line
Regression Equation

COMMENT The straight line that best fits a set of data points according to the least-squares criterion is called the **regression line** whereas the equation of the regression line is called the **regression equation.**

Let us use Formula 11.2 to find the regression equation connecting two variables.

EXAMPLE 1 Fifteen students were asked to indicate how many hours they studied before taking their statistics examination. Their responses were then matched with their grades on the exam, which had a maximum score of 100.

Hours, x	0.50	0.75	1.00	1.25	1.50	1.75	2.00	2.25	2.50	2.75	3.00	3.25	3.50	3.75	4.00
Scores, y	57	64	59	68	74	76	79	83	85	86	88	89	90	94	96

a. Find the regression equation that will predict a student's score if we know how many hours the student studied.

b. If a student studied 0.85 hours, what is the student's predicted grade?

Solution

a. To enable us to perform the computations, we arrange the data in the form of a chart.

x	y	x^2	xy
0.50	57	0.2500	28.5
0.75	64	0.5625	48
1.00	59	1.0000	59
1.25	68	1.5625	85
1.50	74	2.2500	111
1.75	76	3.0625	133
2.00	79	4.0000	158
2.25	83	5.0625	186.75
2.50	85	6.2500	212.5
2.75	86	7.5625	236.5
3.00	88	9.0000	264
3.25	89	10.5625	289.25
3.50	90	12.2500	315
3.75	94	14.0625	352.5
4.00	96	16.0000	384
33.75	1188	93.4375	2863

$$\Sigma\, x = 33.75 \qquad \Sigma\, y = 1188 \qquad \Sigma\, x^2 = 93.4375 \qquad \Sigma\, xy = 2863$$

From the chart we have $n = 15$ so that

$$b_1 = \frac{n(\Sigma\, xy) - (\Sigma\, x)(\Sigma\, y)}{n(\Sigma\, x^2) - (\Sigma\, x)^2}$$

$$= \frac{15(2863) - (33.75)(1188)}{15(93.4375) - (33.75)^2}$$

$$= \frac{42{,}945 - 40{,}095}{1401.5625 - 1139.0625}$$

$$= \frac{2850}{262.5}$$

$$= 10.857$$

and

$$b_0 = \frac{1}{n} (\Sigma\, y - b_1 \cdot \Sigma\, x)$$

$$= \frac{1}{15} (1188 - (10.857) \cdot (33.75))$$

$$= \frac{1}{15} (1188 - 366.424)$$

$$= \frac{1}{15} (821.576)$$

$$= 54.772$$

Thus,*

$$\hat{y} = b_0 + b_1 x$$
$$= 54.772 + 10.857x$$

The equation of the predicted regression line then is

$$\hat{y} = 54.772 + 10.857x$$

b. For $x = 0.85$, we get

$$\hat{y} = 54.772 + 10.857(0.85)$$
$$= 54.772 + 9.22845$$
$$= 64.00045$$

Thus, the predicted grade of a student who studies 0.85 hours is approximately 64.

---

*Values for the variables throughout this chapter are calculated using computer accuracy, even though answers are often rounded to two or three decimal places.

EXAMPLE 2

A West Coast publishing company keeps accurate records of its monthly expenditure for advertising and its total monthly sales. For the first 10 months of 1991, the records showed the following.

Advertising (in thousands), x	43	44	36	38	47	40	41	54	37	46
Sales (in millions), y	74	76	60	68	79	70	71	94	65	78

(Note that units are in dollars.)

a. Find the least-squares prediction equation appropriate for the data.
b. If the company plans to spend $50,000 for advertising next month, what is its predicted sales? Assume that all other factors can be neglected.

Solution

a. We arrange the data in the form of a chart.

x	y	x^2	xy
43	74	1849	3182
44	76	1936	3344
36	60	1296	2160
38	68	1444	2584
47	79	2209	3713
40	70	1600	2800
41	71	1681	2911
54	94	2916	5076
37	65	1369	2405
46	78	2116	3588
426	735	18,416	31,763

$$\Sigma x = 426 \qquad \Sigma y = 735 \qquad \Sigma x^2 = 18,416 \qquad \Sigma xy = 31,763$$

From the chart we have $n = 10$ so that

$$b_1 = \frac{n(\Sigma\ xy) - (\Sigma\ x)(\Sigma\ y)}{n(\Sigma\ x^2) - (\Sigma\ x)^2}$$

$$= \frac{10(31,763) - (426)(735)}{10(18,416) - (426)^2}$$

$$= \frac{317,630 - 313,110}{184,160 - 181,476}$$

$$= \frac{4520}{2684} = 1.684$$

and

$$b_0 = \frac{1}{n} (\Sigma y - b_1 \cdot \Sigma x)$$

$$= \frac{1}{10} (735 - (1.684) \cdot (426))$$

$$= \frac{1}{10} (735 - 717.384)$$

$$= \frac{1}{10} (17.616)$$

$$= 1.762$$

Thus,

$$\hat{y} = b_0 + b_1 x$$
$$= 1.762 + 1.684x$$

The equation of the predicted regression line is

$$\hat{y} = 1.762 + 1.684x$$

b. For $x = 50$, not 50,000, since x is in thousands of dollars, we get

$$\hat{y} = 1.762 + 1.684(50)$$
$$= 1.762 + 84.2$$
$$= 85.962$$

Thus, if the company spends $50,000 next month for advertising, its predicted sales are $85.962 million assuming all other factors can be neglected.

There is an alternate way to compute the equation of the regression line. This involves the sample covariance. We first compute the average of the x-values, denoted as $\bar{x}$, and the average of the y-values denoted as $\bar{y}$. Then we compute the sample standard deviation of the x-values. This is denoted as s_x. Finally, we determine the **Sample Covariance** **sample covariance** of the n data points, which is defined by

$$\text{Sample covariance} = s_{xy} = \frac{\Sigma (x - \bar{x})(y - \bar{y})}{n - 1}$$

We then have the following alternate formula.

FORMULA 11.3   Equation of Regression Line (Alternate version)

The equation of the regression line for a set of n data points is given by

$$\hat{y} = b_0 + b_1 x$$

where

$$b_1 = \frac{s_{xy}}{s_x^2}, \qquad b_0 = \bar{y} - b_1 \bar{x}$$

and s_x is the sample standard deviation of the x-values

COMMENT   In a later exercise the reader will be asked to demonstrate that Formula 11.3 is indeed equivalent to Formula 11.2, which was used earlier to calculate the equation of the regression line.

We will illustrate the use of Formula 11.3 for the data presented in Example 2 of this section.

EXAMPLE 3   Find the equation of the regression line for the data of Example 2.

Solution

We arrange the data in the form of a chart.

x	y	$x - \bar{x}$	$y - \bar{y}$	$(x - \bar{x})(y - \bar{y})$	$(x - \bar{x})^2$
43	74	0.4	0.5	0.2	0.16
44	76	1.4	2.5	3.5	1.96
36	60	−6.6	−13.5	89.1	43.56
38	68	−4.6	−5.5	25.3	21.16
47	79	4.4	5.5	24.2	19.36
40	70	−2.6	−3.5	9.1	6.76
41	71	−1.6	−2.5	4.0	2.56
54	94	11.4	20.5	233.7	129.96
37	65	−5.6	−8.5	47.6	31.36
46	78	3.4	4.5	15.3	11.56
426	735	0	0	452	268.4

$\Sigma x = 426$ $\Sigma y = 735$ $\Sigma(x - \bar{x})(y - \bar{y}) = 452$ $\Sigma(x - \bar{x})^2 = 268.4$

Thus,

$$\bar{x} = \frac{\Sigma x}{n} = \frac{426}{10} = 42.6$$

$$\bar{y} = \frac{\Sigma y}{n} = \frac{735}{10} = 73.5$$

$$s_x^2 = \frac{\Sigma(x - \bar{x})^2}{n - 1} = \frac{268.4}{9} = 29.8222$$

$$s_{xy} = \frac{\Sigma(x - \bar{x})(y - \bar{y})}{n - 1} = \frac{452}{9} = 50.2222$$

so that

$$b_1 = \frac{s_{xy}}{s_x^2} = \frac{50.2222}{29.8222} = 1.684$$

and

$$b_0 = \bar{y} - b_1\bar{x} = 73.5 - (1.684)(42.6) = 1.762$$

The equation of the regression line is

$$\hat{y} = b_0 + b_1 x$$
$$= 1.762 + 1.684x$$

This is indeed the same answer that we obtained earlier.

EXERCISES FOR SECTION 11.6

In each of the following exercises assume that the correlation is high enough to allow for reasonable prediction.

1. As we indicated at the beginning of this chapter, Galton believed that a very close relationship exists between the heights of fathers and the heights of their sons. To test this claim, a scientist selects ten men at random and records their heights and the heights of their sons.

Father's Height (in inches), x	63	64	72	73	74	69	66	72	75	64
Son's Height (in inches), y	64	66	74	75	74	71	69	75	76	67

a. Determine the least-squares prediction equation.
b. If a father is 76 inches tall, what is the predicted height of the son?

2. Consider the following newspaper article:

> ## WEEKEND RAINS ADD WATER TO OUR RESERVOIRS
>
> *Los Angeles (Mar. 4)*—Water department officials estimate that this weekend's downpouring of intermittent rains will add about 100 million gallons of water to the state's reservoir systems. This additional rain should help ease the drought that is currently gripping our state.
>
> March 4, 1991

Based on past experience, the following data have been collected.

Number of Inches of Rainfall, x	Number of Millions of Gallons of Water Added to Reservoirs, y
0.5	80
1.0	93
1.5	101
2.0	112
2.4	118
2.9	124
3.3	139
3.9	145

a. Determine the least-squares prediction equation.

b. If 2.2 inches of rain fall, what is the predicted number of millions of gallons of water to be added to the reservoirs?

3. *Advertising:* A large department store chain is interested in determining the relationship between the amount of money spent on advertising x, and the weekly volume of sales, y. It randomly samples 8 weeks and determines the amount of money spent on advertising and the weekly volume of sales. The following data are available.

Amount of Money Spent on Advertising (in thousands of dollars), x	Weekly Sales Volume (in thousands of dollars), y
46	496
54	499
43	500
59	506
63	513
25	479
75	520
51	500

a. Determine the least-squares prediction equation.

b. If the department store chain spends $70,000 on advertising, what is the predicted volume of sales?

4. Traffic department officials in Chicago believe that the number of parking tickets issued per week is directly related to the number of meter maids hired. The following data have been collected for eight randomly selected weeks to support this claim.

Number of Meter Maids Hired, x	Number of Parking Tickets Issued, y
4	326
8	197
7	409
6	400
10	628
15	709
22	978
9	680

a. Determine the least-squares prediction equation.

b. If 11 meter maids are hired, what is the predicted number of parking tickets to be issued?

5. A sociologist believes that there is a tendency for tall men to marry tall women. To determine if such a relationship exists, 12 couples are randomly selected and the height of the husband and the height of the wife are recorded as shown here.

Wife's Height (in inches), x	Husband's Height (in inches), y
62	66
67	67
61	65
64	66
63	67
63	69
65	67
66	68
68	72
71	74
60	62
64	66

a. Determine the least-squares prediction equation.

b. If a women is 70 inches tall, what is the predicted height of her husband?

6. A chemist for the Barry Plant Food Company undertook a study to determine the relationship that exists between the growth of a plant and the number of units of a chemical given to the plant. The following data are available.

Number of Units of Chemical Given, x	Number of Centimeters of Growth, y
9	9
6	7
4	5
3	4
5	5
7	7
2	2

a. Find the least-squares prediction equation.
b. If eight units of the chemical are added to the plant, what is the predicted number of centimeters of growth for the plant?

7. A medical researcher is interested in determining how coffee with caffeine affects a person's heartbeat. The following data have been collected on eight subjects who were given various dosages of coffee with caffeine added to it over a 24-hour period.

Dosage (units), x	Number of Heartbeats per Minute, y
7	115
4	96
3	90
6	109
2	81
1	72
5	100
9	122

What is the predicted number of heartbeats per minute for a subject who drank coffee with eight units of caffeine added to it?

8. An economist believes that there is a relationship between a family's income and the amount of money spent on entertainment. The following data are available on six randomly selected families.

Annual Family Income (in thousands of dollars), x	Annual Expenditure for Entertainment (in hundreds of dollars), y
25	2
31	8
38	7
23	1
29	6
47	9

a. Find the least-squares prediction equation.
b. What is the predicted annual expenditure for entertainment by a family whose annual income is $33,000?

9. A tax agent believes that the amount of money claimed for charitable deductions on a taxpayer's 1991 federal 1040 income tax form is related to the person's taxable income. To support this claim, the following data have been collected.

Annual Taxable Income (in dollars), x	Amount of Money Claimed for Charitable Deductions (in dollars), y
20,000	220
24,000	290
29,000	400
25,000	470
40,000	550
50,000	650

a. Find the least-squares prediction equation.

b. If a taxpayer's annual taxable income is $33,000, what is the predicted amount of money that will be claimed for charitable deductions?

10. After analyzing the salaries and batting averages of many baseball players in both leagues a sports commentator concluded the higher-paid players will have higher batting averages. To determine if there indeed is a relationship between salary and batting averages, ten players were randomly selected and the following statistics were obtained.

Average Annual Salary (in dollars), x	Batting Average (in prior year), y
670,000	0.290
662,000	0.281
698,000	0.299
825,000	0.302
900,000	0.309
1,400,000	0.345
800,000	0.304
850,000	0.306
685,000	0.295
1,300,000	0.320

a. Determine the least-squares prediction equation.

b. Assuming that the sports commentator's beliefs are accurate, what is the predicted batting average of a baseball player who is earning $175,000 annually?

11. All visitors to a perfume factory on the East Coast receive a free sample of the company's products at the completion of a tour of the factory. The company believes that sales at its on-premises gift shop are related to the number of free samples distributed as follows.

Number of Free Samples Distributed, x	Sales at On-Site Gift Shop (in dollars), y
100,000	800,000
200,000	1,000,000
500,000	1,300,000
700,000	1,600,000
900,000	2,000,000
1,100,000	2,400,000
1,500,000	3,200,000

a. Determine the least-squares prediction equation.
b. If 600,000 free samples are distributed, what is the predicted sales at the on-premises gift shop?

11.7 STANDARD ERROR OF THE ESTIMATE

In Section 11.6 we discussed the least-squares regression line, which predicts a value of y when x has a particular value. Quite often it turns out that the predicted value of y and the observed value of y are different. If the correlation is low, these differences will be large. Only when the correlation is high can we expect the predicted values to be close to the observed values.

In general, the true population regression line is not known, so we use the data to estimate the equation of this true population regression line. For example, consider the data of Example 2 given in Section 11.6 (page 558), which is reproduced here.

x	y
43	74
44	76
36	60
38	68
47	79
40	70
41	71
54	94
37	65
46	78

The equation of the regression line is $\hat{y} = 1.762 + 1.684x$. When $x = 50$ the predicted value of y is 85.962. We cannot expect such predictions to be completely accurate. At different times, when $x = 50$, we may get different y-values. Thus, for each x there is a corresponding population of y-values. The mean of the corresponding y-values lies on some straight line whose equation we do not know but which is of the form $y = \alpha + \beta x$. For each x-value, the distribution of the corresponding population of y-values is normally distributed. Moreover, for each x-value the mean

**Population
Regression Line**

of the corresponding population of y-values lies on a straight line called the **population regression line,** whose equation is of the form $y = \alpha + \beta x$. The population standard deviation, σ, of the population of y-values corresponding to a given x-value is the *same,* regardless of the x-value.

Thus, we assume that the kind of normal distribution occurring when $x = 50$ will also appear at any other value of x. This means that for any x-value the distribution of the population of y-values is a normal distribution and that the variance of that normal distribution is the same for every x. This can be seen in Figure 11.14.

Error Terms

We can then conclude that the **error terms** (the vertical distance between the predicted y-values and the true population values) are normally distributed with mean 0 and the same standard deviation σ. How do we estimate σ, the common standard deviation of the normal distributions given in Figure 11.14? Statisticians have devised a method for measuring σ. This is the **standard error of the estimate.** However, before doing this, let us analyze our least-squares prediction equation, $\hat{y} = 1.762 + 1.684x$. Why bother computing this equation? Why not use the given data to make predictions about the sales by simply ignoring the values of x (the amount of money spent for advertising) and using only the mean value of the sampled y's (the average sales) in making predictions? Thus, we can use

**Standard Error of the
Estimate**

$$\bar{y} = \frac{\Sigma\, y}{n} = \frac{735}{10} = 73.5$$

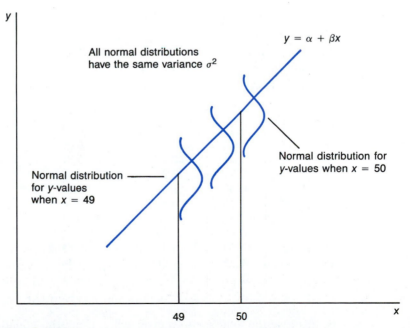

All normal distributions
have the same variance σ^2

$y = \alpha + \beta x$

Normal distribution for
y-values when $x = 50$

Normal distribution
for y-values
when $x = 49$

Figure 11.14
Normal distributions of population y-values about the regression line.

as our predicted sales (in all cases). How large an error have we made? In the following computations we indicate the (squared) error that is made when we predict a value of $\bar{y} = 73.5$ for the observed y-values.

y	$y - \bar{y}$	$(y - \bar{y})^2$
74	$74 - 73.5 = 0.5$	0.25
76	$76 - 73.5 = 2.5$	6.25
60	$60 - 73.5 = -13.5$	182.25
68	$68 - 73.5 = -5.5$	30.25
79	$79 - 73.5 = 5.5$	30.25
70	$70 - 73.5 = -3.5$	12.25
71	$71 - 73.5 = -2.5$	6.25
94	$94 - 73.5 = 20.5$	420.25
65	$65 - 73.5 = -8.5$	72.25
78	$78 - 73.5 = 4.5$	20.25
		780.50

$$\Sigma(y - \bar{y})^2 = 780.50$$

Thus, the total squared error that is made when we predict $\bar{y} = 73.5$ for the observed y-values is

$$\Sigma(y - \bar{y})^2 = 780.50$$

Total Sum of Squares, SST

This is referred to as the **total sum of squares, SST.** Therefore,

$$SST = \Sigma(y - \bar{y})^2 = 780.50$$

Instead of using $\bar{y}$ as our predicted value (in all cases), we can use the equation of the regression line values, $\hat{y}$, for our sales prediction. If we believe that our regression equation can be used to predict sales, then the squared error should be less when we use these values. The actual error obtained when we use these values is as follows:

x	y	$\hat{y}$	$y - \hat{y}$	$(y - \hat{y})^2$
43	74	74.17	-0.17	0.03
44	76	75.85	0.15	0.02
36	60	62.41	-2.41	5.81
38	68	65.77	2.23	4.97
47	79	80.89	-1.89	3.57
40	70	69.13	0.87	0.76
41	71	70.81	0.19	0.04
54	94	92.65	1.35	1.82
37	65	64.09	0.91	0.83
46	78	79.21	-1.21	1.46
				19.31

$$\Sigma(y - \hat{y})^2 = 19.31$$

In this case the total squared error when using the equation of the regression line values $\hat{y}$ for predictions is

$$\Sigma(y - \hat{y})^2 = 19.31$$

Error Sum of Squares, SSE

This is called the **error sum of squares, SSE.** Thus,

$$SSE = \Sigma(y - \hat{y})^2 = 19.31$$

Percentage Reduction

It should be obvious that using the regression equation for prediction reduced the total squared error considerably. The **percentage reduction** is

$$\frac{SST - SSE}{SST} = 1 - \frac{SSE}{SST} = 1 - \frac{19.31}{780.50} = 0.9753$$

Coefficient of Determination

or 97.53%. The percentage reduction in the total squared error obtained by using the regression equation instead of $\bar{y}$ is denoted by r^2 and is called the **coefficient of determination.** Thus, we have Formula 11.4.

FORMULA 11.4	**Coefficient of determination** $= r^2 = 1 - \dfrac{SSE}{SST}$

COMMENT The coefficient of determination, namely, r^2, can be interpreted as representing the percentage of variation in the observed y-values that is explainable by the regression line.

As mentioned earlier, the population of y-values corresponding to the various x-values all have the same (usually unknown) standard deviation, σ. The value of σ can be estimated from the sample data by computing the **standard error of the estimate** also called the **residual standard deviation.**

Residual Standard Deviation

To determine the standard error of the estimate, we first calculate the predicted value of y for each x and then compute the difference between the observed value and the predicted value. We then square these differences and divide the sum of these squares by $n - 2$. The square root of the result is called the standard error of the estimate.

FORMULA 11.5 **Standard Error of the Estimate**	The **standard error of the estimate** is denoted by s_e and is defined as $$s_e = \sqrt{\frac{\Sigma(y - \hat{y})^2}{n - 2}} = \sqrt{\frac{SSE}{n - 2}}$$ where $\hat{y}$ is the predicted value, y is the observed value, and n is the number of pairs of scores.

EXAMPLE 1

Find the standard error of the estimate for the least-squares regression equation of Example 1 of Section 11.6 on page 556.

Solution

The least-squares regression equation was

$$\hat{y} = 54.772 + 10.857x$$

Using this equation, we find the predicted value of y corresponding to each value of x. We arrange our computations in the form of a chart.

x	y	$\hat{y}$	$y - \hat{y}$	$(y - \hat{y})^2$
0.50	57	60.2	−3.2	10.24
0.75	64	62.91	1.09	1.19
1.00	59	65.63	−6.63	43.96
1.25	68	68.34	−0.34	0.12
1.50	74	71.06	2.94	8.64
1.75	76	73.77	2.23	4.97
2.00	79	76.49	2.51	6.30
2.25	83	79.2	3.8	14.44
2.50	85	81.92	3.08	9.49
2.75	86	84.63	1.37	1.88
3.00	88	87.35	0.65	0.42
3.25	89	90.06	−1.06	1.12
3.50	90	92.78	−2.78	7.73
3.75	94	95.49	−1.49	2.22
4.00	96	98.21	−2.21	4.88
				117.60

$$\Sigma(y - \hat{y})^2 = 117.60$$

Applying Formula 11.5, we get

$$s_e = \sqrt{\frac{\Sigma(y - \hat{y})^2}{n - 2}} = \sqrt{\frac{117.60}{15 - 2}} = \sqrt{\frac{117.60}{13}} \approx \sqrt{9.05} \approx 3.01$$

Thus, the standard error of the estimate is approximately 3.01.

EXAMPLE 2   Find the standard error of the estimate for the least-squares regression equation of Example 2 of Section 11.6 (page 558).

Solution

The least-squares regression equation was

$$\hat{y} = 1.762 + 1.684x$$

Using this equation, we find the predicted value of y corresponding to each value of x. We arrange our computations in the form of a chart.

x	y	$\hat{y}$	$y - \hat{y}$	$(y - \hat{y})^2$
43	74	74.17	−0.17	0.03
44	76	75.85	0.15	0.02
36	60	62.41	−2.41	5.81
38	68	65.77	2.23	4.97
47	79	80.89	−1.89	3.57
40	70	69.13	0.87	0.76
41	71	70.81	0.19	0.04
54	94	92.65	1.35	1.82
37	65	64.09	0.91	0.83
46	78	79.21	−1.21	1.46
				19.31

$$\Sigma(y - \hat{y})^2 = 19.31$$

Applying Formula 11.5, we get

$$s_e = \sqrt{\frac{\Sigma(y - \hat{y})^2}{n - 2}}$$

$$= \sqrt{\frac{19.31}{8}} \approx \sqrt{2.41} \approx 1.55$$

Thus, the standard error of the estimate is approximately 1.55.

The goodness of fit of the least-squares regression line is determined by the value of the standard error of the estimate. A relatively small value of s_e indicates that the predicted and observed values of y are fairly close. This means that the regression equation is a good description of the relationship between the two variables. On the other hand, a relatively large value of s_e indicates that there is a large difference between the predicted and observed values of y. When this happens the relationship between x and y as given by the least-squares equation is not a good indication of the relationship between the two variables. Only when the standard error of the estimate is zero can we say for sure that the least-squares regression equation is a perfect description of the relationship between x and y.

COMMENT In computing s_e statisticians often use the following equivalent formula.

$$s_e = \sqrt{\frac{\Sigma y^2 - b_0(\Sigma y) - b_1(\Sigma xy)}{n - 2}}$$

where the values of b_0 and b_1 are the same as those obtained earlier in computing the equation of the regression line. The values for the various summations should have already been obtained, thereby reducing the amount of computation needed.

EXERCISES FOR SECTION 11.7

For each of the following, refer back to the exercise indicated and calculate the standard error of the estimate.

*11.8 USING STATISTICAL INFERENCE FOR REGRESSION

We can use the standard statistical inferential procedures discussed in earlier chapters for regression. Specifically, we mentioned earlier that for each x-value, the corresponding population of y-values is normally distributed with mean $y = \beta_0 + \beta_1 x$ and standard deviation σ. However, if β_1 has a value of 0, then x will be totally worthless in predicting y-values since in that case the regression equation would be

$$\hat{y} = \beta_0 + \beta_1 x$$

$$= \beta_0 + 0x$$

$$= \beta_0$$

Thus, the value of x would have absolutely nothing to do with the distribution of y-values. Hence, it is important for us to determine in advance whether x can be used as a predictor of y, that is, if x and y are linearly related. We can decide this by performing the following hypothesis test.

$$H_0: \beta_1 = 0$$

$$H_1: \beta_1 \neq 0$$

If we conclude that the null hypothesis has to be rejected, then this indicates that x and y are linearly related and that we can proceed to use the equation of the regression line for making predictions.

How do we test the null hypothesis that $\beta_1 = 0$? This can be done by using the value of b_1, which actually represents the slope of the sample regression line. We proceed as follows:

> ### Statistical Inference Concerning β_1
>
> To test the null hypothesis H_0 that the slope β_1 of the population regression line is zero or not (that is, whether x can be used as a predictor of y or not) do the following:
>
> 1. State the null hypothesis and the alternative hypothesis as well as the significance level α.
>
> 2. Find $n - 2$. This gives us the number of degrees of freedom for a t-distribution.
>
> 3. Find the appropriate critical values $\pm t_{\alpha/2}$ by using Table VII in the Appendix.
>
> 4. Compute the value of the test statistic.
>
> $$ t = \frac{b_1}{s_e \Big/ \sqrt{\Sigma x^2 - \dfrac{(\Sigma x)^2}{n}}} $$
>
> 5. If the value of the test statistic falls in the rejection region, reject H_0. Otherwise do not reject H_0.
>
> 6. State the conclusion.

Let us illustrate the above procedure with a few examples.

EXAMPLE 1

Refer back to the data of Example 1 on page 556. Does the data indicate that the value of β_1, that is, the slope of the population regression line, is not zero, which would mean that x (the number of hours studied) can be used as a predictor of y (the score received)? Assume that the level of significance is 5%.

Solution

In this case the null hypothesis is $\beta_1 = 0$, and the alternative hypothesis is $\beta_1 \neq 0$. Since $n = 15$, we will have a t-distribution with $15 - 2$ or 13 degrees of freedom. From Table VII in the Appendix, the appropriate $t_{\alpha/2} = t_{0.05/2} = t_{0.025}$ value is 2.160. Now we compute the value of the test statistic. We have

$$ t = \frac{b_1}{s_e \Big/ \sqrt{\Sigma x^2 - \dfrac{(\Sigma x)^2}{n}}} $$

All the values of the variables needed to use this formula have been found earlier. Thus, we know that $b_1 = 10.857$ (computed on page 557), $\Sigma x^2 = 93.4375$

(computed on page 556), $\Sigma\ x = 33.75$ (computed on page 556), and $s_e = 3.01$ (computed on page 570). Also, $n = 15$ so that

$$t = \frac{10.857}{3.01 \Big/ \sqrt{93.4375 - \dfrac{(33.75)^2}{15}}} = 15.089$$

Since the value of the test statistic 15.089 falls in the rejection region of Figure 11.15, we reject H_0 and conclude that the slope of the population regression line is not 0. Hence, x (number of hours studied) can be used as a predictor of y.

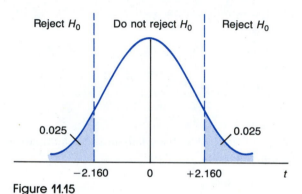

Reject H_0 Do not reject H_0 Reject H_0

0.025   0.025

-2.160 0 $+2.160$ t

Figure 11.15

EXAMPLE 2

Refer back to the data of Example 2 on page 558. Do the data indicate that the value of β_1, that is, the slope of the population regression line, is not zero, which would mean that x (the amount spent on advertising) can be used as a predictor of y (sales)? Use a 5% level of significance.

Solution

The null hypothesis is $\beta_1 = 0$, and the alternative hypothesis is $\beta_1 \neq 0$. Since $n = 10$, we have a t-distribution with $10 - 2$ or 8 degrees of freedom. From Table VII, the $t_{0.025}$ value is 2.306. We then have the acceptance-rejection region shown in Figure 11.16. The value of the test statistic is

$$t = \frac{b_1}{s_e \Big/ \sqrt{\Sigma\ x^2 - \dfrac{(\Sigma\ x)^2}{n}}} = \frac{1.684}{1.55 \Big/ \sqrt{18{,}416 - \dfrac{(426)^2}{10}}} = 17.799$$

where the value of $b_1 = 1.684$ was computed on page 558, the value of $\Sigma\ x^2 = 18{,}416$ was computed on page 558, the value of $\Sigma\ x = 426$ was computed on page 558 and the value of $s_e = 1.55$ was computed on page 571. Since the value of the test statistic 17.799 falls in the rejection region of Figure 11.16, we reject H_0 and conclude

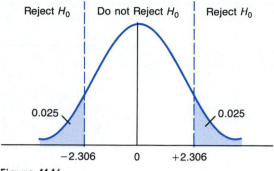

Reject H_0 | Do not Reject H_0 | Reject H_0

0.025

0.025

-2.306 0 $+2.306$

Figure 11.16

that the slope of the population regression line is not 0. Hence, x (the amount spent on advertising) can be used as a predictor of y (sales).

COMMENT Since all the computations have been done already, we can also determine a confidence interval for the slope β_1 of the population regression line. The end points of the confidence interval are

$$b_1 \pm t_{\alpha/2} \cdot \frac{s_e}{\sqrt{\Sigma x^2 - \dfrac{(\Sigma x)^2}{n}}}$$

The interested reader should actually verify that a 95% confidence interval for β_1 of Example 1 is $10.857 \pm 2.160 \left(\dfrac{3.01}{\sqrt{17.5}} \right)$ or from 9.303 to 12.411. Thus, we can be 95% confident that β_1 is somewhere between 9.303 and 12.411.

Prediction Intervals

After we have determined the least-squares prediction equation for some data, we may wish to set up a prediction interval for a population value of y that we are predicting, corresponding to some particular value of x. Under such circumstances we proceed as follows:

1. Find $n - 2$. This gives us the number of degrees of freedom for a t-distribution.
2. Find the appropriate $t_{\alpha/2}$ values by using Table VII in the Appendix.
3. Find the least-squares prediction equation, $\hat{y}$, by using Formula 11.2. Use it to compute the predicted y-value, $\hat{y}$, corresponding to some particular x_p.
4. Find the standard error of the estimate, s_e, by using Formula 11.5.
5. Set up the appropriate prediction interval by using Formula 11.6, as follows.

FORMULA 11.6   A prediction interval for a particular y corresponding to some particular value of $x = x_p$ is

$$\text{Lower boundary:}\quad \hat{y}_p - t_{\alpha/2} \cdot s_e \cdot \sqrt{1 + \frac{1}{n} + \frac{n(x_p - \bar{x})^2}{n(\Sigma\, x^2) - (\Sigma\, x)^2}}$$

$$\text{Upper boundary:}\quad \hat{y}_p + t_{\alpha/2} \cdot s_e \cdot \sqrt{1 + \frac{1}{n} + \frac{n(x_p - \bar{x})^2}{n(\Sigma\, x^2) - (\Sigma\, x)^2}}$$

where $t_{\alpha/2}$ represents the t-distribution value obtained from Table VII using $n - 2$ degrees of freedom, s_e is the standard error of the estimate, and $\hat{y}_p$ is the predicted value of y corresponding to $x = x_p$.

Prediction Intervals   **COMMENT**   The preceding intervals are frequently called **prediction intervals** because they give intervals for future values of y at a specified x.

We illustrate the use of Formula 11.6 with several examples.

EXAMPLE 3   Using the data of Example 1 on page 556, find a 95% prediction interval for the score of a student who studies for 0.85 hours.

Solution

The least-squares prediction equation was already calculated. It is

$$\hat{y} = 54.772 + 10.857x$$

Also, when $x_p = 0.85$, then the predicted value of y is $\hat{y}_p = 64$. This was calculated earlier. The standard error of the estimate was calculated on page 570. It is 3.01. In this case $n = 15$, so we have a t-distribution with $15 - 2$, or 13 degrees of freedom. From Table VII, the appropriate $t_{\alpha/2} = t_{0.05/2} = t_{0.025}$ value is 2.160. Now we apply Formula 11.6. We get

$$\text{Lower boundary} = \hat{y}_p - t_{\alpha/2} \cdot s_e \cdot \sqrt{1 + \frac{1}{n} + \frac{n(x_p - \bar{x})^2}{n(\Sigma\, x^2) - (\Sigma\, x)^2}}$$

$$= 64 - (2.160)(3.01) \cdot \sqrt{1 + \frac{1}{15} + \frac{15(0.85 - 2.25)^2}{15(93.4375) - (33.75)^2}}$$

$$= 64 - (2.160)(3.01)\sqrt{1 + 0.067 + 0.112}$$

$$= 64 - (2.16)(3.01)(1.086)$$

$$= 64 - 7.061 = 56.939$$

$$\text{Upper boundary} = y_p + t_{\alpha/2} \cdot s_e \cdot \sqrt{1 + \frac{1}{n} + \frac{n(x_p - \bar{x})^2}{n(\Sigma \, x^2) - (\Sigma \, x)^2}}$$

$$= 64 + (2.160)(3.01)\sqrt{1 + \frac{1}{15} + \frac{15(0.85 - 2.25)^2}{15(93.4375) - (33.75)^2}}$$

$$= 64 + (2.160)(3.01)(1.086)$$

$$= 64 + 7.061 = 71.061$$

Thus, a 95% prediction interval for the score of a student who studies for 0.85 hours is 56.939 to 71.061.

EXAMPLE 4

Using the data of Example 2 on page 558 find a 95% prediction interval of the predicted sales of the publishing company if it spends $50,000 for advertisement next month.

Solution

The least-squares prediction equation was already calculated. It is

$$\hat{y} = 1.762 + 1.684x$$

When $x_p = 50$, $\hat{y}_p = 85.962$. The standard error of the estimate, s_e, was calculated on page 571 and is 1.55. In this case $n = 10$ so that we have a t-distribution with $10 - 2$, or 8 degrees of freedom. From Table VII the $t_{0.025}$ value is 2.306. Applying Formula 11.6 gives

$$\text{Lower boundary} = \hat{y}_p - t_{\alpha/2} \cdot s_e \cdot \sqrt{1 + \frac{1}{n} + \frac{n(x_p - \bar{x})^2}{n(\Sigma \, x^2) - (\Sigma \, x)^2}}$$

$$= 85.962 - (2.306)(1.55)\sqrt{1 + \frac{1}{10} + \frac{10(50 - 42.6)^2}{10(18,416) - (426)^2}}$$

$$= 85.962 - (2.306)(1.55)\sqrt{1 + 0.1 + 0.204}$$

$$= 85.962 - (2.306)(1.55)(1.142)$$

$$= 85.962 - 4.082 = 81.88$$

$$\text{Upper boundary} = \hat{y}_p + t_{\alpha/2} \cdot s_e \cdot \sqrt{1 + \frac{1}{n} + \frac{n(\Sigma \, x_p - \bar{x})^2}{n(\Sigma \, x^2) - (\Sigma x)^2}}$$

$$= 85.962 + (2.306)(1.55)\sqrt{1 + \frac{1}{10} + \frac{10(50 - 42.6)^2}{10(18,416) - (426)^2}}$$

$$= 85.962 + (2.306)(1.55)(1.142)$$

$$= 85.962 + 4.082 = 90.044$$

Thus, a 95% prediction interval for the predicted sales if the company spends $50,000 for advertisement next month is $81.88 million to $90.044 million (assuming that all other factors can be neglected).

EXERCISES FOR SECTION 11.8

For each of the following, test the null hypothesis that $\beta_1 = 0$ and then find a 95% prediction interval for the indicated value.

1. The predicted height of a son when his father is 76 inches tall in Exercise 1 on page 561
2. The predicted number of millions of gallons of water to be added to the reservoirs when 2.2 inches of rain fall in Exercise 2 on page 561
3. The predicted volume of sales when the department store chain spends $70,000 on advertising in Exercise 3 on page 562
4. The predicted number of parking tickets issued when 11 meter maids are hired in Exercise 4 on page 562
5. The predicted height of a man when his wife is 70 inches tall in Exercise 5 on page 563
6. The predicted number of centimeters of growth for a plant when eight units of the chemical are added to it in Exercise 6 on page 563
7. The predicted number of heartbeats per minutes for a subject who was given coffee with eight units of caffeine added to it in Exercise 7 on page 564
8. The predicted annual expenditure for entertainment by a family whose annual income is $35,000 in Exercise 8 on page 564
9. The predicted amount of money that will be claimed for charitable deductions by a person whose annual income is $33,000 in Exercise 9 on page 564
10. The predicted batting average of a baseball player who is earning $975,000 annually in Exercise 10 on page 565
11. The predicted sales at the on-site premises gift shop when 600,000 free samples are distributed in Exercise 11 on page 565

11.9 THE RELATIONSHIP BETWEEN CORRELATION AND REGRESSION

Although it may seem to you that linear correlation and regression are very similar since many of the computations performed in both are the same, the two ideas are quite different. One uses the correlation coefficient (Formula 11.1) to determine whether two variables are linearly related. The correlation coefficient measures the strength of the linear relationship. Regression analysis, on the other hand, is used when we want to answer questions about the relationship between two variables. Just exactly how the two variables are related (that is, can we find an equation connecting the variables) requires regression analysis. The equation connecting the variables may not necessarily be linear.

Extrapolation

Another factor that has to be considered involves **extrapolation.** Two variables may have a high positive correlation. Yet regression analysis *cannot* be used to make predictions for new x-values that are far removed from the original data. Such uses of regression are called extrapolation. To illustrate, in Example 1 on page 556 we determined that the regression equation or the least-squares prediction equation was $\hat{y} = 54.772 + 10.857x$. This equation *cannot* be used to make predictions for values of x that are far removed from the values of x between 0.50 and 4.00.

*11.10 MULTIPLE REGRESSION

Until now we have been interested in finding a linear equation connecting two variables. However, there are many practical situations where several variables may simultaneously affect a given variable. For example, the yield from an acre of land may depend on, among other things, such variables as the amount of fertilizer used, the amount of rainfall, the amount of sunshine, and so on.

There are many formulas that can be used to express relationships between more than two variables. The most commonly used formulas are linear equations of the form

$$\hat{y} = b_0 + b_1x_1 + b_2x_2 + \cdots + b_mx_m$$

The main difficulty in deriving a linear equation in more than two variables that best describes a given set of data is that of determining the values of b_0, b_1, b_2, ..., b_m. When there are two independent variables x_1 and x_2 and the linear equation connecting them is of the form $y = b_0 + b_1x_1 + b_2x_2$, then we can apply the method of least-squares. This means that we must solve the following equations simultaneously.

FORMULA 11.7
Multiple Regression Formula

Multiple regression formulas

$$\Sigma\, y = n \cdot b_0 + b_1(\Sigma\, x_1) + b_2(\Sigma\, x_2)$$

$$\Sigma\, x_1 y = b_0(\Sigma\, x_1) + b_1(\Sigma\, x_1^2) + b_2(\Sigma\, x_1 x_2)$$

$$\Sigma\, x_2 y = b_0(\Sigma\, x_2) + b_1(\Sigma\, x_1 x_2) + b_2(\Sigma\, x_2^2)$$

Solving these equations usually involves a lot of computation. Nevertheless, we illustrate the procedure with an example.

EXAMPLE 1

The following data give the yield, y, per plot of land depending on the quantity of fertilizer used, x_1, and the number of inches of rainfall, x_2. Compute a linear equation that will enable us to predict the average yield, y, per plot in terms of the quantity of fertilizer used, x_1, and the number of inches, x_2, of rainfall.

Yield per Plot (hundreds of bushels) y	Quantity of Fertilizer Used (units) x_1	Rainfall (inches) x_2
20	2	5
25	3	9
28	5	14
30	7	15
32	11	23

Solution

We arrange the data in the following tabular format.

x_1	x_2	y	x_1y	x_2y	x_1^2	x_1x_2	x_2^2
2	5	20	40	100	4	10	25
3	9	25	75	225	9	27	81
5	14	28	140	392	25	70	196
7	15	30	210	450	49	105	225
11	23	32	352	736	121	253	529
28	66	135	817	1903	208	465	1056

$\Sigma x_1 = 28$ | $\Sigma x_2 = 66$ | $\Sigma y = 135$ | $\Sigma x_1y = 817$ | $\Sigma x_2y = 1903$ | $\Sigma x_1^2 = 208$ | $\Sigma x_1x_2 = 465$ | $\Sigma x_2^2 = 1056$

We now substitute these values into the equations given in Formula 11.7. Here $n = 5$. We get

$$135 = 5b_0 + 28b_1 + 66b_2$$

$$817 = 28b_0 + 208b_1 + 465b_2$$

$$1903 = 66b_0 + 465b_1 + 1056b_2$$

This represents a system of three equations in three unknowns. If we solve these equations simultaneously, we get $b_0 = 17.4459$, $b_1 = -0.7504$, and $b_2 = 1.0422$. Thus, the least-squares prediction equation is

$$\hat{y} = 17.4459 - 0.7504x_1 + 1.0422x_2$$

When the farmer uses four units of fertilizer and there are 9 inches of rain, then the predicted yield per plot is

$$y = 17.4459 - 0.7504(4) + 1.0422(9)$$

or approximately 23.8241 hundreds of bushels.

EXERCISES FOR SECTION 11.10

1. A laundary detergent manufacturer would like to investigate the effects of television (TV) advertising and magazine advertising on weekly gross revenue. The advertising department has analyzed the problem and has presented the following data based on past company records.

Magazine Advertising (in thousands of dollars), x_1	TV Advertising (in thousands of dollars), x_2	Weekly Gross Revenue (in thousands of dollars), y
1.6	5.1	98
2.1	2.1	92
1.6	4.1	97
2.6	2.6	93
3.4	3.1	97
2.4	3.6	96
4.3	2.6	95

a. Determine the least-squares prediction equation.
b. What is the predicted gross revenue when magazine advertising and TV advertising are 1.8 and 2.8, respectively?

2. A real estate appraiser believes that the selling price of a house is related to the size of the building and the size of the lot on which the building stands. The following data are available.

Building Size (in hundreds of sq. ft.), x_1	Lot Size (in thousands of sq. ft.), x_2	Selling Price (in thousands of dollars), y
22	22	48
17	24	40
18	8	29
15	10	35
20	12	37
19	46	52
24	13	56

a. Determine the least-squares prediction equation.
b. What is the predicted selling price of a house whose building size and lot size are 21 hundred square feet and 17 thousand square feet, respectively?

3. The resale value of a car depends among other things on the car's age and the

number of miles indicated on the car's odometer. For one particular model car and geographic area the following selling prices were reported.

Car's Age (in years), x_1	Number of Miles on Odometer, x_2	Selling Price, y
1	14,000	$8800
2	23,000	8000
3	30,000	7100
4	40,000	6100
5	55,000	4900
6	68,000	3700

a. Determine the least-squares prediction equation.
b. What is the predicted selling price of a car that is $2\frac{1}{2}$ years old and whose odometer reading is 45,000 miles?

11.11 USING COMPUTER PACKAGES

Computer packages are ideally suited for drawing scatter diagrams, computing correlation coefficients, and deriving regression equations. The computer output for such data also provides an analysis of variance, which is a topic that we will discuss in a later chapter.

The data from Example 2 of Section 11.6 (page 558) are used to illustrate how MINITAB handles correlation and regression problems.

```
MTB > READ ADVERTISING IN C1, SALES IN C2
DATA > 43    74
DATA > 44    76
DATA > 36    60
DATA > 38    68
DATA > 47    79
DATA > 40    70
DATA > 41    71
DATA > 54    94
DATA > 37    65
DATA > 46    78
DATA > END


       10 ROWS READ


MTB > PLOT SALES IN C2 VS ADVERTISING IN C1
```

This MINITAB program produces the following scatter diagram.

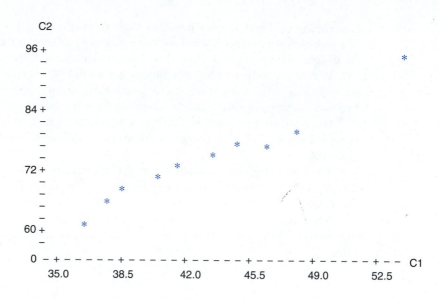

```
MTB > CORRELATION COEFFICIENT BETWEEN ADVERTISING IN C1 AND SALES IN C2
      CORRELATION OF C1 AND C2 = 0.988

MTB > REGRESS SALES IN C2 ON 1 PREDICTOR ADVERTISING IN C1

The regression equation is
C2 = 1.76 + 1.68 C1
Predictor      Coef     Stdev    t-ratio       p
Constant      1.759     4.069       0.43    0.677
C1          1.68405   0.09483      17.76    0.000

s = 1.554   R-sq = 97.5%   R-sq(adj) = 97.2%

Analysis of Variance


SOURCE       DF        SS        MS        F        p
Regression    1    761.19    761.19   315.39    0.000
Error         8     19.31      2.41
Total         9    780.50

MTB > STOP
```

11.12 SUMMARY

In this chapter we analyzed the relationship between two variables. Scatter diagrams were drawn that help us to understand this relationship. We discussed the concept of correlation coefficients, which tell us the extent to which two variables are related.

Correlation coefficients vary between the values of -1 and $+1$. A value of $+1$ or -1 represents a perfect linear relationship between the two variables. A correlation coefficient of 0 means that there is no linear relationship between the two variables.

We indicated how to test whether or not a value of r is significant. Furthermore, we mentioned that even when there is an indication of positive correlation between two variables, great care must be shown in how we interpret this relationship.

Once we determine that there is a significant linear correlation between two variables we find the least-squares equation, which expresses this relation mathematically.

We discussed the standard error of the estimate. This is a way of measuring how well the estimated least-squares regression line really fits the data. The smaller the value of s_e is, the better the estimate will be. We also indicated how to set up prediction intervals for values of y obtained through regression analysis. Finally, we discussed multiple regression where the value of one variable depends upon several other variables.

Study Guide

The following is a chapter summary in capsule form. You should now be able to demonstrate your knowledge of ideas mentioned by giving definitions, descriptions, or specific examples. Page references are given in parentheses.

The **correlation coefficient** is a number that measures the relationship between two sets of scores. (page 535)

To analyze the type of relationship that exists between two variables, we draw a **scatter diagram** where one of the variables is pictured on the horizontal axes and the other is pictured on the vertical axis. (page 537)

If the dots form an approximate straight line, then we say that there is a **linear correlation** between the variables. (page 537)

If the line moves in a direction that is from lower left to upper right, then we have **positive correlation**. If the line moves in a direction that is from upper left to lower right, then we have **negative correlation**. (page 537)

The **coefficient of linear correlation** measures the strength of a relationship between two variables. (page 539)

The **reliability of a particular value of r** can be determined by using Table V in the appendix. (page 549)

The **level of significance** of the reliability of r is indicated by writing r with an appropriate subscript. (page 550)

An **estimated regression line** is a line that best represents the relationship between the two variables. (page 552)

Regression methods involve fitting a straight line or a curve to a set of numbers. (page 552)

Linear regression involves fitting a straight line to a set of numbers. (page 552)

Curvilinear regression involves fitting a curve to a set of numbers. The variable to be predicted is called the **dependent variable** (y) and the known variable is called the **independent variable** (x). (page 552)

The difference between any point and the corresponding point on the regression line is called the (vertical) **deviation** from the line. (page 555)

The **least-squares method** determines the estimated regression line in such a way that the sum of the squares of these vertical deviations is as small as possible. (page 555)

The straight line that best fits a set of data points according to the least-square criterion is called the **regression line**, whereas the equation of the regression line is called the **regression equation**. (page 556)

The equation of the regression line can also be determined by using a formula to calculate the **sample covariance**. (page 559)

For each x-value the mean of the corresponding population of y-values lies on a straight line called the **population regression line**, whose equation is $y = \alpha + \beta x$. (page 567)

The **error terms** (the vertical distance between the predicted y-values and the true population values) are normally distributed with mean 0 and standard deviation σ. The estimated value of σ is called the **standard error of the estimate** or the **residual standard deviation**. (page 567)

The **total sum of squares**, **SST**, representing the total squared error that is made when we predict y for the observed y-values is given by $SST = \Sigma(y - \bar{y})^2$. (page 568)

The **error sum of squares**, **SSE**, representing the total squared error that is made when using the equation of the regression line values for prediction is given by $SSE = \Sigma(y - \hat{y})^2$. (page 569)

The **percentage reduction** in the total squared error obtained by using the regression equation instead of y is denoted by r^2 and is called the **coefficient of determination** (page 569)

After we have determined the least-squares prediction equation for some data we can set up a **prediction interval** for a population value of y that we are predicting, corresponding to some particular value of x. (page 576)

Two variables may have a high positive correlation. Yet regression analysis cannot be used to make predictions using x-values that are far removed from the original data. Such uses of regression are called **extrapolation**. (page 579)

When there are two or more independent variables x_1, x_2, . . ., and the linear equation connecting them is of the form $y = b_0 + b_1x_1 + b_2x_2 + \ldots$, we must use **multiple regression formulas**. (page 579)

Formulas to Remember

You should be able to identify each symbol in the following formulas, understand the relationships among the symbols expressed in each formula, understand the significance of each formula, and use the formulas in solving problems.

1. Coefficient of linear correlation:

$$r = \frac{n(\Sigma\ xy) - (\Sigma\ x)(\Sigma\ y)}{\sqrt{n(\Sigma\ x^2) - (\Sigma\ x)^2}\ \sqrt{n(\Sigma\ y^2) - (\Sigma\ y)^2}}$$

2. Estimated regression line:

$$\hat{y} = b_0 + b_1 x$$

where

$$b_1 = \frac{n(\Sigma\ xy) - (\Sigma\ x)(\Sigma\ y)}{n(\Sigma\ x^2) - (\Sigma\ x)^2} \quad \text{and} \quad b_0 = \frac{1}{n}(\Sigma\ y - b_1 \cdot \Sigma\ x)$$

3. Sample covariance $= s_{xy} = \dfrac{\Sigma(x - \bar{x})(y - \bar{y})}{n - 1}$

4. Equation of regression line (alternate version):

$$\hat{y} = b_0 + b_1 x$$

where

$$b_1 = \frac{s_{xy}}{s_x^2}, \quad b_0 = \bar{y} - b_1\bar{x},$$

and s_x is the sample standard deviation of x-values

5. Coefficient of linear correlation (alternate version):

$$r = \frac{s_{xy}}{s_x s_y}$$

6. Total sum of squares, $SST = \Sigma\ (y - \bar{y})^2$
7. Error sum of squares, $SSE = \Sigma\ (y - \hat{y})^2$
8. Coefficient of determination or percentage reduction:

$$r^2 = \frac{SST - SSE}{SST} = 1 - \frac{SSE}{SST}$$

9. Estimate of the common standard deviation, σ, or standard error of the estimate:

$$s_e = \sqrt{\frac{\Sigma(y - \hat{y})^2}{n - 2}} = \sqrt{\frac{SSE}{n - 2}} \quad \text{or} \quad \sqrt{\frac{\Sigma y^2 - b_0(\Sigma xy) - b_1(\Sigma y)}{n - 2}}$$

10. To test whether or not $\beta_1 = 0$, the test statistic is

$$t = \frac{b_1}{s_e \bigg/ \sqrt{\Sigma x^2 - \frac{(\Sigma x)^2}{n}}}$$

11. End points of prediction interval for β_1:

$$b_1 \pm t_{\alpha/2} \cdot \frac{s_e}{\sqrt{\Sigma x^2 - \frac{(\Sigma x)^2}{n}}}$$

12. Prediction interval for y corresponding to some given value of $x = x_p$:

$$\text{Lower boundary: } \hat{y}_p - t_{\alpha/2} \cdot s_e \sqrt{1 + \frac{1}{n} + \frac{n(x_p - \bar{x})^2}{n(\Sigma x^2) - (\Sigma x)^2}}$$

$$\text{Upper boundary: } \hat{y}_p + t_{\alpha/2} \cdot s_e \sqrt{1 + \frac{1}{n} + \frac{n(x_p - \bar{x})^2}{n(\Sigma x^2) - (\Sigma x)^2}}$$

13. Multiple regression: $\hat{y} = b_0 + b_1 x_1 + b_2 x_2$, where

$$\Sigma y = n \cdot b_0 + b_1(\Sigma x_1) + b_2(\Sigma x_2)$$

$$\Sigma x_1 y = b_0(\Sigma x_1) + b_1(\Sigma x_1^2) + b_2(\Sigma x_1 x_2)$$

$$\Sigma x_2 y = b_0(\Sigma x_2) + b_1(\Sigma x_1 x_2) + b_2(\Sigma x_2^2)$$

Testing Your Understanding of This Chapter's Concepts

1. Where $r = 0$, what is the value of $\sqrt{\dfrac{\Sigma(y - \hat{y})^2}{n - 2}}$?

    a. 1 **b.** -1 **c.** 0 **d.** a very large number **e.** none of these

2. A correlation coefficient of -0.97
    a. indicates a strong negative correlation
    b. indicates a weak negative correlation
    c. is insignificant
    d. is impossible
    e. none of these

3. In many respects the standard error of the estimate very closely resembles the standard deviation. Do you agree with this statement? Explain.
4. When $r = \pm 1.00$, what is the value of

$$\sqrt{\frac{\Sigma(y - \hat{y})^2}{n - 2}}$$

THINKING CRITICALLY

1. Verify that Formulas 11.2 and 11.3 are equivalent. Hence, either can be used to calculate the equation of the regression line.
2. Show that the formula for calculating the linear correlation ocefficient (Formula 11.1) and the following computational formula are equivalent.

$$r = \frac{s_{xy}}{s_x s_y}$$

where

s_{xy} = the sample covariance
s_x = the sample standard deviation of the x-values
s_y = the sample standard deviation of the y-values

(*Hint:* Verify the following identity:

$$\frac{s_{xy}}{s_x s_y} = \frac{n(\Sigma\ xy) - (\Sigma\ x)(\Sigma\ y)}{\sqrt{n(\Sigma\ x^2) - (\Sigma\ x)^2}\ \sqrt{n(\Sigma\ y^2) - (\Sigma\ y)^2}}$$

3. Why is correlation analysis included in the study of linear regression? Can we study these two subjects independently?

Review Exercises for Chapter 11

1. Refer to the information presented in the graphs on top of the next page. A survey was conducted to determine if the sex of a respondent affects the person's response. In each case the person was asked the same question.
   a. Draw a scatter diagram for the data. (Let x = male percentage and y = female percentage.)
   b. Calculate the coefficient of correlation.

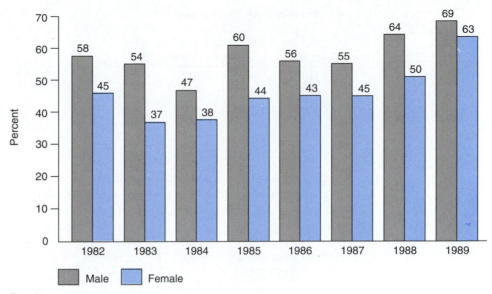

Question:
"Are you in favor of the death penalty for persons convicted of murder?"
Source: Sourcebook of Criminal Justice Statistics, Washington, D.C., 1990.

2. The number of pints of blood in storage on any given day at Willow Hospital depends on the number of operations scheduled for that day as shown here.

Number of Operations Scheduled, x	Number of Pints of Blood in Storage, y
10	98
12	111
13	128
15	175
16	221
17	281

    a. Find the least-squares prediction equation.
    b. If 14 operations are scheduled for tomorrow, find the predicted number of pints of blood in storage.
3. Refer back to Question 2. Find a 95% prediction interval for the predicted number of pints of blood needed when 14 operations are scheduled.
4. A leading psychologist believes that IQ scores and the number of hours per week that a child watches television are related. The following data was collected for

six students to verify this claim:

IQ Scores, x	Number of Hours Spent Watching TV, y
110	12
105	14
115	18
121	8
100	16
90	20

Find the least-squares prediction equation.

5. In a certain school district the absentee rate for kindergarten classes depends on the average outside temperature reading as follows:

Outside Temperature Reading (in degrees), x	Absentee Rate (%), y
30	7
25	9
20	12
15	16
10	20
5	24
0	30

Find the least-squares prediction equation.

6. Mary has determined that the number of miles per gallon of gas that she gets depends on the extra weight of the objects in the trunk of her car as shown below.

Weight of Objects in Trunk (in pounds), x	Miles per Gallon, y
30	24
50	23
80	21
110	19
120	17
140	15

Find the least-squares prediction equation.

7. The following is a list of the grades of five students on a special math aptitude test, the IQ scores of the students, and the number of hours that each prepared for the exam.

Grade on Exam, y	IQ Score, x_1	Preparation for Exam (in hours), x_2
84	120	20
88	130	19
81	118	17
92	142	21
78	116	12

a. Determine the least-squares prediction equation.

b. What is the predicted grade on the exam for a student who has an IQ score of 128 and who studies 18 hours preparing for the exam?

8. What type of correlation do you believe exists between the incidence of heart attacks and overweight?

   a. positive correlation

   c. zero correlation

   b. negative correlation

   d. none of these

9. A fire department official believes that as the temperature decreases the number of fires increases. To support this claim, the official has obtained the following statistics.

Temperature (in degrees), x	40	35	30	25	20	15	10	5
Number of Fires, y	33	37	40	44	56	60	61	71

a. Draw a scatter diagram for the data and then compute the coefficient of correlation.

b. Do these data support the fire official's claim?

10. The value of a car, as given in a used car prices manual, is said to be determined among other things by the age of the car. The following data are available on a particular car:

Age of Car (in years), x	Value of Car (in dollars), y
$\frac{1}{2}$ year	12,000
1 year	9,500
$1\frac{1}{2}$ years	9,100
2 years	8,500
$2\frac{1}{2}$ years	8,100
3 years	7,500
4 years	7,000
5 years	6,000

a. Find the least-squares prediction equation.

b. For this car, what is the predicted value of the car when it is $3\frac{1}{2}$ years old?

Chapter Test

1. An industrial engineer believes that there is a relationship between the number of defective items produced by a machine and the speed at which the machine operates. The following data have been collected.

Speed of Machine (rps), x	Number of Defective Items Produced (per hour), y
8	5
10	7
12	10
14	14
16	18
18	22

a. Compute the least-squares prediction equation.

b. If the machine is operated at a speed of 15 rps (revolutions per second), what is the predicted number of defective items to be produced by the machine?

2. *Teacher's salaries and alcohol:* A student collected the following data for a certain state.

Year	Average Teacher's Salary (in thousands of dollars), x	Sales of Alcohol (in millions of dollars), y
1986	30	15
1987	34	19
1988	35	20
1989	38	22
1990	39	24
1991	42	28

Compute the coefficient of correlation for these data. Comment.

3. The following table shows the age of drivers in a certain town and the average number of automobile accidents in which they were involved in a given month.

Age of Driver (in years), x	Average Number of Accidents, y
18	2
19	3
20	1
21	1
22	0
23	2
24	1
25	2
26	0
27	1

    a. Find the least-squares prediction equation.

    b. Find the predicted average number of accidents per month for a driver who is $21\frac{1}{2}$ years old.

4. The following table shows the average weights of boys in different age groups at a certain school.

Age (in years), x	Weights (in pounds), y
5	54
6	57
7	61
8	63
9	70
10	76
11	84
12	91
13	104
14	121

    a. Compute the coefficient of correlation for the preceding data.

    b. Compute the least-squares prediction equation.

    c. Compute the standard error of the estimate.

    d. What is the predicted weight of a boy who is $10\frac{1}{2}$ years old at this school?

5. Wilt is on the college basketball team. The following chart indicates the number of points that Wilt scored in the last six games and also the number of hours per week that he practiced before each game.

Number of Practice Hours, x	Number of Points Scored, y
9	29
7	25
5	20
10	33
9	28
8	27

Find the least-squares prediction equation.

6. A psychologist believes that how many times humans correctly perform a particular task and the differing levels of noise intensity under which these tasks are performed are related, as shown in the following chart.

Noise Intensity, x	Number of Times Task Was Correctly Performed, y
4	25
9	46
6	34
8	41
12	58
11	50
3	20

   a. Determine the least-squares prediction equation.
   b. What is the predicted number of tasks that can be correctly performed when the noise intensity is 10?

7. Refer back to Question 6. Find a 95% confidence interval for the predicted number of tasks that can be performed correctly when the noise intensity is 10.

8. A city official believes that the number of complaints to the city's heat complaint control board (for the lack of heat) is directly related to the outdoor temperature. The following data have been collected.

Outdoor Temperature (in degrees Fahrenheit) x	Number of Complaints to Heat Control Board y
30	39
25	44
20	65
15	73
10	80
5	89
0	101

   a. Determine the least-squares prediction equation.
   b. What is the predicted number of complaints to the city's heat complaint control board when the outdoor temperature drops to 12° Fahrenheit?

9. The operator of the coliseum claims that the average daily attendance at the automobile show is related to the admission price charged as shown here.

Admission Price (per adult) (in dollars), x	Average Daily Attendance, y
7.00	20,000
8.00	17,500
9.00	16,000
10.00	14,500
12.00	13,000
15.00	12,000

   a. Determine the least-squares prediction equation.
   b. What is the predicted average daily attendance at the auto show when the admission price is $13.00 per adult?

10. An auto mechanic believes that the estimated overall gas mileage, y, depends on the car engine's size and on the number of cylinders in the engine. The following data are available for seven cars:

Engine Size (in liters), x_1	Number of Cylinders x_2	Estimated Gas Mileage (in miles per gallon), y
2.5	4	28
4.2	6	19
4.4	8	21
1.6	4	31
5.8	8	19
3.8	6	23
5.0	8	21

   Determine the least-squares prediction equation.
11. Refer back to Question 10. What is the estimated gas mileage for a car that has 4 cylinders and whose engine size is 2.3?

For questions 12–18 use the following.

   A large corporation has eight branch offices in different parts of the United States. It has gathered the following information concerning the number of secretaries and the number of employees (excluding secretaries) at seven of its branch offices:

Branch	1	2	3	4	5	7	8
Number of employees, x	650	730	810	900	1020	1070	1150
Number of secretaries, y	40	50	54	61	82	110	121

12. Compute the correlation of coefficient for the data.

    a. 0.951   **b.** 0.953   **c.** 0.957   **d.** 0.983   **e.** none of these

13. Determine whether r is significant.

    a. yes   **b.** no   **c.** not enough information given   **d.** none of these

14. Find the least-squares prediction equation.

    a. $y = 0.162 - 72.37x$ **b.** $y = -72.37 + 0.162x$
    c. $y = 72.37 - 0.162x$ **d.** $y = -0.162 + 72.37x$ **e.** none of these

15. If Branch 6 employs 1050 employees, what is the predicted number of secretaries at this branch?

    a. 99.17   **b.** 95.34   **c.** 97.73   **d.** 91.46   **e.** none of these

16. What is the value of the standard error of the estimate?

    a. 8.432   **b.** 3.845   **c.** 6.412   **d.** 9.945   **e.** none of these

17. What is the value of the coefficient of determination for the data?

    a. 0.916   **b.** 0.961   **c.** 0.248   **d.** 0.769   **e.** none of these

18. At the 5% level of significance, does the data indicate that the number of employees can be used as a predictor of the number of secretaries?

    a. yes   **b.** no   **c.** not enough information given   **d.** none of these

1. Draper, N., and H. Smith. *Applied Regression Analysis*, 2nd ed. New York: Wiley, 1981.
2. Hartwig, F., and B. E. Dearing. *Exploratory Data Analysis*. Beverly Hills, CA: Sage, 1979.
3. Mendenhall, W., and J. T. McLave. *A Second Course in Business Statistics: Regression Analysis*, 2nd ed. San Francisco: Dellen, 1986.
4. Neter, J., W. Wasserman and M. H. Kutner. Applied Linear Regression Models. Homewood, IL: Irwin, 1983.
5. Weisberg, S. *Applied Linear Regression*. New York: Wiley, 1980.
6. Younger, M. S. *A Handbook for Linear Regression*, North Scituate, MA: Duxbury, 1979.

12 Analyzing Count Data: The Chi-Square Distribution

Is is true that all booths in a toll booth plaza are used with the same frequency? See discussion on p. 614. (© *Fran Heyl Associates*)

Until now we have been using statistical inference procedures for hypothesis tests and confidence intervals involving means and proportions. In this chapter we apply hypothesis-testing procedures to determine whether two characteristics of a population as given in a contingency table are statistically independent or whether observed data follow some pattern. This is referred to as goodness of fit. The chi-square distribution plays a key role in such analyses.

CHAPTER OBJECTIVES

After studying the material in this chapter, you should be able:

- **To apply** chi-square tests that provide the basis for testing whether more than two population proportions can be considered as equal. (Section 12.2)

- **To discuss** a chi-square test statistic that can be applied to determine whether an observed frequency is in agreement with the expected mathematical distribution. This is known as the goodness of fit. (Section 12.3)

- **To analyze** contingency tables. These are tabular arrangements of data into a two-way classification. The chi-square test statistic tells us whether the two ways of classifying the data are independent. (Section 12.4)

- **To compute** expected frequencies. These are the numbers, that is, the frequencies, that should appear in each of the boxes of a contingency table. (Section 12.4)

Statistics in the News

THE ARMY'S NEW PHYSICAL FITNESS STANDARDS

Age Groups	Push-ups Min.	Push-ups Max.	Sit-ups Min.	Sit-ups Max.	Two-mile run Min.	Two-mile run Max.
17–21	42	82	52	92	15:54	11:45
	18	58	50	90	18:45	14:45
22–26	40	80	47	87	16:36	12:36
	16	56	45	85	19:36	15:36
27–31	38	78	42	82	17:18	13:18
	15	54	40	80	21:00	17:00
32–36	33	73	38	78	18:00	14:00
	14	52	35	75	22:36	18:36
37–41	32	72	33	73	18:42	14:42
	13	48	30	70	22:36	19:36
42–46	26	66	29	69	19:06	15:12
	12	45	27	67	24:00	20:00
47–51	22	62	27	67	19:36	15:36
	10	41	24	64	24:30	20:30
52 & over	16	56	26	66	20:00	16:00
	09	40	22	62	25:00	21:00

■ MALE ■ FEMALE

Source: Office of Deputy Chief of Staff for Operations, Individual Training Branch of the U.S. Army, 1990.

IS GOVERNMENT SPENDING TOO MUCH ON DEFENSE

Washington (March 10)—According to the latest PBC poll, many Americans believe that the defense budget should not be trimmed and that our military preparedness and superiority must be maintained. In a survey of 1000 randomly selected people from four geographical areas of the United States, about 60% of those surveyed believed that in light of the recent Persian Gulf War our military preparedness must be maintained. Each person surveyed was asked the question: "Should the defense budget be cut?" Their responses are summarized below:

Geographic Area

	East	Midwest	West	South
Yes	90	70	102	140
No	141	116	157	184

March 10, 1991

The first newspaper article gives the Army's new physical fitness standards for both men and women in different age groups. By looking at the data, can we conclude that people in different age groups are capable of performing different numbers of push-ups, sit-ups, and so on, or is the age of the person independent of these standards? Such information is very valuable since it enables an individual to measure his or her performance against that of others.

Now consider the second article. Is the geographical area in which a person lives a factor in determining a person's opinion as to whether the defense budget should be cut? Are there ways of comparing the proportions of responses?

12.1 INTRODUCTION

In Chapter 10 we discussed methods for testing whether the observed difference between two sample means is significant. In this chapter we analyze whether differences among two or more sample proportions are significant or whether they are due purely to chance. For example, suppose a college professor distributes a faculty-evaluation form to the 150 students of his Psychology 12 classes. The following are two examples of the multiple-choice questions appearing on the form:

1. What is your grade-point average? (Assume A = 4 points, B = 3 points, C = 2 points, D = 1 point, and F = 0 points.)
   a. 3.0 to 4.0   b. 2.0 to 2.99   c. below 2.0
2. Would you be willing to take another course with this teacher?
   a. Yes   b. No

The results for these two questions are summarized in the following chart:

		Grade Point Average		
		3.0–4.0	*2.0–2.99*	*Below 2.0*
Would You Take This Teacher Again?	*Yes*	28	36	11
	No	22	44	9
	Total	50	80	20

The teacher may be interested in knowing whether these ratings are influenced by the student's grade-point average. In the 3.0–4.0 category the proportion of students who said that they would take this teacher for another course is $\frac{28}{50}$. In the 2.0–2.99 category the proportion is $\frac{36}{80}$, and in the below 2.0 category the proportion is $\frac{11}{20}$. Is it true that students with a higher grade-point average tend to rate the teacher differently from students with a lower grade-point average?

Now consider the magazine clipping (Fig. 12.1), which gives the victimization rates by the type of crime and the age of the victim. Is the type of crime committed related to the age of the intended victim? Such information is vital to law-enforcement officials. Questions of this type occur quite often.

In this chapter we study the chi-square distribution, which is of great help in studying differences between proportions.

PERSONAL CRIMES: VICTIMIZATION RATES FOR PERSONS AGE 12 AND OVER, BY TYPE OF CRIME AND AGE OF VICTIMS, 1980

(Rate per 1,000 population in each age group)

Type of crime	12–15 (16,527,000)	16–19 (15,792,000)	20–24 (17,609,000)	25–34 (29,211,000)	35–49 (33,783,000)	50–64 (30,847,000)	65 and over (20,792,000)
Crimes of violence	52.6	67.9	61.1	38.6	20.8	11.8	9.0
Rape	1.5	2.5	2.1	1.4	0.2[1]	0.3	0.2[1]
Robbery	12.7	11.3	10.7	7.0	5.5	4.1	3.9
Robbery with injury	3.3	3.5	3.3	2.1	2.1	1.5	1.9
From serious assault	1.4	1.9	1.7	1.4	1.3	0.9	1.0
From minor assault	2.0	1.6	1.6	0.6	0.8	0.6	0.9
Robbery without injury	9.4	7.8	7.4	5.0	3.4	2.6	2.0
Assault	38.5	54.1	48.3	30.2	15.2	7.3	4.9
Aggravated assault	12.9	23.7	22.0	12.6	7.0	2.7	1.6
With injury	5.7	7.7	6.9	3.7	2.1	0.7	0.4[1]
Attempted assault with weapon	7.2	16.0	15.2	8.9	4.9	2.0	1.2
Simple assault	25.6	30.4	26.3	17.7	8.2	4.6	3.4
With injury	7.9	8.3	6.5	3.7	1.7	1.0	0.5
Attempted assault without weapon	17.8	22.1	19.7	13.9	6.5	3.6	2.9
Crimes of theft	166.7	159.8	146.3	106.2	79.2	49.4	21.9
Personal larceny with contact	3.1	3.7	3.4	2.6	2.6	3.5	3.4
Purse snatching	0.4[1]	0.6[1]	0.9	0.6	0.7	1.5	1.4
Pocket picking	2.7	3.1	2.4	2.0	1.9	1.9	2.0
Personal larceny without contact	163.6	156.1	143.0	103.5	76.7	45.9	18.5

NOTE: Detail may not add to total shown because of rounding. Numbers in parentheses refer to population in the group.

[1]Estimate, based on about 10 or fewer sample cases, is statistically unreliable.

FIGURE 12.1

Source: Criminal Victimization in the United States, U.S. Department of Justice—Law Enforcement Assistance Administration, Washington, D.C., 1989

12.2 THE CHI-SQUARE DISTRIBUTION

To illustrate the method that is used when analyzing several sample proportions, let us return to the first example discussed in the introduction. Let p_1 be the true proportion of students in the 3.0–4.0 category who will take another course with this teacher. Similarly, let p_2 and p_3 represent the proportion of students in the 2.0–2.99 and below 2.0 categories who will take another course with this teacher. The null hypothesis that we wish to test is

$$H_0: \quad p_1 = p_2 = p_3$$

This means that a student's grade-point average does not affect the student's decision to take this teacher again. The alternative hypothesis is that at least one of p_1, p_2, and p_3 is different. This means that a student's grade-point average does affect the student's decision.

If the null hypothesis is true, the observed difference between the proportions in each of the grade-point average categories is due purely to chance. Under this assumption we combine all the samples into one and consider it as one large sample. We then obtain the following estimate of the true proportion of *all* students in the school who are willing to take another course with the teacher. We get

$$\frac{28 + 36 + 11}{50 + 80 + 20} = \frac{75}{150} = 0.5$$

Thus, our estimate of the true proportion of students who are willing to take another course with this teacher is 0.5.

There are 50 students in the 3.0–4.0 category. We would therefore expect 50(0.5), or 25 of these students to indicate yes they would take another course with this teacher, and we would expect 50(0.5), or 25 to indicate no. Similarly, in the 2.0–2.99 category there are 80 students so that we would expect 80(0.5), or 40 yes answers and 40 no answers. Also, in the below 2.0 category there are 20 students so that we would expect 20(0.5), or 10 yes answers and 10 no answers. The numbers that should appear are called **expected frequencies.** In the following table we have indicated these numbers in parentheses below the ones that were actually observed. We call the numbers that were actually observed **observed frequencies.**

Expected Frequencies

Observed Frequencies

Grade Point Average

	3.0–4.0	2.0–2.99	Below 2.0
Yes	28 (25)	36 (40)	11 (10)
No	22 (25)	44 (40)	9 (10)
Total	50	80	20

Notice that the expected frequencies and the observed frequencies are not the same. If the null hypothesis that $p_1 = p_2 = p_3$ is true, the observed frequencies should be fairly close to the expected frequencies. Since this rarely will happen, we need some way of determining when these differences are significant.

When is the difference between the observed frequencies and the expected frequencies significant? To answer this question, we calculate a test statistic called the **chi-square statistic.**

Chi-square Statistic

FORMULA 12.1

Let E represent the expected frequency and let O represent the observed frequency. Then the **chi-square test statistic,** denoted as χ^2 is defined as

$$\chi^2 = \Sigma \frac{(O - E)^2}{E}$$

COMMENT In using Formula 12.1 we must calculate the square of the difference for each box, that is, cell, of the table. Then we divide the squares of the difference for each cell by the expected frequency for that box. Finally, we add these results together.

Returning to our example, we have

$$\chi^2 = \frac{(28 - 25)^2}{25} + \frac{(36 - 40)^2}{40} + \frac{(11 - 10)^2}{10} + \frac{(22 - 25)^2}{25} + \frac{(44 - 40)^2}{40} + \frac{(9 - 10)^2}{10}$$

$$= \quad 0.36 \quad + \quad 0.40 \quad + \quad 0.10 \quad + \quad 0.36 \quad + \quad 0.40 \quad + \quad 0.10$$

$$= 1.72$$

The value of the χ^2 statistic is 1.72.

It should be obvious from Formula 12.1 that the value of χ^2 will be 0 when there is perfect agreement between the observed frequencies and the expected frequencies since in this case $O - E = 0$. Generally speaking, if the value of χ^2 is small, the observed frequencies and the expected frequencies will be pretty close to each other. On the other hand, if the value of χ^2 is large, this indicates that there is considerable difference between the observed frequency and the expected frequency.

Chi-square
Distribution

To determine when the value of the χ^2 statistic is significant, we use the **chi-square distribution.** This is pictured in Figure 12.2. We reject the null hypothesis when the value of the chi-square statistic falls in the rejection region of Figure 12.2. Table VIII in the Appendix gives us the critical values, that is, the dividing line, depending on the number of degrees of freedom. Thus, $\chi^2_{0.05}$ represents the dividing line that cuts off 5% of the right tail of the distribution. *The number of degrees of freedom is always 1 less than the number of sample proportions that we are testing.*

In our example we are comparing three proportions so that the number of degrees of freedom is $3 - 1$, or 2. Now we look at Table VIII to find the χ^2 value that corresponds to 2 degrees of freedom. We have $\chi^2_{0.05} = 5.991$. Since the test statistic value that we obtained, $\chi^2 = 1.72$, is much less than the table value of 5.991, we do not reject the null hypothesis. The difference between what was expected and what actually happened can be attributed to chance.

Although we will usually use the 5% level of significance, Table VIII in the Appendix also gives us the χ^2 values for the 1% level of significance. We use

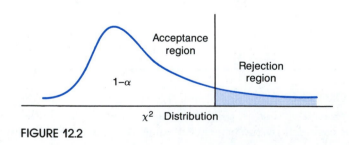

FIGURE 12.2

these values when we wish to find the dividing line that cuts off 1% of the right tail of the distribution.

Let us further illustrate the χ^2 test with several examples.

EXAMPLE 1

There are 10,000 students at a college. 2700 are freshmen, 2300 are sophomores, 3000 are juniors, and 2000 are seniors. Recently a new president was appointed. 2000 students attended the reception party for the president. The attendance breakdown is shown in the following table.

		Freshmen	Sophomores	Juniors	Seniors
	Yes	300	700	650	350
Attended Reception?	No	2400	1600	2350	1650
	Total	2700	2300	3000	2000

Test the null hypothesis that the proportion of freshmen, sophomores, juniors, and seniors that attended the reception is the same. (Use a 5% level of significance.)

Solution

In order to compute the χ^2 test statistic, we must first compute the expected frequency for each box, or cell. To do this we obtain an estimate of the true proportion of students who attended the reception. We have

$$\frac{300 + 700 + 650 + 350}{2700 + 2300 + 3000 + 2000} = \frac{2000}{10,000} = 0.20$$

Out of 2700 freshmen we would expect 2700(0.20) or 540 to attend and 2700 − 540, or 2160, not to attend. Similarly, out of 2300 sophomores we would expect 2300(0.20), or 460 to attend and 2300 − 460, or 1840 not to attend. Also, out of 3000 juniors we would expect 3000(0.20), or 600 to attend and 3000 − 600, or 2400 not to attend. Finally, out of 2000 seniors, we would expect 2000(0.20), or 400 to attend and 2000 − 400, or 1600 not to attend. We have indicated these expected frequencies just below the observed values in the following chart:

		Freshmen	Sophomores	Juniors	Seniors
	Yes	300 (540)	700 (460)	650 (600)	350 (400)
Attended Reception?	No	2400 (2160)	1600 (1840)	2350 (2400)	1650 (1600)
	Total	2700	2300	3000	2000

Now we calculate the value of the χ^2 statistic. We have

$$\chi^2 = \Sigma \frac{(O - E)^2}{E}$$

$$= \frac{(300 - 540)^2}{540} + \frac{(700 - 460)^2}{460} + \frac{(650 - 600)^2}{600}$$

$$+ \frac{(350 - 400)^2}{400} + \frac{(2400 - 2160)^2}{2160} + \frac{(1600 - 1840)^2}{1840}$$

$$+ \frac{(2350 - 2400)^2}{2400} + \frac{(1650 - 1600)^2}{1600}$$

$$= 106.67 + 125.22 + 4.17 + 6.25 + 26.67 + 31.30 + 1.04 + 1.56$$

$$= 302.88$$

There are four proportions that we are testing so that there are $4 - 1$, or 3 degrees of freedom. From Table VIII in the Appendix we find that the $\chi^2_{0.05}$ value with 3 degrees of freedom is 7.815. The value of the χ^2 test statistic ($\chi^2 = 302.88$) is definitely greater than 7.815. Hence, we reject the null hypothesis. The proportions of freshmen, sophomores, juniors, and seniors that attended the reception are not the same.

EXAMPLE 2

A survey of the marital status of the members of three health clubs was taken. The following table indicates the results of the survey.

		Club 1	Club 2	Club 3
	Yes	11	17	8
Married?	No	29	33	22
	Total	40	50	30

Test the null hypothesis that the proportion of members that are married in each of these health clubs is the same. (Use a 5% level of significance.)

Solution

We must first compute the expected frequency for each cell. To do this we obtain an estimate of the true proportion of members who are married. We have

$$\frac{11 + 17 + 8}{40 + 50 + 30} = \frac{36}{120} = 0.3$$

Thus, the estimate of the true proportion is 0.3. Out of the 40 members in Club 1 we would expect 40(0.3), or 12 members to be married and $40 - 12$, or 28 not to be married. In Club 2 we would expect 50(0.3), or 15 members to be married and 50 − 15, or 35 members not to be married. In Club 3 we would expect 30(0.3), or 9 members to be married and 30 − 9, or 21 not to be married. We have indicated these expected frequencies in parentheses in the following table.

		Club 1	Club 2	Club 3
	Yes	11 (12)	17 (15)	8 (9)
Married?				
	No	29 (28)	33 (35)	22 (21)
	Total	40	50	30

Now we calculate the value of the χ^2 statistic. We have

$$\chi^2 = \Sigma \frac{(O - E)^2}{E}$$

$$= \frac{(11 - 12)^2}{12} + \frac{(17 - 15)^2}{15} + \frac{(8 - 9)^2}{9} + \frac{(29 - 28)^2}{28}$$

$$+ \frac{(33 - 35)^2}{35} + \frac{(22 - 21)^2}{21}$$

$$= 0.08 + 0.27 + 0.11 + 0.04 + 0.11 + 0.05$$

$$= 0.66$$

There are three proportions that we are testing so that there are 3 − 1, or 2 degrees of freedom. From Table VIII in the Appendix we find that the $\chi^2_{0.05}$ value with 2 degrees of freedom is 5.991. Since the value of the test statistic, 0.66, is less than 5.991, we do not reject the null hypothesis.

COMMENT Experience has shown us that the χ^2 test can only be used when the expected frequency in each cell is at least 5. If the expected frequency of a cell is not larger than 5, this cell should be combined with other cells until the expected frequency is at least 5. We will not, however, concern ourselves with this situation.

EXERCISES FOR SECTION 12.2

1. A criminologist is interested in determining if a man's age is a factor in whether or not he drinks alcoholic beverages. A random survey of 600 working men in California produced the following results.

		Age Group (in years)			
		Between 20 and 29	*Between 30 and 39*	*Between 40 and 49*	*Over 50*
Drinker?	Yes	82	67	25	12
	No	160	139	74	41

Using a 5% level of significance, test the null hypothesis that there is no significant difference between the corresponding proportion of men in the various age groups who drink alcoholic beverages.

2.

MORE DEAD FISH FOUND WASHING UP ON THE SHORELINE

Dec. 28—Health officials reported yesterday that more than 284 dead fish were found over the weekend on the city's shoreline, bringing to well over 2500 the number of dead fish that were found recently. A health official, who spoke on condition of anonymity, speculated that the fish might have died as a result of the accidental release of highly toxic chemicals into the waterways from a neighboring chemical company.

December 28, 1990

Consider the accompanying newspaper article. Environmental pathologists carefully examined 400 of these dead fish to pinpoint the cause of death. The pathologists found traces of three different chemicals in the bodies of these dead fish as shown below.

		Chemical Found		
		Pesticide A	*Pesticide B*	*Pesticide C*
Traces of Chemical Found?	*Yes*	51	63	35
	No	85	88	78

Using a 1% level of significance, test the null hypothesis that there is no significant difference between the proportion of dead fish with each of the different pesticides in their bodies.

3. A TWA flight attendant reported that after analyzing the 240 people who were on board a particular airplane flight it was found that 41 of the 158 male passengers rented headphones for the in-flight movie and 25 of the 82 female passengers rented headphones for the in-flight movie. Using a 5% level of significance, test

the null hypothesis that there is no significant difference between the corresponding proportion of male or female passengers who rent headphones for in-flight movies.

4. A major refrigerator manfacturer purchases the parts needed for assembly from five different suppliers located in different parts of the country and assembles them locally. Due to the numerous complaints received about defective thermostats, management decides to sample 500 refrigerators assembled using parts supplied by these suppliers and to determine the number of refrigerators with defective thermostats. The following results are available.

		Supplier				
		A	*B*	*C*	*D*	*E*
Does Refrigerator Have Defective Thermostat?	*Yes*	9	11	16	6	9
	No	101	93	71	94	90

Using a 1% level of significance, test the null hypothesis that there is no significant difference between the proportion of refrigerators from all suppliers with defective thermostats.

5. A random survey of 740 families in New Jersey, Texas, and Illinois was taken to determine the proportion of families who own a video cassette recorder (VCR).* The following table presents the results of the survey.

		State		
		New Jersey	*Texas*	*Illinois*
Own a VCR?	*Yes*	160	135	240
	No	70	55	80

Using a 5% level of significance, test the null hypothesis that there is no significant difference between the proportion of families in all three states that own a VCR.

6. A survey of 1876 of the state's motorists was taken throughout the state to determine how many of them were in favor of raising the state gasoline tax so as to obtain the additional funds needed to rebuild and/or improve the highway infrastructure. The following table indicates the results of the survey according to the geographic location of the motorist interviewed.

		Geographic Location Within State			
		North	*South*	*East*	*West*
In Favor of Raising State Gasoline Tax?	*Yes*	139	147	154	168
	No	327	298	312	331

*Public Interest Research Group, "Who Owns A Home Computer System," 1985.

Using a 5% level of significance, test the null hypotheses that there is no significant difference between the proportion of motorists in each of the geographic areas surveyed in favor of raising the state gasoline tax.

7. Consider the accompanying newspaper article. The commission surveyed homes on six streets and obtained the following data.

			Homes Located on			
		Ave. N	*Riggs Blvd.*	*Bailey Drive*	*Lashly Place*	*Hew Ave.*	*7th Street*
Dangerous Levels of Radon Gas Present	*Yes*	14	18	6	16	12	20
	No	62	58	74	52	78	80

Using a 1% level of significance, test the null hypothesis that there is no significant difference between the percentage of homes on all the streets containing excessive levels of radon gas.

MORE HOMES CONTAMINATED WITH RADON

Bergen (Sept. 12)—A new study released today by the state's environmental commission reveals that the number of homes containing excessive amounts of deadly radon gas is larger than expected. The radon gas is entering many homes on the East Coast (particulary those in the states of New Jersey and Pennsylvania) through holes or cracks in the foundation. It is believed that the gas is coming from deep within the earth.

The Commission urged the governor to appropriate additional funds to enable it to expand its monitoring activities.

September 12, 1989

8. A state legislator claims that the percentage of college students who are receiving state tuition assistance and who work part-time is essentially the same for all colleges within the state. A recent survey by the state education department revealed the following information about the students at five colleges within the state.

	College				
Student Receiving	*A*	*B*	*C*	*D*	*E*
State Tuition Assistance and	*Yes* 193	203	128	309	512
Works Part-Time?	*No* 646	716	517	1462	784

Using a 1% level of significance, test the null hypothesis that the percentage of college students receiving state tuition assistance and working part-time is essentially the same at these colleges.

9. Most colleges and universities have annual campaigns in which they ask former alumni to contribute funds for scholarships. For the graduating classes of the years 1987 to 1990 at one college the following information is available on the number of former graduates that contributed money for scholarships.

	Graduation Class			
	1987	*1988*	*1989*	*1990*
Contributed Money? *Yes*	469	582	731	646
No	218	371	589	488

Using a 5% level of significance, test the null hypothesis that the percentage of former graduates who contributed money for scholarships is the same in the different graduating classes.

10. A study was conducted to determine whether an individual's social class has any effect on the individual's willingness to contribute blood to the local hospital or to any Red Cross agency. Each of 450 randomly selected individuals was asked the same question: "Would you be willing to contribute blood to the local hospital or to any Red Cross agency?" A summary of their responses is shown in the accompanying table.

	Social Class			
	Upper	*Upper-Middle*	*Middle-Lower*	*Lower*
Would You Be Willing to *Yes*	27	69	84	98
Contribute Blood? *No*	11	32	53	76

Using a 5% level of significance, test the null hypothesis that the percentage of people in the various social classes who are willing to give blood is the same.

11. Many college libraries add books (or periodicals) to their holdings by either buying them or by receiving donations of books and periodicals from private individuals. Last year one state university added 119 math or math-related books or periodicals to its collection from two sources as shown in the accompanying chart.

		Algebra or Trigonometry	Geometry	Calculus and Analysis	Advanced Theoretical Books
	Statistics				
Was Book or Periodical Purchased? Yes	13	19	10	15	22
No	8	11	4	7	10

Using a 5% level of significance, test the null hypothesis that the percentage of books or periodicals added to the library's holdings is the same for all types of math or math-related books or periodicals.

12. The Washington Savings Bank has five branches, each located in a different section of the city. Management has received numerous complains about the quality of service and the courteousness of the employees at some of these branches. A random survey of 3000 of the banks customers is conducted to determine the extent of this dissatisfaction. The results are summarized below:

Branch Where Account is Located

		Broadway	Main Street	Driggs Avenue	Boynton Blvd.	Forest Mall
Are You Satisfied?	Yes	312	414	375	280	253
	No	248	370	313	242	193

Using a 5% level of significance, test the null hypothesis that the level of customer satisfaction is the same at all of the bank branches.

13. Refer to the newspaper article given at the beginning of this chapter on page 598. Consider the information contained in it. Using a 1% level of significance, test the null hypothesis that the proportion of people in the different geographic areas who believe that the defense budget should be cut is the same.

12.3 GOODNESS OF FIT

In addition to the applications mentioned in the previous section, the chi-square test statistic can also be used to determine whether an observed frequency distribution is in agreement with the expected mathematical distribution. For example, when a die is rolled, we assume that the probability of any one face coming up is $\frac{1}{6}$. Thus, if a die is rolled 120 times, we would expect each face to come up approximately 20 times since $\mu = np = 120 \left(\frac{1}{6} \right) = 20$.

Suppose we actually rolled a die 120 times and obtained the results shown in Table 12.1. In this table we have also indicated the expected frequencies. Are these observed frequencies reasonable? Do we actually have an honest die?

TABLE 12.1 Expected and Observed Frequencies When a Die Was Tossed 120 Times

Die Shows	Expected Frequency	Observed Frequency
1	20	18
2	20	21
3	20	17
4	20	21
5	20	19
6	20	24
	Total = 120	Total = 120

To check whether the differences between the observed frequencies and the expected frequencies are due purely to chance or are significant, we use the chi-square test statistic of Formula 12.1. We reject the null hypothesis that the observed differences are not significant only when the test statistic falls in the rejection region.

In our case the value of the test statistic is

$$\chi^2 = \Sigma \frac{(O - E)^2}{E}$$

$$= \frac{(18 - 20)^2}{20} + \frac{(21 - 20)^2}{20} + \frac{(17 - 20)^2}{20} + \frac{(21 - 20)^2}{20}$$

$$+ \frac{(19 - 20)^2}{20} + \frac{(24 - 20)^2}{20}$$

$$= 1.60$$

There are $6 - 1$, or 5, degrees of freedom. From Table VIII in the Appendix we find that the $\chi^2_{0.05}$ value with 5 degrees of freedom is 11.070. Since the test statistic has a value of only 1.60, which is considerably less than 11.070, we do not reject the null hypothesis. Any differences between the observed frequencies and the expected frequencies are due purely to chance.

Goodness of Fit

The following examples will further illustrate how the chi-square test statistic can be used to test **goodness of fit,** that is, to determine whether the observed frequencies fit with what was expected.

EXAMPLE 1

The number of phone calls received per day by a local chapter of Alcoholics Anonymous is as follows:

Number of Calls Received	M	T	W	T	F
	173	153	146	182	193

Using a 5% level of significance, test the null hypothesis that the number of calls received is independent of the day of the week.

Solution

We first calculate the number of expected calls per day. If the number of calls received is independent of the day of the week, we would expect to receive

$$\frac{173 + 153 + 146 + 182 + 193}{5} = 169.4$$

calls per day. We can then set up the following table:

	M	T	W	T	F
Observed Number of Calls	173	153	146	182	193
Expected Number of Calls	169.4	169.4	169.4	169.4	169.4

Now we calculate the value of the chi-square test statistic. We have

$$\chi^2 = \Sigma \frac{(O - E)^2}{E}$$

$$= \frac{(173 - 169.4)^2}{169.4} + \frac{(153 - 169.4)^2}{169.4} + \frac{(146 - 169.4)^2}{169.4}$$

$$+ \frac{(182 - 169.4)^2}{169.4} + \frac{(193 - 169.4)^2}{169.4}$$

$$= 0.0765 + 1.5877 + 3.2323 + 0.9372 + 3.2878$$

$$= 9.1215$$

There are $5 - 1$, or 4 degrees of freedom. From Table VIII in the Appendix we find that the $\chi^2_{0.05}$ value with 4 degrees of freedom is 9.488. Since the test statistic has a value of 9.1215, which is less than 9.488, we do not reject the null hypothesis and the claim that the number of calls received is independent of the day of the week.

EXAMPLE 2   A scientist has been experimenting with rats. As a result of certain injections, the scientist claims that when two black rats are mated, the offspring will be black, white, and gray in the proportion 5:4:3. (This means that the probability of a black offspring is $\frac{5}{12}$, the probability of a white offspring is $\frac{4}{12}$, and the probability of a gray rat is $\frac{3}{12}$.) Many rats were mated after being injected with the chemical. Of 180 newborn rats 71 were black, 69 were white, and 40 were gray. Can we accept the scientist's claim that the true proportion is 5:4:3? Use a 5% level of significance.

Solution

We first calculate the expected frequencies. Out of 180 rats we would expect $180 \left(\dfrac{5}{12} \right)$, or 75 of them to be black. Similarly, out of 180 rats we would expect $180 \left(\dfrac{4}{12} \right)$, or 60 of them to be white, and we would expect $180 \left(\dfrac{3}{12} \right)$, or 45 of them to be gray. We now set up the following table:

Color of Rat	Expected Frequency	Observed Frequency
Black	75	71
White	60	69
Gray	45	40

The value of the χ^2 test statistic is

$$\chi^2 = \Sigma \frac{(O - E)^2}{E}$$

$$= \frac{(71 - 75)^2}{75} + \frac{(69 - 60)^2}{60} + \frac{(40 - 45)^2}{45}$$

$$= 2.12$$

There are $3 - 1$, or 2 degrees of freedom. From Table VIII in the Appendix we find that the $\chi^2_{0.05}$ value with 2 degrees of freedom is 5.991. Since the test statistic has a value of 2.12, which is less than 5.991, we do not reject the null hypothesis and the scientist's claim.

EXERCISES FOR SECTION 12.3

1. A casino official wants to determine whether or not a die is fair. The official throws the die 120 times with the following results.

Number of Dots Showing on Face of Die	Frequency of Occurrence
1	28
2	32
3	12
4	13
5	21
6	14

Using a 5% level of significance, test the null hypothesis that the die is fair.

2. There are five tolls booths at the entrance to a tunnel. An observer notices that 800 randomly selected cars used these toll booths with the following frequencies.

Toll Booth Number	Number of Cars Using This Toll Booth
1	130
2	194
3	196
4	106
5	174

Using a 5% level of significance, test the null hypothesis that all the toll booths are used with the same frequency.

3. Worker absenteeism affects production schedules. The Marvel Chemical Company analyzed the number of absences occurring daily at its Patchogue plant. The following absences occurred during one week.

Day of Week

	Mon.	Tues.	Wed.	Thurs.	Fri.
Number of Absences	54	38	44	40	24

Using a 5% level of significance, test the null hypothesis that absences occur on the 5 days with equal frequency.

4. The Austrian monk Gregor Mendel performed many experiments with garden peas. In one such experiment the following results were obtained:

148 round and yellow peas.

84 wrinkled and yellow peas.

61 round and green peas.

47 wrinkled and green peas.

Using a 5% level of significance, test the null hypothesis that the frequencies of these types of peas should be in the ratio $8:4:3:2$.

5. Using gene-splicing techniques, a medical researcher claims to have developed a new strain of bacteria with many desirable characteristics. Preliminary indications are that the new bacteria can be used as an antirejection drug to ward off infections. Pending further investigation, the medical researcher has coded these new characteristics as A, B, C, and D. The following results were obtained in several gene-splicing experiments:

261 had characteristic A.

213 had characteristic B.

111 had characteristic C.

105 had characteristic D.

Using a 1% level of significance, test the null hypothesis that the frequencies of these types of characteristics obtained by gene-splicing are in the ratio $9:7:4:3$.

6. Bill Vecker, a spokesperson for the telephone company, has obtained the following information on the number of calls to the business office of the phone company by customers questioning their bills.

Day of the Week	Mon.	Tues.	Wed.	Thurs.	Fri.
Number of Calls Received	276	212	198	246	253

Using a 5% level of significance, test the null hypothesis that the number of calls received by the business office of the phone company is independent of the day of the week.

7. A supermarket manager believes that the number of half gallons of vanilla-flavored, chocolate-flavored, almond-flavored, butter pecan-flavored, and strawberry-flavored ice cream sold per week is in the ratio of $7:3:4:2:5$. In a random sample of 420 purchases of ice cream it was observed that the following flavors were purchased.

Flavor Purchased	Observed Frequency
Vanilla	163
Chocolate	51
Almond	78
Butter Pecan	44
Strawberry	84

Using a 5% level of signficance, test the null hypothesis that the manager's claim is correct.

8. Five candidates have announced their candidacy for governor of a southern state. To determine whether any of the candidates have an early lead in popularity, 5000 randomly selected voters were asked to indicate their gubernatorial preference from among these five candidates. Their responses are summarized below.

	Candidate				
	A	B	C	D	E
Number of Voters Who Prefer This Candidate	1329	876	1021	946	820

Using a 5% level of significance, test the null hypothesis that all the candidates do share the same popularity among the registered voters.

9. There are four television (TV) stations that broadcast news, weather, and sports from 6:00 P.M. to 7:00 P.M. every weekday. An advertising firm is interested in knowing whether there is an unequal breakdown of the viewing audience among these four stations. The advertising firm randomly selects 1000 people who watch the newscasts during these hours. Each is asked the same question: "On which TV station do you watch the news from 6:00 P.M. to 7:00 P.M. every weekday?" Their responses are as follows.

TV Station	Number of Viewers
I	241
II	288
III	263
IV	208

Using a 5% level of significance, test the null hypothesis that these four TV stations do have equal shares of the evening news viewing audience.

10. It has been documented that having an adequate number of fire-fighting personnel on duty at the appropriate time can actually save lives. The number of calls to one particular fire district in New York City during the week of January 13 to January 19, 1991 was as follows.

Sun. Jan. 13	Mon. Jan. 14	Tues. Jan. 15	Wed. Jan. 16	Thurs. Jan. 17	Fri. Jan. 18	Sat. Jan. 19
48	32	66	38	45	44	42

Using a 5% level of significance, test the null hypothesis that the number of calls to this particular fire district is independent of the day of the week.

12.4 CONTINGENCY TABLES

A very useful application of the χ^2 test discussed in Section 12.2 occurs in connection with contingency tables. Contingency tables are used when we wish to determine whether two variables of classification are related or dependent one on the other. For example, consider the following chart, which indicates the eye color and hair color of 100 randomly selected girls.

	Brown Eyes	Blue Eyes
Light Hair	10	33
Dark Hair	44	13

Is eye color independent of hair color or is there a significant relationship between hair color and eye color?

Contingency tables are especially useful in the social sciences where data are collected and often classified into two main groups. We might be interested in determining whether a relationship exists between these two ways of classifying the data or whether they are independent. We have the following definition:

DEFINITION 12.1
Contingency Table

A contingency table is an arrangement of data into a two-way classification. One of the classifications is entered in rows and the other in columns.

When dealing with contingency tables remember that the null hypothesis always assumes that the two ways of classifying the data are independent. We use the χ^2 test statistic as discussed in Section 12.2. The only difference is that we compute the expected frequency for each cell by using the following formula.

FORMULA 12.2 Expected Frequency	The **expected frequency** of any cell in a contingency table is found by multiplying the total of the row with the total of the column to which the cell belongs. The product is then divided by the total sample size.

The following examples will illustrate how we apply the χ^2 test to contingency tables.

EXAMPLE 1

Let us consider the contingency table given at the beginning of this section. Is eye color independent of hair color?

Solution

In order to compute the χ^2 test statistic, we must first compute the row total, column total, and the total sample size. We have

$$\begin{aligned}
\textit{Row Totals:} \quad & 10 + 33 = 43 \\
& 44 + 13 = 57 \\
\textit{Column Totals:} \quad & 10 + 44 = 54 \\
& 33 + 13 = 46 \\
\textit{Total Sample Size:} \quad & 10 + 33 + 44 + 13 = 100
\end{aligned}$$

We indicate these values in the following table:

	Brown Eyes	Blue Eyes	Row Total
Light Hair	10 (23.22)	33 (19.78)	43
Dark Hair	44 (30.78)	13 (26.22)	57
Column Total	54	46	

The expected value for the cell in the first row first column is obtained by multiplying the first row total with the first column total and then dividing the product by the total sample size. We get

$$\frac{54 \times 43}{100} = 23.22$$

For the first row second column we have

$$\frac{46 \times 43}{100} = 19.78$$

For the second row first column we have

$$\frac{54 \times 57}{100} = 30.78$$

For the second row second column we have

$$\frac{46 \times 57}{100} = 26.22$$

These values are entered in parentheses in the appropriate cell. We now use Formula 12.1 of Section 12.2 and calculate the χ^2 statistic. We have

$$\chi^2 = \Sigma \frac{(O - E)^2}{E}$$

$$= \frac{(10 - 23.22)^2}{23.22} + \frac{(33 - 19.78)^2}{19.78} + \frac{(44 - 30.78)^2}{30.78} + \frac{(13 - 26.22)^2}{26.22}$$

$$= 7.53 + 8.84 + 5.68 + 6.67$$

$$= 28.72$$

The χ^2 test statistic has a value of 28.72. *If the contingency table has r rows and c columns, then the number of degrees of freedom is (r − 1)(c − 1).* In this example there are two rows and two columns, so there are $(2 - 1) \cdot (2 - 1)$, or $1 \cdot 1$, which is 1 degree of freedom. From Table VIII in the Appendix we find that the $\chi^2_{0.05}$ value with 1 degree of freedom is 3.841. Since we obtained a value of 28.72, we reject the null hypothesis and conclude that hair color and eye color are *not* independent.

EXAMPLE 2   Criminal Analysis

A sociologist is interested in determining whether the occurrence of different types of crimes varies from city to city. An analysis of 1100 reported crimes produced the following results.

Type of Crime

	Rape	Auto Theft	Robbery and Burglary	Other	Total
City A	76 (61.35)	112 (146.34)	87 (72.66)	102 (96.65)	377
City B	64 (68.83)	184 (164.20)	77 (81.52)	98 (108.44)	423
City C	39 (48.82)	131 (116.45)	48 (57.82)	82 (76.91)	300
Total	179	427	212	282	1100

Do these data indicate that the occurrence of a type of crime is dependent on the location of the city? (Use a 5% level of significance.)

Solution

We first calculate the expected frequency for each cell. We have

$$\textit{First Row:} \quad \frac{(179)(377)}{1100} = 61.35 \quad\quad \frac{(427)(377)}{1100} = 146.34$$

$$\frac{(212)(377)}{1100} = 72.66 \quad\quad \frac{(282)(377)}{1100} = 96.65$$

$$\textit{Second Row:} \quad \frac{(179)(423)}{1100} = 68.83 \quad\quad \frac{(427)(423)}{1100} = 164.20$$

$$\frac{(212)(423)}{1100} = 81.52 \quad\quad \frac{(282)(423)}{1100} = 108.44$$

$$\textit{Third Row:} \quad \frac{(179)(300)}{1100} = 48.82 \quad\quad \frac{(427)(300)}{1100} = 116.45$$

$$\frac{(212)(300)}{1100} = 57.82 \quad\quad \frac{(282)(300)}{1100} = 76.91$$

These numbers appear in parentheses in the preceding chart. Now we calculate the χ^2 test statistic, getting

$$\chi^2 = \Sigma \, \frac{(O - E)^2}{E}$$

$$= \frac{(76 - 61.35)^2}{61.35} + \frac{(112 - 146.34)^2}{146.34} + \cdots + \frac{(82 - 76.91)^2}{76.91}$$

$$= 24.46$$

There are three rows and four columns so that there are 6 degrees of freedom since

$$(3 - 1)(4 - 1) = 2 \cdot 3 = 6$$

From Table VIII in the Appendix we find that the $\chi^2_{0.05}$ value with 6 degrees of freedom is 12.592. Since we obtained a χ^2 value of 24.46, we reject the null hypothesis and conclude that the type of crime and the location of the city are not independent.

COMMENT It should be noted that the chi-square independence test is appropriate only when applied to sample data and not to data for the entire population.

COMMENT For the chi-square independence test to be valid, it is assumed that all expected frequencies are at least 1 and that at most 20% of the expected frequencies are less than 5. Otherwise this test procedure should not be used.

We can summarize the procedures for using the chi-square test for independence as follows:

To perform a chi-square independence test:

1. State the null and alternate hypotheses.

2. Calculate the expected frequency for each cell by using Formula 12.2 so that the entry placed in each cell below the observed frequency is given by

$$\text{Entry} = \frac{(\text{row total}) \cdot (\text{column total})}{\text{total sample size}}$$

3. Assuming the expected frequencies satisfy the necessary preconditions for the chi-square test to be applicable (as given above), determine the significance level.

4. Compute the value of the test statistic $\chi^2 = \dfrac{\Sigma(O - E)^2}{E}$ and compare the results with the critical value χ^2 as given in Table VIII. If the contingency table has r rows and c columns, then the number of degrees of freedom is $(r - 1)(c - 1)$.

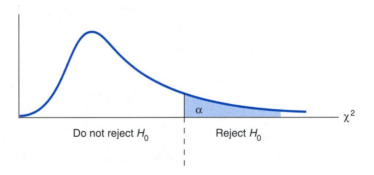

5. If the value of the test statistic falls in the region, reject H_0. Otherwise do not reject H_0. Specify your conclusion in words.

EXERCISES FOR SECTION 12.4

1. A quality control engineer is analyzing the number of defective calculators produced by the different production work shifts. The following data are available.

		Defective	Good
		Number of Calculators Produced That Are	
	Morning	21	347
Shift	Afternoon	19	288
	Night	28	375

Using a 5% level of significance, test the null hypothesis that the number of defective calculators produced is independent of the production work shift.

2. Many colleges require student input in the rehiring and promotion of teachers. This is usually accomplished by having the students complete course and faculty evaluation forms at the end of the semester. One of the questions asks students to rate the course as poor, fair, good, or excellent. Another question asks students to indicate whether the course was taken as an elective or whether it was a required course in the student's academic program. The following data are available for one large lecture course.

		Excellent	Poor	Fair	Good
		How Would You Rate This Course?			
Why Did You Take This Course?	Required	12	16	11	24
	Elective	15	17	22	23

After seeing the data the president of the college would like to know if the rating of the course is independent of the reason for taking the course. Using a 5% level of significance, how do you respond to the president?

3. A Wall Street analysis firm randomly sampled 500 business executives nationwide and classified the executive according to his or her smoking and drinking habits as shown here.

		None	Social Only	Heavy Drinker
		Drinking Habit		
	None	61	38	21
Smoking Habit	Light	57	41	25
	Moderate	46	55	27
	Heavy	29	61	39

Using a 5% level of significance, test the null hypothesis that smoking and drinking habits are independent for this group.

4. A large brokerage firm would like to determine whether the service it provides to its more affluent customers is of the same quality as the service it provides to its lower-income customers. A random sample of 320 customers is selected. Each customer is asked the following two questions: "What is your annual income?" and "How would you rate the service provided by your broker?" The results of the survey are as follows.

		Broker Rating		
		Poor	*Average*	*Good*
	Below $30,000	18	25	36
Annual Income	*$30,000–$70,000*	19	49	61
	Above $70,000	22	33	57

Using a 5% level of significance, test the null hypothesis that the broker rating is independent of the customer's annual income.

5. Pregnancy test kits are now available without a prescription for in-home early determination of pregnancy. Are these pregnancy test kits accurate? A consumer's research group conducted a survey of 591 individuals. The pregnancy test kit results of an individual (a positive reading—indicating pregnancy—or a negative reading) were then compared with the individual's actual clinical condition (pregnant or nonpregnant). The following data are available.

		Clinical Test Result	
		Pregnant	*Not Pregnant*
Pregnancy Test Kit Result	*Positive*	368	29
	Negative	18	176

Using a 5% level of significance, test the null hypothesis that the pregnancy test kit results are independent of the clinical test results.

6. The following is a breakdown of the different types of credit card sales of a large department store for a particular day.

| | Type of Credit Card Used | | | | |
	Master Card	VISA	American Express	Discover Card	Store Card
Under $25	78	82	32	53	46
Between $25 and $75	62	53	18	28	37
Over $75	37	29	11	17	19

Amount of Purchase (row label at left)

Using a 5% level of significance, test the null hypothesis that the type of credit card used is independent of the amount of purchase.

7. Many states have recently begun enacting health care proxy laws whereby an individual can appoint some trustworthy person to make decisions about what medical treatment (for example, artificial nutrition and hydration) to apply should one be unable or incompetent to make such a decision. A newspaper reporter conducted a nationwide survey of 1698 people to find out what they thought about the new health care proxy laws. The results of the survey are shown in the accompanying table.

| | Opinion | | |
	In Favor of Proposal	Against Proposal	Had no Opinion
East	164	204	21
Midwest	110	358	17
South	128	276	9
Far West	146	252	13

Region of Country in Which Respondent Lives (row label at left)

Using a 5% level of significance, test the null hypothesis that the region in which a person lives is independent of the person's opinion about the proposal.

8. A taste test was given to 1800 people at a shopping mall to determine which brand of vanilla ice cream they preferred. The results were as follows.

| | Brand | | | |
	A	B	C	D
Male	256	262	236	203
Female	246	231	218	148

Sex of Consumer (row label at left)

Using a 5% level of significance, test the null hypothesis that the sex of the consumer is independent of the brand preferred.

TIGHTER SECURITY AT AIRPORTS

Washington (July 5)—As a direct result of the TWA hijacking, government officials have announced a series of secutiry checks at airports of all international travelers and their baggage to thwart any possible terrorist attacks. Yesterday, the *Tribune* conducted a random survey of travellers at the nation's airports to determine whether they were satisfied with the new security arrangements and the resulting delays. The results are summarized below:

	Support the checks enthusiastically despite delays	Support the checks moderately	Are opposed to the checks because of delays
Male Frequent traveler	43	67	21
Infrequent traveler	79	99	32
Female Frequent traveler	48	58	17
Infrequent traveler	69	95	27

July 5, 1986

 9. Consider the preceding newspaper article. Using a 5% level of significance, test the null hypotheses that the type of traveler is independent of the degree of satisfaction with the new security arrangement and the resulting delays.

 10. Consider the following newspaper article. A competing drug company conducted a survey to determine the preferences of its customers for various forms of medicines. Their responses as well as their ages are indicated in the accompanying table.

NO MORE TYLENOL CAPSULES

New Jersey (Jan. 19)—As a direct result of the recent drug tampering, Johnson and Johnson, makers of Tylenol, and one of the nation's large drug manufacturers, announced yesterday that it would no longer manufacture any over-the-counter drugs in capsule form.

Wednesday, Jan. 19, 1986

Medicine Preference Form for Pills

		Caplet	*Capsule*	*Tablet*
	Between 20 and 30	75	78	84
Age of User (in years)	*Between 30 and 50*	109	127	89
	Over 50	97	118	72

Using a 1% level of significance, test the null hypothesis that the preference in medicine form is the same for all age groups.

11. An environmental group conducted a survey of 3000 people in different age groups to determine which environmental problems were of most concern to them. The following results were obtained.

Environmental Problem

		Energy Conservation	*Water Pollution*	*Air Pollution*	*Nuclear Energy*
	Under 25 years	187	123	199	432
Age	*25–40 years*	368	221	117	277
	Over 40 years	112	145	637	182

Using a 5% level of significance, test the null hypothesis that age is independent of the environmental problem that most concerns an individual.

12. A college placement officer has compiled the following list on the college grade-point average at the graduation of 630 graduates and the starting salaries of these graduates.

		Starting Salary		
		Under $20,000	*$20,000–$35,000*	*Over $35,000*
Grade-Point Average	*2.00–2.59*	17	53	47
	2.60–2.99	19	58	56
	3.00–3.59	30	71	68
	3.60–4.00	38	90	83

Using a 1% level of significance, test the null hypothesis that the grade-point average of a graduate and the starting salary of the graduate are independent.

12.5 USING COMPUTER PACKAGES

MINITAB is well-suited to perform all calculations needed to test for independence. One only need to enter the cell frequencies to obtain the results. MINITAB is especially useful when the number of rows and columns of a contingency table is large. Let us apply the MINITAB program to the data given in Example 2 of Section 12.4 (page 618).

```
MTB  > READ THE TABLE INTO C1, C2, C3, C4
DATA >   76    112    87    102
DATA >   64    184    77     98
DATA >   39    131    48     82
DATA >   END

MTB  > CHISQUARE ANALYSIS ON TABLE IN C1, C2, C3, C4
Expected counts are printed below observed counts!

              C1        C2        C3        C4     Total
    1         76       112        87       102       377
           61.35    146.34     72.66     96.65

    2         64       184        77        98       423
           68.83    164.20     81.52    108.44

    3         39       131        48        82       300
           48.82    116.45     57.82     76.91

Total        179       427       212       282      1100
```

```
ChiSq =   3.499 +   8.060 +   2.831 +   0.296 +
          0.339 +   2.387 +   0.251 +   1.005 +
          1.975 +   1.817 +   1.667 +   0.337 =   24.465

df = 6

MTB  > STOP
```

12.6 SUMMARY

In this chapter we discussed the chi-square distribution and how it can be used to test hypotheses that differences between expected frequencies and observed frequencies are due purely to chance.

We applied the chi-square test statistic to test whether observed frequency distributions are in agreement with expected mathematical frequencies.

The chi-square test statistic can also be used to analyze whether the two factors of a contingency table are independent. This is very useful, especially in the social sciences, where the data are often grouped according to two factors.

When using the χ^2 test statistic we must take great care in determining the number of degrees of freedom. Also, as we pointed out in a comment on page 605, each expected cell frequency must be at least 5 for the χ^2 test statistic to be applied.

Study Guide

The following is a chapter outline in capsule form. You should be able to demonstrate your knowledge of the ideas mentioned by giving definitions, descriptions, or specific examples. Page references are given in parentheses.

When data have been categorized and arranged in tabular format, the numbers that should appear in each cell (box) are called **expected frequencies** and the numbers that were actually observed are called **observed frequencies**. (page 601)

To determine whether the difference between the observed frequencies and the expected frequencies is significant, we calculate the **chi-square test statistic**. (page 601)

To determine when the value of the χ^2 test statistic is significant, we use the **chi-square distribution.** The number of degrees of freedom is always one less than the number of sample proportions that we are testing. (page 602)

The **goodness-of-fit test** is used to determine whether the observed frequencies fit with what was expected. (page 610)

A **contingency table** is an arrangement of data into a two-way classification. One of the classifications is entered in rows and the other in columns. (page 616)

Contingency tables are used to summarize observed and expected frequencies for a **test of independence of two variables** associated with a population. (page 616)

The **expected frequency** of any cell in a contingency table is found by multiplying the total of the row with the total of the column to which the cell belongs. The product is then divided by the total sample space. (page 617)

If a contingency table has r rows and c columns, then the number of degrees of freedom is $(r - 1) \cdot (c - 1)$. (page 618)

Formulas to Remember

You should be able to identify each symbol in the following formulas, understand the relationships among the symbols expressed in each formula, understand the significance of each formula, and use the formulas in solving problems.

1. $\chi^2 = \Sigma \dfrac{(O - E)^2}{E}$

   where O = observed frequency and E = expected frequency.
2. The expected frequency for any cell of a contingency table:

$$\frac{\text{(total of row to which cell belongs)} \cdot \text{(total of column to which cell belongs)}}{\text{total sample size}}$$

3. The number of degrees of freedom for a contingency table:

   $(r - 1)(c - 1)$

   where c = number of columns and r = number of rows

Testing Your Understanding of This Chapter's Concepts

1. When using the chi-square independence test, the null hypothesis is that the two characteristics under consideration are independent whereas the alternated hypothesis is that they are dependent. Why is it true that this test is always right-tailed?
2. When applying the chi-square test of independence, if the null hypothesis is rejected, we can conclude that the two characteristics are dependent. Does this imply a causal relationship between the two characteristics? Explain your answer. Give an example to support your answer.
3. A school principal is interested in knowing whether the grade level of a child

determines which parent will come (assuming only one parent comes) on open school day to discuss the child's academic progress with the teacher. The following randomly selected data are available.

		Grade Level of Child in School		
		Elementary School	Junior High School	Senior High School
Which Parent Came on Open School Day?	Father	32	47	96
	Mother	78	53	29

The following is a MINITAB printout of this information, showing the expected frequencies as well as the calculated values for chi-square.

a. Verify the values given for the expected frequencies in each cell and also the chi-square values.

b. Find the value of p.

```
MTB > READ TABLE IN C1, C2, C3
DATA> 32   47   96
DATA> 78   53   29
DATA> END
      2 ROWS READ

MTB > CHISQUARE ANALYSIS ON TABLE IN C1, C2, C3
Expected counts are printed below observed counts:
           C1        C2        C3     Total
    1       32        47        96       175
          57.46     52.24     65.30

    2       78        53        29       160
          52.54     47.76     59.70

Total      110       100       125       335

ChiSq =  11.283 +   0.525 + 14.435 +
         12.341 +   0.575 + 15.788 = 54.947

df = 2

MTB > STOP
```

THINKING CRITICALLY

1. In a recent survey conducted by the Acme Insurance Company of 57 cars equipped with some anti-theft device, the following information was obtained.

		Type of Antitheft Device		
		Ignition Shutoff	*Steering Wheel Lock*	*Burglar Alarm*
Size of Car	*Compact*	8	7	13
	Intermediate	2	9	3
	Large	7	2	6

Using a 5% level of significance, test the null hypothesis that the type of antitheft device used is independent of the size of the car.

2. In the previous exercise it was not really possible to perform the chi-square test of independence since the assumptions regarding the expected frequencies were not met for each cell. To overcome this difficulty we can combine rows or columns, eliminate rows or columns, or increase the sample size. Combine the last two rows to form a new contingency table and perform the hypothesis test indicated earlier. Compare the results. What can we conclude?

3. Many people have never flown in an airplane because they are afraid of heights. The following data were obtained by asking 200 people in each of five different age groups whether they had serious fears of heights.

Age Group (in years)	Number of People Who Fear Heights
11–20	121
21–30	93
31–40	88
41–50	76
51–60	69

Using a 5% level of significance, test the null hypothesis that the proportion of people in each age group who fear heights is the same. (*Hint*: The contingency table must account for all 1000 people.)

4. For the following contingency table,

Column

		1	2
Row	1	*a*	*b*
	2	*c*	*d*

verify that the chi-square test statistic is: $\chi^2 = \dfrac{(a + b + c + d)(ad - bc)^2}{(a + b)(c + d)(b + d)(a + c)}$

Review Exercises for Chapter 12

1. A 4 × 3 contingency table has row totals 15, 20, 25, and 15. The column totals are each 25. Determine the expected cell frequencies.

2. Consider the following data.

Had Characteristic

	I	II	III
Yes	30	10	22
No	18	20	25

Find the value of p.

3. Refer back to the previous question. Find the value of the chi-square test statistic *and* the number of degrees of freedom.

4. 180 frequent fliers on the planes of one of the major U.S. airline companies were asked to rate the quality of the meals served on board the aircraft. The following results were obtained.

How Would You Rate the Meal?

Sex of Flier		Below Average	Average	Above Average
	Male	47	27	8
	Female	61	28	9

Using a 5% level of significance, test the null hypothesis that the rating of the meal is independent of the sex of the flier.

5. A recent survey of 1470 drug users and nondrug users contained the following question: "Do you believe that drug use is harmful to your health?" The responses are shown below:

	Drug User	Nondrug User
Harmful	198	712
Not Harmful	452	108

What is the expected number of drug users who believe that drug use is harmful?
a. 402.4   b. 242.1   c. 398.3   d. 474.72   e. none of these

6. According to police department personnel records, the number of police officers who are not working because of job-related disabilities is as follows:

	Married Male	Single Male	Married Female	Single Female
Number of Employees	46	54	76	62

Using a 5% level of significance, test the null hypothesis that the proportion of police officers who are not working because of job-related disabilities is the same for all categories.

7. Many people believe that criminals who plead guilty or plea bargain get lighter jail sentences than those who are convicted by trial. In a survey of 1110 defendants in armed robbery cases the following data were obtained. (*Note*: All the defendants had prior police records.)

		Sent to Jail	Not Sent to Jail
Plea Entered	*Guilty*	403	69
	Not Guilty	612	26

Using a 5% level of significance, test the null hypothesis that the plea entered (guilty or not guilty) is independent of the jail sentence (sent or not sent to jail.)

8. The following table shows the number and type of crimes committed during 1990 for one large city and the police precincts in which they were committed.

		Type of Crime				
		Car Theft	*Murder*	*Robbery*	*Burglary*	*Rape*
	I	68	14	37	57	28
Precinct Where Committed	*II*	46	19	39	61	21
	III	103	28	48	54	32
	IV	49	17	62	40	33
	V	58	31	31	49	27

Using a 5% level of significance, test the null hypothesis that the type of crime committed is independent of the precinct where it was committed.

9. Are female jurors more liberal than male jurors? A recent study of 184 jurors involved in similar cases produced the following results.

		Tendency of Juror to Be		
		Liberal	*Fair*	*Vindictive*
Sex of Juror	*Male*	31	27	35
	Female	42	28	21

Using a 5% level of significance, test the null hypothesis that the sex of the juror is independent of his or her tendency to be liberal, fair, or vindictive.

10. A researcher claims that among people who eat in restaurants, the following preferences (given in percentage instead of ratio form) apply.

Day of Week Preferred	Number of People Indicating Such a Preference
Sunday	25%
Monday	11%
Tuesday	9%
Wednesday	3%
Thursday	5%
Friday	19%
Saturday	28%

Sample results from 400 patrons of the restuarants are given here.

Day	Sun.	Mon.	Tues.	Wed.	Thurs.	Fri.	Sat
Number	87	40	38	10	20	73	132

Using a 5% level of significance, test the null hypothesis that the given percentages are correct.

Chapter Test

1. The United Fruit Corporation operates several fruit stores that are open 24 hours a day. The corporation is interested in knowing whether there is a relationship between the amount of a purchase and the time of the day that it is made. In addition to printing the amount of the purchase, the cash registers used in all of its stores also prints the time of purchase. A random selection of 420 of the purchases made at these stores is taken and the results shown here are obtained.

	Time of Purchase		
	8:00 A.M.–3:59 P.M.	4:00 P.M.–11:59 P.M.	12:00 A.M.–7:59 A.M.
Less than $3.00	62	56	21
$3.01–$8.00	72	48	19
Over $8.00	53	52	37

Amount of Purchase

Using a 5% level of significance, test the null hypothesis that the amount of purchase at stores is independent of the time that it is purchased.

2. A consumer opinion research organization interviewed 124 male viewers and 131 female viewers to determine their opinion about the new police show that is being shown on TV during prime time. The following counts were obtained.

What Do You Think of the New Show?

		Opposed	In Favor	No Opinion
Sex of Viewer	Male	31	71	22
	Female	42	62	27

Using a 5% level of significance, test the null hypothesis that the sex of the viewer is independent of the viewer's opinion.

3. A researcher interviewed 222 people in various occupations to determine whether there is any relation between occupation and smoking. The accompanying table presents the results of the interview.

Occupation

		Teacher	Lawyer	Doctor	Accountant
Smoker?	Yes	21	48	12	18
	No	35	36	25	27

Using a 1% level of significance, test the null hypothesis that the proportion of smokers in the different occupations mentioned is the same.

4. Police department records indicate the following number of arrests for drunken driving on U.S. Highway 1 on the different days of the week.

Mon.	Tues.	Wed.	Thurs.	Fri.	Sat.	Sun.
18	16	14	17	29	27	26

Using a 5% level of significance, test the null hypothesis that the number of arrests for drunken driving is the same for the different days of the week.

5. Many of our nation's lakes and rivers are polluted with mercury, PCV's, or other toxic chemicals. Data gathered at numerous locations along several rivers by the Charleston Environmental Group are shown in the accompanying table.

River

		A	B	C	D
Level of Pollution	Low	3	14	9	4
	Moderate	4	12	11	7
	High	8	10	13	6

Using a 5% level of significance, test the null hypothesis that the level of pollution is independent of the river involved.

6. The Marro Apparel Company sold 200 dresses during January 1991, as follows

79 size 10 dresses

59 size 12 dresses

42 size 8 dresses

20 size 14 dresses

Using a 5% level of significance, test the null hypothesis that the sizes of the dresses sold (in the order given) are in the ratio 8:5:4:3.

7. A heating consultant in the northeast has gathered the following data on how residential homes are heated during the winter season and the family income of the homeowner.

		Family Income	
	Under $30,000	*Between $30,000 and $60,000*	*More Than $60,000*
Gas	31	41	37
Oil	28	34	27
Wood/Coal	19	26	31

How Do You Heat Your Home?

Using a 1% level of significance, test the null hypothesis that family income is independent of the method of heating used.

8. Is our government doing enough for the homeless? Bates and Young[1] (San Francisco, 1991) interviewed 534 randomly selected Americans. The following results were obtained.

	City in Which Respondent Lives				
	Los Angeles	*New York*	*Chicago*	*Detroit*	*Miami*
Yes	56	41	55	48	72
No	57	49	62	53	41

Is Our Government Doing Enough for the Homeless

Using a 5% level of significance, test the null hypothesis that the proportion of Americans who think that our government is doing enough for the homeless is the same for the cities mentioned.

9. Safety engineers often analyze industrial accidents to determine whether the day of the week when an accident occurs is independent of the time of the day when it occurs. For one very large company employing many workers the following data on industrial accidents is available.

[1]Bates and Young, *What is Our Government Doing for the Homeless?*, San Francisco, 1991.

		Mon.	Tues.	Wed.	Thurs.	Fri.
				Day of the Week		
Time of Day When Accident Occurred	Morning Shift	7	6	4	3	9
	Afternoon Shift	5	8	3	2	7
	Evening Shift	10	2	8	4	5

Using a 1% level of significance, test the null hypothesis that the time of day when the accident occurs is independent of the day of the week when the accident occurs.

10. According to a certain theory, the number of flowers of each of four types should be as shown in the accompanying table. The actual numbers obtained as a result of genetic manipulation are also given:

Type of flower	Expected	Actual
Smooth pink	315	325
Wrinkled pink	104	98
Smooth red	88	92
Wrinkled red	72	65
	Total 580	580

At the 5% level of significance, are the data consistent with the theory?

For questions 11–15 use the following information: 1709 voters of a city were asked to indicate whether or not they favored hiring an outside consultant for revamping the city's judicial system. The following results were obtained:

		Opposea	Don't Care	In Favor
Age of Voter	Under 30 Years	124	139	295
	30–50 Years	136	144	266
	Over 50 Years	142	176	287

11. What is the expected number of voters under 30 years of age who are in favor of hiring an outside consultant?
   a. 276.9   b. 149.9   c. 270.9   d. 131.3   e. none of these
12. What is the expected number of voters over 50 years of age who are opposed to hiring an outside consultant?
   a. 146.6   b. 142.3   c. 149.9   d. 162.5   e. none of these
13. The number of degrees of freedom for this problem is
   a. 3   b. 9   c. 4   d. 8   e. none of these
14. The value of χ^2 for this problem is
   a. 4.583   b. 4.982   c. 4.015   d. 4.662   e. none of these

15. Using a 5% level of significance, test the null hypothesis that the age of the voter is independent of the view expressed by the voter.
  a. Do not reject null hypothesis   **b.** Reject null hypothesis   **c.** Not enough information given

Suggested Further Reading

1. Albright, S. Christian. *Statistics for Business and Economics*. New York: Macmillan, 1987.
2. Anderson, T. W., and S. L. Sclove. *An Introduction to the Statistical Analysis of Data*. Boston: Houghton Mifflin, 1978. (Chapter 12)
3. Cochran, W. G. "The χ^2 Test of Goodness of Fit" in *Annals of Mathematical Statistics* Vol. 23 (1952), pp. 315–345.
4. Iman, R. L., and W. J. Conover. *A Modern Approach to Statistics*. New York: Wiley, 1983.
5. Levin, Richard J. *Statistics for Management*, 4th ed. Englewood Cliffs, NJ: Prentice-Hall, 1987.
6. Plane, Donald R., and Edward B. Oppermann. *Business and Economic Statistics*. Plano, TX: Business Publications, 1986.

Analysis of Variance

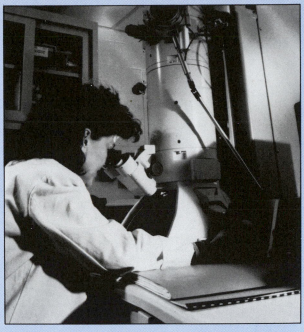

Often a researcher must compare the effect of different diets on the reduction of blood serum cholesterol levels. What analysis is needed? See discussion on p. 662. (© *Herb C. Ohlmeyer, Fran Heyl Associates*)

In previous chapters we presented hypothesis-testing procedures for determining whether the difference between two sample means is significant. In this chapter we test hypotheses about several means. The ANOVA techniques to be discussed allow us to test the null hypothesis that *all* sample means are equal against the alternate hypothesis that at least one mean value is different.

Statistics in the News

COMPARISON OF NEW YORK
STATE AND NATIONAL
AVERAGES ON COLLEGE
BOARD ACHIEVEMENT TEST,
1990

	Average Score	
Achievement Test	NYS	National
English Composition	524	512
Mathematics Level 1	580	539
Biology	588	546
Chemistry	601	571
American History	544	508
Spanish	556	529
Mathematics Level 2	669	654
Physics	611	595
French	561	546
Literature	529	517
European History	591	544
German	587	551
Hebrew	641	602
Latin	579	548
Russian	728	642
All Tests Combined	558	532

Source: New York State Department of Education

An analysis of this article indicates that the average scores of students on many different tests are given. How do we determine if there is any significant difference between the average scores of New York students and the national average on the various tests? A test-by-test comparison is very time-consuming. This is true even if we know the sample sizes.

13.1 INTRODUCTION

Suppose a company is interested in determining whether changing the lighting conditions of its factory will have any effect on the number of items produced by a worker. It can arrange the lighting conditions in four different ways: A, B, C, and D. To determine how lighting affects production, the factory supervisor decides to randomly arrange the lights under each of the four possible conditions for an equal number of days. She will then measure worker productivity under each of these conditions.

After all the data are collected the manager calculates the number of items produced under each of these lighting conditions. She now wishes to test whether there is any significant difference between these sample means. How does she proceed?

The null hypothesis is

$$H_0: \mu_A = \mu_B = \mu_C = \mu_D$$

This tells us that the mean number of items produced is the same for each lighting condition. The alternative hypothesis is that not all the means are the same. Thus, the lighting condition does affect production. The supervisor will reject the null hypothesis if one (or more) of the means is different from the others.

To determine if lighting affects production, she could use the techniques of Section 10.5 for testing differences between means. However, she would have to apply those techniques many times since each time she would be able to test only two means. Thus she would have to test the following null hypotheses:

$$H_0: \mu_A = \mu_B \qquad H_0: \mu_A = \mu_D \qquad H_0: \mu_B = \mu_D$$

$$H_0: \mu_A = \mu_C \qquad H_0: \mu_B = \mu_C \qquad H_0: \mu_C = \mu_D$$

In order to conclude that there is no significant difference between the sample means, she would have to accept each of the six separate null hypotheses previously listed. Performing these tests involves a tremendous amount of computation. Furthermore, in doing it this way, the probability of making a Type-I error is quite large.

Since many problems that occur in applied statistics involve testing whether there is any significant difference between several means, statisticians use a special analysis of variance technique. This is abbreviated as **ANOVA**. These types of problems occur so frequently because many companies often hire engineers to design new techniques or processes for producing products. The company must then compare the sample means of these new processes with the sample mean of the old process. Using the ANOVA technique, we have to test only one hypothesis in order to determine when to reject or not reject a null hypothesis.

ANOVA techniques can be applied to many different types of problems. In this chapter we first discuss one simple application of ANOVA techniques. This is the

ANOVA

case in which the data are classified into groups on the basis of one single property. Then we discuss two-way ANOVA techniques, where the data are classified into groups on the basis of two properties.

13.2 DIFFERENCES AMONG r SAMPLE MEANS

Suppose that a chemical researcher is interested in comparing the average number of months that four different paints will last on an exterior wall before beginning to blister. The researcher has available the following data on the lasting time of three paint samples for each of the four different brands of paint.

Lasting Time (in months)

	Paint A	Paint B	Paint C	Paint D
	13.4	14.8	12.9	13.7
	11.8	13.9	14.3	14.1
	14.4	12.4	13.3	12.1
Average:	13.2	13.7	13.5	13.3

The average lasting time for these paints is 13.2, 13.7, 13.5, and 13.3 months, respectively. Since not all the paints had the same average lasting time, the chemical researcher is interested in knowing whether the observed differences between the sample means is significant or whether they can be attributed purely to chance. Is there a rule available for determining when observed differences between sample means are significant?

Before proceeding further, we wish to emphasize the fact that our discussion will be based on certain assumptions. These are the following:

1. We will always assume that the populations from which the samples are obtained are normally distributed and that the samples are obtained independently of one another.
2. We will also assume that the populations from which the samples are obtained all have the same (often unknown) variance, σ^2.

When these assumptions are satisfied, we can obtain an estimate of σ^2 in two different ways. First we note that if the paints have the same mean, then the average lasting time of the different paints (in our example) is

$$\bar{x} = \frac{13.2 + 13.7 + 13.5 + 13.3}{4} = 13.425$$

The averages will be normally distributed with a variance of σ^2. This can be estimated by the following procedure:

$$s_{\bar{x}}^2 = \frac{(13.2 - 13.425)^2 + (13.7 - 13.425)^2 + (13.5 - 13.425)^2 + (13.3 - 13.425)^2}{4 - 1}$$

$$= \frac{0.1475}{3} = 0.0491667$$

From our earlier work we already know that

$$s_{\bar{x}}^2 \approx \left(\frac{\sigma}{\sqrt{n}}\right)^2 = \frac{\sigma^2}{n}$$

Here n = the number of samples of each brand of paint. Thus, $n \cdot s_{\bar{x}}^2$ can be used as an estimate of σ^2. In our case we get $3(0.0491667) = 0.1475$. Our first estimate of σ^2, which is based on the variation among the sample means (assuming that all the groups have the same mean), has a value of 0.1475.

Since under our second assumption the populations from which the samples are obtained all have the same (often unknown) variance, σ^2, we can obtain another estimate of σ^2 by selecting any one of the sample variances. For each brand of paint we can obtain an estimate of σ^2 based on 2 degrees of freedom. We can then pool the estimates of the σ from each of the brands of paint to obtain an estimate of σ. This estimate of σ is not affected by variation of the population means among the paints. We have the following computations.

Paint Brand	Variance
A	$s_1^2 = \dfrac{(13.4 - 13.2)^2 + (11.8 - 13.2)^2 + (14.4 - 13.2)^2}{3 - 1} = \dfrac{3.44}{2} = 1.72$
B	$s_2^2 = \dfrac{(14.8 - 13.7)^2 + (13.9 - 13.7)^2 + (12.4 - 13.7)^2}{3 - 1} = \dfrac{2.94}{2} = 1.47$
C	$s_3^2 = \dfrac{(12.9 - 13.5)^2 + (14.3 - 13.5)^2 + (13.3 - 13.5)^2}{3 - 1} = \dfrac{1.04}{2} = 0.52$
D	$s_4^2 = \dfrac{(13.7 - 13.3)^2 + (14.1 - 13.3)^2 + (12.1 - 13.3)^2}{3 - 1} = \dfrac{2.24}{2} = 1.12$

Taking the average of these gives

$$\frac{s_1^2 + s_2^2 + s_3^2 + s_4^2}{4} = \frac{1.72 + 1.47 + 0.52 + 1.12}{4}$$

$$= \frac{4.83}{4} = 1.2075$$

Thus, our second estimate of σ^2, which is based on the variation within the samples, is 1.2075.

How do we compare these two estimates of σ^2? Our first estimate of σ^2, which is based on the variation among the sample means, had a value of 0.1475, whereas our second estimate, which is based on the variation within the samples (or on the fact that the variation is due purely to chance), had a value of 1.2075. It seems reasonable that the variation among the sample means should be larger than that which is due purely to chance.

Fortunately, statisticians use a special F-distribution that facilitates such comparisons. It tells us under what conditions to reject the null hypothesis assuming the variation of the sample means is simply too great to be attributed to chance. This, of course, implies that the differences among the sample means is significant. The exact technique will be discussed in the next section after we introduce a convenient tabular arrangement for displaying our computations.

COMMENT   The straight averaging technique discussed until now works since the example given is a case with equal sample sizes.

13.3 ONE-WAY OR SINGLE-FACTOR ANOVA

Let us return to the factory supervisor problem discussed in the introduction. The manager repeated each of the lighting conditions on five different days. Table 13.1 indicates the number of items produced under each of the conditions.

The means for each of these lighting conditions are 12.4, 12.8, 11.2, and 13. Since the average number of items produced is not the same for all the lighting conditions, the question becomes: "Is the variation among individual sample means due purely to chance or are these differences due to the different lighting conditions?" The null hypothesis is H_0: $\mu_A = \mu_B = \mu_C = \mu_D$. If the null hypothesis is true, then the lighting condition does not affect production. We can then consider our

TABLE 13.1   Number of Items Produced Under Different Lighting Conditions

		Day					
		1	*2*	*3*	*4*	*5*	*Average*
	A	12	10	15	12	13	12.4
Lighting	*B*	16	14	9	10	15	12.8
Conditions	*C*	11	15	8	12	10	11.2
	D	15	14	12	11	13	13

results as a listing of the number of items produced on 20 randomly selected days under one of the lighting conditions.

As mentioned earlier, when working with such problems we will assume that the number of items produced is normally distributed and that the variance is the same for each of the lighting conditions. We will also assume that the experiments with the different lighting conditions are independent of each other. (Usually the experiments are conducted in random order so that we have independence.)

Let us now apply the ANOVA techniques to this problem. Notice that we have included the row totals in Table 13.2. In applying the ANOVA technique, we analyze the reasons for the difference among the means. What is the source of the variance?

Total Sum of Squares

The first number that we calculate is called the **Total Sum of Squares** and is abbreviated as SS (total). To obtain the SS (total), we first square each of the numbers in the table and add these squares together. Then we divide the square of the grand total (total of all the rows or the total of all the columns) by the number in the sample. (In our case there are 20 in the sample.) Subtracting this result from the sum of the squares, we get in our case

$$[12^2 + 10^2 + 15^2 + 12^2 + 13^2 + 16^2 + 14^2 + 9^2 + 10^2$$

$$+ 15^2 + 11^2 + 15^2 + 8^2 + 12^2 + 10^2 + 15^2 + 14^2 + 12^2 + 11^2 + 13^2] - \left[\frac{(247)^2}{20}\right]$$

$$= 3149 - 3050.45$$

$$= 98.55$$

*Thus, SS (total) $= 98.55$.   (An * will precede important results.)

To find the next important result, we first square each of the row totals and divide the sum of these squares by the number in each row, which is 5. Then we divide the square of the grand total by the number in the sample. We subtract this

		Day						
		1	*2*	*3*	*4*	*5*	*Row Total*	
	A	12	10	15	12	13	62	
Lighting Conditions	*B*	16	14	9	10	15	64	
	C	11	15	8	12	10	56	
	D	15	14	12	11	13	65	
							247	Grand Total

result from the sum of squares and get

$$\left(\frac{62^2 + 64^2 + 56^2 + 65^2}{5}\right) - \frac{(247)^2}{20} = 3060.2 - 3050.45 = 9.75$$

Sum of Squares Due to the Factor

This result is called the **Sum of Squares Due to the Factor.** The factor in our example is the lighting conditions. Thus,

$$*SS(\text{lighting}) = 9.75$$

Sum of Squares of the Error

Now we calculate a number that is called the **Sum of Squares of the Error** abbreviated as SS(error). We first square each entry of the table and add these squares together. Then we square each row total and divide the sum of these squares by the number in each row. Subtracting this result from the sum of squares, we get

$$\left(12^2 + 10^2 + 15^2 + 12^2 + \cdots + 11^2 + 13^2\right) - \frac{62^2 + 64^2 + 56^2 + 65^2}{5}$$

$$= 3149 - 3060.2$$

$$= 88.8$$

Thus,

$$* \; SS(\text{error}) = 88.8$$

COMMENT   Actually, SS(error) = SS(total) − SS(factor).

ANOVA Table

We enter these results along with the appropriate number of degrees of freedom, which we will determine shortly, in a table known as an **ANOVA table.** The general form of an ANOVA table is shown in Table 13.3.

TABLE 13.3   ANOVA Table

Source of Variation	Sum of Squares (SS)	Degrees of Freedom (df)	Mean Square (MS)	F-Ratio
Factors of experiment				
Error				
Total				

In our case we have the ANOVA table shown in Table 13.4. We have entered the important results that we obtained in the preceding paragraphs in the appropriate space of the table. The degrees of freedom are obtained according to the following rules.

TABLE 13.4 ANOVA Table

Source of Variation	Sum of Squares	Degrees of Freedom	Mean Square	F-Ratio
Factors (lighting)	9.75	3	3.25	$\dfrac{3.25}{5.55} = 0.59$
Error	88.8	16	5.55	
Total	98.55	19		

1. The number of degrees of freedom for the factor tested is one less than the number of possible levels at which the factor is tested. If there are r levels of the factor, the number of degrees of freedom is $r - 1$.
2. The number of degrees of freedom for the error is one less than the number of repetitions of each condition multiplied by the number of possible levels of the factor. If each condition is repeated c times, then the number of degrees of freedom is $r(c - 1)$.
3. Finally, the number of degrees of freedom for the total is one less than the total number in the sample. If the total sample consists of n things, the number of degrees of freedom is $n - 1$.

In our case there are four experimental conditions and each condition is repeated five times so that $r = 4$, $c = 5$, and $n = 20$.

Thus,

$$df(\text{factor}) = r - 1 = 4 - 1 = 3$$

$$df(\text{error}) = r(c - 1) = 4(5 - 1) = 4 \cdot 4 = 16$$

$$df(\text{total}) = n - 1 = 20 - 1 = 19$$

We enter these values in an ANOVA table as shown in Table 13.4.

Mean Square

Now we calculate the values that belong in the **mean square** column. We divide the sum of squares for the row by the number of degrees of freedom for that row. Thus,

$$MS(\text{lighting}) = \frac{SS(\text{lighting})}{df(\text{lighting})} = \frac{9.75}{3} = 3.25$$

$$MS(\text{error}) = \frac{SS(\text{error})}{df(\text{error})} = \frac{88.8}{16} = 5.55$$

Although we are comparing the sample means for the different lighting conditions, the ANOVA technique compares variances under the assumption that the variance among all the levels is 0. If the variance is 0, then all the means are the same.

Suppose we are given two samples of size n_1 and n_2 for two populations that have roughly the shape of a normal distribution. Then we base tests on the equality of the two population standard deviations (or variances) on the ratio $\dfrac{s_1^2}{s_2^2}$ or $\dfrac{s_2^2}{s_1^2}$, where s_1 and s_2 are the standard deviations of both samples. The distribution of such a ratio is a continuous distribution called the **F-distribution.** This distribution depends on the number of degrees of freedom of both sample estimates, $n_1 - 1$ and $n_2 - 1$. When using the distribution **we reject the null hypotheses** of equal variances ($\sigma_1 = \sigma_2$) when the test statistic value of F exceeds $F_{\alpha/2}$, where α is the level of significance.

When comparing variances we use the **F-distribution.** The test statistic is

$$F = \frac{MS(\text{lighting})}{MS(\text{error})}$$

COMMENT   When using the F-distribution we are testing whether or not the variances among the factors is zero (in which case the means are equal). These situations are pictured in Figure 13.1 for the case involving three means.

F-distribution (margin note)

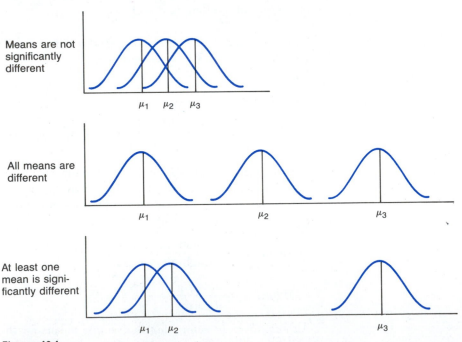

Means are not significantly different

μ_1 μ_2 μ_3

All means are different

μ_1 μ_2 μ_3

At least one mean is significantly different

μ_1 μ_2 μ_3

Figure 13.1

As mentioned earlier, the values of the F-distribution depend on the number of degrees of freedom of the numerator, that of the denominator, and on the level of significance. Table IX in the Appendix gives us the different values corresponding to different degrees of freedom. From Table IX we find that the F-value with 3 degrees of freedom for the numerator and 16 degrees of freedom for the denominator at the 5% level of significance is 3.24. We reject the null hypothesis if the F-value is greater than 3.24.

In our case the value of the F statistic is

$$\frac{MS(\text{lighting})}{MS(\text{error})} = \frac{3.25}{5.55} = 0.59$$

Since 0.59 is less than 3.24, we do not reject the null hypothesis. Thus, the data do not indicate that the lighting condition affects production.

ANOVA Table

Let us summarize the procedure to be used when testing several sample means by the ANOVA technique. First draw an ANOVA table such as shown in Table 13.5. In this table we have placed numbers in parentheses in the various cells. The values that belong in each of these cells are obtained by using the following formulas.

Source of Variation	Sum of Squares	Degrees of Freedom	Mean Square	F-Ratio
Factor of the experiment	(1)	(4)	(7)	(9)
Error	(2)	(5)	(8)	
Total	(3)	(6)		

Cell (1): $\dfrac{\Sigma(\text{each row total})^2}{\text{number in each row}} - \dfrac{(\text{grand total})^2}{\text{total sample size}}$

Cell (2): $\Sigma(\text{each original number})^2 - \dfrac{\Sigma(\text{each row total})^2}{\text{number in each row}}$

Cell (3): $\Sigma(\text{each original number})^2 - \dfrac{(\text{grand total})^2}{\text{total sample size}}$

Cell (4): there are r levels of the factor, $r - 1$

Cell (5): If there are c repetitions of each of the r levels, $r(c - 1)$

Cell (6): If the total sample consists of n things, $n - 1$

Cell (7): $\dfrac{\text{cell (1) value}}{\text{cell (4) value}}$

Cell (8): $\dfrac{\text{cell (2) value}}{\text{cell (5) value}}$

Cell (9): $\dfrac{\text{cell (7) value}}{\text{cell (8) value}}$

Although these formulas seem complicated, they are easy to use as the following examples illustrate.

EXAMPLE 1 Testing the Tar Content of Cigarettes

Three brands of cigarettes, six cigarettes from each brand, were tested for tar content. Do the following data indicate that there is a significant difference in the average tar content for these three brands of cigarettes? (Use a 5% level of significance.)

								Row Total
	X	14	16	12	18	11	13	84
Brand	Y	10	11	22	19	9	18	89
	Z	24	22	19	18	20	19	122
								295 Grand Total

Solution

The null hypothesis is that the sample mean tar content is the same for the three brands of cigarettes. To solve this problem, we set up the following ANOVA table.

Source of Variation	Sum of Squares	Degrees of Freedom	Mean Square	F-Ratio
Cigarettes	(1)	(4)	(7)	(9)
Error	(2)	(5)	(8)	
Total	(3)	(6)		

Now we calculate the values that belong in each of the cells by using the formulas given on pages 649–650. We have

Cell (1): $\dfrac{\Sigma(\text{each row total})^2}{\text{number in each row}} - \dfrac{(\text{grand total})^2}{\text{total sample size}}$

$= \dfrac{84^2 + 89^2 + 122^2}{6} - \dfrac{(295)^2}{18}$

$$= 4976.83 - 4834.72$$

$$= 142.11$$

Cell (2): $\Sigma(\text{each original number})^2 - \dfrac{\Sigma(\text{each row total})^2}{\text{number in each row}}$

$$= \left(14^2 + 16^2 + 12^2 + \cdots + 19^2 \right) - \dfrac{(84^2 + 89^2 + 122^2)}{6}$$

$$= 5187 - 4976.83$$

$$= 210.17$$

Cell (3): $\Sigma(\text{each original number})^2 - \dfrac{(\text{grand total})^2}{\text{total sample size}}$

$$= \left(14^2 + 16^2 + 12^2 + \cdots + 19^2 \right) - \dfrac{(295)^2}{18}$$

$$= 5187 - 4834.72$$

$$= 352.28$$

Cell (4): $df = r - 1$

$$= 3 - 1 = 2$$

Cell (5): $df = r(c - 1)$

$$= 3(6 - 1) = 3(5) = 15$$

Cell (6): $df = n - 1$

$$= 18 - 1 = 17$$

Cell (7): $\dfrac{\text{cell (1) value}}{\text{cell (4) value}} = \dfrac{142.11}{2} = 71.06$

Cell (8): $\dfrac{\text{cell (2) value}}{\text{cell (5) value}} = \dfrac{210.17}{15} = 14.01$

Cell (9): $\dfrac{\text{cell (7) value}}{\text{cell (8) value}} = \dfrac{71.06}{14.01} = 5.07$

Now we enter these numbers on the ANOVA table.

Source of Variation	Sum of Squares	Degrees of Freedom	Mean Square	F-Ratio
Cigarettes	142.11 (1)	2 (4)	71.06 (7)	5.07 (9)
Error	210.17 (2)	15 (5)	14.01 (8)	
Total	352.28 (3)	17 (6)		

From Table IX in the Appendix we find that the F-value with 2 degrees of freedom for the numerator and 15 degrees of freedom for the denominator at the 5% level of significance is 3.68. Since we obtained an *F*-value of 5.07, which is larger than 3.68, we reject the null hypothesis that the sample mean tar content is the same for the three brands of cigarettes.

COMMENT   Rejection of the null hypothesis $H_0: \mu_1 = \mu_2 = \mu_3$ does not tell us where the difference lies.

EXAMPLE 2

Four groups of five students each were taught a skill by four different teaching techniques. At the end of a specified time the students were tested and their scores were recorded. Do the following data indicate that there is a significant difference in the mean achievement for the four teaching techniques? (Use a 1% level of significance.)

						Row Total
A	64	73	69	75	78	359
B	73	82	71	69	74	369
C	61	79	71	73	66	350
D	63	69	68	74	75	349

Teaching Technique

1427   Grand Total

Solution

The null hypothesis is that the mean achievement for the four teaching techniques is the same. To solve this problem, we set up the following ANOVA table.

Source of Variation	Sum of Squares	Degrees of Freedom	Mean Square	F-Ratio
Teaching method	(1) 52.15	(4) 3	(7) 17.38	(9) 0.56
Error	(2) 500.4	(5) 16	(8) 31.28	
Total	(3) 552.55	(6) 19		

Now we calculate the values that belong in each of the cells by using the formulas given on pages 649–650. We have

$$\text{Cell (1):} \quad \frac{\Sigma(\text{each row total})^2}{\text{number in each row}} - \frac{(\text{grand total})^2}{\text{total sample size}}$$

$$= \frac{359^2 + 369^2 + 350^2 + 349^2}{5} - \frac{(1427)^2}{20}$$

$$= 101{,}868.6 - 101{,}816.45$$

$$= 52.15$$

Cell (2): $\Sigma(\text{each original number})^2 - \dfrac{\Sigma(\text{each row total})^2}{\text{number in each row}}$

$$= (64^2 + 73^2 + 69^2 + 75^2 + \cdots + 74^2 + 75^2)$$

$$\quad - \left(\frac{359^2 + 369^2 + 350^2 + 349^2}{5} \right)$$

$$= 102{,}369 - 101{,}868.6$$

$$= 500.4$$

Cell (3): $\Sigma(\text{each original number})^2 - \dfrac{(\text{grand total})^2}{\text{total sample size}}$

$$= \left(64^2 + 73^2 + 69^2 + 75^2 + \cdots + 74^2 + 75^2 \right) - \frac{(1427)^2}{20}$$

$$= 102{,}369 - 101{,}816.45$$

$$= 552.55$$

Cell (4): $df = r - 1 = 4 - 1 = 3$

Cell (5): $df = r(c - 1) = 4(5 - 1) = 4(4) = 16$

Cell (6): $df = n - 1 = 20 - 1 = 19$

Cell (7): $\dfrac{\text{cell (1) value}}{\text{cell (4) value}} = \dfrac{52.15}{3} = 17.38$

Cell (8): $\dfrac{\text{cell (2) value}}{\text{cell (5) value}} = \dfrac{500.4}{16} = 31.28$

Cell (9): $\dfrac{\text{cell (7) value}}{\text{cell (8) value}} = \dfrac{17.38}{31.28} = 0.56$

We enter all these values on the ANOVA table as shown.

From Table IX in the Appendix we find that the F-value with 3 degrees of freedom for the numerator and 16 degrees of freedom for the denominator at the 1% level of significance is 5.29. Since we obtained an F-value of 0.56, we do not reject the null hypothesis that the mean achievement for the four teaching techniques is the same.

*13.4 TWO-WAY ANOVA

Let us return to the example given in Section 13.2 and construct an analysis-of-variance table for the data that are reproduced here except that we have now rewritten the chart in a slightly different manner. (The reason for this will be apparent shortly.)

	Lasting Time (in months)			Row Total
Paint A	13.4	11.8	14.4	39.6
Paint B	14.8	13.9	12.4	41.1
Paint C	12.9	14.3	13.3	40.5
Paint D	13.7	14.1	12.1	39.9
Column Total	54.8	54.1	52.2	161.1 *Grand Total*

We set up the following ANOVA table to which we have added the values that belong in each of the cells by using the formulas given on pages 649–650.

Source of Variation	Sum of Squares	Degrees of Freedom	Mean Square	F-Ratio
Different Paints	0.4425 (1)	3 (4)	0.1475 (7)	0.1222 (9)
Error	9.66 (2)	8 (5)	1.2075 (8)	
Total	10.1025 (3)	11 (6)		

Thus, we have

Cell (1): $\left(\dfrac{(39.6)^2 + (41.1)^2 + (40.5)^2 + (39.9)^2}{3} \right) - \dfrac{(161.1)^2}{12}$

$$= 2163.21 - 2162.7675 = 0.4425$$

Cell (2): $[(13.4)^2 + (11.8)^2 + \cdots + (12.1)^2]$

$$- \left[\dfrac{(39.6)^2 + (41.1)^2 + (40.5)^2 + (39.9)^2}{3} \right]$$

$$= 2172.87 - 2163.21 = 9.66$$

Cell (3): $\left[(13.4)^2 + (11.8)^2 + \cdots + (12.1)^2 \right] - \dfrac{(161.1)^2}{4.3}$

$$= 2172.87 - 2162.7675 = 10.1025$$

Cell (4): $4 - 1 = 3$

Cell (5): $4(3 - 1) = 8$

Cell (6): $12 - 1 = 11$

Cell (7): $\dfrac{0.4425}{3} = 0.1475$

Cell (8): $\dfrac{9.66}{8} = 1.2075$

Cell (9): $\dfrac{0.1475}{1.2075} = 0.1222$

From Table IX in the Appendix we find that the F-value with 3 degrees of freedom for the numerator and 8 degrees of freedom for the denominator at the 5% level of significance is 4.07. Therefore, since we obtained an F-value of 0.1222, we do not reject the null hypothesis that there is a significant difference in the average lasting times of the four brands of paint.

After analyzing the data carefully the chemical engineer discovers that the test conditions for each of the brands of paint were not the same and that the samples for each brand of paint were subject to different temperature and humidity conditions. This factor, which was not considered earlier, definitely has to be considered. Thus, instead of just having to determine if the average lasting time for the four different brands of paint is the same, we also have to consider the various test conditions as a possible cause of the variation as well as pure chance. Under these new circumstances the data must be arranged as follows:

Test Conditions

	Test Condition I	Test Condition II	Test Condition III	Row Total
Paint A	13.4	11.8	14.4	39.6
Paint B	14.8	13.9	12.4	41.1
Paint C	12.9	14.3	13.3	40.5
Paint D	13.7	14.1	12.1	39.9
Column Total	54.8	54.1	52.2	161.1 *Grand Total*

Blocks
Treatments
Two-way Analysis of Variance

When discussing this second factor it is customary to refer to it as **blocks** as opposed to the original factor, which is referred to as the **treatments.** We then have the following ANOVA table for such a **two-way analysis of variance.**

Source of Variation	Sum of Squares	Degrees of Freedom	Mean Square	F-Ratio
Treatments	(1)	(5)	(9)	(12)
Blocks	(2)	(6)	(10)	(13)
Error	(3)	(7)	(11)	
Total	(4)	(8)		

The values that belong in each of these cells for a two-way analysis of variance table are obtained by using the following formulas.

Cell (1): $\dfrac{\Sigma(\text{each row total})^2}{\text{number in each row}} - \dfrac{(\text{grand total})^2}{r \cdot c}$

where r = number of levels of one factor (row)

c = number of levels of second factor (column)

Cell (2): $\dfrac{\Sigma(\text{each column total})^2}{\text{number in each column}} - \dfrac{(\text{grand total})^2}{r \cdot c}$

Cell (3): $\Sigma(\text{each original number})^2 - \dfrac{\Sigma(\text{each row total})^2}{\text{number in each row}}$

$- \dfrac{\Sigma(\text{each column total})^2}{\text{number in each column}} + \dfrac{(\text{grand total})^2}{r \cdot c}$

Cell (4): $\Sigma(\text{each original number})^2 - \dfrac{(\text{grand total})^2}{r \cdot c}$

Cell (5): $r - 1$

Cell (6): $c - 1$

Cell (7): $(r - 1)(c - 1)$

Cell (8): $rc - 1$

Cell (9): $\dfrac{\text{cell (1) value}}{\text{cell (5) value}}$

Cell (10): $\dfrac{\text{cell (2) value}}{\text{cell (6) value}}$

Cell (11): $\dfrac{\text{cell (3) value}}{\text{cell (7) value}}$

Cell (12): $\dfrac{\text{cell (9) value}}{\text{cell (11) value}}$

Cell (13): $\dfrac{\text{cell (10) value}}{\text{cell (11) value}}$

Let us apply these formulas to our example. We have the following two-way ANOVA table to which we have added the values that belong in each of the cells by using the formulas given previously.

Source of Variation	Sum of Squares	Degrees of Freedom	Mean Square	F-Ratio
Paints	0.4425 (1)	3 (5)	0.1475 (9)	0.1011(12)
Condition	0.905 (2)	2 (6)	0.4525 (10)	0.3101(13)
Error	8.755 (3)	6 (7)	1.4592 (11)	
Total	10.1025 (4)	11 (8)		

We have

Cell (1) value:
$$\left[\frac{(39.6)^2 + (41.1)^2 + (40.5)^2 + (39.9)^2}{3}\right] - \frac{(161.1)^2}{4 \cdot 3}$$

$$= 2163.21 - 2162.7675 = 0.4425$$

Cell (2) value:
$$\left[\frac{(54.8)^2 + (54.1)^2 + (52.2)^2}{4}\right] - \frac{(161.1)^2}{4 \cdot 3}$$

$$= 2163.6725 - 2162.7675 = 0.905$$

Cell (3) value:
$$\left[(13.4)^2 + (11.8)^2 + \cdots + (12.1)^2\right]$$
$$- \left[\frac{(39.6)^2 + (41.1)^2 + (40.5)^2 + (39.9)^2}{3}\right]$$
$$- \left[\frac{(54.8)^2 + (54.1)^2 + (52.2)^2}{4}\right] + \frac{(161.1)^2}{4 \cdot 3}$$

$$= 2172.87 - 2163.21 - 2163.6725 + 2162.7675$$

$$= 8.755$$

Cell (4) value: $= \left[(13.4)^2 + (11.8)^2 + \cdots + (12.1)^2\right] - \frac{(161.1)^2}{4 \cdot 3}$

$$= 2172.87 - 2162.7675 = 10.1025$$

Cell (5) value: $4 - 1 = 3$

Cell (6) value: $3 - 1 = 2$

Cell (7) value: $(4 - 1)(3 - 1) = 6$

Cell (8) value: $4 \cdot 3 - 1 = 12 - 1 = 11$

Cell (9) value: $\dfrac{0.4425}{3} = 0.1475$

Cell (10) value: $\dfrac{0.905}{2} = 0.4525$

Cell (11) value: $\dfrac{8.755}{6} = 1.4592$

Cell (12) value: $\dfrac{0.1475}{1.4592} = 0.1011$

Cell (13) value: $\dfrac{0.4525}{1.4592} = 0.3101$

From Table IX in the Appendix we find that the F-value with 3 degrees of freedom for the numerator and 6 degrees of freedom for the denominator is 4.76 at the 5% level of significance. Similarly, the F-value with 2 degrees of freedom for the numerator and 6 degrees of freedom for the denominator is 5.14. In both cases we obtained values of F equal to 0.1011 and 0.3101, respectively, that are less than 4.76 and 5.14. Thus, we do not reject the null hypothesis that the average drying time for the several brands of paint is significantly different nor do we reject the null hypothesis that the different testing conditions were a factor in determining average drying time.

COMMENT Our analysis of two-way ANOVA is by no means a complete discussion of the topic. A thorough two-way ANOVA problem and solution depends on how the treatments were assigned to the experimental units (i.e., how the randomization was conducted) as well as on the number of replications of each experimental condition. Our intention was merely to introduce you to the idea of two-way ANOVA. A more detailed analysis can be found by consulting any of the references listed in the Suggested Further Reading for this chapter.

EXERCISES FOR SECTION 13.4

 1. A consumer's public interest research group samples four different gas stations, each owned and operated by the four major oil companies, in an effort to determine if there is a significant difference in the average price charged for a gallon of regular unleaded gas. The following results are obtained.

	Price Charged for Gallon of Gas (in cents)			
Company A	101	107	106	103
Company B	102	105	109	107
Company C	101	104	111	105
Company D	102	107	106	104

Station Operated By

Do the data indicate that there is a significant difference in the average price charged by these gas stations for a gallon of unleaded gas? (Use a 5% level of significance.)

 2. The U.S. Federal Bureau of Investigation collects and publishes data on the frequency of various crimes in different cities of the United States. For the first 5 months of 1991 the following data are available on the number of murders reported monthly in four different cities.

Number of Murders Reported Monthly

A	27	35	28	41	32
B	26	21	39	47	30
C	29	21	28	32	36
D	26	28	19	47	37

City

Using a 1% level of significance, test the null hypothesis that the average number of murders reported over the 5-month period is not significantly different for all four cites.

3. A medical researcher is interested in determining which of four different blood pressure pills are most effective in controlling a person's blood pressure. The researcher gives each blood pressure pill to five different people over a period of a month and records the following average blood pressure loss for these people.

Average Systolic Blood Pressure Loss (in mm Hg)

A	22	11	17	15	10
B	12	16	18	13	15
C	19	11	14	15	17
D	16	3	9	10	12

Blood Pressure Pill

Using a 1% level of significance, test the null hypothesis that the average blood pressure loss is not significantly different for the various blood pressure pills.

4. The Commissioner of Health believes that the average daily hospital charge for a routine (noncaesarean) delivery is significantly different for the various hospitals where the child is delivered and is related to the geographic location of the hospital within the state. In support of this claim, the following data are available concerning charges for a routine delivery from four hospitals in different geographic regions within the state.

	Daily Hospital Charge (in dollars)			
Northern	427	382	502	476
Southern	391	402	427	501
Eastern	517	378	476	409
Western	501	499	404	428

Geographic Location of Hospital Within Part of the State

At a 5% level of significance, is the commissioner's claim justified?

 5. An auto engineer is interested in comparing the mileage performance of cars using three gasolines to which four different additives have been added. The engineer originally planned to repeat each determination six times. Not having enough time, however, the engineer has to base any conclusion on the following data.

	Miles Per Gallon Obtained					
A	28.0	26.4	25.7	29.6	26.5	28.2
B	31.1	27.8	28.6	30.8	27.9	29.0
C	26.5	27.8	26.1	22.5	23.6	
D	24.3	29.2	34.1	30.7	27.6	21.4

Gasoline With Additive

Using a 5% level of significance, test the null hypothesis that the differences among the sample means are significant.

 6. The shipping department of the Stahl Book Publishing Co. wants to compare the breaking strength of cartons that it uses to ship books. It has available cartons from four different companies. The following breaking strengths were obtained in laboratory analyses:

	Breaking Strength (in pounds)					
A	88	94	62	86	74	89
B	76	81	74	79	75	
C	69	83	95	77	80	
D	82	68	70	71		

Carton Manufactured by Company

Using a 5% level of significance, test the null hypothesis that the average breaking strength of the cartons produced by the different companies is not significantly different.

7. The price of a gallon of regular unleaded gas varies both according to the neighborhood in which it is purchased and the brand of gasoline purchased, as shown here.

Brand of Gasoline Purchased		Neighborhood of City in Which Purchased		
		Upper-Class	*Middle-Class*	*Lower-Class*
	A	\$1.21	\$1.25	\$1.33
	B	1.19	1.23	1.29
	C	1.23	1.27	1.27
	D	1.18	1.21	1.28
	E	1.25	1.27	1.22

Using a two-way analysis of variance, do the data indicate that the average price of a gallon of unleaded gas is not significantly different for all brands and is not significantly different for all neighborhoods. (Use a 5% level of significance.)

8. A record was kept on the number of defective items produced by three machines on each day of the week. The list is as follows.

Day of Week		Machine		
		A	B	C
	Mon	36	21	31
	Tues	32	29	32
	Wed	33	31	36
	Thurs	19	32	22
	Fri	24	38	27

Do these data indicate that there is a significant difference in the average number of defective items produced by these machines? (Use a 5% level of significance.)

9. A consumer's group is interested in determining how long it takes a teller to complete the identical transaction at each of four banks. The following results were obtained when numerous tellers at the four banks were surveyed.

Time Required to Complete Identical Transaction at Different Banks

Bank A (in minutes)	Bank B (in minutes)	Bank C (in minutes)	Bank D (in minutes)
12	9	23	12
14	12	8	18
11	15	17	19
16	11	9	
10	8		

Do the data indicate that there is a significant difference in the average time required to complete the transaction at these banks? (Use a 1% level of significance.)

 10. The number of angiograms performed daily at four hospitals that specialize in cardiovascular problems during a week is as follows.

		Hospital			
		A	B	C	D
	Mon	42	28	45	36
	Tues	29	49	24	40
Day of Week	Wed	45	31	29	47
	Thurs	24	37	32	45
	Fri	37	29	35	36

Do the data indicate that there is a significant difference between the average number of angiograms performed at these hospitals? (Use a 5% level of significance.)

11. *Reducing blood serum cholesterol levels:* An experimenter is interested in comparing the effect of four different diets on the reduction of the blood serum cholesterol levels. 16 men were randomly assigned (four each) to one of the four treatment (diet) groups. After three months had passed the blood serum cholesterol of each participant was determined and is given here.

Diet A	Diet B	Diet C	Diet D
268	280	261	232
211	251	278	280
235	245	255	270
240	253	249	261

Do these data indicate that the type of diet used affects the average blood serum cholesterol level? (Use a 1% level of significance.)

TOLL COLLECTORS PROTEST UNHEALTHY WORKING CONDITIONS

New Dorp (May 17):—Toll collectors staged a 3-hour protest yesterday against the unhealthy working conditions occurring at different times of the day at the toll plazas of the city's bridges and tunnels. A spokesperson for the toll collectors indicated that the level of sulfur oxide pollutants in the air at the plazas was far in excess of the recommended level. Between the hours of 10 A.M. and 1 P.M., motorists were able to use the tunnel without paying any toll. In an effort to resolve the dispute, Bill Sigowsky has been appointed as mediator.

May 17, 1991

---

 12. Consider the accompanying article above. After taking samples of the air on different days at the various tunnels and bridges the following levels of pollutants were obtained.

	Day of the Week				
	Mon.	Tues.	Wed.	Thurs.	Fri.
Bridge 1	12.8	13.3	12.6	12.9	13.5
Bridge 2	12.1	12.3	12.8	13.7	12.7
Bridge 3	13.8	13.4	13.5	13.1	12.9
Tunnel 1	12.8	12.7	13.1	13.2	13.3
Tunnel 2	12.8	13.3	12.9	13.0	13.1

Location Where Sample was Obtained

Using a two-way analysis of variance, test the null hypothesis that the average amount of sulfur oxide pollutants is not significantly different at all the bridges and tunnels and is the same for each day of the week. (Use a 5% level of significance.)

 13. An experimenter is interested in knowing which method yields the greatest amount of usable oil from shale rock. The experimenter gathers shale rock from five different parts of the country and applies one of four conversion techniques. The quantity of oil obtained (on a certain scale) is as follows.

Quantity of Oil Obtained

	Region A	Region B	Region C	Region D	Region E
I	42	38	46	44	43
II	34	37	41	49	58
III	46	37	26	35	41
IV	26	29	25	28	27

Conversion Technique Used (rows I–IV)

Using a two-way analysis of variance, test the null hypothesis that the average quantity of oil obtained is not significantly different for all the conversion techniques and is also not significantly different for shale rock obtained from each of the regions. (Use a 1% level of significance.)

 14. A lawyer has gathered the following statistics from several states for the length of a jail term for people with the same prior record convicted of different types of crimes.

Jail Term by Crime (years)

	Burglary	Armed Robbery	Assault	Murder
1	3	6	1	34
2	2	5	2	36
3	4	7	1	35
4	3	8	2	37

State (rows 1–4)

Using a two-way analysis of variance, test the null hypothesis that the average length of a jail term is the same for the various crimes mentioned and that this average is not significantly different for the several states. (Use a 1% level of significance.)

13.5 USING COMPUTER PACKAGES

MINITAB greatly simplifies the computations involved when working with ANOVA problems. We illustrate the output from MINITAB when applied to the data of Example 2 of Section 13.3 (page 652).

```
MTB  > READ THE FOLLOWING DATA INTO C1, C2, C3, C4
DATA > 64    73    61    63
DATA > 73    82    79    69
DATA > 69    71    71    68
DATA > 75    69    73    74
DATA > 78    74    66    75
DATA > END

MTB  > AOVONE WAY ON C1-C4

ANALYSIS OF VARIANCE
SOURCE    DF        SS        MS       F          P
FACTOR     3       52.2      17.4     0.56      0.652
ERROR     16      500.4      31.3
TOTAL     19      552.6
                                INDIVIDUAL 95 PCT CI'S FOR MEAN
                                BASED ON POOLED STDEV
LEVEL      N      MEAN      STDEV  --------+---------+---------+-------
C1         5     71.800     5.450        (-------------*------------)
C2         5     73.800     4.970           (-------------*------------)
C3         5     70.000     6.856   (-----------*------------)
C4         5     69.800     4.868  (-------------*------------)
          --------+---------+---------+-------
POOLED STDEV = 5.592            68.0       72.0       76.0

MTB  > STOP
```

COMMENT Although the MINITAB analysis of variance for this problem will produce additional output (involving confidence intervals), we will not analyze such information here, as our main objective is merely to indicate how MINITAB handles simple ANOVA problems.

13.6 SUMMARY

In this chapter we briefly introduced the important statistical technique known as analysis of variance or ANOVA. This technique is used when we wish to test a hypothesis about the equality of several means. We limited our discussion to normal populations with equal variances.

When applying the ANOVA technique, we do not test the differences between the means directly. Instead, we test the variances. If the variances are zero, there is no difference between the means. We therefore analyze the source of the variation. Is it due purely to chance or is the difference in the variation significant? When analyzing variation by means of analysis of variance techniques, we use an ANOVA table.

Although we discussed only single-factor (one-way) and two-way ANOVA, it is worth noting that ANOVA techniques can be applied to more complicated situations or when we have replications for each test condition. However, these are beyond the scope of this book.

Study Guide

The following is a chapter outline in capsule form. You should now be able to demonstrate your knowledge of the ideas mentioned by giving definitions, descriptions, or specific examples. Page references are given in parentheses.

We can analyze whether there is any significant difference between several means by using analysis of variance techniques. This is abbreviated as **ANOVA.** (page 641)

When setting up an ANOVA table the first number that we calculate is called the **Total Sum of Squares** and is abbreviated as SS(total). This number is obtained by squaring each number in the table and then adding these squares together. (page 645)

The second number that we calculate is the **Sum of Squares Due to the Factor.** We square each of the row totals and divide the sum of these squares by the number in each row. From this result we subtract the square of the grand total divided by the number in the sample. (page 646)

Then we calculate the **Sum of Squares of the Error,** abbreviated as SS(error). We find the sum of the squares of each entry in the table. From this sum we subtract the sum of the squares of each row total divided by the number in each row. (page 646)

The **number of degrees for the factor,** df(factor), is one less than the number of possible levels at which the factor is tested. (page 647)

The **number of degrees for the error,** df(error), is one less the number of repetitions of each condition multiplied by the number of possible levels of the factor. (page 647)

The **number of degrees for the total,** df(total), is one less than the total number in the sample. (page 647)

To determine the entries that belong in the **mean square** column, we divide the sum of squares for the row by the number of degrees of freedom for that row. (page 647)

If we are given two samples of size n_1 and n_2 for two populations that are roughly normally distributed with sample standard deviations of s_1 and s_2, respectively, then we base tests on the equality of the two population standard deviations on the ratio $\dfrac{s_1}{s_2}$ by using the F-distribution. (page 648)

The **F-distribution** depends on the number of degrees of freedom of both sample estimates, $n_1 - 1$ and $n_2 - 1$. We **reject the null hypotheses** of equal variances when the test statistic value of F exceeds the chart value of F at the appropriate level of significance. (page 648)

For a single-factor analysis of variance we have the following **ANOVA table.**

Source of Variation	Source of Squares	Degrees of Freedom	Mean Square	F-Ratio
Factors of the experiment				
Error				
Total				

When performing an analysis of variance, if we are discussing only one factor, it is referred to as **treatments.** (*Note*: MINITAB uses "FACTOR" instead of "Treatments.") We then have the single-factor ANOVA table given above. If a second factor is involved, it is referred to as **blocks** and we have a two-way ANOVA. The two-way ANOVA table is slightly different. Instead of having a single line for factors, it has two lines, one for treatments and one for blocks. (page 654)

Formulas to Remember

The most important thing to remember when testing several means is how to set up an ANOVA table and how to find the appropriate values for each cell of the ANOVA table. Both of these ideas are summarized on pages 649–650 for single-factor ANOVA and on pages 656–657 for a two-way ANOVA.

Testing Your Understanding of This Chapter's Concepts

1. We cannot use one-way ANOVA techniques without first satisfying certain assumptions. Name three of them.
2. For an F-distribution, find the probability that $F \leq 3.37$ when the degrees of freedom for the numerator is 10 and the degrees of freedom for the denominator is 20.
3. The ANOVA techniques discussed in this chapter actually analyze whether the difference between several sample means is significant. Explain how variance plays a key role in such analyses.

THINKING CRITICALLY

1. A researcher designed an experiment that utilizes one qualitative factor with five levels (A, B, C, D, and E). What are the treatments?
2. If the Total Sum of Squares for a completely randomized experiment with five treatments and $n = 25$ measurements (five per treatment) is 400 and with error sum of squares equal to 150, perform an F-test of the null hypothesis that the five sample means are equal. (Use a 5% level of significance.)
3. If our analysis of variance leads us to conclude that the sample means are not all the same, then a confidence interval estimate of the difference between the two treatment means is given by

$$\bar{x}_1 - \bar{x}_2 \pm t_{\alpha/2} \sqrt{MSE \left(\frac{1}{n_1} + \frac{1}{n_2} \right)}$$

where $\bar{x}_1$ and $\bar{x}_2$ are the sample means for the first and second treatments respectively, n_1 and n_2 are the sample sizes, MSE = mean square error as given in the ANOVA table, and $t_{\alpha/2}$ = t-value (using the t-distribution) based on the number of degrees of freedom for the error as given in the ANOVA table. Using a 5% level of significance ($\alpha = 0.05$), find a confidence interval for the difference between the average tar content of Brand X and Y cigarettes as given in Example 1 of Section 13.3. on page 650).
4. Refer back to Question 3. Find a confidence interval for the difference between the average tar content of
   a. Brand X and Z cigarettes.
   b. Brand Y and Z cigarettes.

Review Exercises for Chapter 13

For Questions 1–9, use the following information:

Twelve plots of land were randomly divided into three groups. The first 4 plots were used as a control group, whereas the 4 plots of land in the second group were treated with fertilizer A and the 4 plots of land in the third group were treated with fertilizer B. Yields, in bushels, were as follows.

Control Group	Group Treated with Fertilizer A	Group Treated ith Fertilizer B
29	24	25
33	27	24
26	33	33
27	31	32

In an effort to determine whether the type of fertilizer used significantly affects the yield, the following ANOVA has been set up.

ANOVA Table

Source of Variation	Sum of Squares	Degrees of Freedom	Mean Square	F-Ratio
Different workers	(1)	(4)	(7)	(9)
Error	(2)	(5)	(8)	
Total	(3)	(6)		

1. The appropriate entry for cell (1) is —————.

2. The appropriate entry for cell (2) is —————.

3. The appropriate entry for cell (3) is —————.

4. The appropriate entry for cell (4) is —————.

5. The appropriate entry for cell (5) is —————.

6. The appropriate entry for cell (6) is —————.

7. The appropriate entry for cell (7) is —————.

8. The appropriate entry for cell (8) is —————.

9. The appropriate entry for cell (9) is —————.

10. It is appropriate to use the *F*-distribution when comparing
    a. differences between sample means.
    b. differences between variances.
    c. differences resulting from the use of different sample sizes.
    d. variances from non–normally distributed populations.
    e. none of these.

11. Which of the following assumptions are made when we use the ANOVA techniques?
    a. The populations from which the samples are selected are normally distributed.
    b. The sample sizes are large.

c. The Central Limit Theorem is not applicable.
d. There is a correlation between the factors of the experiment.
e. None of these.

12. The total number of degrees of freedom when using a two-way ANOVA, where there are r levels of one factor and c levels of another factor, is
a. $r - 1$ b. $c - 1$ c. $rc - 1$ d. $r(c - 1)$ e. none of these

---

FALSE FIRE ALARMS ON THE RISE

East Brighton (Dec. 12)—Fire Department officials disclosed yesterday that the number of false alarms called in to the department has reached epidemic proportions. Last month, 59% of all fire alarms received were false alarms. Cited one official, "While our men are out on a false alarm, a true alarm may come in. Our equipment cannot be 2 places at 1 time."

December 12, 1989

---

13. Consider the previous newspaper article. After reading it the mayor orders the fire commissioner to submit a detailed report on the number of false alarms for one week covering five different sections of the city. The following report was submitted to the mayor.

Number of False Alarms Reported
Daily over a Five-Day Period

Section of City					
A	26	42	35	25	45
B	16	19	27	14	22
C	40	34	29	33	37
D	23	14	16	20	12
E	17	14	16	13	18

Using a 1% level of significance, do the data indicate that the average number of false alarms reported is significantly different for the various sections of the city?

14. Accidental breakage of glass items by customers as they push their shopping carts through the aisles of a supermarket can be dangerous to other customers as well as costly to the management. A large supermarket chain reported the

following number of reported cases of the breakage of glass items on a weekly basis for a period of four weeks and the store location.

Number of Cases of
Glass Breakage

Brooks Mall	49	36	40	44
Weaver Mall	41	38	50	38
Tanner Mall	40	42	40	49
Acres Mall	46	43	35	57

Store Location

Using a 5% level of significance, test the null hypothesis that the average number of accidental glass breakage cases is not significantly different for all the stores, irrespective of location.

Chapter Test

 1. Hearing-aid batteries produced by four different companies were tested to determine which brand lasts longer. Five batteries of each brand were tested. Do the following data indicate that there is no significant difference between the average life of these batteries? (Use a 1% level of significance.)

How Many Weeks Did It Last

A	22	29	19	16	18
B	33	12	17	26	17
C	18	27	19	17	32
D	34	17	19	32	29

Batteries

 2. A record was kept on the number of items completed by three workers on each day of the week. Do the following data indicate that there is a significant difference in the average production of those employees? (Use a 5% level of significance.)

Day of Week

	Mon	*Tues*	*Wed*	*Thurs*	*Fri*
A	17	14	28	16	11
B	12	19	18	16	24
C	14	13	18	22	19

Worker

 3. The number of cases handled by each of five judges each day of the week is as follows.

	Day of Week				
	Mon	*Tues*	*Wed*	*Thurs*	*Fri*
A	22	17	32	16	19
B	19	36	24	18	31
C	25	29	14	30	26
D	20	33	26	11	17
E	25	17	22	34	27

Judge

Do the data indicate that there is a significant difference in the average number of cases handled by these judges? (Use a 5% level of significance.)

 4. 16 children with speech defects were divided into four groups of four students each. Each group was taught how to overcome its speech impediment, each by different method. The following chart indicates the number of practice sessions needed by each student to overcome the same type of problem. Do the data indicate that there is a significant difference in the average time needed by each student to learn to overcome the particular speech defect by the different methods? (Use a 5% level of significance.)

Method

A	55	48	49	58
B	42	39	59	50
C	46	32	51	48
D	38	47	54	39

 5. 24 animals were divided into groups of four each. Each animal in a group was injected with a chemical to increase its weight. Each group was given a different chemical. Do the following data indicate that there is a significant difference in the average weight gain of an animal over a fixed length of time as a result of the different chemicals? (Use a 1% level of significance.)

Chemical

A	10	13	7	9
B	6	14	8	11
C	13	6	8	11
D	10	9	15	13
E	18	7	8	14
F	12	13	8	9

6. Last year a publishing company published 12 math books. Five were typeset by Company A, four were typeset by Company B, and three were typeset by Company C. The number of errors in these books is as follows.

Company A	Company B	Company C
84	75	87
60	81	78
78	60	90
90	86	
61		

Do these data indicate that the average number of errors in the books typeset by each of these companies is significantly different? (Use a 1% level of significance.)

 7. Three professors have been teaching different sections of a Statistics I course. The final exam was the same for all three sections. The grades of the students on the final are as follows. (Each section had only seven students.)

Prof. Randolph	Prof. Frankel	Prof. Lichtenfeld
72	67	49
66	79	69
58	59	92
82	88	83
91	72	75
85	91	71
75	69	83

Do the preceding data indicate that the students in each of these teacher's class are not significantly different in their preparation for Statistics II? (Use a 5% level of significance.)

 8. 25 people were divided into five groups of five each. Each group was subjected to a different stimulus and the reaction time (in seconds) to the stimulus was recorded. The results are as follows.

		Stimulus		
A	B	C	D	E
2.2	1.8	1.1	1.9	1.7
3.3	3.2	2.7	2.0	2.9
1.7	1.9	2.8	1.9	1.2
3.4	1.7	1.6	1.1	2.8
1.9	3.4	2.6	1.8	3.1

Do the preceding data indicate that the average reaction time to the different stimuli is not significantly different? (Use a 1% level of significance.)

9. An analysis of the cause of hospitalization of patients at Maimonides Medical Center reveals the following information:

Reason for Hospitalization

		Cancer	Disease of Heart and Blood Vessels	Accidents	Pneumonia and Influenza	AIDS Related
Age of Person (in years)	Under 30	29	17	13	21	74
	Between 30–60	37	68	41	32	47
	Over 60	41	84	79	68	37

Using a two-way analysis of variance, do the data indicate that the average age of people hospitalized is not significantly different for all the age groups and is not significantly different for all the reasons for hospitalization? (Use a 1% level of significance.)

10. A large automobile insurance company operates drive-in claim centers where drivers involved in automobile accidents can bring their cars for immediate claims processing. The company is interested in determining whether the particular claims examiner involved in a settlement has any effect on the amount of money that the company must pay. The company decides to sample identical cases resolved by five different claims examiners in each of the four drive-in centers that it operates in a large northeastern state. The car damage is identical in all claims examined. The following are the reported settlement amounts.

Drive-in Center

		A	B	C	D
Claim Examiner	I	$1225	$1240	$1175	$1205
	II	$1420	$1395	$1105	$1295
	III	$1095	$1600	$1245	$1265
	IV	$1525	$1305	$1420	$1150
	V	$1240	$1310	$1290	$1430

Using a two-way analysis of variance, do the data indicate that the average amount of money in a settled claim is not significantly different for all the centers and is not significantly different for all the employees? (Use a 5% level of significance.)

For questions 11–16 use the following information: A large insurance company in the west processes numerous car accident reports daily at 5 different locations.

A record on the number of reports processed daily at the five different branches for one week is shown below:

Number of reports processed

Location						
	I	58	61	73	46	39
	II	43	52	29	35	63
	III	49	55	62	57	54
	IV	48	63	71	51	68
	V	47	69	50	70	67

In an effort to determine whether there is a significant difference in the average number of reports processed daily at the different locations, the following one-way ANOVA table has been set up.

ANOVA Table

Source of Variation	Sum of Squares	Degrees of Freedom	Mean Square	F-Ratio
Different workers	(1)	(4)	(7)	(9)
Error	(2)	(5)	(8)	
Total	(3)	(6)		

11. The appropriate entry for cell (1) is
    a. 854.4   b. 2441.6   c. 3296   d. None of these
12. The appropriate entry for cell (2) is
    a. 854.4   b. 2441.6   c. 3296   d. None of these
13. The appropriate entry for cell (3) is
    a. 854.4   b. 2441.6   c. 3296   d. None of these
14. The appropriate entry for cell (7) is
    a. 4   b. 20   c. 213.6   d. 122.08   e. None of these
15. The appropriate entry for cell (8) is
    a. 24   b. 4   c. 122.08   d. 213.6   e. None of these
16. Do the data indicate that there is a significant difference between the average number of reports processed daily at the five different locations? (Use a 5% level of significance.)
    a. yes   b. No   c. not enough information given   d. none of these

Suggested Further Reading

1. Dunn, Olive and Virginia Clark. *Applied Statistics: Analysis of Variance and Regression*, 2nd ed., New York: John Wiley, 1987.
2. Lapin, Lawrence. *Statistics*. New York: Harcourt, Brace, Jovanovich, 1975.
3. Neter, J. and W. Wasserman. *Applied Linear Statistics Models*, 2nd ed. Homewood, IL: Irwin, 1985.
4. Ott, Lyman. *An Introduction to Statistical Methods and Data Analysis*, 2nd ed. North Scituate, MA: Duxbury Press, 1984.
5. Snedecor, George and William C. Cochran. *Statistical Methods*, 7th ed. Ames, IA: Iowa State University Press, 1980.
6. Winer, B. J. *Statistical Principles in Experimental Design*, 2nd ed. New York: McGraw-Hill, 1971.
7. Wonnacott, Thomas and Ronald Wonnacott. *Introductory Statistics for Business and Economics*, 3rd ed. New York: John Wiley, 1984.

14 Nonparametric Statistics

It is often claimed that railroad ridership decreases as fares are raised. How do we determine whether the effects of fare increases are significant? See the newspaper article on p. 678 and the discussion on p. 689. (© *Barbara Rios, Photo Researchers, Inc.*)

In recent years much emphasis has been placed on nonparametric statistics. Since the concepts behind nonparametric statistics (or distribution-free methods as they are called) are based on elementary probability theory, many of the formulas and tests used can be derived quite easily by employing only a little algebra. Once we develop a table of critical values for a given situation, the nonparametric techniques are quite easy to use.

In this chapter we discuss some of the more popular nonparametric techniques.

CHAPTER OBJECTIVES

After studying the material in this chapter, you should be able:

- **To analyze** several tests that can be used when assumptions about normally distributed populations or sample size cannot be satisfied. These are called nonparametric statistics or distribution-free methods. (Sections 14.1 and 14.2)

- **To apply** the sign test that is used in the "before and after" type study. We test whether or not $\mu_1 = \mu_2$ when we know that the samples are not independent. We can also use the Wilcoxon signed-rank test. (Sections 14.3 and 14.4)

- **To discuss** an alternative to the standard significance tests for the difference between two sample means that is used when we have nonnormally distributed populations. This is the rank-sum test. We can also use this test when the population variances are not equal. (Section 14.5)

- **To understand** the Spearman rank correlation test. (Section 14.6)

- **To point out** how we use the runs test when we wish to test for randomness. (Section 14.7)

- **To work with** the Kruskal-Wallis H-test as a nonparametric test that can be used to test whether the difference between numerous sample means is significant. (Section 14.8)

NEW STATISTICS INVOLVING COMMUTER RIDERSHIP RELEASED YESTERDAY

Long Island (Feb. 18)—Officials of the Regional Transportation Commission released the following statistics indicating how commuter ridership has been affected by the 75% increase in fares that took effect February 4. The data are for riders who use the 8:05 A.M. train daily between Pleasure Point and the City.

Number of Riders (in thousands) Using the 8:05 A.M. Daily

	Before Feb. 4	After Feb. 4
Monday	183	179
Tuesday	177	175
Wednesday	169	168
Thursday	180	179
Friday	179	179

A spokesperson for the railroad said that the effects of the rate increase on commuter ridership was under study and that no comments would be forthcoming at this time.

Monday—February 18, 1991

Referring to this article, we can conclude that a technique is needed for measuring the effects of a new policy like the one described here. Such "before and after" types of situations often occur when new products are introduced or when a new technique is tried.

Are such statistical tests available?

14.1 INTRODUCTION

In previous chapters we discussed procedures for testing various hypotheses involving means, proportions, variances, and the like. In almost all the cases discussed we assumed that the populations from which the samples were taken were approximately normally distributed. Only when we applied the chi-square distribution in comparing observed frequencies with expected frequencies did we not specify the normal distribution.

Since there are many situations where this requirement cannot be satisfied, statisticians have developed techniques to be used in such cases. These techniques are known as **nonparametric statistics** or **distribution-free methods.** As the names imply, these methods are not dependent upon the distribution or parameters involved.

There are advantages and disadvantages associated with using nonparametric statistics. The advantages in using these methods as opposed to the **standard methods** are as follows:

1. They are easier to understand.
2. They often involve much less computation.
3. They are less demanding in their assumptions about the nature of the sampled populations.

For these reasons many people often refer to nonparametric statistics as shortcut statistics. The disadvantages associated with nonparametric statistics are that they usually waste information, as we will see shortly, and that they tend to result in the acceptance of null hypotheses more often than they should.

Nonparametric statistical methods are frequently used when samples are small since most of the **standard tests** require that the sample sizes be reasonably large.

In this chapter we discuss only briefly some of the more commonly used nonparametric tests. A complete discussion of all these methods would require many chapters or perhaps even several volumes.

14.2 THE ONE-SAMPLE SIGN TEST

All the standard sample tests involving means that we have discussed so far in this book are based on the assumption that the populations are approximately normally distributed. This often may not be the case.

When the preceding assumption is not necessarily true, then we can replace the standard tests by one of the numerous nonparametric tests. By far the simplest of all the nonparametric tests is the **one-sample sign test.** It is used as an alternative to the *t*-test with one mean that was discussed in Section 10.5. We apply this test when the hypothesis we are testing concerns the value of the mean or median of the population and we are sampling a continuous population in the vicinity of the unknown mean or median *M*. Under these assumptions the probability that a sample

Nonparametric Statistics

Distribution-free Methods

Standard Methods

Standard Tests

One-sample Sign Test

value is less than the mean or median and the probability that a sample value is greater than the mean or median are both $\frac{1}{2}$.

If the sample size is small, we can perform the sign test by referring to a table of binomial probabilities (Table III in the Appendix.) When the sample size is large we can perform the sign test by using the normal curve approximation to the binomial. We illustrate the techniques with several examples.

EXAMPLE 1

A trucking industry spokesperson claims that the median weight of a load carried by a truck traveling on State Highway No. 7 is 15,000 pounds (lb). A transportation official believes that this figure is much too low. To verify the trucking industry claim, a random survey of 15 trucks is taken and the weight of each truck's load is determined. It is found that 5 trucks have loads with weights below 15,000 lbs, 2 trucks have loads equal to 15,000 lbs, and 8 trucks have loads above 15,000 lbs. Using the one-sample sign test, test the null hypothesis that the median weight of a load carried by a truck is 15,000 lbs against the alternative hypothesis that the median weight is more than 15,000 lbs (Use a 5% level of significance.)

Solution

We replace each truck's load weight above 15,000 lbs with a plus sign and each truck's load weight below 15,000 lbs with a minus sign. We discard those trucks whose load weight equals 15,000 lbs. We then test the null hypothesis that the plus signs and minus signs are values of a random variable having the binomial distribution with $p = \frac{1}{2}$. In our case we have 5 minus signs and 8 plus signs. Here $n = 13$ and not 15 since we disregard those trucks whose load weight is exactly 15,000 lbs. We then determine whether 8 plus signs in 13 observed trials agrees with the null hypothesis that $p = \frac{1}{2}$ or with the alternative hypothesis that $p > \frac{1}{2}$. Using Table III in the Appendix, we find that for $n = 13$ and $p = \frac{1}{2}$ the probability of obtaining 8 or more successes is

$$
\begin{aligned}
\text{Prob (8 successes)} \;&=\; 0.157 \\
\text{Prob (9 successes)} \;&=\; 0.087 \\
\text{Prob (10 successes)} \;&=\; 0.035 \\
\text{Prob (11 successes)} \;&=\; 0.010 \\
\text{Prob (12 successes)} \;&=\; 0.002 \\
\text{Prob (13 successes)} \;&=\; \text{-------} \\
\hline
\text{Prob (8 or more successes)} \;&=\; 0.291
\end{aligned}
$$

Since this value is more than $\alpha = 0.05$, we do not reject the null hypothesis. The sample data do not contradict the trucking industry's claim.

The previous example was a one-sided hypothesis problem. However, the same procedure can be used when testing two-sided alternative hypotheses problems. In essence, what we do is calculate the number of plus signs and the number of minus signs. We then determine whether a random variable having a binomial distribution can have the calculated number of plus signs or minus signs and still have $p = \dfrac{1}{2}$.

TABLE 14.1   Critical Values for Sign Test

n	Level of Significance ($\alpha =$)				n	Level of Significance ($\alpha =$)			
	0.01	0.05	0.10	0.25		0.01	0.05	0.10	0.25
1					21	4	5	6	7
2					22	4	5	6	7
3				0	23	4	6	7	8
4				0	24	5	6	7	8
5			0	0	25	5	7	7	9
6		0	0	1	26	6	7	8	9
7		0	0	1	27	6	7	8	10
8	0	0	1	1	28	6	8	9	10
9	0	1	1	2	29	7	8	9	10
10	0	1	1	2	30	7	9	10	11
11	0	1	2	3	31	7	9	10	11
12	1	2	2	3	32	8	9	10	12
13	1	2	3	3	33	8	10	11	12
14	1	2	3	4	34	9	10	11	13
15	2	3	3	4	35	9	11	12	13
16	2	3	4	5	36	9	11	12	14
17	2	4	4	5	37	10	12	13	14
18	3	4	5	6	38	10	12	13	14
19	3	4	5	6	39	11	12	13	15
20	3	5	5	6	40	11	13	14	15

By referring to the binomial probability chart (Table III in the Appendix), we then specify the critical rejection region.

Since applications using the sign test occur often, statisticians have constructed a chart that enables us to determine whether the number of plus signs or minus signs is significant. This eliminates the need to use the binomial distribution probability chart and to add probabilities. One such chart is given in Table 14.1.

> **RULE**
> When using Table 14.1 for a two-sided test, the following applies:
>
> 1. n represents the total number of plus signs and negative signs, disregarding any zeros.
> 2. The test statistic is the number of the less frequent sign. This means that we first determine the number of plus signs and the number of minus signs. We select the smaller number between the number of plus signs or minus signs. This represents the test statistic.
> 3. We reject the null hypothesis if the test statistic value is less than or equal to the chart value. If the test statistic is larger than the chart value, we do not reject the null hypothesis.

COMMENT   Table 14.1 gives us the critical values for a two-sided test. When working with a one-sided test, double the value of α specified in the problem.

Use of the preceding rule is illustrated in the following example.

EXAMPLE 2   A doctor suspects that the median annual cost for malpractice insurance in her specialty is approximately $18,000. Her nurse believes that the median annual cost is not equal to $18,000. She samples 13 insurance companies and obtains the following quotes for the identical cost of malpractice insurance.

$14,350 $17,010 $13,936 $17,073 $17,985 $18,840 $17,240
$18,000 $19,420 $17,840 $16,090 $17,360 $17,053

Using the one-sample sign test, test the null hypothesis that the median cost for malpractice insurance is $18,000 against the alternative hypothesis that the median cost is not $18,000. (Use a 5% level of significance.)

Solution

We replace each price quote with a minus sign if it is less than $18,000 and with a plus sign if it is above $18,000. We neglect those quotes that equal $18,000. We then have

$$-\quad-\quad-\quad-\quad-\quad+\quad-\quad+\quad-\quad-\quad-\quad-$$

or a total of 10 minus signs and 2 plus signs in 12 trials. Here we are testing whether the number of plus signs (two in our case) supports the null hypothesis that $p = \dfrac{1}{2}$ or the alternative hypothesis that $p \neq \dfrac{1}{2}$. The number of plus signs is 2 and the number of minus signs is 10. The smaller of these two numbers is 2. This represents the test statistic. Now we look in Table 14.1 to find the appropriate critical value for $n = 12$ and $\alpha = 0.05$. The chart value is 2. Since the test statistic value that we obtained is less than or equal to the chart value of 2 (in our case it equals 2), we reject the null hypothesis and conclude that the median cost of malpractice insurance is not $18,000.

14.3 THE PAIRED-SAMPLE SIGN TEST

Paired Data

Two Dependent Samples

Before and After Type Study

The sign test can also be used when working with **paired data** that occur when we deal with **two dependent samples.** This often happens when we measure the same sample twice, as is done in the **before and after type study.** We proceed in a manner similar to what was done in Section 14.2.

To illustrate, suppose a college administrator is interested in knowing how a particular 3-week math minicourse affects a student's grade. Twenty students are selected and are given a math test. Then these students attend the minicourse and are retested. The administration would like to use the results of these two tests to determine whether the minicourse actually improves a student's score.

Table 14.2 contains the scores of the 20 students on the precourse test and the postcourse test. In this table we have taken each student's precourse test score and subtracted it from the student's postcourse test score to obtain the change score. An increase in score is assigned a plus $(+)$ sign and a decrease in score is assigned a minus $(-)$ sign. No sign is indicated when the two scores are identical.

The null hypothesis in this case is $\mu_1 = \mu_2$. This means that the minicourse does not significantly affect a student's score. Under this assumption we would expect an equal number of plus signs and minus signs. Thus, if $\hat{p}$ is the proportion of plus signs, we would expect $\hat{p}$ to be around 0.5 (subject only to chance error).

Since we think that the minicourse does increase a student's score, the alternative hypothesis would be $\mu_2 > \mu_1$. Thus,

H_0: $\mu_1 = \mu_2$

H_1: $\mu_2 > \mu_1$

We can now apply the methods discussed in Section 10.8 (pages 515–518) for

TABLE 14.2   Scores of Twenty Students on Precourse and Postcourse Tests

Student	Precourse Score	Postcourse Score	Sign of Difference
1	68	71	+
2	63	65	+
3	82	88	+
4	70	79	+
5	65	57	−
6	66	77	+
7	64	62	−
8	69	73	+
9	72	70	−
10	74	76	+
11	71	68	−
12	80	80	
13	59	71	+
14	85	80	−
15	57	65	+
16	83	87	+
17	43	48	+
18	94	94	
19	82	93	+
20	91	94	+

testing a proportion. Recall that for testing a proportion we use the test statistic

$$z = \frac{\hat{p} - p}{\sqrt{\dfrac{p(1 - p)}{n}}}$$

In applying this test statistic, we let p be the true proportion of plus signs as specified in the null hypothesis. Thus, if the minicourse does not affect a student's scores, we would expect as many plus signs as minus signs. There should be no more students obtaining higher scores than students obtaining lower scores as a result of this mini-course. Therefore, $p = 0.50$. We now count the number of plus signs. There are 13 plus signs out of a possible 18 sign changes. We ignore the cases that involve no change. So, $\hat{p} = \dfrac{13}{18}$, or 0.72, $n = 18$, and $p = 0.50$. Applying the test statistic, we have

$$z = \frac{\hat{p} - p}{\sqrt{\dfrac{p(1 - p)}{n}}}$$

$$= \frac{0.72 - 0.50}{\sqrt{\dfrac{(0.5)(1 - 0.5)}{18}}}$$

$$= \frac{0.22}{\sqrt{0.013889}} = \frac{0.22}{0.118}$$

$$= 1.86$$

We use the one-tail rejection region of Figure 14.1. Since the value of $z = 1.86$ falls in the critical region, we reject the null hypothesis at the 5% level of significance. Thus, the minicourse seems to have improved a student's score.

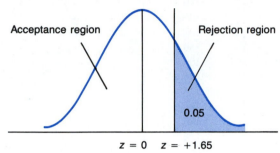

Figure 14.1

The following example further illustrates the paired-sign test technique.

EXAMPLE 1

A new weight-reducing pill is given to 15 people once a week for 3 months to determine its effectiveness in reducing weight. The following data indicate the before and after weights (in pounds) of these 15 people.

Weight Before Taking Pill	Weight After Taking Pill
131	125
127	128
116	118
153	155
178	179
202	200
192	195
183	180
171	180
182	180
169	174
155	150
163	169
171	172
208	200

Using a 5% level of significance, test the null hypothesis that the pill is not effective in reducing weight.

Solution

We arrange the data as follows.

Weight Before	Weight After	Sign of Difference
131	125	−
127	128	+
116	118	+
153	155	+
178	179	+
202	200	−
192	195	+
183	180	−
171	180	+
182	180	−
169	174	+
155	150	−
163	169	+
171	172	+
208	200	−

For each person we determine the change in weight. A plus sign indicates a gain and a minus sign indicates a loss. If the weight-reducing pill is not effective, the average weight should be the same before and after taking this pill. Since we are testing whether a person's weight remains the same or is reduced, we have

$$H_0: \mu_1 = \mu_2$$

$$H_1: \mu_2 < \mu_1$$

Out of the 15 sign changes 6 are minus so that

$$\hat{p} = \frac{6}{15} = 0.4 \quad \text{and} \quad n = 15$$

Also, $p = 0.50$. Applying the test statistic, we get

$$z = \frac{\hat{p} - p}{\sqrt{\dfrac{p(1 - p)}{n}}}$$

$$= \frac{0.4 - 0.5}{\sqrt{\dfrac{(0.5)(1 - 0.5)}{15}}}$$

$$= \frac{-0.1}{\sqrt{0.01667}} = \frac{-0.1}{0.129}$$

$$= -0.78$$

We use the one-tail rejection region of Figure 14.2. Since the value of $z = -0.78$ falls in the acceptance region, we cannot reject the null hypothesis. The weight-reducing pill does not seem to be effective in reducing weight. It may even cause an increase in weight.

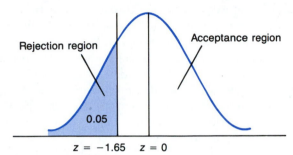

Figure 14.2

COMMENT   We mentioned earlier that nonparametric methods are wasteful. A plus or minus sign merely tells us that a person gained or lost weight. It does not specify whether the gain was 1 pound, 10 pounds, or even 100 pounds. The same is true for the minus signs.

COMMENT   Since the sign test is so easy to use, many people use it even when the *standard* tests can be used.

COMMENT   The paired-sample sign test is actually a nonparametric alternative to the paired difference *t*-test.

EXERCISES FOR SECTION 14.3

1. The dean of Marx College believes that the median starting salary for an accountant graduating from the college's business school is $27,000. To verify the accuracy of this claim, a reporter for the student newspaper randomly selects 12 of the college's recent graduates who had similar backgrounds and who recently obtained accountant jobs. Their starting salaries were $25,000, $23,638, $30,360, $26,350, $27,430, $28,000, $27,000, $28,500, $26,000, $27,850, $26,800, $28,400. Using the one-sample sign test, test the null hypothesis that the median starting salary for an accountant who graduates from this college is $27,000 against the

alternate hypothesis that the median starting salary is not $27,000. (Use a 5% level of significance.)

2. Dr. Griffith, a leading pediatrician, believes that infants and young children will develop chicken pox approximately 10 days after being exposed to it. To test the accuracy of this claim, the parents of 15 young children who recently developed chicken pox were randomly selected. These parents report that their children developed the chicken pox 8, 12, 11, 16, 9, 10, 7, 11, 15, 10, 14, 13, 8, 14, and 15 days after being exposed to it. Using the one-sample sign test, test the null hypothesis that the median time needed for a young child to develop chicken pox after being exposed to it is 10 days against the alternative hypothesis that the median time is more than 10 days. (Use a 5% level of significance.)

3. A highway patrol officer believes that the median age of a person arrested for driving while intoxicated along the Roosevelt Expressway near the State University Campus is 19 years. A random sample of 13 people arrested for driving while intoxicated along the Roosevelt Expressway is taken. The ages of these drivers at the time of arrest is found to be 17, 23, 18, 17, 19, 34, 28, 26, 20, 24, 22, 18, and 16 years. Use the one-sample sign test to test the null hypothesis that the highway patrol officer is correct against the alternative hypothesis that the median age is higher. (Use a 5% level of significance.)

4. *School violence:* After analyzing the numerous complaints that the Board of Education has received about muggings, racial altercations, and general violence incidents at the city's public high schools, the Board decides to hire more security personnel (both uniformed and nonuniformed). The following table gives the number of weekly arrests per school for unlawful activity before and after more security personnel were added.

Number of Weekly Arrests for Unlawful
Activity at Various Schools

		Before New Personnel Hired	After New Personnel Hired
	1	7	2
	2	4	3
	3	8	6
	4	5	4
	5	9	9
School	6	9	7
	7	9	8
	8	5	3
	9	6	7
	10	8	8
	11	9	8
	12	10	9

Using a 5% level of significance, test the null hypothesis that the new security personnel hired have no effect on the number of weekly arrests for unlawful activity at the schools against the alternative hypothesis that the weekly number of arrests has decreased.

5. Refer back to the newspaper article on page 678. Using a 5% level of significance, test the null hypothesis that the new fares have no effect on the subway ridership against the alternative hypothesis that the number of commuters using the 8:05 A.M. train has decreased.

6. Many patients with heart conditions are given medications known as beta blockers. Some patients report having side effects such as dizziness or tingling in the arms. In one study the number of different times that dizziness or tingling in the arms was reported per month by patients prior to and after taking a certain beta blocker medicine is shown here.

<table>
<tr><th></th><th></th><th colspan="2">Reported Dizziness or Tingling in Arms</th></tr>
<tr><th></th><th></th><th>Before Taking Medicine</th><th>After Taking Medicine</th></tr>
<tr><td></td><td>A</td><td>18</td><td>24</td></tr>
<tr><td></td><td>B</td><td>19</td><td>16</td></tr>
<tr><td></td><td>C</td><td>22</td><td>25</td></tr>
<tr><td></td><td>D</td><td>16</td><td>19</td></tr>
<tr><td></td><td>E</td><td>12</td><td>15</td></tr>
<tr><td>Patient</td><td>F</td><td>7</td><td>7</td></tr>
<tr><td></td><td>G</td><td>8</td><td>4</td></tr>
<tr><td></td><td>H</td><td>9</td><td>12</td></tr>
<tr><td></td><td>I</td><td>18</td><td>21</td></tr>
<tr><td></td><td>J</td><td>19</td><td>17</td></tr>
</table>

Using a 5% level of significance, test the null hypothesis that the beta blocker medicine has no effect on the number of times that dizziness or tingling in the arms is reported against the alternative hypothesis that the medicine increases the number of times that dizziness or tingling in the arms is reported.

7. Most states, as well as the Internal Revenue Service, have phone numbers that taxpayers can call for information about specific tax questions. However, many taxpayers complain to government authorities or to their elected representatives that the information received from these agents is often wrong and results in the assessment of additional taxes or penalties. One regional taxpayer assistance office has instituted a new policy whereby it randomly "listens in" and records all agent conversations with taxpayers. The number of complaints received about incorrect tax information given by tax agents on ten different days both before and after the new policy was instituted is shown in the accompanying table.

Daily Number of Complaints Received
for a Two-Week Period

	Before New Policy	After New Policy
Mon	23	17
Tues	23	21
Wed	14	10
Thurs	23	39
Fri	20	20
Mon	21	16
Tues	28	16
Wed	29	24
Thurs	27	18
Fri	37	28

Using a 5% level of significance, test the null hypothesis that the new policy has no effect on the number of complaints received about wrong tax information given by the agents against the alternative hypothesis that the number of complaints has decreased.

8. In an effort to combat the increasing number of cab robberies many cabbies are installing two-way radios in their taxis. See the article on the top of the next page. These radios keep the cab drivers in direct contact with the police department and their dispatcher. The following statistics are available on the number of cab robberies for one city.

Number of Weekly Robberies of Cab Drivers

Before Installation of Two-way Radios	After Installation of Two-way Radios
18	11
14	13
16	16
19	20
22	16
15	18
17	17
11	11
15	8
13	11

Using a 5% level of significance test the null hypothesis that the two-way radios have no effect on the number of robberies of cab drivers.

ANOTHER CAB DRIVER ROBBED AND SHOT

New York (March 12)—Police report that José Rodriguez was shot by a fare that he had picked up in the Times Square area. The shooting of a cab driver, the 21st this year, occurred on Broadway after the passenger pulled a gun on Rodriguez and robbed him. The driver was then ordered to drive on Broadway where he was shot. The assailant escaped on foot.

March 12, 1991

---

9. The U.S. Postal Service recently installed a new high-technology mail sorting machine at one of its branches to cut down on the number of damaged pieces of mail. The following data on the number of pieces of damaged mail at this post office branch both before and after installation of the new machine are available.

Daily Number of Damaged Pieces of Mail
for a Two-Week Period

Before New Equipment Installed	After New Equipment Installed
11	11
28	18
22	17
19	5
20	10
22	20
16	17
12	12
8	7
9	10
18	19
16	16
21	20
17	15

Using a 5% level of significance, test the null hypothesis that the new equipment has no effect in reducing the number of damaged pieces of mail.

10. In a certain city the telephone company has instituted a charge for directory assistance. The following data on the number of calls for directory assistance over a 2-week period both before and after the imposition of the charge are available:

Number of Calls for Directory Assistance

Before Charge	After Charge
512	512
642	601
397	388
760	758
490	500
601	585
454	400
768	601
491	448
587	508

Using a 5% level of significance, test the null hypothesis that the number of calls for directory assistance is not affected by the charge.

ONE OFFICER PATROL CARS A FIASCO

New York (May 10)—Police department officials have refused to comment on surveys conducted by civic groups which indicate that the number of police arrests for various crimes including robberies, felonies, burglaries, etc. has dropped sharply ever since the department's new one police officer per patrol car policy went into effect. Although this new policy was started as an economy measure, its effectiveness is being questioned.

The department officials steadfastly refuse to admit that this new policy has had any effect on the number of polic arrests.

May 10, 1989

11. Refer to the accompanying newspaper article. The number of felony arrests over a comparable period for ten precincts both before and after the one-police-officer-per-patrol-car policy was instituted is as follows:

		Number of Felony Arrests	
		Before One-Police-Officer-per-Car Policy	*After One-Police-Officer-per-Car Policy*
	7	12	8
	9	6	4
	12	10	9
	13	8	8
Precinct	16	10	6
	18	14	9
	22	11	5
	47	10	6
	66	9	7
	68	13	12

Using a 5% level of significance, test the null hypothesis that the daily number of felony arrests is not affected by the number of police officers in the patrol car against the alternative hypothesis that the number of felony arrests has decreased.

A 16-year-old driver getting a "talking-to" instead of a ticket from a policeman after a minor traffic violation. (© *Spencer Grant/Photo Researchers, Inc.*)

14.4 THE WILCOXON SIGNED-RANK TEST

Paired-sample Sign Test

Wilcoxon Signed-rank Test

As we indicated in the last section, the **paired-sample sign test** merely utilizes information concerning whether the differences between pairs of numbers is positive or negative. Quite often the sign test will accept a null hypothesis simply because too much information is "thrown away." The **Wilcoxon signed-rank test** is less likely to accept a null hypothesis since it considers both the *magnitude* as well as the *direction* (positive or negative) of the differences between pairs.

To illustrate the use of the Wilcoxon signed-rank test, let us consider the following exam scores of 12 students who were tested both before and after receiving special instruction.

Student	Pre-instruction Score X_B	Post-instruction Score X_A	Difference $D = X_B - X_A$	Absolute Value of Difference $\lvert D \rvert$	Rank of $\lvert D \rvert$	Signed Rank
1	46	81	$46 - 81 = -35$	35	9	-9
2	58	73	$58 - 73 = -15$	15	8	-8
3	69	72	$69 - 72 = -3$	3	2	-2
4	72	77	$72 - 77 = -5$	5	4	-4
5	82	82	$82 - 82 = 0$	---	---	---
6	65	72	$65 - 72 = -7$	7	6	-6
7	69	63	$69 - 63 = 6$	6	5	$+5$
8	72	68	$72 - 68 = 4$	4	3	$+3$
9	73	85	$73 - 85 = -12$	12	7	-7
10	87	88	$87 - 88 = -1$	1	1	-1

Notice that we have added several new columns.

To apply the Wilcoxon signed-rank test we proceed as follows:

1. Find the entry for the difference column by subtracting the new value from the corresponding old value. These differences may be positive, negative, or zero.
2. Form the absolute value of these differences.
3. Rank these absolute values in order from the lowest (1) to the highest.
4. Give each rank a plus (+) sign or a minus (−) sign, which is the same sign as in the column for D.
5. The test statistic is the sum of the ranks with the smaller sum. If the null hypothesis is correct, then we would expect the sum of the positive ranks and the sum of the negative ranks to balance each other. If the sum of the ranks is considerably more positive or considerably more negative, then we would be more likely to reject the null hypothesis.

In our case the sum of the ranks is

Positive sign ranks $= (+5) + (+3) = +8$

Negative sign ranks $= (-9) + (-8) + (-2) + (-4)$
$$+ (-6) + (-7) + (-1) = -37$$

TABLE 14.3 Critical Values for Wilcoxon Signed-Rank Test

	One-Tailed Test Level of Significance ($\alpha =$)					Two-Tailed Test Level of Significance ($\alpha =$)			
n	0.005	0.01	0.025	0.05	n	0.01	0.02	0.05	0.10
5	—	—	—	0	5	—	—	—	0
6	—	—	0	2	6	—	—	0	2
7	—	0	2	3	7	—	0	2	3
8	0	1	3	5	8	0	1	3	5
9	1	3	5	8	9	1	3	5	8
10	3	5	8	10	10	3	5	8	10
11	5	7	10	13	11	5	7	10	13
12	7	9	13	17	12	7	9	13	17
13	9	12	17	21	13	9	12	17	21
14	12	15	21	25	14	12	15	21	25
15	15	19	25	30	15	15	19	25	30
16	19	23	29	35	16	19	23	29	35
17	23	27	34	41	17	23	27	34	41
18	27	32	40	47	18	27	32	40	47
19	32	37	46	53	19	32	37	46	53
20	37	43	52	60	20	37	43	52	60
21	42	49	58	67	21	42	49	58	67
22	48	55	65	75	22	48	55	65	75
23	54	62	73	83	23	54	62	73	83
24	61	69	81	91	24	61	69	81	91
25	68	76	89	100	25	68	76	89	100
26	75	84	98	110	26	75	84	98	110
27	83	92	107	119	27	83	92	107	119
28	91	101	116	130	28	91	101	116	130
29	100	110	126	140	29	100	110	126	140
30	109	120	137	151	30	109	120	137	151

We select $+8$ since this is the sum of the ranks with the smaller sum. We now compare this test statistic value with the critical value given in Table 14.3 using $n = 9$ and $\alpha = 0.05$. The chart value for a two-tailed test is 5. Since our test statistic value of $+8$ is larger than the chart value, we do not reject the null hypothesis. If, on the other hand, the test statistic value is less than or equal to the chart value, then we reject the null hypothesis.

COMMENT   In performing the calculations in the previous example we did not use the fifth student's scores since, as with the sign test, a difference of zero is not considered as positive or negative for our purposes.

COMMENT   All entries given in Table 14.3 are for absolute values of the test statistic.

COMMENT   Occasionally, two differences will have the same rank. For example, if two differences are tied for the fourth place, then each is assigned a rank of 4.5. We then assign the next value a rank of 6. The same procedure is followed when there is a tie for any rank.

We illustrate the use of the Wilcoxon signed-rank test with another example.

EXAMPLE 1

A new cholesterol-lowering pill is given to 15 people once a day for 2 months to determine its effectiveness in lowering blood serum cholesterol levels. The following data indicate the before and after cholesterol levels (in milligrams) of these 15 people.

Cholesterol Level Before Taking Pill (in milligrams)	Cholesterol Level After Taking Pill (in milligrams)
240	227
261	238
283	257
276	276
220	208
186	193
195	198
198	199
247	233
238	227
220	210
250	241
263	255
298	276
317	269

Using a 5% level of significance, test the null hypothesis that the pill has no effect on a person's blood serum cholesterol level; it neither lowers nor raises it.

Solution

We arrange the data as follows.

Cholesterol Level Before X_B	Cholesterol Level After X_A	Difference $D = X_B - X_A$	Absolute Value of Difference $\lvert D \rvert$	Rank of $\lvert D \rvert$	Signed Rank
240	227	240 − 227 = 13	13	9	+9
261	238	261 − 238 = 23	23	12	+12
283	257	283 − 257 = 26	26	13	+13
276	276	276 − 276 = 0	---	---	---
220	208	220 − 208 = 12	12	8	+8
186	193	186 − 193 = −7	7	3	−3
195	198	195 − 198 = −3	3	2	−2
198	199	198 − 199 = −1	1	1	−1
247	233	247 − 233 = 14	14	10	+10
238	227	238 − 227 = 11	11	7	+7
220	210	220 − 210 = 10	10	6	+6
250	241	250 − 241 = 9	9	5	+5
263	255	263 − 255 = 8	8	4	+4
298	276	298 − 276 = 22	22	11	+11
317	269	317 − 269 = 48	48	14	+14

For each person we determine the difference in the cholesterol level and the absolute value of the difference. Then we assign ranks from the lowest to the highest and assign a plus sign or a minus sign to each of these ranks. The sign is the same as the sign in the column for D. Now we calculate the sum of the ranks. We have

$$\text{Positive sign ranks} = (+9) + (+12) + (+13) + (+8) + (+10) + (+7)$$
$$+ (+6) + (+5) + (+4) + (+11) + (+14) = +99$$

$$\text{Negative sign ranks} = (-3) + (-2) + (-1) = -6$$

We select -6 since this is the sum of the ranks with the smaller sum. The absolute value of -6 is 6. We now compare this test statistic value with the critical value given in Table 14.3 using $n = 14$ and $\alpha = 0.05$. The chart value for a two-tailed test is 21. Since our test statistic value of 6 is less than the chart value, we reject the null hypothesis and conclude that the pill does affect the blood serum cholesterol level of an individual.

COMMENT   When using the Wilcoxon signed-rank test, it is assumed that the sample data are continuous and that the sampled population is symmetric. Furthermore, the data results can be arranged into relationships of "greater than" or "less than." In an applied example it is often difficult to verify that the sampled population is symmetric.

EXERCISES FOR SECTION 14.4

1. Many junior and senior high schools have been experimenting with having famous sports athletes speak at their school about the problems encountered should one choose to use or push drugs. Students sometimes identify with such celebrities. In one study of eight public schools it was found that over a given time period the number of reported incidents of drug use both before and after the new policy was instituted is as follows.

<table>
<tr><td></td><td></td><td colspan="2">Number of Incidents of
Drug Use</td></tr>
<tr><td></td><td></td><td>*Before New
Policy*</td><td>*After New
Policy*</td></tr>
<tr><td></td><td>*A*</td><td>29</td><td>28</td></tr>
<tr><td></td><td>*B*</td><td>26</td><td>23</td></tr>
<tr><td></td><td>*C*</td><td>21</td><td>19</td></tr>
<tr><td>School</td><td>*D*</td><td>25</td><td>26</td></tr>
<tr><td></td><td>*E*</td><td>26</td><td>19</td></tr>
<tr><td></td><td>*F*</td><td>25</td><td>25</td></tr>
<tr><td></td><td>*G*</td><td>26</td><td>18</td></tr>
<tr><td></td><td>*H*</td><td>27</td><td>27</td></tr>
</table>

Using the Wilcoxon signed-rank test at a 5% level of significance, test the null hypothesis that the new policy has no effect on drug use at the schools.

2. In response to consumer concern about cholesterol and saturated fat in baked goods, a large bakery on the East Coast has decided to reformulate the ingredients used in making a particular cake. It will bake this cake without using any egg yolks. To determine consumer reaction to this new product, the company randomly selects 12 people and asks them to rate the cake (on a particular taste appeal scale) both before and after the egg yolks are removed.

	Ratings by Tasters	
	Before Egg Yolks Removed	After Egg Yolks Removed
1	112	109
2	120	121
3	139	141
4	117	116
5	121	121
Subject 6	138	141
7	129	137
8	118	130
9	121	105
10	119	130
11	137	125
12	128	150

Using the Wilcoxon signed-rank test, do the data indicate that there is no significant difference in the ratings of the cake before and after the egg yolks are removed? (Use a 5% level of significance.)

3. In an effort to combat the recent increase in the incidence of burglaries many police officials recommend that silverware and jewelry be engraved with a code by using a special engraving tool that often can be borrowed from local police stations. A burglar cannot easily dispose of silverware or jewelry with such identification codes. The number of reported cases of burglaries (where silverware or jewelry was taken) both before and after the implementation of this new policy for one city is shown here:

	Number of Reported Burglaries	
	Before New Policy	After New Policy
Brighton	39	35
Queens	43	37
Boro Hall	47	44
Gravesend	45	35
Togo	42	43
Braverly	41	41
Stapelton	57	49
Beaver	51	57
Kensington	59	57
Pleasant Plains	50	47

Section of City

Using the Wilcoxon signed-rank test at a 5% level of significance, test the null hypothesis that the new policy has no effect on reducing burglaries.

4. Many camp directors often complain that a lot of food is wasted simply because campers do not finish meals that have been prepared for them. One camp director has hired a new cook and a nutritionist. Their job is to make the meals more nutritious and visually appealing. After several weeks the following data are available.

		Number of Campers Completing Their Meals	
		Before New Cook Began Preparing Meals	After New Cook Began Preparing Meals
	1	24	28
	2	23	27
	3	26	25
	4	25	26
Division	5	28	27
	6	29	28
	7	31	30
	8	21	21
	9	24	27
	10	28	33

Using the Wilcoxon signed-rank test, test the null hypothesis that the number of campers completing their meals has not changed significantly even after the new cook was hired. (Use a 5% level of significance.)

14.5 THE MANN-WHITNEY TEST

Wilcoxon Rank-sum Test

Mann-Whitney Test

An important nonparametric test that is used as an alternative to the standard significance tests for the difference between two sample means is the **Wilcoxon rank-sum test** or **the Mann-Whitney Test.** We can use this test when the assumption about normality is not satisfied.

To illustrate how this test is used, we consider the following data on the number of minutes needed by two independent groups of music students to learn to play a particular song. Group A received special instruction whereas Group B did not.

									Average
Group A	35	39	51	63	48	31	29	41	55	43.56
Group B	85	28	42	37	61	54	36	57		50

The means of these two samples are 43.56 and 50. In this case we wish to decide whether the difference between the means is significant.

The two samples are arranged jointly, as if they were one sample, in order of increasing time. We get

Time	Group	Rank
28	B	1
29	A	2
31	A	3
35	A	4
36	B	5
37	B	6
39	A	7
41	A	8
42	B	9
48	A	10
51	A	11
54	B	12
55	A	13
57	B	14
61	B	15
63	A	16
85	B	17

We indicate each value, whether it belongs to Group A or to Group B. Then we assign the ranks 1, 2, 3, 4, . . . , 17 to the scores, in this order, as shown.

Notice that the Group A scores occupy the ranks of 2, 3, 4, 7, 8, 10, 11, 13, and 16. The Group B scores occupy the ranks of 1, 5, 6, 9, 12, 14, 15, and 17. Now we sum the ranks of the group with the *smaller* sample size, in this case Group B, getting

$$1 + 5 + 6 + 9 + 12 + 14 + 15 + 17 = 79$$

The sum of the ranks is denoted by R. In this case $R = 79$.

We always let n_1 and n_2 denote the sizes of the two samples where n_1 represents the smaller of the two sample sizes. Thus, R represents the sum of the ranks of this smaller group. (If both groups are of equal sizes, then either one is called n_1, and R represents the sum of the ranks of this group.) Statistical theory tells us that if both n_1 and n_2 are large enough, each equal to 8 or more, then the distribution of R can be approximated by a normal distribution. The test statistic is given by Formula 14.1.

FORMULA 14.1

$$z = \frac{R - \mu_R}{\sigma_R}$$

where

$$\mu_R = \frac{n_1(n_1 + n_2 + 1)}{2}$$

and

$$\sigma_R = \sqrt{\frac{n_1 n_2(n_1 + n_2 + 1)}{12}}$$

Using a 5% level of significance, we reject the null hypothesis of equal means if $z > 1.96$ or $z < -1.96$.

In our case $R = 79$, $n_1 = 8$, and $n_2 = 9$ so that

$$\mu_R = \frac{n_1(n_1 + n_2 + 1)}{2}$$

$$= \frac{8(8 + 9 + 1)}{2}$$

$$= 72$$

and

$$\sigma_R = \sqrt{\frac{n_1 n_2 (n_1 + n_2 + 1)}{12}}$$

$$= \sqrt{\frac{8(9)(8 + 9 + 1)}{12}} = \sqrt{108}$$

$$= 10.39$$

The test statistic then becomes

$$z = \frac{R - \mu_R}{\sigma_R} = \frac{79 - 72}{10.39} = 0.67$$

Since the value of $z = 0.67$ falls in the acceptance region of Figure 14.3, we *do not* reject the null hypothesis. There is no significant difference between the means of these two groups.

COMMENT   The test that we have just described is the Wilcoxon rank-sum test with the normal approximation of the test statistic. **Mann-Whitney's test** is equivalent, but the test statistic is calculated in a slightly different way. Statisticians have constructed tables that give the appropriate critical values when both sample sizes, n_1 and n_2, are smaller than 8. The interested reader can find such tables in many books

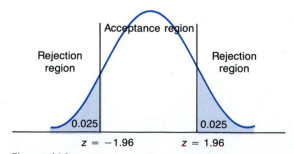

Figure 14.3

Mann-Whitney U Test

on nonparametric statistics. The corresponding exact statistic is called the **Mann-Whitney U test.**

EXAMPLE 1

An animal trainer in a circus is teaching 20 lions to perform a special trick. The lions have been divided into two groups, A and B. Group A gets positive reinforcement of food and favorable comments during the learning session whereas Group B does not. The following table indicates the number of days needed by each lion to learn the trick.

Group A	78	95	82	69	111	65	73	84	92	110
Group B	121	132	101	79	94	88	102	93	98	127

Using a 5% level of significance, test the null hypothesis that the mean time for both groups is the same.

Solution

The two samples are first arranged jointly, as if they were one large sample, in order of increasing size. We get

Days	Group	Rank
65	A	1
69	A	2
73	A	3
78	A	4
79	B	5
82	A	6
84	A	7
88	B	8
92	A	9
93	B	10
94	B	11
95	A	12
98	B	13
101	B	14
102	B	15
110	A	16
111	A	17
121	B	18
127	B	19
132	B	20

Since both groups are of equal size, we will work with Group A. The sum of the ranks of Group A is

$$1 + 2 + 3 + 4 + 6 + 7 + 9 + 12 + 16 + 17 = 77$$

Thus, $R = 77$. Now we apply Formula 14.1. We have $R = 77$, $n_1 = 10$, and $n_2 = 10$ so that

$$\mu_R = \frac{n_1(n_1 + n_2 + 1)}{2}$$

$$= \frac{10(10 + 10 + 1)}{2}$$

$$= 105$$

and

$$\sigma_R = \sqrt{\frac{n_1 n_2(n_1 + n_2 + 1)}{12}}$$

$$= \sqrt{\frac{10 \cdot 10(10 + 10 + 1)}{12}}$$

$$= \sqrt{175}$$

$$\approx 13.23$$

The test statistic then becomes

$$z = \frac{R - \mu_R}{\sigma_R}$$

$$= \frac{77 - 105}{13.23}$$

$$= -2.12$$

We use the two-tail rejection of Figure 14.3. Since the value of $z = -2.12$ falls in the rejection region, we reject the null hypothesis and conclude that the number of minutes needed by each group is not the same. Positive reinforcement affects learning time.

EXAMPLE 2   Many airlines are forced to cancel scheduled flights for a variety of reasons. Records submitted to aviation officials indicate that the weekly number of canceled flights reported by two large airline companies over a consecutive period of weeks is as follows.

Airline A	Airline B
7	13
9	5
12	4
14	8
8	11
6	3
11	17
	15

Using a 5% level of significance, test the null hypothesis that the average number of canceled flights is the same for both airlines.

Solution

The null hypothesis is that the average number of canceled flights is the same for both airlines. We arrange both samples jointly, as if they were one large sample, in order of increasing size. We get

Number of Canceled Flights	Airline	Rank
3	B	1
4	B	2
5	B	3
6	A	4
7	A	5
8	A	6.5
8	B	6.5
9	A	8
11	A	9.5
11	B	9.5
12	A	11
13	B	12
14	A	13
15	B	14
17	B	15

You will notice that there are some ties for several ranks. Whenever a tie comes up each of the tied observations is assigned the mean of the ranks they occupy. Thus, we assign a rank of 6.5 when the number of canceled flights was 8, and a rank of 9.5 when the number of canceled flights was 11. For Airline A we have 7 weeks of data and for Airline B we have 8 weeks of data. Thus, we will work with the smaller group, which is A. The sum of the ranks for Airline A is

$$4 + 5 + 6.5 + 8 + 9.5 + 11 + 13 = 57$$

Thus, $R = 57$. Applying Formula 14.1 with $n_1 = 7$ and $n_2 = 8$ gives

$$\mu_R = \frac{n_1(n_1 + n_2 + 1)}{2}$$

$$= \frac{7(7 + 8 + 1)}{2}$$

$$= 56$$

and

$$\sigma_R = \sqrt{\frac{n_1 n_2 (n_1 + n_2 + 1)}{12}}$$

$$= \sqrt{\frac{7(8)(7 + 8 + 1)}{12}}$$

$$= \sqrt{74.6667}$$

$$\approx 8.641$$

The test statistic then becomes

$$z = \frac{R - \mu_R}{\sigma_R}$$

$$= \frac{57 - 56}{8.641}$$

$$= -0.116$$

The value of $z = -0.116$ falls in the acceptance region of Figure 14.3. Thus, we cannot reject the null hypothesis. There are not sufficient data to conclude that the average number of cancellations of scheduled flights is significantly different for both airlines.

EXERCISES FOR SECTION 14.5

1. Two groups of college graduates were given pretests for the National Teacher's Examination (NTE). Group A graduates were from colleges located in a northeastern state, whereas Group B graduates were from colleges located in a midwestern state. The following data indicate the results of the exam.

Group A	Group B
45	52
50	41
48	30
54	42
47	43
58	38
53	41
44	43
49	37
55	42
57	44
	45
	41
	40
	46

Using the rank-sum test, test the null hypothesis that the mean score for both groups is the same. (Use a 5% level of significance.)

2. Nine female volunteers and ten male volunteers were each given the same amount of alcohol and then asked to perform a special task correctly. The number of trials required by each volunteer before completing the task is shown below:

Male Volunteer	Female Volunteer
35	21
37	29
34	33
29	18
20	25
30	20
25	26
28	22
24	24
30	

Using the rank-sum test, test the null hypothesis that the average number of trials required by a member of each group to complete the task after being given alcohol is the same. (Use a 5% level of significance.)

3. The number of patients treated in the Emergency Room for complications of the flu at Spovik Hospital during the first 8 days of January and in the Emergency Room at Mercy Hospital for the first 9 days of January are as follows.

Spovik Hospital	57	60	72	49	55	82	69	32	
Mercy Hospital	61	46	83	95	3	13	88	70	53

Using a 5% level of significance, test the null hypothesis that the average number of patients treated at both hospitals is the same.

4. Environmentalists have taken 20 samples of the air during rush hour at two tunnels to determine the level of pollutants in the air (in parts per million). The following results are available.

Tunnel 1	24	22	26	20	21	22	28	29			
Tunnel 2	27	21	25	20	23	26	20	25	22	29	28	26

Using a 5% level of significance, test the null hypothesis that the average amount of pollutants in the air (in parts per million) at both tunnels is the same.

5. The number of home equity loans approved by two banks per day during a 2-week period is as follows. (One bank is closed on weekends.)

Bank A	5	8	4	4	6	5	3	8	7	5				
Bank B	5	6	2	6	5	7	2	8	8	4	7	3	4	5

Using a 5% level of significance, test the null hypothesis that the average number of home equity loans approved by both banks is the same.

6. The number of minor traffic accidents occurring on the Brooks Expressway for an 8-day period and on the Stockton Parkway for a 10-day period is as follows.

Brooks Expressway	23	63	75	68	51	47	69	54		
Stockton Parkway	86	57	59	65	71	38	61	43	51	76

Using a 5% level of significance, test the null hypothesis that the average number of accidents occurring on both highways is the same.

7. The number of abandoned cars towed away daily by the city's two disposal agencies is as follows:

Towing Agency 1	43	57	80	68	40	42	56	48	31	20
Towing Agency 2	32	49	44	49	92	69	47	53		

Using a 5% level of significance, test the null hypothesis that the average number of abandoned cars towed away by the two disposal agencies is the same.

14.6 THE SPEARMAN RANK CORRELATION TEST

For many years the most widely used nonparametric statistical test was the rank correlation test developed by C. Spearman in the early 1900s. Although originally devised as a shortcut method for computing the coefficient of correlation discussed in Chapter 11, the Spearman rank correlation test has the advantage that it uses rankings only, it makes no assumptions about the distribution of the underlying populations, and it does not assume normality. We simply arrange some data in rank order and then apply the following formula.

FORMULA 14.2   The Spearman Rank Correlation Coefficient

$$R = 1 - \frac{6 \, \Sigma(x - y)^2}{n(n^2 - 1)}$$

where $\Sigma(x - y)^2$ represents the sum of the squares of the difference in ranks and n stands for the number of individuals who have been ranked.

COMMENT   When using Formula 14.2 the value of R will be between -1 and $+1$. It is used in much the same way that we used the correlation coefficient in Chapter 11.

COMMENT   When using Formula 14.2 the null hypothesis to be tested is that there is no significant correlation between the two rankings as opposed to the alternative hypothesis, which assumes that there is a significant correlation between the rankings.

We illustrate the use of this formula with several examples.

EXAMPLE 1   Beer Tasting

Two judges are testing five different brands of beer for their taste appeal, and the judges rate the beers as follows.

	Judge 1	Judge 2
Brand A	3	4
Brand B	4	3
Brand C	5	1
Brand D	2	2
Brand E	1	5

Using a 5% level of significance, test the null hypothesis that there is no significant (linear) correlation between the two judge's ratings.

Solution

We first rewrite the preceding rankings by letting x represent Judge 1's rankings and by letting y represent Judge 2's rankings. We then have the following.

x	y	$x - y$	$(x - y)^2$
3	4	-1	1
4	3	1	1
5	1	4	16
2	2	0	0
1	5	-4	16
			34

$$\Sigma(x - y)^2 = 34$$

Now we apply Formula 14.2. We have

$$R = 1 - \frac{6\,\Sigma(x - y)^2}{n(n^2 - 1)}$$

$$= 1 - \frac{6(34)}{5(25 - 1)}$$

$$= 1 - 1.7 = -0.7$$

Table XI in the Appendix allows us to interpret the value of the Spearman rank coefficient correctly. We use this table in the following way:

1. First compute the value of R using Formula 14.2.
2. Then look in the chart for the appropriate R-value corresponding to some given value of n where n is the number of pairs of scores.
3. The value of R is *not* statistically significant if it is between $-R_{0.025}$ and $R_{0.025}$ for a particular value of n at the 5% level of significance.

Returning to the previous example, we conclude that since the value of R is -0.7, it is not significant at the 5% level of significance. We cannot conclude that there is a significant correlation between the two rankings.

EXAMPLE 2

Several 9-year-old children recently competed in the Brown Bowling League playoffs and the Southview Bowling League games. The scores of these contestants were as follows.

	Score	
Contestant	*Brown League Playoff Game*	*Southview Bowling League Game*
Kim	180	162
Heather	176	157
Jason	198	176
Cassandra	197	183
Wilfredo	171	188

Using a 5% level of significance, test the null hypothesis that there is no significant correlation between the Brown League playoff game scores and the Southview scores.

Solution

We first rewrite each of the scores in terms of rankings from the highest score, which is assigned a ranking of 1 to the lowest score, which is assigned a ranking of 5. Now we let x represent the Brown score rankings and let y represent the Southview score rankings. We then have the following.

x	y	$x - y$	$(x - y)^2$
3	4	-1	1
4	5	-1	1
1	3	-2	4
2	2	0	0
5	1	4	16
			22

$$\Sigma(x - y)^2 = 22$$

Applying Formula 14.2 gives

$$R = 1 - \frac{6 \, \Sigma(x - y)^2}{n(n^2 - 1)}$$

$$= 1 - \frac{6(22)}{5(25 - 1)}$$

$$= 1 - 1.1 = -0.1$$

By referring to Table XI in the Appendix, we conclude that since our value of R is -0.1, it is not significant at the 5% level of significance. We cannot conclude that there is a significant correlation between the two bowling league scores.

14.7 THE RUNS TEST

All the samples discussed so far in this book were assumed to be random samples. How does one test for randomness?

Runs Test

In recent years mathematicians have developed a **runs test** for determining the randomness of samples. This test is based on the order in which the observations are made. For example, suppose 25 people are waiting in line for admission to a theater and they are arranged as follows where m denotes male and f denotes female:

f, f, f, f, m, f, m, m, m, m, f, f, f, f, f, f, m, m, m, m, f, m, m, f, m

Is this a random arrangement of the m's and f's?

Theory of runs

In order to answer this question, statisticians use the **theory of runs.** We first have the following definition.

DEFINITION 14.1
Run

A **run** is a succession of identical letters or symbols that is followed and preceded by a different letter or by no letter at all.

There are ten runs in the preceding sequence of m's and f's. These are

Run	Letters
1	ffff
2	m
3	f
4	mmmm
5	ffffff
6	mmmm
7	f
8	mm
9	f
10	m

Many runs would indicate that the data occur in definite cycles according to some pattern. The same is true for data with few runs. In either case we do not have a random sample. We still need some way of determining when the number of runs is reasonable.

When using the runs test note that the length of each individual run is not important. What is important is the number of times that each letter appears in the entire sequence of letters. Thus, in our example there are 25 people waiting in line; 13 are female and 12 are male so that f appears 13 times and m appears 12 times. We have n_1 samples of one kind and n_2 samples of another kind. We now wish to test whether this sample is random.

Table X in the Appendix gives us the critical values for the total number of runs. To use this table we first determine the larger of n_1 and n_2. In our case there are 13 f's and 12 m's so that the larger is 13 and the smaller is 12. We now move across the top of the chart until we reach 13. Then we move down until we get to the 12 row. Notice that there are two numbers in the box corresponding to larger 13, smaller 12. These are the numbers 8 and 19. These are also the critical values. If the number of runs is between 8 and 19, we do not reject the null hypothesis. This would mean that we have a random sample. If the number of runs is less than 8 or more than 19, we no longer have a random sample. In our case since we had ten runs we do not reject the null hypotheses of randomness.

Let us further illustrate the runs test with several examples.

EXAMPLE 1   Twenty people are waiting on line in a bank. These people will either deposit or withdraw money. Let d represent a customer who makes a deposit and let w represent a customer who is making a withdrawal. If the people are arranged in the following order, test for randomness. (Use a 5% level of significance.)

d, d, d, w, w, w, w, w, d, d, d, d, d, d, d, w, w, w, d, d

Solution

There are five runs as shown in the following chart:

Run	Letters
1	ddd
2	wwwww
3	ddddddd
4	www
5	dd

There are also 12 d's and 8 w's in this succession of letters, so that the larger is 12 and the smaller is 8. From Table X in the Appendix we find that the critical values, where the larger number is 12 and the smaller number is 8, are 6 and 16 runs. Since we obtained only 5 runs, we reject the null hypotheses and conclude that these people are not arranged in random order.

EXAMPLE 2

Thirty dresses are arranged on a rack as follows, where r represents a red dress and b represents a blue dress. Using a 5% level of significance, determine if the dresses are arranged in a random order.

r, b, r, b, r, r, b, r, b, b, b, r, r, r, r, r, b, r, b, b, r, b, r, b, r, b, b, b, b, r, r

Solution

There are 16 r's and 14 b's so that the larger number is 16 and the smaller number is 14. There are also 19 runs as shown.

Run	Letters
1	r
2	b
3	r
4	b
5	rr
6	b
7	r
8	bb
9	rrrr
10	b
11	r
12	bb
13	r
14	b
15	r
16	b
17	r
18	bbbb
19	rr

From Table X we note that the critical values, where the larger number is 16 and the smaller number is 14, are 10 and 22. Since we obtained 19 runs, we do not reject the null hypothesis of randomness.

The theory of runs can also be applied to any set of numbers to determine whether or not these numbers appear in a random order. In such cases we first calculate the median of these numbers (see page 109) since approximately one-half the numbers are below the median and one-half the numbers are above the median. We then go through the sequence of numbers and replace each number with the letter a if it is above the median and with the letter b if it is below the median. We omit any values that equal the median. Once we have a sequence of a's and b's, we proceed in exactly the same way as we did in Examples 1 and 2 of this section.

EXAMPLE 3

The number of defective items produced by a machine per day over a period of a month is

13, 17, 14, 20, 18, 16, 14, 19, 21, 20, 14, 17, 12, 14, 19, 20, 17, 18, 14, 20, 17, 19, 17, 14, 19, 21, 16, 12, 15, 22

Using a 5% level of significance, test for randomness.

Solution

We first arrange the numbers in order, from the smallest to the largest, to determine the median. We get 12, 12, 13, 14, 14, 14, 14, 14, 14, 15, 16, 16, 17, 17, 17, 17, 17, 18, 18, 19, 19, 19, 19, 20, 20, 20, 20, 21, 21, 22. The median of these numbers is 17. We now replace all the numbers of the original sequence with the letter b if the number is below 17 and with the letter a if the number is above 17. We do not replace the 17 with any letter. The new sequence then becomes

b, b, a, a, b, b, a, a, a, b, b, b, a, a, a, b, a, a, b, a, a, b, b, b, a

There are now 13 a's and 12 b's, and there are 12 runs as follows:

Run	Letters
1	bb
2	aa
3	bb
4	aaa
5	bbb
6	aaa
7	b
8	aa
9	b
10	aa
11	bbb
12	a

From Table X we note that the critical values, where the larger number is 13 and the smaller number is 12, are 8 and 19. Since we obtained 12 runs, we do not reject the null hypothesis.

COMMENT Table X can be used only when n_1 and n_2 are not greater than 20. If either is larger than 20, we use the normal curve approximation. In this case the test statistic is

$$z = \frac{X - \mu_R}{\sigma_R}$$

where

$$\mu_R = \frac{2n_1n_2}{n_1 + n_2} + 1$$

and

$$\sigma_R = \sqrt{\frac{2n_1n_2(2n_1n_2 - n_1 - n_2)}{(n_1 + n_2)^2(n_1 + n_2 - 1)}}$$

EXERCISES FOR SECTION 14.7

1. Four baseball scouts have rated five potential baseball prospects (Bill, Thomas, Roger, Mack, and John) in terms of their overall fielding and batting abilities according to the following rankings.

Prospect Rankings

Baseball Scout 1	Baseball Scout 2	Baseball Scout 3	Baseball Scout 4
Bill	Bill	Thomas	Thomas
John	Thomas	Roger	John
Roger	Mack	Mack	Mack
Mack	John	John	Roger
Thomas	Roger	Bill	Bill

Using a 5% level of significance, test the null hypothesis that there is no correlation between the rankings of
   a. baseball scouts 1 and 2.
   b. baseball scouts 1 and 3.
   c. baseball scouts 2 and 4.
2. The following list gives the final exam grade for ten students in both Calculus I and Calculus II.

Student	Calculus I Grade	Calculus II Grade
Bill	89	82
Mary	67	45
Joe	78	84
Francine	93	95
Edward	96	96
Jason	71	57
Marilyn	76	56
George	50	81
Bruce	73	39
Lisa	68	83

Using a 5% level of significance, test the null hypothesis that there is no correlation between the Calculus I final exam grade and the Calculus II final exam grade; that is, they are independent.

3. *Teacher preference:* There are seven teachers who teach advanced math elective courses at Stony University. Two math majors were asked to rank these teachers from the best to the worst. Their rankings were as follows:

Teachers	Ratings by Student 1	Ratings by Student 2
Prof. Lichtenfeld	5	6
Prof. Kuperman	4	4
Prof. Randolph	6	5
Prof. McMillan	3	3
Prof. Rodriguez	1	7
Prof. Mathews	2	1
Prof. Hobson	7	2

Using a 5% level of significance, test the null hypothesis that there is no correlation between the student's rankings of the teachers.

4. Two sportscasters were asked to rate the following baseball teams in terms of overall performance and in terms of who would win the pennant. The following are their rankings.

Team	Sportscaster 1	Sportscaster 2
Chicago Cubs	7	8
Pittsburgh Pirates	4	3
Montreal Expos	5	5
New York Mets	3	7
San Francisco Giants	2	2
Los Angeles Dodgers	1	1
Philadelphia Phillies	6	4
St. Louis Cardinals	8	6

Using a 5% level of significance, test the null hypothesis that there is no correlation between the two sportscaster's rankings of the teams.

If we are given the attendance figures for numerous games, how do we test for randomness? (© *Art Stein/Photo Researchers, Inc.*)

5. The first 25 cars that entered a parking lot were rated as either compact or intermediate. The order was

C, I, C, C, I, C, C, C, I, I, C, I, I, I, C, C, C, I. C, C, I C, C, I, C

where C = Compact and I = Intermediate.

Test for randomness.

6. The sex of the first 24 people arriving at a disco is as follows (M = male and F = female):

M, F, F, M, F, M, M, F, F, F, M, M, F, M, F, F, F, M, M, M, F, F, F, M

Test for randomness.

7. A quality control engineer rates packages as either acceptable (A) or nonacceptable (N). In one day's production she rated the first 24 cases examined as follows:

A, A, N, N, N, N, A, A, A, A, A, A, A, A, N, N, A, A, A, N, N, A, A, N

Test for randomness.

8. An interviewer asks people as they are getting off an elevator whether or not they are in favor of lowering the drinking age. They answer yes (Y) or no (N) as follows:

Y, Y, Y, N, N, N, Y, Y, N, Y, N, N, Y, Y, Y, Y, N, N, N, Y, N, Y, N, Y, N, Y

Test for randomness.

9. The number of counterfeit bills confiscated last month by the Treasury Department on a daily basis is as follows:

16, 18, 19, 23, 9, 43, 25, 21, 17, 16, 14, 10, 11, 17, 36, 29

Test for randomness.

10. The number of daily calls for an ambulance in a particular city over the last few weeks is as follows:

68, 40, 45, 65, 87, 69, 73, 68, 30, 97, 65, 90, 89, 65, 97, 68, 80, 83, 96, 94, 75

Test for randomness.

11. A newspaper reported covering a rally asks students if they are Democrats (D) or Republicans (R). Their answers are as follows:

D, D, R, D, R, D, D, R, R, R, R, D, R, R, R, D, R, D, R, R, D, R, R, D, D, R, D, R

Test for randomness.

12. The number of minutes that a commuter's train was late in arriving over the past 19 days is as follows:

10, 17, 3, 19, 15, 16, 14, 5, 23, 17, 18, 16, 19, 28, 20, 18, 16, 14, 12

Test for randomness.

13. The ages of the 20 girls of the Alpha Sorority in order of their height are as follows:

19, 23, 20, 21, 19, 18, 19, 21, 17, 19, 21, 20, 18, 17, 19, 18, 19, 20, 19, 18

Test for randomness.

14. The number of people attending a baseball game during the 22 June home games of a baseball team is as follows:

42,371	40,191	36,089	51,376	39,057	23,792
26,421	38,112	40,037	22,769	33,007	41,326
39,602	37,504	45,691	43,069	21,003	19,792
16,001	39,056	48,886	37,691		

Test for randomness.

15. Five women were asked to rate seven different brands of perfurme for its appeal and lasting ability. Their rankings were as follows.

Brand of Perfume	Woman 1	Woman 2	Woman 3	Woman 4	Woman 5
A	2	2	4	5	7
B	3	4	6	6	5
C	7	7	3	4	3
D	1	5	5	3	4
E	5	1	7	2	6
F	6	3	1	1	1
G	4	6	2	7	2

Using a 5% level of significance, test the null hypothesis that there is no correlation between the women's ratings of the perfumes.

14.8 THE KRUSKAL-WALLIS H-TEST

In Chapter 13 we discussed ANOVA techniques for determining whether the sample means of several populations are equal. In order to apply the ANOVA techniques and the F-distribution we assumed that the samples were randomly selected from independent populations that were approximately normally distributed.

Kruskal-Wallis H-test

A nonparametric statistical test that does not require that these assumptions be satisfied is the **Kruskal-Wallis H-test**. This rank-sum test is used when we wish to test the null hypothesis that r independent random samples were obtained from identical (not necessarily normal) populations. Of course, the alternative hypothesis is that the means of these populations are not the same. The only requirement is that each sample have at least five observations.

When using the Kruskal-Wallis H-test we combine all the data and rank them jointly as if they represent one single sample. If the null hypothesis is true, then the sampling distribution of these numbers can be approximated by a chi-square distribution with $r - 1$ degrees of freedom. We accept the null hypothesis that the sample means are equal whenever the test statistic value is less than the χ^2 value for $r - 1$ degrees of freedom at a level of significance of α. Otherwise we reject the null hypothesis. The test statistic value is calculated using the following formula.

Test Statistic When Using the Kruskal-Wallis H-Test

> ### Test Statistic When Using the Kruskal-Wallis H-Test
>
> $$\frac{12}{n(n + 1)} \sum_{i=1}^{r} \frac{R_i^2}{n_i} - 3(n + 1)$$

When using the preceding test statistic, we have

$n = $ the total number in the entire sample
(when the data are joined together)

that is,

$$n = n_1 + n_2 + \cdots + n_r$$

R_i = the sum of the ranks assigned to n_i values of the ith sample.

We illustrate the use of this formula with several examples.

EXAMPLE 1

A consumer's group tested numerous cans of paint from three different companies to determine whether there is any significant difference between these brands in average drying time. The number of minutes needed for each of these brands of paint to dry when applied to identical walls was as follows:

Brand A	Brand B	Brand C
38	52	47
32	48	30
27	39	37
29	42	41
23	46	44
43	53	49

Use the Kruskal-Wallis H-test to determine if there is any significant difference in the average drying time for these brands of paint. (Use a 5% level of significance.)

Solution

The null hypothesis is that the means of these different brands are the same. The three samples are first arranged jointly, as if they were one large sample, in order of increasing size. We get

Drying Time	Brand	Rank
23	A	1
27	A	2
29	A	3
30	C	4
32	A	5
37	C	6
38	A	7
39	B	8
41	C	9
42	B	10
43	A	11
44	C	12
46	B	13
47	C	14
48	B	15
49	C	16
52	B	17
53	B	18

The sum of the rankings for Brand A is

$$R_1 = 1 + 2 + 3 + 5 + 7 + 11 = 29$$

The sum of the rankings for Brand B is

$$R_2 = 8 + 10 + 13 + 15 + 17 + 18 = 81$$

The sum of the rankings for Brand C is

$$R_3 = 4 + 6 + 9 + 12 + 14 + 16 = 61$$

In this case the number of cans of paint tested for each brand is 6 so that $n_1 = n_2 = n_3 = 6$. Also, the total number in the entire joined sample is $n = 18$. Substituting these values into the test statistic formula gives

$$\text{Test statistic} = \frac{12}{18(18 + 1)} \left(\frac{29^2}{6} + \frac{81^2}{6} + \frac{61^2}{6} \right) - 3(18 + 1)$$

$$= 8.048$$

From Table VIII in the Appendix the χ^2 value with $3 - 1$ or 2 degrees of freedom at the 5% level of significance is 5.991. Since our test statistic value of 8.048 exceeds this value, we reject the null hypothesis and conclude that the average drying time for these brands of paint is not the same.

EXAMPLE 2

An independent agency is interested in determining how long (in minutes) it takes a teller to complete the identical transaction at each of three banks. The following results were obtained when numerous tellers at the three banks were surveyed.

Time to Transaction
Completion (minutes)

Bank 1	Bank 2	Bank 3
12	23	15
14	8	20
11	17	21
16	9	7
10	13	22
18	24	
19		

Do the data indicate that there is a significant difference in the average time required to complete the transaction at these banks? (Use the Kruskal-Wallis H-test at the 5% level of significance.)

Solution

The null hypothesis is that the means for the three banks are the same. The three samples are first arranged jointly, as if they constitute one large sample, in order of increasing size. We get

Time	Bank	Rank
7	3	1
8	2	2
9	2	3
10	1	4
11	1	5
12	1	6
13	2	7
14	1	8
15	3	9
16	1	10
17	2	11
18	1	12
19	1	13
20	3	14
21	3	15
22	3	16
23	2	17
24	2	18

The sum of the rankings for Bank 1 is

$$R_1 = 4 + 5 + 6 + 8 + 10 + 12 + 13 = 58$$

The sum of the rankings for Bank 2 is

$$R_2 = 2 + 3 + 7 + 11 + 17 + 18 = 58$$

The sum of the rankings for Bank 3 is

$$R_3 = 1 + 9 + 14 + 15 + 16 = 55$$

In this case the number of tellers from Bank 1 is 7, so $n_1 = 7$; the number of tellers from Bank 2 is 6, so $n_2 = 6$; and the number of tellers from Bank 3 is 5, so $n_3 = 5$. Also, the total number in the entire joined sample is $n = 18$. Substituting these values into the test statistic formula gives

$$\text{Test statistic} = \frac{12}{18(18 + 1)} \left(\frac{58^2}{7} + \frac{58^2}{6} + \frac{55^2}{5} \right) - 3(18 + 1)$$

$$= 0.7627$$

From Table VIII in the Appendix the χ^2 value with $3 - 1$, or 2 degrees of freedom at the 5% level of significance is 5.991. Since our test statistic value of 0.7627 is less than this value, we do not reject the null hypothesis. We conclude that there is no significant difference in the average time required to complete the transaction at these banks.

EXERCISES FOR SECTION 14.8

 1. After a home computer has been purchased and removed from the packing material, the various components have to be assembled and connected for the machine to become functional. Many home computer stores provide this service at no charge. One large midwestern computer store employs five people who perform this function. The number of minutes required by each to complete the assembly task for 26 identical machines is shown as follows:

Time to Complete Task (in minutes)

Employee								
	Bill	65	47	51	72	62	81	
	Mary	53	58	62	49	53		
	Bob	61	57	49	58	57	64	72
	Sue	65	63	59	56			
	Sandra	49	58	51	45			

Using a 1% level of confidence, test the null hypothesis that there is no significant difference between the average time required to assemble the computer by the various employees.

 2. A consumer's group wishes to verify the claim made by several paint manufacturers that the average gallon of a certain type of paint will cover a 420-square-foot wall when applied according to specifications. Fifteen walls are divided into groups of five each. Each wall in a subgroup is then painted with one of three brands of paint. The coverage (in square feet) by each of these brands of paint is as follows.

Paint Coverage (in square feet)

Brand A	400	390	402	407	410
Brand B	422	416	421	413	420
Brand C	415	422	425	410	411

Using a 5% level of significance, test the null hypothesis that there is no significant difference between the average coverage by each of these brands of paint.

 3. To compare the effectiveness of four new pills in lowering blood serum cholesterol levels, 20 people with high cholesterol levels were carefully selected in order to make them as comparable as possible. The 20 people were divided into four groups of 5 each. Each person in a subgroup was then given a pill. The decrease in the blood cholesterol level of each person taking the pill is as follows.

Decrease in Blood Cholesterol Level
(in milligrams)

Pill A	12	16	18	13	15
Pill B	10	15	22	11	17
Pill C	16	3	9	12	10
Pill D	17	19	11	14	15

Using a 1% level of significance, test the null hypothesis that there is no significant difference between the average decrease in the blood serum cholesterol levels of people taking each of these types of pills.

 4. *Water pollution:* Environmentalists have accused a chemical company of polluting the waters of a particular river by dumping untreated chemical wastes in it. To test this charge, a judge orders that water samples from four different locations on the river be taken and the quantity of dissolved oxygen contained in the river at each location be determined. (The quantity of dissolved oxygen contained in water is often used to determine the extent of water pollution. The lower the dissolved oxygen content in the water is, the higher the level of water pollution will be.) The locations to be used are:

1. upstream above the chemical plant.
2. adjacent to the chemical plant's discharge pipe.
3. one-half mile downstream from the chemical plant.
4. a considerable distance downstream from the chemical plant.

It is decided that at least five samples will be taken at each location. The results of the experiment are as follows:

Average Quantity of Oxygen in Water

Location								
1	4.7	4.2	6.4	4.1	5.1	6.6		
2	6.2	6.3	3.8	5.7	6.1	5.4	3.9	4.6
3	5.2	5.8	5.9	5.5	5.6	7.3		
4	6.5	6.0	5.3	4.8	3.7			

Do these data indicate that there is a significant difference in the average dissolved oxygen content at these four locations? (Use a 1% level of significance.)

14.9 COMPARISON OF PARAMETRIC AND NONPARAMETRIC STATISTICS

In this chapter we have discussed only some of the nonparametric statistical techniques. Actually, there are many other such techniques that can be used. The question that is often asked is: Why should one bother using the standard statistical techniques discussed in the remainder of this book when the nonparametric statistical techniques discussed in this chapter are easier to use? As a matter of fact, many statisticians actually recommend the use of such nonparametric techniques in many different situations.

The decision as to which statistical technique to use depends on the particular situation. Generally speaking, if you are sure that the data come from a population that is approximately normally distributed, you should use a parametric test. On the other hand, if you are not sure, then use the appropriate nonparametric test. Other factors, such as the possible error generated by using one test as opposed to another, have to be considered. While nonparametric calculations are easier, the widespread availability and use of the computer today for parametric calculations should make such techniques easier to use.

For the benefit of the reader, we summarize some of the nonparametric and parametric statistical techniques discussed and the cases in which each is used.

When Performing Tests Involving	Parametric Test to Use	Nonparametric Test to Use
One mean (median)	t-test (page 494)	One-sample or paired-sample sign test (pages 679 and 683)
Two means (independent samples)	t-test (page 507)	Mann-Whitney test (page 700)
Two means (paired samples)	Paired t-test (page 507)	Sign test or Wilcoxon signed-rank test (page 694)
More than two means (independent samples)	ANOVA (page 644)	Kruskal-Wallis H-test (page 719)
Correlation	Linear correlation (page 541)	Spearman's rank correlation test (page 708)
Randomness		Runs test (page 711)

14.10 USING COMPUTER PACKAGES

The MINITAB statistical package is well suited to perform many of the nonparametric statistical tests discussed in this chapter. We illustrate the use of the MINITAB for one such nonparametric test. Let us return to Example 1 of Section 14.6 (page 709). In that example two judges were asked to rate five different brands of beer on the basis of taste appeal. The actual test results were

	Judge A	Judge B
Brand 1	72	68
Brand 2	68	70
Brand 3	62	79
Brand 4	81	77
Brand 5	87	63

Although we deleted the actual test results and presented only the judge's rankings, when using MINITAB we do not have to do this. MINITAB does the ranking and then computes the Spearman rank coefficient. This is shown in the following printout:

```
MTB > READ JUDGE A INTO C1, JUDGE B INTO C3
DATA > 72   68
DATA > 68   70
DATA > 62   79
DATA > 81   77
DATA > 87   63
DATA > END
       5 ROWS READ
MTB > RANK THE VALUES IN C1, PUT RANKS INTO C2
MTB > RANK THE VALUES IN C3, PUT THE RANKS INTO C4
MTB > PRINT C1 - C4

 ROW   C1   C2   C3   C4

   1   72    3   68    2
   2   68    2   70    3
   3   62    1   79    5
   4   81    4   77    4
   5   87    5   63    1

MTB > CORRELATION COEFFICIENT BETWEEN RANKS IN C2 AND C4

CORRELATION OF C2 AND C4 = -0.700

MTB > STOP
```

14.11 SUMMARY

In this chapter we discussed several of the nonparametric statistical methods that are often used when we cannot use the standard tests. By far the easiest and most popular of these methods is the sign test. This test is used when we wish to compare two sample means and we know that the samples are not independent. Because of its simplicity, the sign test is used by many people even when a standard test can be used. However, this method wastes much information.

Another important nonparametric test is the rank-sum test (Mann-Whitney test), which is used when the normality assumption is not satisfied or when the variances are not equal. The sum of the ranks is normally distributed when the sample size is large enough, in which case we can use an appropriate z statistic.

We also mentioned the Spearman rank coefficient test, which was originally devised as a shortcut method for computing the coefficient of correlation.

Then we discussed the runs test, which is used to test for randomness or a lack of it. In determining whether or not we have a random sample, we use Table X in the Appendix to find the appropriate critical values.

Finally, we presented the Kruskal-Wallis H-test, which we can use instead of the ANOVA techniques discussed in Chapter 13.

Study Guide

The following is a chapter summary in capsule form. You should now be able to demonstrate your knowledge of the ideas mentioned by giving definitions, descriptions, or specific examples. Page references are given in parentheses.

Statistical techniques that are not dependent on the distribution or parameters involved are known as **nonparametric statistics** or **distribution-free methods** as opposed to **standard methods.** (page 679)

Nonparametric tests are often used in place of the **standard tests** because their assumptions are less demanding. (page 679)

The **one-sample sign test** can be used as an alternative to the *t*-test with one mean when the assumption about normally distributed populations is not satisfied. (page 679)

The sign test can also be used when working with **paired data** that occur when we deal with **two dependent samples.** This can happen when we measure the same sample twice, as is done in the **before and after type study.** (page 683)

The **paired-sample sign test** is a nonparametric alternative to the paired difference *t*-test (page 683)

The **Wilcoxon signed-rank test** utilizes information concerning whether the differences between pairs of numbers is positive or negative. It considers both the magnitude as well as the direction of the differences between pairs. (page 694)

The **Wilcoxon rank-sum test** or the **Mann-Whitney test** is a nonparametric test that can be used to test whether the difference between two observed means is significant when the assumption about normality is not satisfied. The two samples are arranged jointly as if they were one sample. (page 700)

The **Spearman rank correlation test** is a nonparametric shortcut method that can be used to compute the coefficient of correlation between data. It uses rankings only. (page 708)

A **run** is a succession of identical letters or symbols that is followed and preceded by a different letter or no letter at all. (page 711)

The **theory of runs** allows us to test for randomness. (page 711)

The **Kruskal-Wallis H-test** is a nonparametric test that can be used to test whether the sample means of several populations are equal. It is the counterpart to the ANOVA techniques discussed in Chapter 13. When using the Kruskal-Wallis H-test, we combine all the data and rank them jointly as if they represent one single sample. (page 719)

Formulas to Remember

You should be able to identify each symbol in the following formulas, understand the relationship among the symbols expressed in the formulas, understand the significance of the formulas, and use the formulas in solving problems.

1. When using the Mann-Whitney test (rank-sum test):

$$z = \frac{R - \mu_R}{\sigma_R}$$

where

$$\mu_R = \frac{n_1(n_1 + n_2 + 1)}{2}$$

and

$$\sigma_R = \sqrt{\frac{n_1 n_2(n_1 + n_2 + 1)}{12}}$$

and n_1 is the smaller sample size.

2. Spearman rank coefficient test:

$$R = 1 - \frac{6 \Sigma(x - y)^2}{n(n^2 - 1)}$$

3. Normal curve approximation for the runs test:

$$z = \frac{X - \mu_R}{\sigma_R}$$

$$\mu_R = \frac{2n_1 n_2}{n_1 + n_2} + 1$$

$$\sigma_R = \sqrt{\frac{2n_1 n_2(2n_1 n_2 - n_1 - n_2)}{(n_1 + n_2)^2(n_1 + n_2 - 1)}}$$

4. Kruskal-Wallis H-test statistic:

$$\frac{12}{n(n + 1)} \sum_{i=1}^{r} \frac{R_i^2}{n_i} - 3(n + 1)$$

where n = the total number in the entire sample (when the data are joined together) and R_i = the sum of the ranks assigned to n_i values of the ith sample.

Testing Your Understanding of This Chapter's Concepts

1. Assume that we are performing a Wilcoxon rank-sum test for independent random variables. Is there any difference between a one-tailed and a two-tailed test? Explain your answer.
2. In order to apply the Wilcoxon signed-rank test, we assume that the probability distribution of the differences is continuous. Explain why this is necessary.
3. When using the Kruskal-Wallis H-test only large values of H will lead to rejection of the null hypothesis that the samples come from identical populations. Can we say that the Kruskal-Wallis H-test is a right-tailed test? Explain your answer.
4. When using the Kruskal-Wallis H-test, if we have three samples, each of size 5, what are the smallest and the largest values of H? Explain your answer.

THINKING CRITICALLY

1. When testing for randomness the null hypothesis is that the sampling distribution is random so that each sequence position has the same prior chance of being assigned an a as any other. The following expression can be used as a test statistic:

$$z = \frac{R_a - \left(\dfrac{2n_a n_b}{n_a + n_b} + 1\right)}{\sqrt{\dfrac{2n_a n_b(2n_a n_b - n_a - n_b)}{(n_a + n_b)^2(n_a + n_b - 1)}}} \qquad R_a = \text{number of runs for category a.}$$

where n_a and n_b represent the number of a's and b's in the sample. Applying this formula, test the data of Example 3 on page 714 for randomness.

2. Refer back to the newspaper clipping given on page 187 concerning the 1970 draft lottery. The U.S. Selective Service was vehemently criticized about the way in which the capsules were selected due to the wide differences in the draft priority numbers. People whose birthdays were later in the year had a higher priority. Apply the runs test to justify the null hypothesis of randomness.

3. It is true that for the case of two samples the Kruskal-Wallis H-test is equivalent to the Wilcoxon rank-sum test? (*Hint:* For the case of two samples the Kruskal-Wallis test statistic equals the square of the test statistic used in the Wilcoxon rank-sum test. Also, note that when we have 1 degree of freedom the critical values of χ^2 correspond to the square of the z-score.)

Review Exercises for Chapter 14

1. The following data indicate the number of minutes needed by two groups of joggers to run a 3-mile marathon.

Group A	30	20	17	16	14	12	15	19		
Group B	25	16	19	31	15	18	12	19	17	16

   Test the null hypothesis that the average time needed to run the 3-mile marathon by both groups is the same. (Use a 5% level of significance.)

2. A gas station attendant makes a list of the first 28 drivers who buy gas for their cars as to whether the gas is high test (H) or regular (R). The following is the list.

   R, R, H, H, H, R, H, H, R, R, R, R, H, H, H, R, H,
   R, H, H, R, R, R, R, H, H, R, H

   Test for randomness.

3. A sociologist asked 20 engaged couples how many children they planned to have. The answers of the men and the answers of the women are as follows.

Couple	Man	Woman
1	3	2
2	2	1
3	1	1
4	2	2
5	3	2
6	4	1
7	3	0
8	0	0
9	2	5
10	3	2
11	5	4
12	6	6
13	8	8
14	2	1
15	1	2
16	2	2
17	3	3
18	5	6
19	6	2
20	3	0

   Using the sign test, do the men and women differ significantly on this issue?

4. A stock analyst has compiled the following list on the number of stock transactions conducted per day and the day of the week. The list is as follows.

Day	Number of Transactions Conducted
Mon	57
Tues	61
Wed	48
Thurs	39
Fri	73
Mon	65
Tues	56
Wed	69
Thurs	41
Fri	50
Mon	59
Tues	39
Wed	83
Thurs	69
Fri	53

By using the Spearman rank coefficient test, test the hypothesis that there is no correlation between the day on which the transactions took place and the number of transactions conducted. (Using a 5% level of significance.)

5. Ten identical cars were tested with one brand of gasoline. Twelve other cars, identical to the first cars, were tested with a second brand of gasoline. The number of miles per gallon of gas with each of these brands is as follows. (The test conditions were the same for all cars.)

Brand A	23	25	27	19	18	24	22	27	21	25	
Brand B	19	25	17	29	27	24	27	29	26	28	24	23

Using a 5% level of significance, test the null hyothesis that the average number of miles per gallon is the same for both cars.

6. The number of false alarms phoned in to the fire department of a large city on a daily basis over a 22-day period is as follows.

79 62 53 49 59 27 69 53 46 80 29
61 53 39 51 46 21 51 72 66 82 12

Test for randomness.

7. Two teachers have ranked six students according to their intellectual ability as follows.

Student	Teacher 1	Teacher 2
Pat	2	3
Dick	3	2
Nat	1	1
Sol	4	4
Ted	6	5
Ann	5	6

Using a 5% level of significance, test the null hypothesis that the rankings are not related.

8. A group of students was given a manual dexterity test. After the test the students were trained by a special technique and then were retested. The following are the results.

Student	Score Before Training	Score After Training
A	58	65
B	61	65
C	79	82
D	63	63
E	52	52
F	78	70
G	82	81
H	42	45
I	39	41
J	75	77
K	85	88

Using the sign test, test the null hypothesis that the special teaching technique is effective.

9. An anxiety test was given to a group of 15 college students while not on drugs. The same test was then given to these students while they were on drugs. The following list gives the results of the two tests.

Subject	Rating While Not on Drugs	Rating While on Drugs
1	92	89
2	70	48
3	85	91
4	91	93
5	55	42
6	55	12
7	70	73
8	90	90
9	70	61
10	91	87
11	91	97
12	91	88
13	65	65
14	93	92
15	55	20

Using the sign, test the null hypothesis that drugs have no effect on the performance on the anxiety test.

10. The number of marriage licenses issued by the Marriage Bureau of a city during 25 business days is as follows:

    22, 12, 8, 22, 15, 17, 16, 14, 11, 5, 13, 11, 10, 12, 10, 10, 10, 11, 6, 18, 5, 8, 34, 28, 17

    Test for randomness.

11. The following chart indicates the tolerance level of 12 females and 10 males to a certain stimuli.

Female	101	115	110	96	130	112	106	116	94	117	116	134
Male		97	110	92	129	111	95	116	90	125	113	

    Using the rank-sum test, test the null hypothesis that the average tolerance level is the same for females and males. (Use a 5% level of significance.)

12. *At what age are we the smartest?* A psychologist administered an intelligence test to ten 25-year-old and to twelve 40-year-old business executives. The results were as follows.

    Results

25 Years Old	40 Years Old
110	112
99	101
87	93
92	115
112	116
114	121
114	100
105	106
103	91
119	92
	87
	99

    Using the rank-sum test, test the null hypothesis that the average results at both ages is the same. (Use a 5% level of significance.)

13. Twenty-five people are lined up and waiting for their unemployment insurance claim to be processed. Their sex is as follows:

    M, W, M, M, W, W, M, M, M, M, W, W, W, M, M, W, W, M, W, M, W, W, M, M, M

    Test for randomness.

14. The weights (in pounds) of ten people before they stopped drinking and 3 months after they stopped drinking are shown below.

Before	147	165	182	173	177	180	162	169	175	155
After	153	160	175	168	170	175	158	161	171	150

Using the sign test, test the null hypothesis that stopping drinking has no effect on a person's weight.

15. The ages of the 21 drivers involved in auto accidents on U.S. Highway #10 during March in the order in which they occurred are as follows:

19, 58, 42, 63, 24, 31, 39, 22, 67, 28, 37, 53, 32, 29, 18,
22, 34, 46, 18, 19, 21

Test for randomness.

Chapter Test

1. A panel of movie critics was asked to indicate its preferences for ten movies. The following rankings were obtained.

Movie	Ranking by Male Reviewers	Ranking by Female Reviewers
1	7	7
2	2	2
3	9	3
4	8	10
5	1	4
6	3	6
7	4	9
8	6	1
9	10	5
10	5	8

Using a 5% level of significance, test the null hypothesis that there is no correlation between the rankings of the male and female reviewers.

2. The sex of the first 27 people arriving to buy a ticket in the state's $41 million dollar drawing is as follows: (M = male and F = female)

F, M, M, F, F, F, M, M, F, M, F, F, F, M, M, M,
F, F, F, F, M, M, F, F, F, F, F

Test for randomness.

3. Twenty students prepared for the Graduate Record Examination (GRE) administered by the Educational Testing Service. Each received special training by one of three different programs. The scores obtained by these students is as follows.

Candidate Received Special Training Using		
Method 1	*Method 2*	*Method 3*
580	500	610
480	450	640
520	580	520
640	380	580
485	490	480
630	550	510

Using the Kruskal-Wallis Test, test the null hypothesis that there is no significant difference in the scores obtained by students using any of these three programs. (Use a 5% level of significance.)

4. A comparison shopper obtained the following prices for the same brand and model number computer printer at 12 computer supplies stores in New York City and 13 computer supplies stores in Los Angeles.

Prices in New York City (in dollars)	Prices in Los Angeles (in dollars)
468	488
459	501
444	479
451	484
479	499
498	469
482	476
470	484
493	487
488	494
438	498
456	446
	488

Using a 5% level of significance, test the null hypothesis that the average price for the computer printer is not significantly different in both cities.

5. Many patients often feel drowsy and/or dizzy when taking certain medications. This often affects their ability to complete tasks requiring mental alertness, such as driving. The following data are available for ten people.

		Number of Tasks Completed	
		Before Taking Medication	After Taking Medication
Subject	A	20	15
	B	17	15
	C	27	27
	D	13	16
	E	14	12
	F	8	6
	G	12	14
	H	11	9
	I	17	13
	J	15	11

Using a 5% level of significance, test the null hypothesis that the medication has no effect on the number of tasks completed. (*Hint:* Use the sign test)

6. Twenty-nine people are lined up at the motor vehicle office waiting to have their picture taken for their new photo driver's license. The ages of these people are as follows:

24, 38, 71, 62, 49, 19, 17, 22, 39, 31, 18, 19, 27, 29, 32, 37, 48, 61, 57, 41, 58, 32, 40, 21, 19, 25, 26, 29, 38

Test for randomness.

7. On one popular television show contestants are asked to guess the manufacturer's suggested selling price or the value of various items. In one particular show ten contestants were asked to estimate the value of two different vacations. The contestants provided the following estimates.

		Value of Vacation 1 (in dollars)	Value of Vacation 2 (in dollars)
Contestant	A	875	912
	B	795	725
	C	999	845
	D	820	820
	E	795	880
	F	845	1099
	G	920	750
	H	800	900
	I	700	700
	J	825	865

Using a 5% level of significance, test the null hypothesis that the contestants estimate the value of both vacations to be the same.

8. 18 randomly selected people were asked to complete a difficult and lengthy questionnaire. Six of the people were given instructions by one pollster, six by a second pollster, and six by a third pollster. Each pollster used his or her own instructions. The times (in minutes) required by each of the people to complete the questionnaire are as follows.

Instructions Given by

Pollster 1	Pollster 2	Pollster 3
28	22	18
34	33	27
25	35	29
36	28	17
37	32	19
32	30	22

Using a 5% level of significance, test the null hypothesis that the average time needed to complete the questionnaire is the same regardless of which pollster gives the instructions.

9. Two professional tasters were asked to rank six different brands of beer in terms of taste appeal and ability to retain its head. The following are their ratings.

Ranking by

Brand of Beer	Taster 1	Taster 2
A	2	4
B	6	3
C	3	6
D	5	2
E	1	1
F	4	5

Using a 5% level of significance, test the null hypothesis that there is no correlation between the rankings. (*Hint:* Use the Spearman-rank correlation test.)

10. Customers of a large bank were asked to rate the services provided by the bank. The following performance ratings (on a certain scale) were obtained by a panel of 21 customers.

Branch Location	Performance Rating						
Downtown	46	53	39	48	50	62	60
Midtown	68	62	47	40	53	39	42
Uptown	39	51	47	59	61	53	78

Using a 5% level of significance, test the null hypothesis that there is a significant difference in the performance evaluations. (*Hint:* Use the Kruskal-Wallis-H-test.)

11. The chairperson of the physical education department at Shoerville College is interested in knowing whether the average number of push-ups performed by students in Professor Brier's class is the same as in Professor Silvernail's class. The following statistics are available on the number of push-ups performed by students in both classes.

Prof. Brier's class	23	46	35	52	58	47	49	39	53	
Prof. Silvernail's class	38	40	43	59	53	49	51	54	21	26	45

Using a 5% level of significance, test the null hypothesis that the average number of push-ups performed by students in both classes is the same.

12. The ages of the 19 women filing for divorce on January 12, 1988 in a particular city (in the order in which they were filed) are as follows:

19, 24, 58, 42, 32, 31, 28, 22, 67, 44, 37, 53, 32, 19, 18, 22, 34, 26 and 23.

Test for randomness.

a. Random **b.** Not random **c.** Not enough information given **d.** None of these

13. There are 25 people waiting in line to cash their payroll check. Their order is as follows (where F = female and M = male):

F, M, F, F, M, F, F, M, M, M, F, F, M, M, M, F, F, M, M, F, F, F, F, M, M

Test for randomness.

a. Random **b.** Not random **c.** Not enough information given **d.** None of these

14. Ten male musicians and 13 female musicians were tested on their ability to learn a particularly difficult musical selection. The number of practice minutes needed by each musician to learn it is as follows:

Female	52	57	61	48	56	50	44	63	67	55	53	59	60
Male	43	54	49	71	62	58	64	69	70	72			

Using the rank-sum test, test the null hypothesis that the mean number of minutes needed by both groups is not significantly different. (Use a 5% level of significance).

a. Reject null hypothesis **b.** Do not reject null hypothesis **c.** Not enough information given **d.** None of these

15. Ten men were given a special experimental blood pressure pill. The following blood pressure readings were obtained:

Reading before pill	116	121	113	145	163	172	162	182	192	111
Reading after pill	114	118	110	140	157	177	162	175	173	109

Using the sign test, test the null hypothesis at the 5% level of significance that the blood pressure pill does not significantly reduce the blood pressure reading.

a. Reject null hypothesis b. Do not reject null hypothesis c. Not enough information given d. None of these

16. Three judges were asked to rate six different newspapers for their overall accuracy. Their ratings are as follows:

	Judge 1	Judge 2	Judge 3
Newspaper A	1	6	4
Newspaper B	5	2	3
Newspaper C	2	3	2
Newspaper D	3	5	1
Newspaper E	4	4	6
Newspaper F	6	1	5

Using a 5% level of significance, test the null hypothesis that there is no significant correlation between the rankings of Judge 1 and Judge 2.

a. No significant correlation b. Significant correlation c. Not enough information given d. None of these

Suggested Further Reading

1. Albright, S. Christian. *Statistics for Business and Economics*. New York: Macmillan Publishing Company, 1987.
2. Conover, W. J. *Practical Nonparametric Statistics*, 2nd ed. New York: Wiley, 1980.
3. Daniel, Wayne. *Applied Non-parametric Statistics*. Boston: Houghton Mifflin, 1978.
4. Gibbons, J. D. *Non-parametric Statistical Inference*. New York: Marcel Dekker, Inc., 1985.
5. Gibbons, J. D., I. Olkin, and M. Sobel. *Selecting and Ordering Populations: A New Statistical Methodology*. New York: Wiley, 1977.
6. Lehmann, E. L. *Non-parametrics: Statistical Methods Based on Ranks*. San Francisco: Holden-Day, 1975.
7. Meek, Gary E., Howard L. Taylor, Kenneth A. Dunning, and Keith A. Klajehn. *Business Statistics*. Boston: Allyn and Bacon, Inc., 1987.
8. Noether, G. E. *Introduction to Statistics: A Nonparametric Approach*. Boston: Houghton Mifflin, 1976.
9. Pfaffenberger, Roger C. and James H. Paterson. *Statistical Methods for Business and Economics*, 3rd ed. Homewood, IL: Irwin, 1987.

Statistical Tables

Table I Factorials

n	n!
0	1
1	1
2	2
3	6
4	24
5	120
6	720
7	5,040
8	40,320
9	362,880
10	3,628,800
11	39,916,800
12	479,001,600
13	6,227,020,800
14	87,178,291,200
15	1,307,674,368,000
16	20,922,789,888,000
17	355,687,428,096,000
18	6,402,373,705,728,000
19	121,645,100,408,832,000
20	2,432,902,008,176,640,000

TABLE II BINOMIAL COEFFICIENTS **A.3**

Table II Binomial Coefficients $\dfrac{n!}{x!(n-x!)}$

n \ x	2	3	4	5	6	7	8	9	10
2	1								
3	3	1							
4	6	4	1						
5	10	10	5	1					
6	15	20	15	6	1				
7	21	35	35	21	7	1			
8	28	56	70	56	28	8	1		
9	36	84	126	126	84	36	9	1	
10	45	120	210	252	210	120	45	10	1
11	55	165	330	462	462	330	165	55	11
12	66	220	495	792	924	792	495	220	66
13	78	286	715	1,287	1,716	1,716	1,287	715	286
14	91	364	1,001	2,002	3,003	3,432	3,003	2,002	1,001
15	105	455	1,365	3,003	5,005	6,435	6,435	5,005	3,003
16	120	560	1,820	4,368	8,008	11,440	12,870	11,440	8,008
17	136	680	2,380	6,188	12,376	19,448	24,310	24,310	19,448
18	153	816	3,060	8,568	18,564	31,824	43,758	48,620	43,758
19	171	969	3,876	11,628	27,132	50,388	75,582	92,378	92,378
20	190	1,140	4,845	15,504	38,760	77,520	125,970	167,960	184,756

Table III   Binomial Probabilities

							p					
n	x	0.05	0.1	0.2	0.3	0.4	0.5	0.6	0.7	0.8	0.9	0.95
2	0	0.902	0.810	0.640	0.490	0.360	0.250	0.160	0.090	0.040	0.010	0.002
	1	0.095	0.180	0.320	0.420	0.480	0.500	0.480	0.420	0.320	0.180	0.095
	2	0.002	0.010	0.040	0.090	0.160	0.250	0.360	0.490	0.640	0.810	0.902
3	0	0.857	0.729	0.512	0.343	0.216	0.125	0.064	0.027	0.008	0.001	
	1	0.135	0.243	0.384	0.441	0.432	0.375	0.288	0.189	0.096	0.027	0.007
	2	0.007	0.027	0.096	0.189	0.288	0.375	0.432	0.441	0.384	0.243	0.135
	3		0.001	0.008	0.027	0.064	0.125	0.216	0.343	0.512	0.729	0.857
4	0	0.815	0.656	0.410	0.240	0.130	0.062	0.026	0.008	0.002		
	1	0.171	0.292	0.410	0.412	0.346	0.250	0.154	0.076	0.026	0.004	
	2	0.014	0.049	0.154	0.265	0.346	0.375	0.346	0.265	0.154	0.049	0.014
	3		0.004	0.026	0.076	0.154	0.250	0.346	0.412	0.410	0.292	0.171
	4			0.002	0.008	0.026	0.062	0.130	0.240	0.410	0.656	0.815
5	0	0.774	0.590	0.328	0.168	0.078	0.031	0.010	0.002			
	1	0.204	0.328	0.410	0.360	0.259	0.156	0.077	0.028	0.006		
	2	0.021	0.073	0.205	0.309	0.346	0.312	0.230	0.132	0.051	0.008	0.001
	3	0.001	0.008	0.051	0.132	0.230	0.312	0.346	0.309	0.205	0.073	0.021
	4			0.006	0.028	0.077	0.156	0.259	0.360	0.410	0.328	0.204
	5				0.002	0.010	0.031	0.078	0.168	0.328	0.590	0.774
6	0	0.735	0.531	0.262	0.118	0.047	0.016	0.004	0.001			
	1	0.232	0.354	0.393	0.303	0.187	0.094	0.037	0.010	0.002		
	2	0.031	0.098	0.246	0.324	0.311	0.234	0.138	0.060	0.015	0.001	
	3	0.002	0.015	0.082	0.185	0.276	0.312	0.276	0.185	0.082	0.015	0.002
	4		0.001	0.015	0.060	0.138	0.234	0.311	0.324	0.246	0.098	0.031
	5			0.002	0.010	0.037	0.094	0.187	0.303	0.393	0.354	0.232
	6				0.001	0.004	0.016	0.047	0.118	0.262	0.531	0.735
7	0	0.698	0.478	0.210	0.082	0.028	0.008	0.002				
	1	0.257	0.372	0.367	0.247	0.131	0.055	0.017	0.004			
	2	0.041	0.124	0.275	0.318	0.261	0.164	0.077	0.025	0.004		
	3	0.004	0.023	0.115	0.227	0.290	0.273	0.194	0.097	0.029	0.003	
	4		0.003	0.029	0.097	0.194	0.273	0.290	0.227	0.115	0.023	0.004
	5			0.004	0.025	0.077	0.164	0.261	0.318	0.275	0.124	0.041
	6				0.004	0.017	0.055	0.131	0.247	0.367	0.372	0.257
	7					0.002	0.008	0.028	0.082	0.210	0.478	0.698

TABLE III BINOMIAL PROBABILITIES **A.5**

Table III Binomial Probabilities (*continued*)

n	x	0.05	0.1	0.2	0.3	0.4	0.5	0.6	0.7	0.8	0.9	0.95
						p						
8	0	0.663	0.430	0.168	0.058	0.017	0.004	0.001				
	1	0.279	0.383	0.336	0.198	0.090	0.031	0.008	0.001			
	2	0.051	0.149	0.294	0.296	0.209	0.109	0.041	0.010	0.001		
	3	0.005	0.033	0.147	0.254	0.279	0.219	0.124	0.047	0.009		
	4		0.005	0.046	0.136	0.232	0.273	0.232	0.136	0.046	0.005	
	5			0.009	0.047	0.124	0.219	0.279	0.254	0.147	0.033	0.005
	6			0.001	0.010	0.041	0.109	0.209	0.296	0.294	0.149	0.051
	7				0.001	0.008	0.031	0.090	0.198	0.336	0.383	0.279
	8					0.001	0.004	0.017	0.058	0.168	0.430	0.663
9	0	0.630	0.387	0.134	0.040	0.010	0.002					
	1	0.299	0.387	0.302	0.156	0.060	0.018	0.004				
	2	0.063	0.172	0.302	0.267	0.161	0.070	0.021	0.004			
	3	0.008	0.045	0.176	0.267	0.251	0.164	0.074	0.021	0.003		
	4	0.001	0.007	0.066	0.172	0.251	0.246	0.167	0.074	0.017	0.001	
	5		0.001	0.017	0.074	0.167	0.246	0.251	0.172	0.066	0.007	0.001
	6			0.003	0.021	0.074	0.164	0.251	0.267	0.176	0.045	0.008
	7				0.004	0.021	0.070	0.161	0.267	0.302	0.172	0.063
	8					0.004	0.018	0.060	0.156	0.302	0.387	0.299
	9						0.002	0.010	0.040	0.134	0.387	0.630
10	0	0.599	0.349	0.107	0.028	0.006	0.001					
	1	0.315	0.387	0.268	0.121	0.040	0.010	0.002				
	2	0.075	0.194	0.302	0.233	0.121	0.044	0.011	0.001			
	3	0.010	0.057	0.201	0.267	0.215	0.117	0.042	0.009	0.001		
	4	0.001	0.011	0.088	0.200	0.251	0.205	0.111	0.037	0.006		
	5		0.001	0.026	0.103	0.201	0.246	0.201	0.103	0.026	0.001	
	6			0.006	0.037	0.111	0.205	0.251	0.200	0.088	0.011	0.001
	7			0.001	0.009	0.042	0.117	0.215	0.267	0.201	0.057	0.010
	8				0.001	0.011	0.044	0.121	0.233	0.302	0.194	0.075
	9					0.002	0.010	0.040	0.121	0.268	0.387	0.315
	10						0.001	0.006	0.028	0.107	0.349	0.599

Table III   Binomial Probabilities (*continued*)

n	x	0.05	0.1	0.2	0.3	0.4	0.5	0.6	0.7	0.8	0.9	0.95
11	0	0.569	0.314	0.086	0.020	0.004						
	1	0.329	0.384	0.236	0.093	0.027	0.005	0.001				
	2	0.087	0.213	0.295	0.200	0.089	0.027	0.005	0.001			
	3	0.014	0.071	0.221	0.257	0.177	0.081	0.023	0.004			
	4	0.001	0.016	0.111	0.220	0.236	0.161	0.070	0.017	0.002		
	5		0.002	0.039	0.132	0.221	0.226	0.147	0.057	0.010		
	6			0.010	0.057	0.147	0.226	0.221	0.132	0.039	0.002	
	7			0.002	0.017	0.070	0.161	0.236	0.220	0.111	0.016	0.001
	8				0.004	0.023	0.081	0.177	0.257	0.221	0.071	0.014
	9				0.001	0.005	0.027	0.089	0.200	0.295	0.213	0.087
	10					0.001	0.005	0.027	0.093	0.236	0.384	0.329
	11							0.004	0.020	0.086	0.314	0.569
12	0	0.540	0.282	0.069	0.014	0.002						
	1	0.341	0.377	0.206	0.071	0.017	0.003					
	2	0.099	0.230	0.283	0.168	0.064	0.016	0.002				
	3	0.017	0.085	0.236	0.240	0.142	0.054	0.012	0.001			
	4	0.002	0.021	0.133	0.231	0.213	0.121	0.042	0.008	0.001		
	5		0.004	0.053	0.158	0.227	0.193	0.101	0.029	0.003		
	6			0.016	0.079	0.177	0.226	0.177	0.079	0.016		
	7			0.003	0.029	0.101	0.193	0.227	0.158	0.053	0.004	
	8			0.001	0.008	0.042	0.121	0.213	0.231	0.133	0.021	0.002
	9				0.001	0.012	0.054	0.142	0.240	0.236	0.085	0.017
	10					0.002	0.016	0.064	0.168	0.283	0.230	0.099
	11						0.003	0.017	0.071	0.206	0.377	0.341
	12							0.002	0.014	0.069	0.282	0.540
13	0	0.513	0.254	0.055	0.010	0.001						
	1	0.351	0.367	0.179	0.054	0.011	0.002					
	2	0.111	0.245	0.268	0.139	0.045	0.010	0.001				
	3	0.021	0.100	0.246	0.218	0.111	0.035	0.006	0.001			
	4	0.003	0.028	0.154	0.234	0.184	0.087	0.024	0.003			
	5		0.006	0.069	0.180	0.221	0.157	0.066	0.014	0.001		
	6		0.001	0.023	0.103	0.197	0.209	0.131	0.044	0.006		
	7			0.006	0.044	0.131	0.209	0.197	0.103	0.023	0.001	
	8			0.001	0.014	0.066	0.157	0.221	0.180	0.069	0.006	
	9				0.003	0.024	0.087	0.184	0.234	0.154	0.028	0.003
	10				0.001	0.006	0.035	0.111	0.218	0.246	0.100	0.021
	11					0.001	0.010	0.045	0.139	0.268	0.245	0.111
	12						0.002	0.011	0.054	0.179	0.367	0.351
	13							0.001	0.010	0.055	0.254	0.513

TABLE III BINOMIAL PROBABILITIES   **A.7**

Table III Binomial Probabilities (*continued*)

n	x						p					
		0.05	0.1	0.2	0.3	0.4	0.5	0.6	0.7	0.8	0.9	0.95
14	0	0.488	0.229	0.044	0.007	0.001						
	1	0.359	0.356	0.154	0.041	0.007	0.001					
	2	0.123	0.257	0.250	0.113	0.032	0.006	0.001				
	3	0.026	0.114	0.250	0.194	0.085	0.022	0.003				
	4	0.004	0.035	0.172	0.229	0.155	0.061	0.014	0.001			
	5		0.008	0.086	0.196	0.207	0.122	0.041	0.007			
	6		0.001	0.032	0.126	0.207	0.183	0.092	0.023	0.002		
	7			0.009	0.062	0.157	0.209	0.157	0.062	0.009		
	8			0.002	0.023	0.092	0.183	0.207	0.126	0.032	0.001	
	9				0.007	0.041	0.122	0.207	0.196	0.086	0.008	
	10				0.001	0.014	0.061	0.155	0.229	0.172	0.035	0.004
	11					0.003	0.022	0.085	0.194	0.250	0.114	0.026
	12					0.001	0.006	0.032	0.113	0.250	0.257	0.123
	13						0.001	0.007	0.041	0.154	0.356	0.359
	14							0.001	0.007	0.044	0.229	0.488
15	0	0.463	0.206	0.035	0.005							
	1	0.366	0.343	0.132	0.031	0.005						
	2	0.135	0.267	0.231	0.092	0.022	0.003					
	3	0.031	0.129	0.250	0.170	0.063	0.014	0.002				
	4	0.005	0.043	0.188	0.219	0.127	0.042	0.007	0.001			
	5	0.001	0.010	0.103	0.206	0.186	0.092	0.024	0.003			
	6		0.002	0.043	0.147	0.207	0.153	0.061	0.012	0.001		
	7			0.014	0.081	0.177	0.196	0.118	0.035	0.003		
	8			0.003	0.035	0.118	0.196	0.177	0.081	0.014		
	9			0.001	0.012	0.061	0.153	0.207	0.147	0.043	0.002	
	10				0.003	0.024	0.092	0.186	0.206	0.103	0.010	0.001
	11				0.001	0.007	0.042	0.127	0.219	0.188	0.043	0.005
	12					0.002	0.014	0.063	0.170	0.250	0.129	0.031
	13						0.003	0.022	0.092	0.231	0.267	0.135
	14							0.005	0.031	0.132	0.343	0.366
	15								0.005	0.035	0.206	0.463

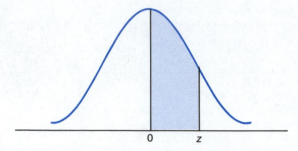

The entries in Table IV are the probabilities that a random variable having the standard normal distribution takes on a value between 0 and z; they are given by the area under the curve shaded in the diagram.

Table IV   Areas Under the Standard Normal Curve

z					Second decimal place in z					
	0.00	0.01	0.02	0.03	0.04	0.05	0.06	0.07	0.08	0.09
0.0	0.0000	0.0040	0.0080	0.0120	0.0160	0.0199	0.0239	0.0279	0.0319	0.0359
0.1	0.0398	0.0438	0.0478	0.0517	0.0557	0.0596	0.0636	0.0675	0.0714	0.0753
0.2	0.0793	0.0832	0.0871	0.0910	0.0948	0.0987	0.1026	0.1064	0.1103	0.1141
0.3	0.1179	0.1217	0.1255	0.1293	0.1331	0.1368	0.1406	0.1443	0.1480	0.1517
0.4	0.1554	0.1591	0.1628	0.1664	0.1700	0.1736	0.1772	0.1808	0.1844	0.1879
0.5	0.1915	0.1950	0.1985	0.2019	0.2054	0.2088	0.2123	0.2157	0.2190	0.2224
0.6	0.2257	0.2291	0.2324	0.2357	0.2389	0.2422	0.2454	0.2486	0.2517	0.2549
0.7	0.2580	0.2611	0.2642	0.2673	0.2704	0.2734	0.2764	0.2794	0.2823	0.2852
0.8	0.2881	0.2910	0.2939	0.2967	0.2995	0.3023	0.3051	0.3078	0.3106	0.3133
0.9	0.3159	0.3186	0.3212	0.3238	0.3264	0.3289	0.3315	0.3340	0.3365	0.3389
1.0	0.3413	0.3438	0.3461	0.3485	0.3508	0.3531	0.3554	0.3577	0.3599	0.3621
1.1	0.3643	0.3665	0.3686	0.3708	0.3729	0.3749	0.3770	0.3790	0.3810	0.3830
1.2	0.3849	0.3869	0.3888	0.3907	0.3925	0.3944	0.3962	0.3980	0.3997	0.4015
1.3	0.4032	0.4049	0.4066	0.4082	0.4099	0.4115	0.4131	0.4147	0.4162	0.4177
1.4	0.4192	0.4207	0.4222	0.4236	0.4251	0.4265	0.4279	0.4292	0.4306	0.4319
1.5	0.4332	0.4345	0.4357	0.4370	0.4382	0.4394	0.4406	0.4418	0.4429	0.4441
1.6	0.4452	0.4463	0.4474	0.4484	0.4495	0.4505	0.4515	0.4525	0.4535	0.4545
1.7	0.4554	0.4564	0.4573	0.4582	0.4591	0.4599	0.4608	0.4616	0.4625	0.4633
1.8	0.4641	0.4649	0.4656	0.4664	0.4671	0.4678	0.4686	0.4693	0.4699	0.4706
1.9	0.4713	0.4719	0.4726	0.4732	0.4738	0.4744	0.4750	0.4756	0.4761	0.4767

0.4732

TABLE IV AREAS UNDER THE STANDARD NORMAL CURVE **A.9**

Table IV Areas Under the Standard Normal Curve (*continued*)

z	Second decimal place in z									
	0.00	0.01	0.02	0.03	0.04	0.05	0.06	0.07	0.08	0.09
2.0	0.4772	0.4778	0.4783	0.4788	0.4793	0.4798	0.4803	0.4808	0.4812	0.4817
2.1	0.4821	0.4826	0.4830	0.4834	0.4838	0.4842	0.4846	0.4850	0.4854	0.4857
2.2	0.4861	0.4864	0.4868	0.4871	0.4875	0.4878	0.4881	0.4884	0.4887	0.4890
2.3	0.4893	0.4896	0.4898	0.4901	0.4904	0.4906	0.4909	0.4911	0.4913	0.4916
2.4	0.4918	0.4920	0.4922	0.4925	0.4927	0.4929	0.4931	0.4932	0.4934	0.4936
2.5	0.4938	0.4940	0.4941	0.4943	0.4945	0.4946	0.4948	0.4949	0.4951	0.4952
2.6	0.4953	0.4955	0.4956	0.4957	0.4959	0.4960	0.4961	0.4962	0.4963	0.4964
2.7	0.4965	0.4966	0.4967	0.4968	0.4969	0.4970	0.4971	0.4972	0.4973	0.4974
2.8	0.4974	0.4975	0.4976	0.4977	0.4977	0.4978	0.4979	0.4979	0.4980	0.4981
2.9	0.4981	0.4982	0.4982	0.4983	0.4984	0.4984	0.4985	0.4985	0.4986	0.4986
3.0	0.4987	0.4987	0.4987	0.4988	0.4988	0.4989	0.4989	0.4989	0.4990	0.4990
3.1	0.4990	0.4991	0.4991	0.4991	0.4992	0.4992	0.4992	0.4992	0.4993	0.4993
3.2	0.4993	0.4993	0.4994	0.4994	0.4994	0.4994	0.4994	0.4995	0.4995	0.4995
3.3	0.4995	0.4995	0.4995	0.4996	0.4996	0.4996	0.4996	0.4996	0.4996	0.4997
3.4	0.4997	0.4997	0.4997	0.4997	0.4997	0.4997	0.4997	0.4997	0.4997	0.4998
3.5	0.4998	0.4998	0.4998	0.4998	0.4998	0.4998	0.4998	0.4998	0.4998	0.4998
3.6	0.4998	0.4998	0.4999	0.4999	0.4999	0.4999	0.4999	0.4999	0.4999	0.4999
3.7	0.4999	0.4999	0.4999	0.4999	0.4999	0.4999	0.4999	0.4999	0.4999	0.4999
3.8	0.4999	0.4999	0.4999	0.4999	0.4999	0.4999	0.4999	0.4999	0.4999	0.4999
3.9	0.5000†									

† For $z \geq 3.90$, the areas are 0.5000 to four decimal places.

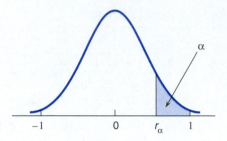

Table V Critical Values of *r*

n	$r_{0.025}$	$r_{0.005}$	n	$r_{0.025}$	$r_{0.005}$
3	0.997		18	0.468	0.590
4	0.950	0.999	19	0.456	0.575
5	0.878	0.959	20	0.444	0.561
6	0.811	0.917	21	0.433	0.549
7	0.754	0.875	22	0.423	0.537
8	0.707	0.834	27	0.381	0.487
9	0.666	0.798	32	0.349	0.449
10	0.632	0.765	37	0.325	0.418
11	0.602	0.735	42	0.304	0.393
12	0.576	0.708	47	0.288	0.372
13	0.553	0.684	52	0.273	0.354
14	0.532	0.661	62	0.250	0.325
15	0.514	0.641	72	0.232	0.302
16	0.497	0.623	82	0.217	0.283
17	0.482	0.606	92	0.205	0.267

This table is abridged from Table VI of R. A. Fisher and F. Yates: *Statistical Tables for Biological, Agricultural, and Medical Research*, published by Longman Group, Ltd., London (previously published by Oliver & Boyd, Edinburgh), by permission of the authors and publishers.

Table VI Table of Random Digits

Line	Col (1)	(2)	(3)	(4)	(5)	(6)	(7)	(8)	(9)	(10)	(11)	(12)	(13)	(14)
1	10480	15011	01536	02011	81647	91646	69179	14194	62590	36207	20969	99570	91291	90700
2	22368	46573	25595	85393	30995	89198	27982	53402	93965	34095	52666	19174	39615	99505
3	24130	48360	22527	97265	76393	64809	15179	24830	49340	32081	30680	19655	63348	58629
4	42167	93093	06243	61680	07856	16376	39440	53537	71341	57004	00849	74917	97758	16379
5	37570	39975	81837	16656	06121	91782	60468	81305	49684	60672	14110	06927	01263	54613
6	77921	06907	11008	42751	27756	53498	18602	70659	90655	15053	21916	81825	44394	42880
7	99562	72905	56420	69994	98872	31016	71194	18738	44013	48840	63213	21069	10634	12952
8	96301	91977	05463	07972	18876	20922	94595	56869	69014	60045	18425	84903	42508	32307
9	89579	14342	63661	10281	17453	18103	57740	84378	25331	12566	58678	44947	05585	56941
10	85475	36857	53342	53988	53060	59533	38867	62300	08158	17983	16439	11458	18593	64952
11	28918	69578	88231	33276	70997	79936	56865	05859	90106	31595	01547	85590	91610	78188
12	63553	40961	48235	03427	49626	69445	18663	72695	52180	20847	12234	90511	33703	90322
13	09429	93969	52636	92737	88974	33488	36320	17617	30015	08272	84115	27156	30613	74952
14	10365	61129	87529	85689	48237	52267	67689	93394	01511	26358	85104	20285	29975	89868
15	07119	97336	71048	08178	77233	13916	47564	81056	97735	85977	29372	74461	28551	90707
16	51085	12765	51821	51259	77452	16308	60756	92144	49442	53900	70960	63990	75601	40719
17	02368	21382	52404	60268	89368	19885	55322	44819	01188	65255	64835	44919	05944	55157
18	01011	54092	33362	94904	31273	04146	18594	29852	71585	85030	51132	01915	92747	64951
19	52162	53916	46369	58586	23216	14513	83149	98736	23495	64350	94738	17752	35156	35749
20	07056	97628	33787	09998	42698	06691	76988	13602	51851	46104	88916	19509	25625	58104
21	48663	91245	85828	14346	09172	30168	90229	04734	59193	22178	30421	61666	99904	32812
22	54164	58492	22421	74103	47070	25306	76468	26384	58151	06646	21524	15227	96909	44592
23	32639	32363	05597	24200	13363	38005	94342	28728	35806	06912	17012	64161	18296	22851
24	29334	27001	87637	87308	58731	00256	45834	15398	46557	41135	10367	07684	36188	18510
25	02488	33062	28834	07351	19731	92420	60952	61280	50001	67658	32586	86679	50720	94953
26	81525	72295	04839	96423	24878	82651	66566	14778	76797	14780	13300	87074	79666	95725
27	29676	20591	68086	26432	46901	20849	89768	81536	86645	12659	92259	57102	80428	25280
28	00742	57392	39064	66432	84673	40027	32832	61362	98947	96067	64760	64584	96096	98253
29	05366	04213	25669	26422	44407	44048	37937	63904	45766	66134	75470	66520	34693	90449
30	91921	26418	64117	94305	26766	25940	39972	22209	71500	64568	91402	42416	07844	69618
31	00582	04711	87917	77341	42206	35126	74087	99547	81817	42607	43808	76655	62028	76630
32	00725	69884	62797	56170	86324	88072	76222	36086	84637	93161	76038	65855	77919	88006
33	69011	65795	95876	55293	18988	27354	26575	08625	40801	59920	29841	80150	12777	48501
34	25976	57948	29888	88604	67917	48708	18912	82271	65424	69774	33611	54262	85963	03547
35	09763	83473	73577	12908	30883	18317	28290	35797	05998	41688	34952	37888	38917	88050
36	91567	42595	27958	30134	04024	86385	29880	99730	55536	84855	29080	09250	79656	73211
37	17955	56349	90999	49127	20044	59931	06115	20542	18059	02008	73708	83517	36103	42791
38	46503	18584	18845	49618	02304	51038	20655	58727	28168	15475	56942	53389	20562	87338
39	92157	89634	94824	78171	84610	82834	09922	25417	44137	48413	25555	21246	35509	20468
40	14577	62765	35605	81263	39667	47358	56873	56307	61607	49518	89656	20103	77490	18062
41	98427	07523	33362	64270	01638	92477	66969	98420	04880	45585	46565	04102	46880	45709
42	34914	63976	88720	82765	34476	17032	87589	40836	32427	70002	70663	88863	77775	69348
43	70060	28277	39475	46473	23219	53416	94970	25832	69975	94884	19661	72828	00102	66794
44	53976	54914	06990	67245	68350	82948	11398	42878	80287	88267	47363	46634	06541	97809
45	76072	29515	40980	07391	58745	25774	22987	80059	39911	96189	41151	14222	60697	59583
46	90725	52210	83974	29992	65831	38857	50490	83765	55657	14361	31720	57375	56228	41546
47	64364	67412	33339	31926	14883	24413	59744	92351	97473	89286	35931	04110	23726	51900
48	08962	00358	31662	25388	61642	34072	81249	35648	56891	69352	48373	45578	78547	81788
49	95012	68379	93526	70765	10592	04542	76463	54328	02349	17247	28865	14777	62730	92277
50	15664	10493	20492	38391	91132	21999	59516	81652	27195	48223	46751	22923	32261	85653

Page 1 of *Table of 105,000 Random Decimal Digits*, Statement No. 4914, May 1949, File No. 261-A-1, Interstate Commerce Commission, Washington, D.C.

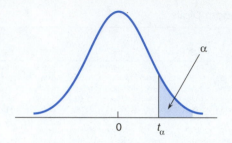

Table VII The t-distribution

df	$t_{0.050}$	$t_{0.025}$	$t_{0.010}$	$t_{0.005}$	df
1	6.314	12.706	31.821	63.657	1
2	2.920	4.303	6.965	9.925	2
3	2.353	3.182	4.541	5.841	3
4	2.132	2.776	3.747	4.604	4
5	2.015	2.571	3.365	4.032	5
6	1.943	2.447	3.143	3.707	6
7	1.895	2.365	2.998	3.499	7
8	1.860	2.306	2.896	3.355	8
9	1.833	2.262	2.821	3.250	9
10	1.812	2.228	2.764	3.169	10
11	1.796	2.201	2.718	3.106	11
12	1.782	2.179	2.681	3.055	12
13	1.771	2.160	2.650	3.012	13
14	1.761	2.145	2.624	2.977	14
15	1.753	2.131	2.602	2.947	15
16	1.746	2.120	2.583	2.921	16
17	1.740	2.110	2.567	2.898	17
18	1.734	2.101	2.552	2.878	18
19	1.729	2.093	2.539	2.861	19
20	1.725	2.086	2.528	2.845	20
21	1.721	2.080	2.518	2.831	21
22	1.717	2.074	2.508	2.819	22
23	1.714	2.069	2.500	2.807	23
24	1.711	2.064	2.492	2.797	24
25	1.708	2.060	2.485	2.787	25
26	1.706	2.056	2.479	2.779	26
27	1.703	2.052	2.473	2.771	27
28	1.701	2.048	2.467	2.763	28
29	1.699	2.045	2.462	2.756	29
inf.	1.645	1.960	2.326	2.576	inf.

This table is abridged from Table IV of R. A. Fisher and F. Yates: *Statistical Tables for Biological, Agricultural, and Medical Research*, published by Longman Group, Ltd., London (previously published by Oliver & Boyd, Edinburgh), by permission of the authors and publishers.

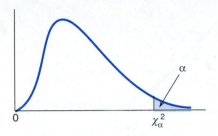

TABLE VIII   The χ^2 distribution

df	$\chi^2_{0.05}$	$\chi^2_{0.01}$	df
1	3.841	6.635	1
2	5.991	9.210	2
3	7.815	11.345	3
4	9.488	13.277	4
5	11.070	15.086	5
6	12.592	16.812	6
7	14.067	18.475	7
8	15.507	20.090	8
9	16.919	21.666	9
10	18.307	23.209	10
11	19.675	24.725	11
12	21.026	26.217	12
13	22.362	27.688	13
14	23.685	29.141	14
15	24.996	30.578	15
16	26.296	32.000	16
17	27.587	33.409	17
18	28.869	34.805	18
19	30.144	36.191	19
20	31.410	37.566	20
21	32.671	38.932	21
22	33.924	40.289	22
23	35.172	41.638	23
24	36.415	42.980	24
25	37.652	44.314	25
26	38.885	45.642	26
27	40.113	46.963	27
28	41.337	48.278	28
29	42.557	49.588	29
30	43.773	50.892	30

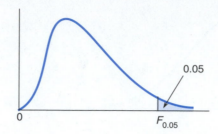

Table IX Critical Values of the *F*-distribution ($\alpha = 0.05$)

		Degrees of Freedom for Numerator									
		1	2	3	4	5	6	7	8	9	10
	1	161	200	216	225	230	234	237	239	241	242
	2	18.5	19.0	19.2	19.2	19.3	19.3	19.4	19.4	19.4	19.4
	3	10.1	9.55	9.28	9.12	9.01	8.94	8.89	8.85	8.81	8.79
	4	7.71	6.94	6.59	6.39	6.26	6.16	6.09	6.04	6.00	5.96
	5	6.61	5.79	5.41	5.19	5.05	4.95	4.88	4.82	4.77	4.74
	6	5.99	5.14	4.76	4.53	4.39	4.28	4.21	4.15	4.10	4.06
	7	5.59	4.74	4.35	4.12	3.97	3.87	3.79	3.73	3.68	3.64
	8	5.32	4.46	4.07	3.84	3.69	3.58	3.50	3.44	3.39	3.35
	9	5.12	4.26	3.86	3.63	3.48	3.37	3.29	3.23	3.18	3.14
	10	4.96	4.10	3.71	3.48	3.33	3.22	3.14	3.07	3.02	2.98
Degrees	11	4.84	3.98	3.59	3.36	3.20	3.09	3.01	2.95	2.90	2.85
of	12	4.75	3.89	3.49	3.26	3.11	3.00	2.91	2.85	2.80	2.75
Freedom	13	4.67	3.81	3.41	3.18	3.03	2.92	2.83	2.77	2.71	2.67
for	14	4.60	3.74	3.34	3.11	2.96	2.85	2.76	2.70	2.65	2.60
Denominator	15	4.54	3.68	3.29	3.06	2.90	2.79	2.71	2.64	2.59	2.54
	16	4.49	3.63	3.24	3.01	2.85	2.74	2.66	2.59	2.54	2.49
	17	4.45	3.59	3.20	2.96	2.81	2.70	2.61	2.55	2.49	2.45
	18	4.41	3.55	3.16	2.93	2.77	2.66	2.58	2.51	2.46	2.41
	19	4.38	3.52	3.13	2.90	2.74	2.63	2.54	2.48	2.42	2.38
	20	4.35	3.49	3.10	2.87	2.71	2.60	2.51	2.45	2.39	2.35
	21	4.32	3.47	3.07	2.84	2.68	2.57	2.49	2.42	2.37	2.32
	22	4.30	3.44	3.05	2.82	2.66	2.55	2.46	2.40	2.34	2.30
	23	4.28	3.42	3.03	2.80	2.64	2.53	2.44	2.37	2.32	2.27
	24	4.26	3.40	3.01	2.78	2.62	2.51	2.42	2.36	2.30	2.25
	25	4.24	3.39	2.99	2.76	2.60	2.49	2.40	2.34	2.28	2.24
	30	4.17	3.32	2.92	2.69	2.53	2.42	2.33	2.27	2.21	2.16
	40	4.08	3.23	2.84	2.61	2.45	2.34	2.25	2.18	2.12	2.08
	60	4.00	3.15	2.76	2.53	2.37	2.25	2.17	2.10	2.04	1.99
	120	3.92	3.07	2.68	2.45	2.29	2.18	2.09	2.02	1.96	1.91
	∞	3.84	3.00	2.60	2.37	2.21	2.10	2.01	1.94	1.88	1.83

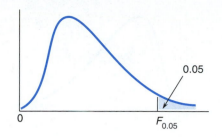

Table IX Critical Values of the *F*-distribution (*α* = 0.05) (*continued*)

| | | \multicolumn{9}{c}{Degrees of Freedom for Numerator} |
		12	15	20	24	30	40	60	120	∞
	1	6,106	6,157	6,209	6,235	6,261	6,287	6,313	6,339	6,366
	2	99.4	99.4	99.4	99.5	99.5	99.5	99.5	99.5	99.5
	3	27.1	26.9	26.7	26.6	26.5	26.4	26.3	26.2	26.1
	4	14.4	14.2	14.0	13.9	13.8	13.7	13.7	13.6	13.5
	5	9.89	9.72	9.55	9.47	9.38	9.29	9.20	9.11	9.02
	6	7.72	7.56	7.40	7.31	7.23	7.14	7.06	6.97	6.88
	7	6.47	6.31	6.16	6.07	5.99	5.91	5.82	5.74	5.65
	8	5.67	5.52	5.36	5.28	5.20	5.12	5.03	4.95	4.86
	9	5.11	4.96	4.81	4.73	4.65	4.57	4.48	4.40	4.31
	10	4.71	4.56	4.41	4.33	4.25	4.17	4.08	4.00	3.91
Degrees	11	4.40	4.25	4.10	4.02	3.94	3.86	3.78	3.69	3.60
of	12	4.16	4.01	3.86	3.78	3.70	3.62	3.54	3.45	3.36
Freedom	13	3.96	3.82	3.66	3.59	3.51	3.43	3.34	3.25	3.17
for	14	3.80	3.66	3.51	3.43	3.35	3.27	3.18	3.09	3.00
Denominator	15	3.67	3.52	3.37	3.29	3.21	3.13	3.05	2.96	2.87
	16	3.55	3.41	3.26	3.18	3.10	3.02	2.93	2.84	2.75
	17	3.46	3.31	3.16	3.08	3.00	2.92	2.83	2.75	2.65
	18	3.37	3.23	3.08	3.00	2.92	2.84	2.75	2.66	2.57
	19	3.30	3.15	3.00	2.92	2.84	2.76	2.67	2.58	2.49
	20	3.23	3.09	2.94	2.86	2.78	2.69	2.61	2.52	2.42
	21	3.17	3.03	2.88	2.80	2.72	2.64	2.55	2.46	2.36
	22	3.12	2.98	2.83	2.75	2.67	2.58	2.50	2.40	2.31
	23	3.07	2.93	2.78	2.70	2.62	2.54	2.45	2.35	2.26
	24	3.03	2.89	2.74	2.66	2.58	2.49	2.40	2.31	2.21
	25	2.99	2.85	2.70	2.62	2.53	2.45	2.36	2.27	2.17
	30	2.84	2.70	2.55	2.47	2.39	2.30	2.21	2.11	2.01
	40	2.66	2.52	2.37	2.29	2.20	2.11	2.02	1.92	1.80
	60	2.50	2.35	2.20	2.12	2.03	1.94	1.84	1.73	1.60
	120	2.34	2.19	2.03	1.95	1.86	1.76	1.66	1.53	1.38
	∞	2.18	2.04	1.88	1.79	1.70	1.59	1.47	1.32	1.00

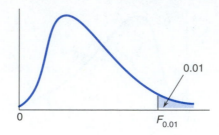

Table IX Critical Values of the *F*-distribution ($\alpha = 0.01$) (*continued*)

					Degrees of Freedom for Numerator					
	1	*2*	*3*	*4*	*5*	*6*	*7*	*8*	*9*	*10*
1	4,052	5,000	5,403	5,625	5,764	5,859	5,928	5,982	6,023	6,056
2	98.5	99.0	99.2	99.2	99.3	99.3	99.4	99.4	99.4	99.4
3	34.1	30.8	29.5	28.7	28.2	27.9	27.7	27.5	27.3	27.2
4	21.2	18.0	16.7	16.0	15.5	15.2	15.0	14.8	14.7	14.5
5	16.3	13.3	12.1	11.4	11.0	10.7	10.5	10.3	10.2	10.1
6	13.7	10.9	9.78	9.15	8.75	8.47	8.26	8.10	7.98	7.87
7	12.2	9.55	8.45	7.85	7.46	7.19	6.99	6.84	6.72	6.62
8	11.3	8.65	7.59	7.01	6.63	6.37	6.18	6.03	5.91	5.81
9	10.6	8.02	6.99	6.42	6.06	5.80	5.61	5.47	5.35	5.26
10	10.0	7.56	6.55	5.99	5.64	5.39	5.20	5.06	4.94	4.85
11	9.65	7.21	6.22	5.67	5.32	5.07	4.89	4.74	4.63	4.54
12	9.33	6.93	5.95	5.41	5.06	4.82	4.64	4.50	4.39	4.30
13	9.07	6.70	5.74	5.21	4.86	4.62	4.44	4.30	4.19	4.10
14	8.86	6.51	5.56	5.04	4.70	4.46	4.28	4.14	4.03	3.94
15	8.68	6.36	5.42	4.89	4.56	4.32	4.14	4.00	3.89	3.80
16	8.53	6.23	5.29	4.77	4.44	4.20	4.03	3.89	3.78	3.69
17	8.40	6.11	5.19	4.67	4.34	4.10	3.93	3.79	3.68	3.59
18	8.29	6.01	5.09	4.58	4.25	4.01	3.84	3.71	3.60	3.51
19	8.19	5.93	5.01	4.50	4.17	3.94	3.77	3.63	3.52	3.43
20	8.10	5.85	4.94	4.43	4.10	3.87	3.70	3.56	3.46	3.37
21	8.02	5.78	4.87	4.37	4.04	3.81	3.64	3.51	3.40	3.31
22	7.95	5.72	4.82	4.31	3.99	3.76	3.59	3.45	3.35	3.26
23	7.88	5.66	4.76	4.26	3.94	3.71	3.54	3.41	3.30	3.21
24	7.82	5.61	4.72	4.22	3.90	3.67	3.50	3.36	3.26	3.17
25	7.77	5.57	4.68	4.18	3.86	3.63	3.46	3.32	3.22	3.13
30	7.56	5.39	4.51	4.02	3.70	3.47	3.30	3.17	3.07	2.98
40	7.31	5.18	4.31	3.83	3.51	3.29	3.12	2.99	2.89	2.80
60	7.08	4.98	4.13	3.65	3.34	3.12	2.95	2.82	2.72	2.63
120	6.85	4.79	3.95	3.48	3.17	2.96	2.79	2.66	2.56	2.47
∞	6.63	4.61	3.78	3.32	3.02	2.80	2.64	2.51	2.41	2.32

Degrees of Freedom for Denominator

A.16

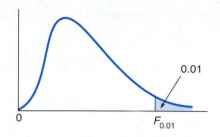

0.01

0

$F_{0.01}$

Table IX Critical Values of the F-distribution ($\alpha = 0.01$) (*continued*)

		Degrees of Freedom for Numerator								
		12	15	20	24	30	40	60	120	∞
	1	244	246	248	249	250	251	252	253	254
	2	19.4	19.4	19.4	19.5	19.5	19.5	19.5	19.5	19.5
	3	8.74	8.70	8.66	8.64	8.62	8.59	8.57	8.55	8.53
	4	5.91	5.86	5.80	5.77	5.75	5.72	5.69	5.66	5.63
	5	4.68	4.62	4.56	4.53	4.50	4.46	4.43	4.40	4.37
	6	4.00	3.94	3.87	3.84	3.81	3.77	3.74	3.70	3.67
	7	3.57	3.51	3.44	3.41	3.38	3.34	3.30	3.27	3.23
	8	3.28	3.22	3.15	3.12	3.08	3.04	3.01	2.97	2.93
	9	3.07	3.01	2.94	2.90	2.86	2.83	2.79	2.75	2.71
	10	2.91	2.85	2.77	2.74	2.70	2.66	2.62	2.58	2.54
Degrees	11	2.79	2.72	2.65	2.61	2.57	2.53	2.49	2.45	2.40
of	12	2.69	2.62	2.54	2.51	2.47	2.43	2.38	2.34	2.30
Freedom	13	2.60	2.53	2.46	2.42	2.38	2.34	2.30	2.25	2.21
for	14	2.53	2.46	2.39	2.35	2.31	2.27	2.22	2.18	2.13
Denominator	15	2.48	2.40	2.33	2.29	2.25	2.20	2.16	2.11	2.07
	16	2.42	2.35	2.28	2.24	2.19	2.15	2.11	2.06	2.01
	17	2.38	2.31	2.23	2.19	2.15	2.10	2.06	2.01	1.96
	18	2.34	2.27	2.19	2.15	2.11	2.06	2.02	1.97	1.92
	19	2.31	2.23	2.16	2.11	2.07	2.03	1.98	1.93	1.88
	20	2.28	2.20	2.12	2.08	2.04	1.99	1.95	1.90	1.84
	21	2.25	2.18	2.10	2.05	2.01	1.96	1.92	1.87	1.81
	22	2.23	2.15	2.07	2.03	1.98	1.94	1.89	1.84	1.78
	23	2.20	2.13	2.05	2.01	1.96	1.91	1.86	1.81	1.76
	24	2.18	2.11	2.03	1.98	1.94	1.89	1.84	1.79	1.73
	25	2.16	2.09	2.01	1.96	1.92	1.87	1.82	1.77	1.71
	30	2.09	2.01	1.93	1.89	1.84	1.79	1.74	1.68	1.62
	40	2.00	1.92	1.84	1.79	1.74	1.69	1.64	1.58	1.51
	60	1.92	1.84	1.75	1.70	1.65	1.59	1.53	1.47	1.39
	120	1.83	1.75	1.66	1.61	1.55	1.50	1.43	1.35	1.25
	∞	1.75	1.67	1.57	1.52	1.46	1.39	1.32	1.22	1.00

From E. S. Pearson and H. O. Hartley, *Biometrika Tables for Statisticians*, 1 (1958), 159–163. Reprinted by permission of the Biometrika Trustees.

Table X Critical Values for Total Number of Runs (Table Shows Critical Values for Two-tailed Test at $\alpha = 0.05$)

The Smaller of n_1 and n_2	\multicolumn{16}{c}{The Larger of n_1 and n_2}

	5	6	7	8	9	10	11	12	13	14	15	16	17	18	19	20
2								2/6	2/6	2/6	2/6	2/6	2/6	2/6	2/6	2/6
3		2/8	2/8	2/8	2/8	2/8	2/8	2/8	2/8	2/8	3/8	3/8	3/8	3/8	3/8	3/8
4	2/9	2/9	2/10	3/10	3/10	3/10	3/10	3/10	3/10	3/10	3/10	4/10	4/10	4/10	4/10	4/10
5	2/10	3/10	3/11	3/11	3/12	3/12	4/12	4/12	4/12	4/12	4/12	4/12	4/12	5/12	5/12	5/12
6		3/11	3/12	3/12	4/13	4/13	4/13	4/13	5/14	5/14	5/14	5/14	5/14	5/14	6/14	6/14
7			3/13	4/13	4/14	5/14	5/14	5/14	5/15	5/15	6/15	6/16	6/16	6/16	6/16	6/16
8				4/14	5/14	5/15	5/15	6/16	6/16	6/16	6/16	6/17	7/17	7/17	7/17	7/17
9					5/15	5/16	6/16	6/16	6/17	7/17	7/18	7/18	7/18	8/18	8/18	8/18
10						6/16	6/17	7/17	7/18	7/18	7/18	8/19	8/19	8/19	8/20	9/20
11							7/17	7/18	7/19	8/19	8/19	8/20	9/20	9/20	9/21	9/21
12								7/19	8/19	8/20	8/20	9/21	9/21	9/21	10/22	10/22
13									8/20	9/20	9/21	9/21	10/22	10/22	10/23	10/23
14										9/21	9/22	10/22	10/23	10/23	11/23	11/24
15											10/22	10/23	11/23	11/24	11/24	12/25
16												11/23	11/24	11/25	12/25	12/25
17													11/25	12/25	12/26	13/26
18														12/26	13/26	13/27
19															13/27	13/27
20																14/28

From C. Eisenhart and F. Swed, "Tables for testing randomness of grouping in a sequence of alternatives," *The Annals of Statistics*, 14(1943), 66–87. Reprinted by permission.

Table XI Critical Value of Spearman's Rank Correlation Coefficient

n	Level of Significance for One-Tailed Test			
	0.05	0.025	0.01	0.005
	Level of Significance for Two-Tailed Test			
	0.10	0.05	0.02	0.01
5	0.900	1.000	1.000	——
6	0.829	0.886	0.943	1.000
7	0.714	0.786	0.893	0.929
8	0.643	0.738	0.833	0.881
9	0.600	0.683	0.783	0.833
10	0.564	0.648	0.745	0.794
11	0.523	0.623	0.736	0.818
12	0.497	0.591	0.703	0.780
13	0.475	0.566	0.673	0.745
14	0.457	0.545	0.646	0.716
15	0.441	0.525	0.623	0.689
16	0.425	0.507	0.601	0.666
17	0.412	0.490	0.582	0.645
18	0.399	0.476	0.564	0.625
19	0.388	0.462	0.549	0.608
20	0.377	0.450	0.534	0.591
21	0.368	0.438	0.521	0.576
22	0.359	0.428	0.508	0.562
23	0.351	0.418	0.496	0.549
24	0.343	0.409	0.485	0.537
25	0.336	0.400	0.475	0.526
26	0.329	0.392	0.465	0.515
27	0.323	0.385	0.456	0.505
28	0.317	0.377	0.448	0.496
29	0.311	0.370	0.440	0.487
30	0.305	0.364	0.432	0.478

Table XII   The Exponential Function

x	e^{-x}	x	e^{-x}	x	e^{-x}	x	e^{-x}
0.00	1.00000	**0.40**	0.67032	**0.80**	0.44933	**1.20**	0.30119
0.01	0.99005	0.41	0.66365	0.81	0.44486	1.21	0.29820
0.02	0.98020	0.42	0.65705	0.82	0.44043	1.22	0.29523
0.03	0.97045	0.43	0.65051	0.83	0.43605	1.23	0.29229
0.04	0.96079	0.44	0.64404	0.84	0.43171	1.24	0.28938
0.05	0.95123	0.45	0.63763	0.85	0.42741	1.25	0.28650
0.06	0.94176	0.46	0.63128	0.86	0.42316	1.26	0.28365
0.07	0.93239	0.47	0.62500	0.87	0.41895	1.27	0.28083
0.08	0.92312	0.48	0.61878	0.88	0.41478	1.28	0.27804
0.09	0.91393	0.49	0.61263	0.89	0.41066	1.29	0.25727
0.10	0.90484	**0.50**	0.60653	**0.90**	0.40657	**1.30**	0.27253
0.11	0.89583	0.51	0.60050	0.91	0.40252	1.31	0.26982
0.12	0.88692	0.52	0.59452	0.92	0.39852	1.32	0.26714
0.13	0.87810	0.53	0.58860	0.93	0.39455	1.33	0.26448
0.14	0.86936	0.54	0.58275	0.94	0.39063	1.34	0.26185
0.15	0.86071	0.55	0.57695	0.95	0.38674	1.35	0.25924
0.16	0.85214	0.56	0.57121	0.96	0.38289	1.36	0.25666
0.17	0.84366	0.57	0.56553	0.97	0.37908	1.37	0.25411
0.18	0.83527	0.58	0.55990	0.98	0.37531	1.38	0.25158
0.19	0.82696	0.59	0.55433	0.99	0.37158	1.39	0.24908
0.20	0.81873	**0.60**	0.54881	**1.00**	0.36788	**1.40**	0.24660
0.21	0.81058	0.61	0.54335	1.01	0.36422	1.41	0.24414
0.22	0.80252	0.62	0.53794	1.02	0.36059	1.42	0.24171
0.23	0.79453	0.63	0.53259	1.03	0.35701	1.43	0.23931
0.24	0.78663	0.64	0.52729	1.04	0.35345	1.44	0.23693
0.25	0.77880	0.65	0.52205	1.05	0.34994	1.45	0.23457
0.26	0.77105	0.66	0.51685	1.06	0.34646	1.46	0.23224
0.27	0.76338	0.67	0.51171	1.07	0.34301	1.47	0.22993
0.28	0.75578	0.68	0.50662	1.08	0.33960	1.48	0.22764
0.29	0.74826	0.69	0.51058	1.09	0.33622	1.49	0.22537
0.30	0.74082	**0.70**	0.49659	**1.10**	0.33287	**1.50**	0.22313
0.31	0.73345	0.71	0.49164	1.11	0.32956	1.51	0.22091
0.32	0.72615	0.72	0.48675	1.12	0.32628	1.52	0.21871
0.33	0.71892	0.73	0.48191	1.13	0.32303	1.53	0.21654
0.34	0.71177	0.74	0.47711	1.14	0.31982	1.54	0.21438
0.35	0.70469	0.75	0.47237	1.15	0.31664	1.55	0.21225
0.36	0.69768	0.76	0.46767	1.16	0.31349	1.56	0.21014
0.37	0.69073	0.77	0.46301	1.17	0.31037	1.57	0.20805
0.38	0.68386	0.78	0.45841	1.18	0.30728	1.58	0.20598
0.39	0.67706	0.79	0.45384	1.19	0.30422	1.59	0.20393

TABLE XII THE EXPONENTIAL FUNCTION   A.21

Table XII   The Exponential Function (*continued*)

x	e^{-x}	x	e^{-x}	x	e^{-x}	x	e^{-x}
1.60	0.20190	**2.00**	0.13534	**2.40**	0.09072	**2.80**	0.06081
1.61	0.19989	2.01	0.13399	2.41	0.08982	2.81	0.06020
1.62	0.19790	2.02	0.13266	2.42	0.08892	2.82	0.05961
1.63	0.19593	2.03	0.13134	2.43	0.08804	2.83	0.05901
1.64	0.19398	2.04	0.13003	2.44	0.08716	2.84	0.05843
1.65	0.19205	2.05	0.12873	2.45	0.08629	2.85	0.05784
1.66	0.19014	2.06	0.12745	2.46	0.08543	2.86	0.05727
1.67	0.18825	2.07	0.12619	2.47	0.08458	2.87	0.05670
1.68	0.18637	2.08	0.12493	2.48	0.08374	2.88	0.05613
1.69	0.18452	2.09	0.12369	2.49	0.08291	2.89	0.05558
1.70	0.18268	**2.10**	0.12246	**2.50**	0.08208	**2.90**	0.05502
1.71	0.18087	2.11	0.12124	2.51	0.08127	2.91	0.05448
1.72	0.17907	2.12	0.12003	2.52	0.08046	2.92	0.05393
1.73	0.17728	2.13	0.11884	2.53	0.07966	2.93	0.05340
1.74	0.17552	2.14	0.11765	2.54	0.07887	2.94	0.05287
1.75	0.17377	2.15	0.11648	2.55	0.07808	2.95	0.05234
1.76	0.17204	2.16	0.11533	2.56	0.07730	2.96	0.05182
1.77	0.17033	2.17	0.11418	2.57	0.07654	2.97	0.05130
1.78	0.16864	2.18	0.11304	2.58	0.07577	2.98	0.05079
1.79	0.16696	2.19	0.11192	2.59	0.07502	2.99	0.05029
1.80	0.16530	**2.20**	0.11080	**2.60**	0.07427	**3.00**	0.04979
1.81	0.16365	2.21	0.10970	2.61	0.07353	3.01	0.04929
1.82	0.16203	2.22	0.10861	2.62	0.07280	3.02	0.04880
1.83	0.16041	2.23	0.10753	2.63	0.07208	3.03	0.04832
1.84	0.15882	2.24	0.10646	2.64	0.07136	3.04	0.04783
1.85	0.15724	2.25	0.10540	2.65	0.07065	3.05	0.04736
1.86	0.15567	2.26	0.10435	2.66	0.06995	3.06	0.04689
1.87	0.15412	2.27	0.10331	2.67	0.06925	3.07	0.04642
1.88	0.15259	2.28	0.10228	2.68	0.06856	3.08	0.04596
1.89	0.15107	2.29	0.10127	2.69	0.06788	3.09	0.04550
1.90	0.14957	**2.30**	0.10026	**2.70**	0.06721	**3.10**	0.04505
1.91	0.14808	2.31	0.09926	2.71	0.06654	3.11	0.04460
1.92	0.14661	2.32	0.09827	2.72	0.06587	3.12	0.04416
1.93	0.14515	2.33	0.09730	2.73	0.06522	3.13	0.04372
1.94	0.14370	2.34	0.09633	2.74	0.06457	3.14	0.04328
1.95	0.14227	2.35	0.09537	2.75	0.06393	3.15	0.04285
1.96	0.14086	2.36	0.09442	2.76	0.06329	3.16	0.04243
1.97	0.13946	2.37	0.09348	2.77	0.06266	3.17	0.04200
1.98	0.13807	2.38	0.09255	2.78	0.06204	3.18	0.04159
1.99	0.13670	2.39	0.09163	2.79	0.06142	3.19	0.04117

Table XII The Exponential Function (continued)

x	e^{-x}	x	e^{-x}	x	e^{-x}	x	e^{-x}
3.20	0.04076	**3.60**	0.02732	**4.00**	0.01832	**4.40**	0.01228
3.21	0.04036	3.61	0.02705	4.01	0.01813	4.41	0.01216
3.22	0.03996	3.62	0.02678	4.02	0.01795	4.42	0.01203
3.23	0.03956	3.63	0.02652	4.03	0.01777	4.43	0.01191
3.24	0.03916	3.64	0.02625	4.04	0.01760	4.44	0.01180
3.25	0.03877	3.65	0.02599	4.05	0.01742	4.45	0.01168
3.26	0.03839	3.66	0.02573	4.06	0.01725	4.46	0.01156
3.27	0.03801	3.67	0.02548	4.07	0.01708	4.47	0.01145
3.28	0.03763	3.68	0.02522	4.08	0.01691	4.48	0.01133
3.29	0.03725	3.69	0.02497	4.09	0.01674	4.49	0.01122
3.30	0.03688	**3.70**	0.02472	**4.10**	0.01657	**4.50**	0.01111
3.31	0.03652	3.71	0.02448	4.11	0.01641	4.51	0.01100
3.32	0.03615	3.72	0.02423	4.12	0.01624	4.52	0.01089
3.33	0.03579	3.73	0.02399	4.13	0.01608	4.53	0.01078
3.34	0.03544	3.74	0.02375	4.14	0.01592	4.54	0.01067
3.35	0.03508	3.75	0.02352	4.15	0.01576	4.55	0.01057
3.36	0.03474	3.76	0.02328	4.16	0.01561	4.56	0.01046
3.37	0.03439	3.77	0.02305	4.17	0.01545	4.57	0.01036
3.38	0.03405	3.78	0.02282	4.18	0.01530	4.58	0.01025
3.39	0.03371	3.79	0.02260	4.19	0.01515	4.59	0.01015
3.40	0.03337	**3.80**	0.02237	**4.20**	0.01500	**4.60**	0.01005
3.41	0.03304	3.81	0.02215	4.21	0.01485	4.61	0.00995
3.42	0.03271	3.82	0.02193	4.22	0.01470	4.62	0.00985
3.43	0.03239	3.83	0.02171	4.23	0.01455	4.63	0.00975
3.44	0.03206	3.84	0.02149	4.24	0.01441	4.64	0.00966
3.45	0.03175	3.85	0.02128	4.25	0.01426	4.65	0.00956
3.46	0.03143	3.86	0.02107	4.26	0.01412	4.66	0.00947
3.47	0.03112	3.87	0.02086	4.27	0.01398	4.67	0.00937
3.48	0.03081	3.88	0.02065	4.28	0.01384	4.68	0.00928
3.49	0.03050	3.89	0.02045	4.29	0.01370	4.69	0.00919
3.50	0.03020	**3.90**	0.02024	**4.30**	0.01357	**4.70**	0.00910
3.51	0.02990	3.91	0.02004	4.31	0.01343	4.71	0.00900
3.52	0.02960	3.92	0.01984	4.32	0.01330	4.72	0.00892
3.53	0.02930	3.93	0.01964	4.33	0.01317	4.73	0.00883
3.54	0.02901	3.94	0.01945	4.34	0.01304	4.74	0.00874
3.55	0.02872	3.95	0.01925	4.35	0.01291	4.75	0.00865
3.56	0.02844	3.96	0.01906	4.36	0.01278	4.76	0.00857
3.57	0.02816	3.97	0.01887	4.37	0.01265	4.77	0.00848
3.58	0.02788	3.98	0.01869	4.38	0.01253	4.78	0.00840
3.59	0.02760	3.99	0.01850	4.39	0.01240	4.79	0.00831

TABLE XII THE EXPONENTIAL FUNCTION **A.23**

Table XII The Exponential Function (continued)

x	e^{-x}	x	e^{-x}	x	e^{-x}
4.80	0.00823	**6.00**	0.00248	**9.00**	0.00012
4.81	0.00815	6.10	0.00244	9.10	0.00011
4.82	0.00807	6.20	0.00203	9.20	0.00010
4.83	0.00799	6.30	0.00184	9.30	0.00009
4.84	0.00791	6.40	0.00166	9.40	0.00008
4.85	0.00783	6.50	0.00150	9.50	0.00007
4.86	0.00775	6.60	0.00136	9.60	0.00007
4.87	0.00767	6.70	0.00123	9.70	0.00006
4.88	0.00760	6.80	0.00111	9.80	0.00006
4.89	0.00752	6.90	0.00101	9.90	0.00005
4.90	0.00745	**7.00**	0.00091	**10.00**	0.00005
4.91	0.00737	7.10	0.00083	10.10	0.00004
4.92	0.00730	7.20	0.00075	10.20	0.00004
4.93	0.00723	7.30	0.00068	10.30	0.00003
4.94	0.00715	7.40	0.00061	10.40	0.00003
4.95	0.00708	7.50	0.00055	10.50	0.00003
4.96	0.00701	7.60	0.00050	10.60	0.0002
4.97	0.00694	7.70	0.00045	10.70	0.00002
4.98	0.00687	7.80	0.00041	10.80	0.00002
4.99	0.00681	7.90	0.00037	10.90	0.00002
5.00	0.00674	**8.00**	0.00034	**11.00**	0.00002
5.10	0.00610	8.10	0.00030	11.10	0.00002
5.20	0.00552	8.20	0.00027	11.20	0.00001
5.30	0.00499	8.30	0.00025	11.30	0.00001
5.40	0.00452	8.40	0.00022	11.40	0.00001
5.50	0.00409	8.50	0.00020	11.50	0.00001
5.60	0.00370	8.60	0.00018	11.60	0.00001
5.70	0.00335	8.70	0.00017	11.70	0.00001
5.80	0.00303	8.80	0.00015	11.80	0.00001
5.90	0.00274	8.90	0.00014	11.90	0.00001

Answers to Selected Exercises

CHAPTER 1

Section 1.5 (pages 12–14)

1. Although it should be descriptive only, in reality it is used as both descriptive and inferential statistics.
3. a. Sex, eye color, etc. . . .
   b. Number of brothers or sisters that each college student has, etc. . . .
   c. Weight or height of each college student, etc. . . .
5. a. The part that indicates that the number of Americans over 65 years of age increased by 6.3% over last year at this time.
   b. The part that predicts that by the end of the century approximately 25% of the American population will be over 65 years of age.
7. Not necessarily. Other factors have to be considered.
9. Not necessarily. The subscribers might have been healthy anyway, without the vitamin C.

Testing Your Understanding of This Chapter's Concepts (pages 15–16)

1. No. The 10% increase represents 10% on a lower base pay.
2. It indicates past performance and possibly represents a prediction of future performance.
3. Not necessarily. Other factors have to be considered.
4. Not necessarily. Other factors have to be considered.
5. No. The sample is likely to be nonrepresentative of the population.

A.25

Thinking Critically (pages 16–17)

1. No
2. Not necessarily; there may not be many dolphins left.
3. No. The sample is likely to be nonrepresentative of the population.
5. No   6. No

Review Exercises (pages 17–19)

1. Choice (d). The conclusion involves statistical inference.
2. The part that states that the blood levels at the various hospitals stood at 20% of their normal levels and that 837 pints of blood were donated last week.
3. The part that claims that the drop in blood donations is due to the fear of AIDS and that the findings have far-reaching implications for individuals needing surgery.
4. Choice (d)   5. No. There are fewer cars on the road.
6. Statistics seem to indicate that this conclusion may be valid.
7. True   8. True   9. True

Chapter Test (pages 19–22)

1. No
2. Not necessarily; there may be fewer highway patrol officers.
3. a. Statistical inference   b. Descriptive statistics
   c. Statistical inference
4. True   5. True
6. Not necessarily; the percentage of the population may not have increased.
7. The part that indicates that 14 of the 527 homes inspected were contaminated with radon gas.
8. The part that estimates that about 15% of the homes in the region are contaminated.
9. Not necessarily; people may be more selective of the foods they eat.
10. Not necessarily.   11. Choice (d)   12. Choice (b)
13. Choice (c)   14. Choice (b)   15. Choice (d)

CHAPTER 2

Section 2.2 (pages 35–39)

1.

Number of Parking Tickets Issued	Tally	Frequency
23–28	ЦНI ЦНI ЦНI II	17
29–34	ЦНI ЦНI ЦНI I	16
35–40	ЦНI ЦНI ЦНI III	18
41–46	ЦНI ЦНI	10
47–52	ЦНI ЦНI	10
53–58	ЦНI II	7
59–64	ЦНI	5
65–70	ЦНI	5
71–76	ЦНI ЦНI	10
77–82	II	2
		100

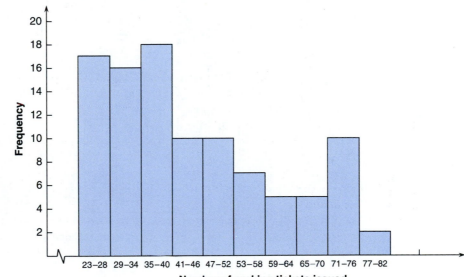

3.

Age of Applicants	Tally	Frequency
17–22	IIII I	6
23–28	IIII III	8
29–34	IIII IIII	9
35–40	III	3
41–46	IIII I	6
47–52	IIII	4
53–58	III	3
59–64	III	3
65–70	IIII I	6
71–76	II	$\underline{2}$
		50

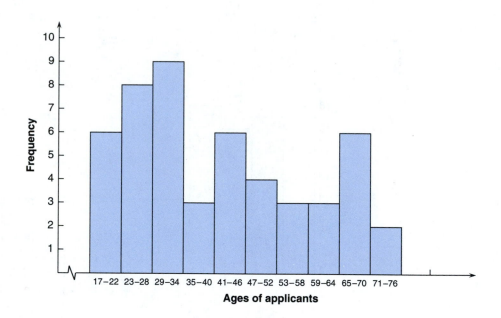

Ages of applicants

5.

Weight of Luggage	Tally	Frequency
74–80	ⅢⅠ Ⅰ	6
81–87	ⅢⅠ	5
88–94	ⅢⅠ ⅢⅠ	10
95–101	ⅢⅠ	5
102–108	ⅢⅠ ⅢⅠ	10
109–115	ⅢⅠ ⅢⅠ Ⅰ	11
116–122	ⅢⅠ	5
123–129	‖	2
130–136	‖	2
137–144	‖‖	4
		60

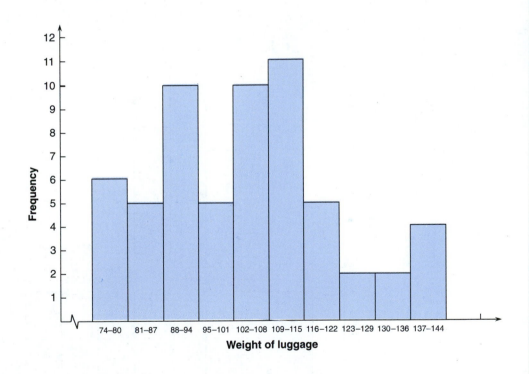

7. a. The frequency distribution using 10 classes.

Number of Pounds Collected	Tally	Frequency
12–31	‖‖	4
32–51	‖‖	4
52–71	‖‖ ‖‖	10
72–91	‖‖ ‖‖ ‖‖	13
92–111	‖‖ ‖‖ ‖	11
112–131	‖‖	5
132–151	‖‖ ‖	6
152–171	‖‖	4
172–191	‖	1
192–211	‖‖	2
		60

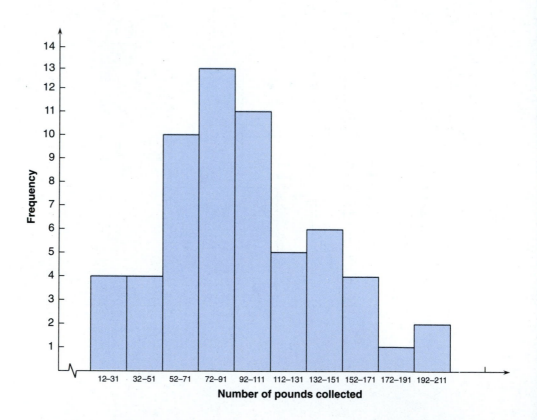

b. The frequency distribution using 5 classes.

Number of Pounds Collected	Tally	Frequency
12–51	ЖН |||	8
52–91	ЖН ЖН ЖН ЖН |||	23
92–131	ЖН ЖН ЖН		16
132–171	ЖН ЖН	10
172–211	|||	$\frac{3}{60}$

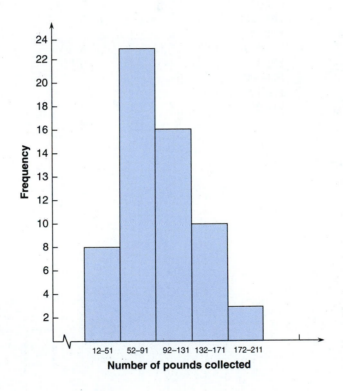

c. The frequency distribution using 15 classes.

Number of Pounds Collected	Tally	Frequency
12–24	||	2
25–37	|||	3
38–50	|||	3
51–63	卌	5
64–76	卌 ||	7
77–89	卌 卌	10
90–102	卌 |	6
103–115	卌 ||	7
116–128	|||	3
129–141	||||	4
142–154	卌 |	6
155–167		0
168–180	||	2
181–193		0
194–206	||	2
		60

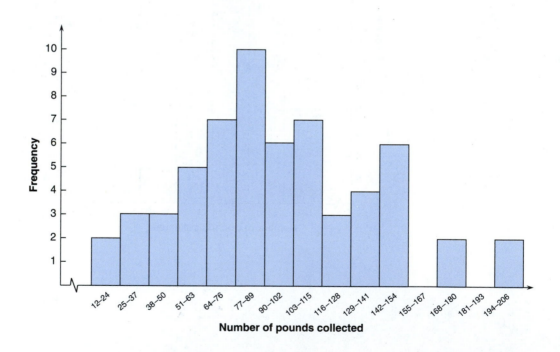

9. a.

Number of Products Completed	Tally	Frequency
4–7		0
8–11	|	1
12–15	||	2
16–19	ЖЖ |	6
20–23	||||	4
24–27	|	1
28–31	||||	4
32–37	||	2
		20

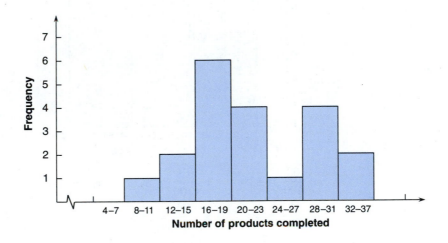

11. To avoid misinterpreting the data.

13.

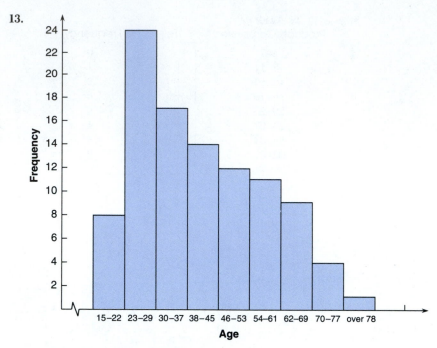

a. $\dfrac{63}{100} = 63\%$ b. $\dfrac{51}{100} = 51\%$ c. $\dfrac{43}{100} = 43\%$

d. $\dfrac{4}{100} = 4\%$ e. $\dfrac{96}{100} = 96\%$

15. a. (i)

Number of Stocks	Tally	Frequency
1–6	~~IIII~~ III	8
7–12	~~IIII~~	5
13–18	~~IIII~~ ~~IIII~~ III	13
19–24	~~IIII~~ ~~IIII~~ ~~IIII~~ IIII	19
25–30	~~IIII~~	5
		50

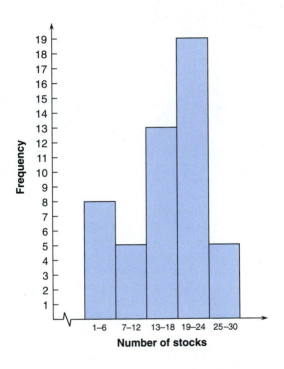

(ii)

Number of Stocks	Tally	Frequency
1–2	||	2
3–4	|||	3
5–6	|||	3
7–8	|||	3
9–10	||	2
11–12		0
13–14	|||	3
15–16	ⅢⅡ ||	7
17–18	|||	3
19–20	||	2
21–22	ⅢⅡ ||	7
23–24	ⅢⅡ ⅢⅡ	10
25–26		0
27–28	||	2
29–30	|||	3
		50

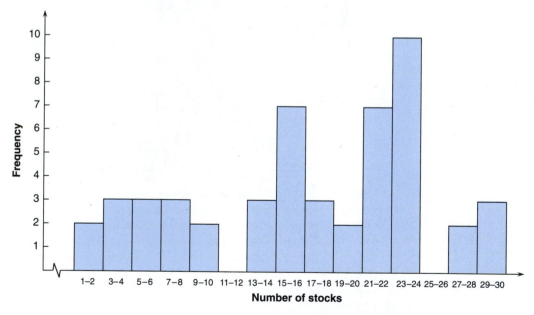

b. When too few or too many classes are used, much information is lost, or the information that is presented is hard to interpret.

Section 2.3 (pages 61–67)

1.

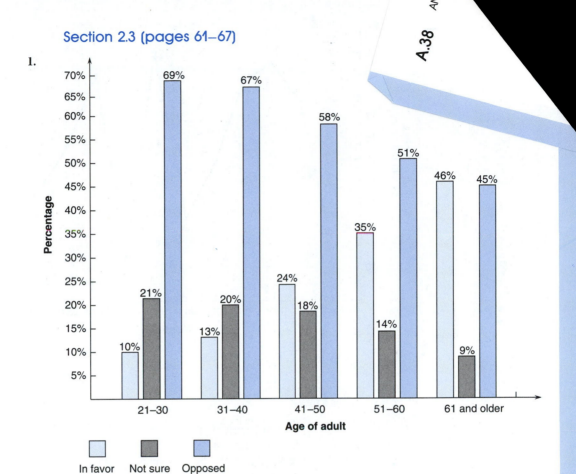

3 a. Humber Bridge: 1400 meters
       Verrazano Narrows Bridge: 1300 meters
       Golden Gate Bridge: 1200 meters
       Mackinac Straits Bridge: 1100 meters
       Bosporus Bridge: 1100 meters
       George Washington Bridge: 1000 meters
    b. 1420 − 1260 or 160 meters
    c. 1240 − 1070 or 170 meters

5.

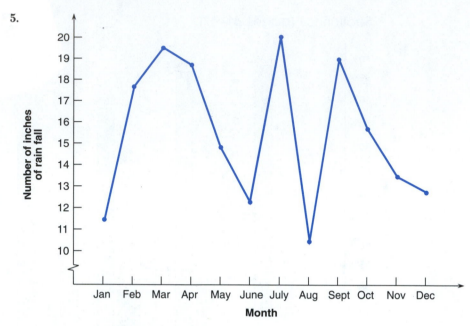

7.

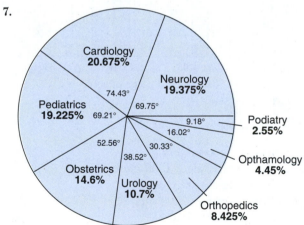

9. a. $1.5 \times 0.26 = \$0.39$ million

b. $1.5 \times 0.07 = \$0.105$ million

c. $1.5 \times (0.32 + 0.26) = \$0.87$ million

d. $1.5 \times (0.26 + 0.32 + 0.07 + 0.05) = \$1.05$ million

11. a.

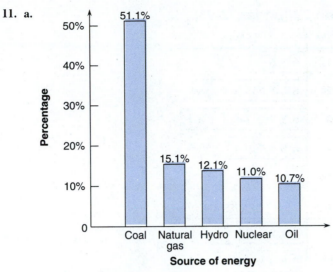

b.

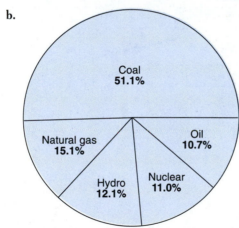

c. They both are useful.

13. a. China: 850 million
India: 625 million
Soviet Union: 250 million
United States: 225 million

b. 850 − 250 = 600 million

15. a. Normally distributed   **b.** Normally distributed
c. Normally distributed   **d.** Not normally distributed
e. Normally distributed   **f.** Not normally distributed

Section 2.4 (pages 72–73)

1.

STEM	LEAVES
1	9 8 6 6
2	8 5 8 4 8 9 9 8 8
3	6 8 7 9 7 6 3 4 9 1
4	5 3 5 7 9 0
5	1

3.

STEM	LEAVES
18	8 4 9 7 9
19	7 8 9 9 6
20	0 1 4 5 3 1 8 5 8
21	1 8 1 9 6
22	7 1 2 1 3 1 7 7
23	2 8 2 9 8
24	5 1 9 1
25	6 7 1
26	9
27	2
28	3 4 7
29	7

5.

STEM	LEAVES
(3900–3949) 39	45 .01 45 45
(3950–3999) 39	98 89 75 76 84 98 87
(4000–4049) 40	00 00 04 00 49
(4050–4099) 40	50
(4100–4149) 41	49 49 45
(4150–4199) 41	98 50
(4200–4249) 42	49 00 49

Section 2.5 (pages 77–78)

1. Both graphs are statistically correct, but because different spacings are used on the vertical scale, the graphs appear different.
3. The graph is truncated. Also the scale is misleading.

Section 2.6 (pages 81–83)

1. $\dfrac{0.75}{0.43} \times 100 = 174.42$ or an increase of approximately 74.42% when compared to 1986.

3. $\dfrac{0.43}{0.75} \times 100 = 57.33$ or that prices were approximately $57.33 - 100$ or 42.67% cheaper in 1986 when compared to 1991.

5.

Year	Cost Index	Interpretation
1985	$\dfrac{2.75}{2.75} \times 100 = 100$	Base year price
1986	$\dfrac{3.25}{2.75} \times 100 = 118.18$	Price increased by approximately 18.18% when compared to 1985
1987	$\dfrac{4.00}{2.75} \times 100 = 145.45$	Price increased by approximately 45.45% when compared to 1985
1988	$\dfrac{4.25}{2.75} \times 100 = 154.55$	Price increased by approximately 54.55% when compared to 1985
1989	$\dfrac{5.00}{2.75} \times 100 = 181.82$	Price increased by approximately 81.82% when compared to 1985
1990	$\dfrac{6.00}{2.75} \times 100 = 218.18$	Price increased by approximately 118.18% when compared to 1985

7.

Year	Cost Index
1984	$\dfrac{1420}{1992} \times 100 = 71.285$
1985	$\dfrac{1605}{1992} \times 100 = 80.572$
1986	$\dfrac{1840}{1992} \times 100 = 92.369$
1987	$\dfrac{1992}{1992} \times 100 = 100 \leftarrow$ Base year price
1988	$\dfrac{2079}{1992} \times 100 = 104.367$
1989	$\dfrac{2250}{1992} \times 100 = 112.952$
1990	$\dfrac{2408}{1992} \times 100 = 120.884$

9. The cost of paper supplies decreased by 1% in February 1989 when compared to January 1988.

Testing Your Understanding of This Chapter's Concepts (pages 88–89)

1. The interval endpoints are overlapping. In which category should we put a monthly bill of $15? Also, the interval lengths are not the same.

2. Yes   **3.** True

4. $\dfrac{112 - 0}{12} = 9.33$. Use 10 as the class width.

Thinking Critically (pages 89–90)

1. Yes. The next interval is supposed to begin where the previous interval ends. No gaps. Also, the interval lengths are not the same.

2. a. $4 + 6 + 9 + 8 + 4 + 3 + 7 + 5 + 1 = 47$
   b. Frequency histogram

3. Nothing is wrong mathematically, although it is misleading.

4. a. No. Vertical scale is inaccurate.
   b. Should not be twice as tall.
   c. Although the conclusion may be true, the graph presenting this information is inaccurate.

Review Exercises (pages 90–93)

1.

Class No.	Interval	Midpoint	Tally	Frequency
1	2–16	9	|||	3
2	17–31	24	||||| ||||	9
3	32–46	39	||||	4
4	47–61	54	|||	3
5	62–76	69	|||||	5
6	77–91	84	|||||		6/30

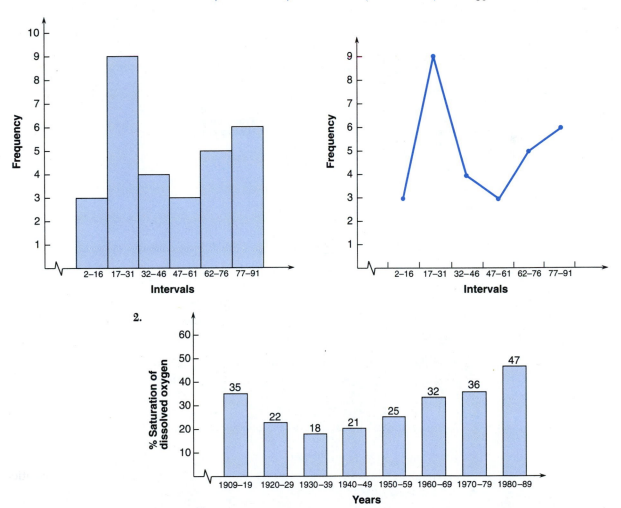

2.

3. **a.** $8 \times 0.31 = 2.48$ million
   b. $8 \times 0.23 = 1.84$ million
   c. $8(0.23 + 0.01 + 0.08) = 2.56$ million

4.

Mileage	Tally	Frequency	Relative Frequency
36–37	|||	3	$\dfrac{3}{16}$
38–39	||||	4	$\dfrac{4}{16}$
40–41	|||	3	$\dfrac{3}{16}$
42–43	||||	4	$\dfrac{4}{16}$
44–45	||	2	$\dfrac{2}{16}$

5.

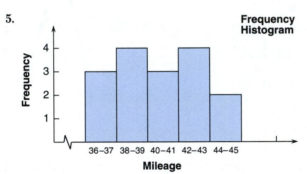

Frequency Histogram

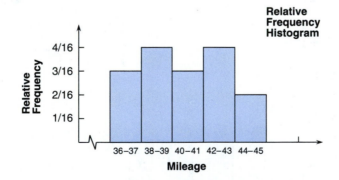

Relative Frequency Histogram

6.

STEM	LEAVES
3	
3	8 9 7 8 8 6 6
4	3 2 1 4 3 0 3 1
4	5

7.

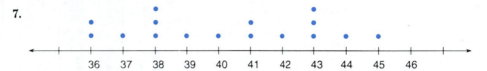

8.

Number of Housing Permits Issued	Tally	Frequency
13–15	|	1
16–18	ⵏ |	6
19–21	||||	4
22–24	||	2
25–27	||	2
28–30	||||	4
31–33	ⵏ	5
34–36	||	2
37–39	||	2
40–42		0
43–45		0
46–48	|	$\frac{1}{29}$

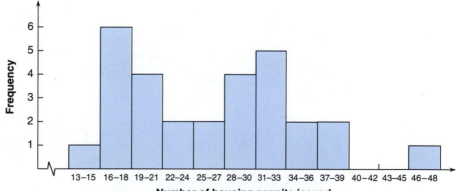

9. a.

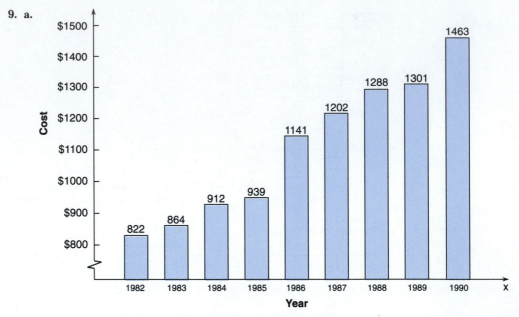

b.

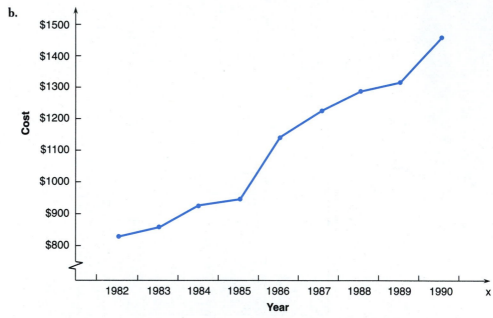

10.

Year	Cost Index
1982	$\dfrac{822}{822} \times 100 = 100$
1983	$\dfrac{864}{822} \times 100 = 105.109$
1984	$\dfrac{912}{822} \times 100 = 110.949$
1985	$\dfrac{939}{822} \times 100 = 114.234$
1986	$\dfrac{1141}{822} \times 100 = 138.808$
1987	$\dfrac{1202}{822} \times 100 = 146.229$
1988	$\dfrac{1288}{822} \times 100 = 156.691$
1989	$\dfrac{1301}{822} \times 100 = 158.273$
1990	$\dfrac{1463}{822} \times 100 = 177.981$

Chapter Test (pages 93–98)

1.

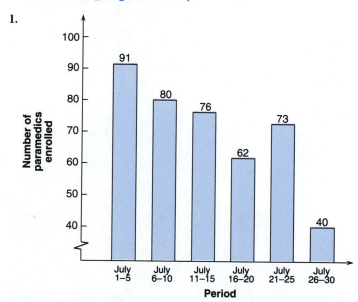

2.

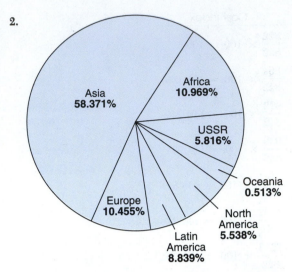

3.

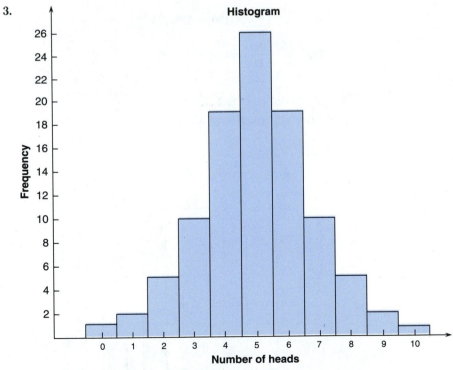

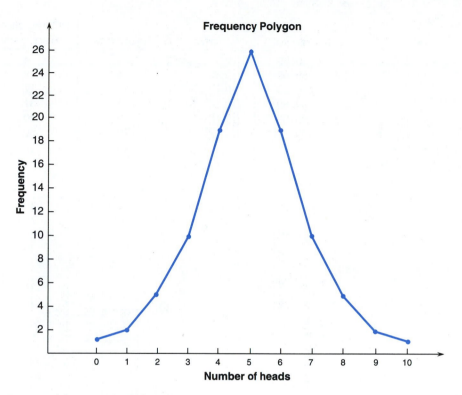

It resembles a normal distribution.

4. a.

STEM	LEAVES
5	8 4 3 5
6	3 9 3 4
7	3 6 9 1 4 4 1 3 6
8	4 6 3 0 8 5 0 9 5 8 2 3 5 6
9	8 1 2 2 7 3 1 6 8
10	4 7 5 4 4 1 6 2 5 9

b.

Interval	Tally	Frequency	Cumulative Frequency
50–59	||||	4	4
60–69	||||	4	8
70–79	卌 ||||	9	17
80–89	卌 卌 ||||	14	31
90–99	卌 ||||	9	40
100–109	卌 卌	10	50
		50	

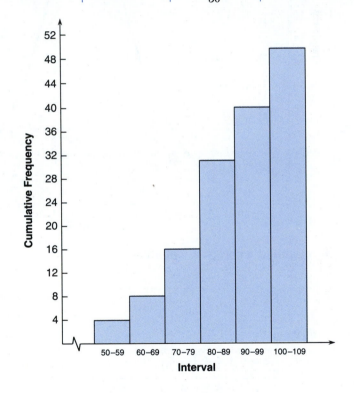

5.

Year	Cost Index
1984	$\frac{341}{341} \times 100 = 100$
1985	$\frac{363}{341} \times 100 = 106.452$
1986	$\frac{387}{341} \times 100 = 113.4897$
1987	$\frac{399}{341} \times 100 = 117.009$
1988	$\frac{407}{341} \times 100 = 119.355$
1989	$\frac{419}{341} \times 100 = 122.874$
1990	$\frac{439}{341} \times 100 = 128.739$

6. a. $130 - 50 = 80$ million (approximately)
   b. $205 - 150 = 55$ million (approximately)

7.

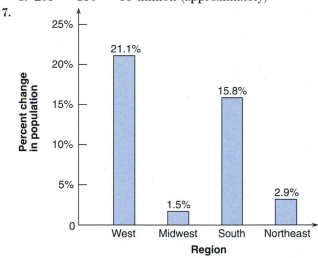

8. The number of people per family living in the same household has been decreasing. The number of housing units has been increasing. Thus there may be more housing units but there are fewer people per household.

9. The coins should be of the same size, and there should be twice as many coins stacked vertically for 1989 as compared to the number of coins stacked vertically for 1980.

10. Probably all are.   **11.** Choice (c)   **12.** Choice (a)   **13.** Choice (b)
14. Choice (e)   **15.** Choice (c)

CHAPTER 3

Section 3.2 (pages 106–107)

1. a. $\sum_{i=1}^{8} y_i$ b. $\sum_{i=1}^{8} y_i^2$ c. $\sum_{i=1}^{8} y_i f_i$ d. $\sum_{i=1}^{7} 19 y_i$ e. $\sum_{i=1}^{n} (3 y_i + x_i)$

3. a. $\Sigma x = 86$ b. $(\Sigma x)^2 = 7396$ c. $\Sigma x^2 = 1916$ d. $\Sigma(x + 1) = 91$
   e. $\Sigma(x + 1)^2 = 2093$ f. $\Sigma(2x + 3) = 187$ g. $\Sigma(2x + 3)^2 = 8741$

5. a. $\sum_{i=1}^{10} (x_1 + 9) = \sum_{i=1}^{10} x_1 + \sum_{i=1}^{10} 9 = 27 + 10 \cdot 9 = 117$

   b. $\sum_{i=1}^{10} (x_1 + 9)^2 = \sum_{i=1}^{10} (x_1^2 + 18 x_1 + 81) = \sum_{i=1}^{10} x_1^2 + 18 \sum_{i=1}^{10} x_1 + \sum_{i=1}^{10} 81$
$$= 50 + 18(27) + 10 \cdot 81 = 1346$$

7. a. The total number of students submitting essays.
   b. The total number of essays (of all types) that she has received.

Section 3.3 (pages 118–124)

1.   Mean $= 31.58$ inches
    Median $=$ between 31 and 35 or 33.
     Mode $= 25$ and 39

3.   Mean $= \$57,375$
    Median $= \$59,000$
     Mode $= \$61,000$
    Extreme scores have a significant effect on the mean.

5.   Mean $= \$6124.17$
    Median $=$ between 5995 and 6100 or $\$6047.50$
     Mode $=$ none

7.   Mean $= 11.906$
    Median $= 12$
     Mode $= 12$
    The mode is more important to the management of the store.

9.   Mean $= \$1.74$
    Median $= \$1.75$
     Mode $= \$1.75$

11.   Mean $= \dfrac{\Sigma x f}{\Sigma f} = \dfrac{2079975}{50} = \$41,599.50$

    Median $= \$42,499.50$ using the class mark or $40,000 + \dfrac{4.5}{14}(5000) = \$41,607.14$

    assuming entries are evenly distributed in each class.
     Mode $=$ In the $\$40,000$–$\$44,999$ interval

13. No. Both classes may not have the same number of students.
15.   Mean = 4.78
      Median = 4
      Mode = 3
      Probably the mode.
17.   Harmonic mean = 6.6359
      Geometric mean = $\sqrt[4]{2160}$
19. **a.** Midrange of bank failures = $\dfrac{2 + 14}{2} = 8$

    b. Midrange of scholastic aptitude test scores
    $$= \dfrac{482 + 623}{2} = 552.5$$

Section 3.5 (pages 133–136)

1. **a.** Range = 38
   b. Sample variance = 146.26667
      Sample standard deviation ≈ 12.094
3. Mean = 2536
   Sample variance = 681468.857
   Sample standard deviation = $\sqrt{681468.857} \approx 825.51$
   Average deviation = 649.33
5. $\mu = 52$
   Population variance = 93.2857
   Population standard deviation = $\sqrt{93.2857}$
   $$\approx 9.658$$
   Average deviation = 7.571
7. New range = 15
   New sample mean = 12
   New sample variance = 22.5
   New standard deviation = $\sqrt{22.5} \approx 4.743$
   New average deviation = 3.333
   The range, sample mean, sample standard deviation, and average deviation are 3 times as great as they were originally. The sample variance is 9 times as great as it was originally.
9. The new mean will be 3400 + 500 or $3900. The new standard deviation will remain at $375.
11. **a.** Mean = $\dfrac{0.44}{5} + 1.29 = \$1.378$

    Population standard deviation = $\sqrt{\dfrac{0.0530}{5} - \dfrac{(0.44)^2}{5^2}} \approx 0.053$

    b. Mean = $\dfrac{0.09}{5} + 1.36 = \$1.378$

$$\text{Population standard deviation} = \sqrt{\frac{0.0159}{5} - \frac{(0.09)^2}{25}} \approx 0.053$$

c. In either case, the population mean and standard deviation turn out to be the same.

13. We will use the class mark for each interval.

$$\text{Mean} = \frac{\Sigma xf}{\Sigma f} = \frac{3140}{50} = 62.8$$

$$\text{Sample variance} = 156.2857$$

$$\text{Sample standard deviation} = \sqrt{156.2857} \approx 12.501$$

15. Although the average cost per gallon of gas in both neighborhoods is the same, the standard deviation is smaller in the poorer neighborhoods.

Section 3.6 (pages 139–140)

1. We must first calculate the sample standard deviation. We have

$$\text{Sample mean} = 51.6$$
$$\text{Sample standard deviation} = s = 15.25$$

When $k = 2$, at least $\frac{3}{4}$ of the measurements fall within $51.6 \pm 2(15.25)$ or between 21.1 and 82.1. Since $n = 10$, $\frac{3}{4}$ of the sample is 7.5. Actually all 10 of the numbers are between 21.1 and 82.1, which is greater than 7.5. When $k = 3$, at least $\frac{8}{9}$ of the measurements fall within $51.6 \pm 3(15.25)$ or between 5.85 and 97.35.

3. $k = 1.5$ At least 55.56%.
5. a. Sample mean $= 72.4$
   Sample standard deviation $= 12.828$
   b. At least 84%. At least 93.75%.
   c. When $k = 2.5$, at least 84% of the measurements will fall within $72.4 \pm 2.5(12.828)$ or between 40.33 and 104.47. In our case, 100% of the observations fall in this interval.

   When $k = 4$, at least 93.75% of the measurements will fall within $72.4 \pm 4(12.828)$ or between 21.088 and 123.712. The percentage of measurements falling within these intervals is indeed true. In our case, 100% of the observations fall in this interval.
   d. Chebyshev's theorem is valid.

Section 3.7 (pages 150–153)

1. Percentile rank of Alfred = 29.16 percentile
   Percentile rank of Bruce = 8.33 percentile
3. 73.81 percentile
5. 76 percentile
7. **a.** Smallest value = 50; largest value = 98

   When arranged in order, the data becomes 50, 55, 59, 65, 68, 69, 73, 76, 78, 80, 80, 80, 80, 83, 84, 85, 85, 86, 90, 90, 93, 95, 96, 97, 98.

   Median or middle quartile = 80

   Lower half of data: 50 55 59 65 68 69 73 76 78 80 80 80

   Median of lower half or lower quartile = $\dfrac{69 + 73}{2} = 71$

   b. Upper half of data: 83 84 85 85 86 90 90 93 95 96 97 98

   Median of upper half or upper quartile = 90

   c.

   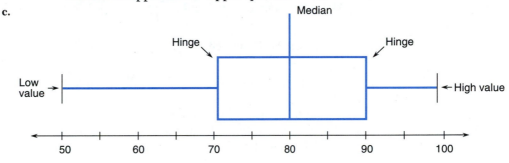

9. **a.** Smallest value = 7; largest value = 21

   Median or middle quartile = $\dfrac{15 + 16}{2} = 15.5$

   Median of lower half or lower quartile = 11

   b. Median of upper half or upper quartile = 18
   c. Interquartile range = 18 − 11 = 7
   d.

   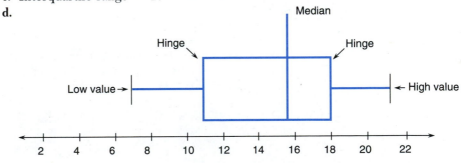

Section 3.8 (pages 156–159)

1. **a.** $z = 1$ **b.** $z = -4.33$ **c.** $z = 0$ **d.** $z = 3.67$
3. **a.** Lake D since it has the highest z-value ($z = 1.54$).
   b. Lake S since it has the lowest z-value ($z = -1.72$).

5. Alan's z-score $= \dfrac{75 - 82}{4} = -1.75$

   Derek's z-score $= \dfrac{120 - 137}{12} = -1.42$

   Since Derek's z-score is better, he has a higher math aptitude.

7. **a.**
   Accounting: $z = \dfrac{49 - 45}{4.7} = 0.85$

   Finance: $z = \dfrac{81 - 80}{8.3} = 0.12$

   Marketing/management: $z = \dfrac{62 - 79}{11.2} = -1.52$

   Real estate/insurance: $z = \dfrac{19 - 20}{3.1} = -0.32$

   Retailing/sales: $z = \dfrac{40 - 41}{2.6} = -0.38$

   b. Accounting since it has the highest z-score.
   c. Marketing/management since it has the lowest z-score.
9. **a.** 84.13%   **b.** $100 - 15.87 = 84.13\%$ **c.** 84.13%

Testing Your Understanding of This Chapter's Concepts (pages 165–166)

1. He may not necessarily be accepted to college. The high school average has nothing to do with percentile rank.
2. Yes. Especially when you have extreme scores.
3. $\Sigma(x - \mu) = \Sigma x - \Sigma \mu$

   $= \Sigma x - n\mu$

   $= \Sigma x - n\left(\dfrac{\Sigma x}{n}\right)$ since $\mu = \dfrac{\Sigma x}{n}$

   $= \Sigma x - \Sigma x = 0$

4. $\sqrt{\dfrac{\Sigma(x-\bar{x})^2}{n-1}} = \sqrt{\dfrac{\Sigma(x^2-2x\bar{x}+\bar{x}^2)}{n-1}}$

$= \sqrt{\dfrac{\Sigma x^2 - \Sigma 2x\bar{x} + \Sigma(\bar{x}^2)}{n-1}}$

$= \sqrt{\dfrac{\Sigma x^2}{n-1} - \dfrac{2\bar{x}\Sigma x}{n-1} + \dfrac{n(\bar{x}^2)}{n-1}}$ Note: $\bar{x} = \dfrac{\Sigma x}{n}$

$= \sqrt{\dfrac{\Sigma x^2}{n-1} - \dfrac{2n(\bar{x}^2)}{n-1} + \dfrac{n(\bar{x}^2)}{n-1}}$

$= \sqrt{\dfrac{\Sigma x^2}{n-1} - \dfrac{n(\bar{x}^2)}{n-1}}$

$= \sqrt{\dfrac{\Sigma x^2}{n-1} - \dfrac{(\Sigma x)(\Sigma x)}{n(n-1)}}$

$= \sqrt{\dfrac{n\Sigma x^2 - (\Sigma x)^2}{n(n-1)}}$

5. b.

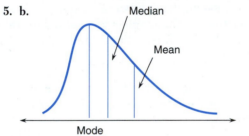

c.

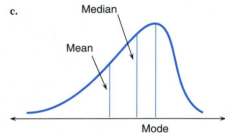

6. You will have 99.7% of the terms of the distribution falling between $z = -3$ and $z = +3$.

Thinking Critically (page 166)

1. a. $\Sigma x = 750$ b. $\Sigma x^2 = 56,890$
2. Using the results of the graphs given in Exercise 9 of Section 3.8, we convert each of Cindy's percentile ranks into z-scores.

	Percentile Rank	Approximate z-score
Accounting	99	+2.33
Finance	72	+0.58
Marketing/management	87	+1.13
Real estate/insurance	50	0
Retailing/sales	33	−0.44

Higher z-scores reflect higher talent: Thus, Cindy has a higher relative talent for accounting, Heather has a higher relative talent for finance, Cindy has a higher relative talent for marketing/management, both girls have the same talent for real estate/insurance, and Cindy has a higher relative talent for retailing/sales.

3. Yes, if all the terms of the distribution are the same.
4. Zero.
5. Not necessarily true. See the data given in Exercise 4 of Section 3.7.
6. False. Positive z-scores will occur for any terms (positive or negative) that are above the mean.
7. Since $\sigma = 0$, all the terms are the same.
8. No. Both teams may have different numbers of members. A weighted average is needed.

Review Exercises (pages 167–168)

1. Probably brand A because of its smaller standard deviation. Others may disagree.
2. a. $\Sigma x = 82$ b. $\Sigma x^2 = 1436$.
3. The standard deviation for the number of windows on Trafalagar Ave. is less than the standard deviation for the number of windows on Billings Lane.
4. No. If the average depth is 3 feet, there can be parts of the pool that are more than 6 feet in depth.
5. $686.36
6. Yes. Although both dating services have the same average age of 20 years, the standard deviation for dating service A is considerably larger.
7. We use the class mark.

$$\text{Sample mean} = 367.5$$

Sample standard deviation ≈ 148.65

8. $P_{55} =$ in the 300–399 category as $12 + 28 + 15 = 55$. Thus $P_{55} = 349.5$ (using the class mark).
9. $z = 1.08$

10. Using the class marks,
    a. $P_{70} = 524.5$ b. $P_{90} = 624.5$
11. 25% as the third quartile is the same as the upper quartile and thus 75% of the scores are below 83.

Chapter Test (pages 168–171)

1. a. The mode is 15.
   b. The median is between 24 and 35 or equals 29.5.
   c. The mean is 40.417.
   d. The lower quartile or P_{25} is the median of the lower half of numbers 9, 13, 15, 15, 15, 24. Thus the lower quartile is 15.
   e. The upper quartile or P_{75} is the median of the upper half of numbers 35, 42, 64, 75, 82, 96. Thus, the upper quartile is 69.5.

2. Midterm z-score $\dfrac{80 - 85}{8} = -0.625$. Final z-score $\dfrac{60 - 68}{12} = -0.666$. When compared to the rest of the class, she did better on the midterm as it has a higher z-score.

3. Using Chebyshev's theorem, at least $1 - \dfrac{1}{(1.5)^2}$ of the terms will lie within 1.5 standard deviation units of the mean, so at least 55.56% of the terms will fall in this category.

4. We use the class mark.
                Sample mean = 21.8062
   Sample standard deviation ≈ 4.8077

5. a. $Q_1 = 3$ b. $Q_3 = 4$ c. $P_{80} = 5$ d. $P_{95} = 6$
6. a. 29
   b. Lower quartile = 23 and upper quartile = 31
   c. Interquartile range = $31 - 23 = 8$

7. We will group the data.

Interval	Frequency	Relative Frequency
28–29	6	0.20
30–31	3	0.10
32–33	4	0.13
34–35	4	0.13
36–37	1	0.03
38–39	4	0.13
40–41	3	0.10
42–43	3	0.10
44–45	2	0.07
	30	≈1.00

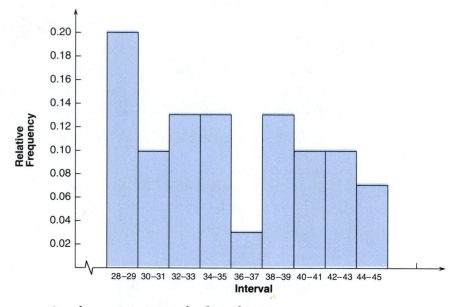

8. $\mu = 48$ and $\sigma = 9.1$. Using the formula $x = \mu + z\sigma$:
Bob: $48 + (3.3)(9.1) = 78.03$ seconds
Mark: $48 + (-0.45)(9.1) = 43.905$ seconds
Chris: $48 + (1.43)(9.1) = 61.013$ seconds

9. Average age $= 30.36$ years

10. $\bar{x} = 19.2$ and $\Sigma(x - \bar{x}) = 0$

11. Maureen's percentile rank $= 71$

12. a. Marketing/sales z-score $= 0.15$
    Plant maintenance z-score $= 0.73$
    Inventory/shipping z-score $= 1.08$
    Personnel z-score $= 0.19$
b. Most talent: Inventory/shipping
c. Least talent: Marketing/sales

13. Sample mean $= 10$
    Sample standard deviation $= \approx 2.898$
14. Choice (a)   15. Choice (a)   16. Choice (d)   17. Choice (c)
18. Choice (c)   19. Choice (d)   20. Choice (b)

CHAPTER 4

Section 4.2 (pages 188–191)

1. a. $\dfrac{2619}{10,672}$ b. $\dfrac{7270}{10,672} = \dfrac{3635}{5336}$ c. 0 d. 1

3. a. $\dfrac{12}{274} = \dfrac{6}{137}$ b. $\dfrac{97}{274}$ c. $\dfrac{60}{274} = \dfrac{30}{137}$ d. $\dfrac{123}{274}$

5. a. $\dfrac{4}{20} = \dfrac{1}{5}$ b. $\dfrac{7}{20}$ c. $\dfrac{9}{20}$ d. $\dfrac{16}{20} = \dfrac{4}{5}$

7. By listing the elements of the sample space, one finds that there are 10 possible combinations of 3 winners from the 5 nominated. Thus the probability that the winners are Jeremy, Mary, and Ahamad is $\dfrac{1}{10}$.

9. $p(2 \text{ boys and 2 girls}) = \dfrac{6}{16} = \dfrac{3}{8}$

11. a. $\dfrac{533}{1000}$ b. $\dfrac{17}{1000}$ c. $\dfrac{525}{1000} = \dfrac{21}{40}$ d. $\dfrac{97}{1000}$

13. a. $\dfrac{192}{656}$ b. $\dfrac{270}{656}$ c. $\dfrac{65}{256}$

15. Probability is $\dfrac{1}{6}$.

17. a. Let L = lemon, C = cherry, and A = apple. The sample space is LLL, LLC, LCL, LCC, LCA, LLA, LAA, LAC, LAL, CLL, CLC, CCL, CCC, CCA, CLA, CAA, CAC, CAL, ALL, ALC, ACL, ACC, ACA, ALA, AAA, AAC, AAL. There are 27 possible outcomes.
    b. $\dfrac{3}{27} = \dfrac{1}{9}$

Section 4.3 (pages 195–198)

1. $3 \times 2 \times 2 = 12$ different ways
3. $4 \times 3 \times 5 = 60$ possible meals
5. a. 120 possible ways   b. 3125 possible ways

7. 1,000,000 possible combinations

9. $3 \cdot 3 \cdot 2 \cdot 1 = 18$ possible numbers. Note: There are only 3 choices for the first digit since the number must be greater than 3000.

11.

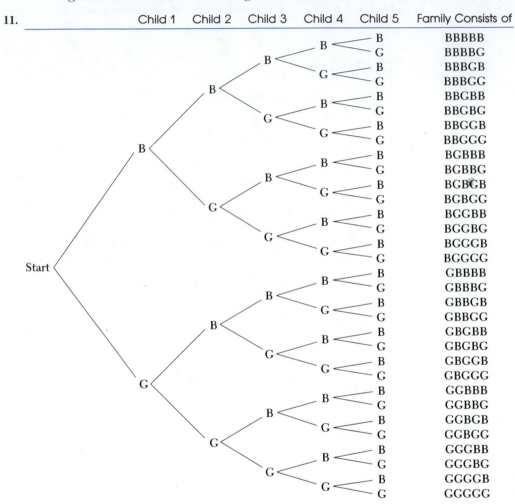

	Child 1	Child 2	Child 3	Child 4	Child 5	Family Consists of

13. a. 12,654,720 possible plates   b. 17,576,000 possible plates

15. a. 5040 possible ways   b. 3600   c. $\dfrac{1440}{5040} = \dfrac{2}{7}$

17.

Slacks	Blouses	Shoes	Outfit Consists of
Black	White print	White	Black slacks, white print blouse, white shoes
		Black	Black slacks, white print blouse, black shoes
	Red print	White	Black slacks, red print blouse, white shoes
		Black	Black slacks, red print blouse, black shoes
	Black	White	Black slacks, black blouse, white shoes
		Black	Black slacks, black blouse, black shoes
White	White print	White	White slacks, white print blouse, white shoes
		Black	White slacks, white print blouse, black shoes
	Red print	White	White slacks, red print blouse, white shoes
		Black	White slacks, red print blouse, black shoes
	Black	White	White slacks, black blouse, white shoes
		Black	White slacks, black blouse, black shoes
Blue	White print	White	Blue slacks, white print blouse, white shoes
		Black	Blue slacks, white print blouse, black shoes
	Red print	White	Blue slacks, red print blouse, white shoes
		Black	Blue slacks, red print blouse, black shoes
	Black	White	Blue slacks, black blouse, white shoes
		Black	Blue slacks, black blouse, black shoes
Red	White print	White	Red slacks, white print blouse, white shoes
		Black	Red slacks, white print blouse, black shoes
	Red print	White	Red slacks, red print blouse, white shoes
		Black	Red slacks, red print blouse, black shoes
	Black	White	Red slacks, black blouse, white shoes
		Black	Red slacks, black blouse, black shoes

Start

Section 4.4 (pages 205–207)

1. **a.** $8! = 40{,}320$ **b.** $9! = 362{,}880$ **c.** $3! = 6$ **d.** $\dfrac{6!}{5!} = 6$

e. $\dfrac{0!}{4} = \dfrac{1}{4}$ **f.** $\dfrac{5!}{3!2!} = 10$ **g.** $\dfrac{6!}{4!2!} = 15$ **h.** $\dfrac{7!}{5!2!} = 21$

i. $_8P_5 = 6720$ **j.** $_7P_4 = 840$ **k.** $_9P_5 = 15{,}120$ **l.** $_0P_0 = 1$

m. $_7P_0 = 1$ **n.** $_7P_7 = 7! = 5040$ **o.** $_6P_4 = 360$ **p.** $_3P_2 = 6$

3. $_8P_5 = 6720$

5. **a.** $_8P_4 = 1680$ **b.** $\dfrac{1}{8}$ **c.** $\dfrac{1}{2}$

7. The letter C is repeated twice.

If the first letter is an E, there are $\dfrac{8!}{2!2!} = 10{,}080$ possible ways.

If the first letter is an O, there are $\dfrac{2 \cdot 8!}{2!} = 40{,}320$ possible ways.

If the first letter is an I, there are $\dfrac{8!}{2!2!} = 10,080$ possible ways.

Thus, there are $10,080 + 40,320 + 10,080 = 60,480$ possible ways.

9. $\dfrac{10!}{2!2!} = 907,200$

11. a. $_{10}P_{10} = 10! = 3,628,800$

    b. The men can be seated in $_5P_5 = 120$ possible ways and the women can also be seated in $_5P_5$ or 120 possible ways. However, the men can be seated on either the left side or right side, so that there are $120 \times 120 \times 2 = 28,800$ possible ways to seat the guests.

    c. The couples can be seated in $_5P_5 = 5!$ ways. However, each man can sit on the left or right side of his wife. Thus, there are $2^5 \cdot 120 = 3840$ possible ways.

13. a. $1 \cdot 10 \cdot 10 \cdot 10 \cdot 10 = 10,000$

    b. $10 \cdot 9 \cdot 8 \cdot 7 \cdot 6 = 30,240$

    c. $10 \cdot 10 \cdot 10 \cdot 10 \cdot 10 = 100,000$

15. a. Assuming there is no "head" of the circle, we can think of anyone as being arbitrarily designated as the "head" and the other $n - 1$ people then can be arranged in $(n - 1)!$ ways.

    b. $5! = 120$

17. $_{10}P_4 = 5040$

19. $_7P_7 = 7! = 5040$ assuming that order counts.

6. 4023737 /5
1. 46 31321 11

Section 4.5 (pages 217–219)

1. a. $_7C_3 = 35$ b. $_6C_5 = 6$ c. $_8C_5 = 56$ d. $_7C_0 = 1$

    e. $_7C_1 = 7$ f. $_8C_4 = 70$ g. $_7C_4 = 35$ h. $\dbinom{9}{5} = {}_9C_5 = 126$

    i. $\dbinom{5}{5} = {}_5C_5 = 1$ j. $\dbinom{8}{9} = {}_8C_9 =$ impossible

3. $_{18}C_8 = 43,758$ 5. $_{10}C_4 = 210$ 7. $_{20}C_5 = 15,504$

9. a. $_8C_3 \cdot {}_9C_3 = 56 \cdot 84 = 4704$

    b. $\dfrac{1 \text{ woman}}{5 \text{ men}} + \dfrac{2 \text{ women}}{4 \text{ men}} + \dfrac{3 \text{ women}}{3 \text{ men}} + \dfrac{4 \text{ women}}{2 \text{ men}} + \dfrac{5 \text{ women}}{1 \text{ man}} + \dfrac{6 \text{ women}}{0 \text{ men}}$

        $_9C_1 \cdot {}_8C_5 + {}_9C_2 \cdot {}_8C_4 + {}_9C_3 \cdot {}_8C_3 + {}_9C_4 \cdot {}_8C_2 + {}_9C_5 \cdot {}_8C_1 + {}_9C_6 \cdot {}_8C_0$

        $= 9 \cdot 56 + 36 \cdot 70 + 84 \cdot 56 + 126 \cdot 28 + 126 \cdot 8 + 84 \cdot 1$

        $= 12,348$

    c. $\dfrac{1 \text{ man}}{5 \text{ women}} + \dfrac{2 \text{ men}}{4 \text{ women}} + \dfrac{3 \text{ men}}{3 \text{ women}} + \dfrac{4 \text{ men}}{2 \text{ women}} + \dfrac{5 \text{ men}}{1 \text{ woman}}$

$$_8C_1 \cdot _9C_5 + _8C_2 \cdot _9C_4 + _8C_3 \cdot _9C_3 + _8C_4 \cdot _9C_2 + _8C_5 \cdot _9C_1$$

$$= 8 \cdot 126 + 28 \cdot 126 + 56 \cdot 84 + 36 \cdot 70 + 56 \cdot 9$$

$$= 12{,}264$$

d. $\dfrac{0 \text{ men}}{6 \text{ women}} + \dfrac{1 \text{ man}}{5 \text{ women}} + \dfrac{2 \text{ men}}{4 \text{ women}}$

$$= _8C_0 \cdot _9C_6 + _8C_1 \cdot _9C_5 + _8C_2 \cdot _9C_4$$

$$= 1 \cdot 84 + 8 \cdot 126 + 28 \cdot 126$$

$$= 4620$$

11. a. $_{11}C_4 = 330$ **b.** $_{11}C_7 = 330$ **c.** They are equal.

13. a. $_8C_1 \cdot _5C_1 \cdot _4C_1 \cdot _2C_1$

$$= 8 \cdot 5 \cdot 4 \cdot 2 = 320$$

13. b.
$\begin{pmatrix} 1 \text{ cardiologist} \\ 1 \text{ surgeon} \\ 1 \text{ anesthesiologist} \\ 1 \text{ obstetrician} \end{pmatrix} + \begin{pmatrix} 1 \text{ cardiologist} \\ 2 \text{ surgeons} \\ 1 \text{ anesthesiologist} \end{pmatrix} +$

$\begin{pmatrix} 1 \text{ cardiologist} \\ 2 \text{ surgeons} \\ 1 \text{ obstetrician} \end{pmatrix} + \begin{pmatrix} 1 \text{ cardiologist} \\ 3 \text{ surgeons} \end{pmatrix} + \begin{pmatrix} 2 \text{ cardiologists} \\ 1 \text{ surgeon} \\ 1 \text{ anesthesiologist} \end{pmatrix} +$

$\begin{pmatrix} 2 \text{ cardiologists} \\ 1 \text{ surgeon} \\ 1 \text{ obstetrician} \end{pmatrix} + \begin{pmatrix} 2 \text{ cardiologists} \\ 2 \text{ surgeons} \end{pmatrix} + \begin{pmatrix} 3 \text{ cardiologists} \\ 1 \text{ surgeon} \end{pmatrix} +$

$\begin{pmatrix} 1 \text{ cardiologist} \\ 1 \text{ surgeon} \\ 2 \text{ anesthesiologists} \end{pmatrix} + \begin{pmatrix} 1 \text{ cardiologist} \\ 1 \text{ surgeon} \\ 2 \text{ obstetricians} \end{pmatrix}$

$$= _8C_1 \cdot _5C_1 \cdot _4C_1 \cdot _2C_1 + _8C_1 \cdot _5C_2 \cdot _4C_1 + _8C_1 \cdot _5C_2 \cdot _2C_1 + _8C_1 \cdot _5C_3$$

$$+ _8C_2 \cdot _5C_1 \cdot _4C_1 + _8C_2 \cdot _5C_1 \cdot _2C_1 + _8C_2 \cdot _5C_2 + _8C_3 \cdot _5C_1$$

$$+ _8C_1 \cdot _5C_1 \cdot _4C_2 + _8C_1 \cdot _5C_1 \cdot _2C_2$$

$$= 2560$$

15. $\dbinom{n}{r} = \dbinom{n}{n-r}$

$$\dfrac{n!}{r!(n-r)!} = \dfrac{n!}{(n-r)![n-(n-r)]!}$$

$$= \dfrac{n!}{(n-r)!r!}$$

$$= \dfrac{n!}{r!(n-r)!}$$

Section 4.6 (pages 224–226)

1. $160,000

3. $2\left(\dfrac{1}{8}\right) + 16\left(\dfrac{3}{8}\right) + (-8)\left(\dfrac{4}{8}\right) = \dfrac{18}{8} = \$2.25$

5. $1\left(\dfrac{9}{60}\right) + 5\left(\dfrac{8}{60}\right) + 10\left(\dfrac{12}{60}\right) + 25\left(\dfrac{16}{60}\right) + 50\left(\dfrac{11}{60}\right) + 100\left(\dfrac{4}{60}\right) = 25.32$ cents

7. 372:628 or 93:157   9. 5:95 or 1:19

11.   Account I: $(3975)(0.61) = \$2424.75$
Account II: $(3300)(0.82) = \$2706.00$
Account III: $(4100)(0.59) = \$2419.00$
Account IV: $(4880)(0.51) = \$2488.80$
Account V: $(5705)(0.46) = \$2624.30$

13. a. $7(0.85) + (-2.5)(0.15) = \$5.575$ million   b. 85:15 or 17:3

Testing Your Understanding of This Chapter's Concepts (pages 228–229)

1.

	Toss 1	Toss 2	Toss 3	Toss 4	Toss 5	Possible Outcome
					H	HHHHH
				H	T	HHHHT
			H		H	HHHTH
				T	T	HHHTT
		H			H	HHTHH
				H	T	HHTHT
			T		H	HHTTH
	H			T	H	HTHHH
			H		T	HTHHT
				T	H	HTHTH
		T				

Start

2. a. There are 1000 possible identification tags. This is not sufficient for 17,225 students.
   b. There are 17,576 possible identification tags. This is sufficient for 17,225 students.

3. a. Assuming no restrictions, $10 \cdot 10 \cdot 10 \cdot 10 \cdot 10 \cdot 10 \cdot 10 \cdot 10 \cdot 10 = 1{,}000{,}000{,}000$ possible numbers
   b. 10,000,000,000 possible numbers
   c. 26,000,000,000 possible numbers

4. 5184 possible cases

Thinking Critically (pages 229–230)

1. No, the events do not have the same probability.
2. Yes, the sample space has been reduced.
3. No, the events do not have the same probability.

Review Exercises (pages 230–231)

1. $\dfrac{3}{5}$ 2. 35

3. $p(\text{at least } 4 \text{ X's}) = p(4 \text{ X's}) + p(5 \text{ X's})$

$$= 5\left(\frac{1}{26}\right)\left(\frac{1}{26}\right)\left(\frac{1}{26}\right)\left(\frac{1}{26}\right)\left(\frac{25}{26}\right) + \left(\frac{1}{26}\right)\left(\frac{1}{26}\right)\left(\frac{1}{26}\right)\left(\frac{1}{26}\right)\left(\frac{1}{26}\right)$$

$$= \frac{126}{11,881,376}$$

4. The sample space consists of 16 possible outcomes. Of these, the outcomes GBBG, GGBG, GBGG, and GGGG are favorable. Thus probability is $\dfrac{4}{16} = \dfrac{1}{4}$.

5. $_8P_8 = 40,320$ 6. $2 \cdot 3 \cdot 2 = 12$ 7. $_7P_3 = 210$

8. $4 \cdot 6 \cdot 2 = 48$ 9. 3:4   10. $\dfrac{1}{2^6} = \dfrac{1}{64}$

11. $4\left(\dfrac{6}{36}\right) + (-1)\left(\dfrac{30}{36}\right) = \dfrac{-6}{36} = \$-0.17$

Chapter Test (pages 231–232)

1. Choice (b)
2. $_7P_7 = 7! = 5040$ assuming order counts.

3. $_6C_4 \cdot {}_5C_3 = 15 \cdot 10 = 150$ 4. $4 \cdot 5 \cdot 4 \cdot 3 = 240$ 5. $\dfrac{10!}{3!} = 604,800$

6. a. $\dfrac{_{12}C_4 \cdot {}_{38}C_0}{_{50}C_4} = \dfrac{495}{230,300}$

   b. $\dfrac{_{12}C_4 \cdot {}_{38}C_0 + {}_{12}C_3 \cdot {}_{38}C_1 + {}_{12}C_2 \cdot {}_{38}C_2}{_{50}C_4} = \dfrac{55,253}{230,300}$

   c. $\dfrac{_{12}C_0 \cdot {}_{38}C_4}{_{50}C_4} = \dfrac{73,815}{230,300}$

7. Four letters followed by three numbers yields
$26 \cdot 26 \cdot 26 \cdot 26 \cdot 10 \cdot 10 \cdot 10 = 456{,}976{,}000$ codes
Three letters followed by four numbers yields
$26 \cdot 26 \cdot 26 \cdot 10 \cdot 10 \cdot 10 \cdot 10 = 175{,}760{,}000$ codes
Use the code of four letters followed by three numbers.

8. $\dfrac{1}{6}$

9.

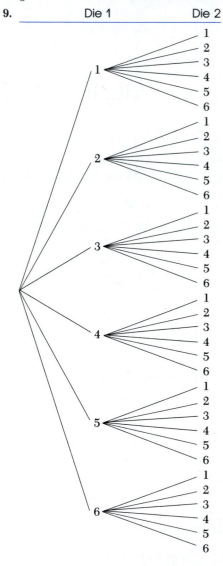

10. $_{11}C_5 = \dfrac{11!}{5!\ 6!} = 462$ assuming order does not count.

11. $_{11}C_6 \cdot {}_{13}C_7 = 462 \cdot 1716 = 792{,}792$

12. $_4P_4 \cdot {}_3P_3 \cdot {}_2P_2 \cdot {}_3P_3 = 24 \cdot 6 \cdot 2 \cdot 6 = 1728$. The groups of soda can be arranged in 4! ways, so that there are $24 \cdot 1728$ or $41{,}472$ possible ways of arranging these cans if the cans of the same brand are to stay together.

13. **a.** 3:3 or 1:1

    b. $9\left(\dfrac{2}{6}\right) + 7\left(\dfrac{1}{6}\right) + (-2)\left(\dfrac{2}{6}\right) + 0\left(\dfrac{1}{6}\right) = \dfrac{21}{6} = \$3.50$

14. 2.3:97.7 or 23:977   15. $_7P_3 = 210$

16. Choice (b)   17. Choice (d)   18. Choice (b)   19. Choice (b)

20. Choice (a)

CHAPTER 5

Section 5.2 (pages 247–250)

1. **a.** Mutually exclusive.   **b.** Not mutually exclusive.

   c. Mutually exclusive.   **d.** Not mutually exclusive.

   e. Not mutually exclusive.   **f.** Not mutually exclusive.

   g. Not mutually exclusive.   **h.** Mutually exclusive.

3. p(mugged or house burglarized) $= 0.41$

5. p(needs remediation or has college-bound brother or sister) $= 0.83$

7. p(has both pet dog and pet cat) $= 0.11$

9. p(both defects) $= 0.08$

11. p(either wrong form used or not filled out completely) $= \dfrac{25}{144}$

13. p(all three courses) $= \dfrac{28}{120}$

Section 5.3 (pages 256–259)

1. **a.** $\dfrac{3}{5}$ **b.** $\dfrac{19}{40}$ **c.** $\dfrac{27}{100}$ **d.** $\dfrac{39}{200}$ **e.** $\dfrac{93}{200}$

3. p(develops reaction | takes medication) $= \dfrac{9}{14}$

5. p(hunter | mountain climber) $= \dfrac{23}{92}$

7. p(accepts Discover Card | accepts Master Card) $= \dfrac{53}{74}$

9. a. $\dfrac{221}{1000}$ b. $\dfrac{221}{461}$ c. $\dfrac{156}{1000}$

11. a. $\dfrac{141}{353}$ b. $\dfrac{188}{611}$ c. $\dfrac{47}{141} = \dfrac{1}{3}$

Section 5.4 (pages 264–266)

1. 0.4806   3. 0.5696   5. 0.4028
7. p(only Janet's application is approved) $= (0.15)(0.92)(0.21) = 0.02898$
9. p(stopped by neither) $= (0.26)(0.01) = 0.0026$
11. 0.3717   13. 0.001232

Section 5.5 (pages 275–277)

1. $\dfrac{95}{158}$ 3. $\dfrac{9}{43}$ 5. $\dfrac{3}{11}$ 7. 0.0167   9. 0.5228

Testing Your Understanding of This Chapter's Concepts (pages 278–279)

1. Two events A and B are mutually exclusive if they both cannot occur at the same time. The events A and B are independent if the likelihood of the occurrence of event B is in no way affected by the occurrence or nonoccurrence of event A.
2. a. Not mutually exclusive.   b. Not mutually exclusive.
3. a. Not independent.   b. Not independent.
4. False. The probability of selecting the ace of spades plus the probability of selecting the queen of spades does not add up to 1.
5. No. Events are not necessarily independent. (One could be present while the other is absent.)
6. No. The number of names listed in the phone directory under each letter is not the same.

Thinking Critically (pages 279–280)

1. No. Probability is $\dfrac{1}{2}$.

2. p(second different from first) $= \dfrac{364}{365}$ (assuming no leap year).

3. p(all different) $= p$(second different from first) $\cdot$ p(third different from first 2 | first and second are different)

$$= \left(\dfrac{364}{365}\right)\left(\dfrac{363}{365}\right) = 0.9918 \quad \text{(assuming no leap year)}$$

4. p(only 1 delivered) = p(machine delivered in morning and carpet in afternoon)
   + p(machine delivered in afternoon and carpet in morning)
   = $(0.57)(0.32) + (0.43)(0.68) = 0.4748$
5. a. p(at least one subscribes)
      = $1 - p$(no one subscribes to closed-circuit TV)
      = $1 - (0.83)^6 = 0.6731$
   b. p(at most one subscribes)
      = p(0 subscribes) + p(1 subscribes)
      = $(0.83)^6 + 6(0.17)(0.83)^5 = 0.7287$
6. No, they are dependent since $p(B \mid A) = 0$ and is not equal to $p(B)$.

Review Exercises (pages 280–282)

1. a. 0.07 b. 0.195
   c. p(at least one succeeds) = 0.805
2. $\frac{1}{5}$ 3. 0.1575 4. 0.35
5. a. $\frac{1}{13}$ b. $\frac{1}{4}$ c. $\frac{4}{40} = \frac{1}{10}$
6. $\frac{3}{15} = \frac{1}{5}$
7. p(Paul contacted and Juliana not contacted) + p(Paul not contacted and Juliana contacted) + p(both contacted) = $(0.75)(0.45) + (0.25)(0.55) + (0.75)(0.55) = 0.8875$
8. p(at least one will default) = $1 - p$(none will default
   $= 1 - (0.6)(0.4)(0.25)$
   $= 0.94$
9. $\left(\frac{2}{9}\right)\left(\frac{1}{8}\right) = \frac{1}{36}$
10. a. $\frac{397}{1600}$ b. $\frac{36}{397}$ c. $\frac{36}{431}$
11. p(all different) = p(second different from first) $\cdot$ p(third different from first 2 | first and second are different)
    $= \left(\frac{364}{365}\right)\left(\frac{363}{365}\right) = 0.9918$ (assuming no leap year)
12. Similar to last problem:
    p(all different) $= \left(\frac{364}{365}\right)\left(\frac{363}{365}\right)\left(\frac{362}{365}\right) = 0.9836$

Chapter Test (pages 282–285)

1. $\dfrac{1}{7}$

2. a. $\dfrac{1722}{2450}$ b. $\dfrac{56}{2450}$ c. $\dfrac{672}{2450}$

3. p (at least two have same birthday)
   $= 1 - p(\text{all birthdays are different})$
   $= 1 - 0.9918$
   $= 0.0082$
   See question 11 in Review Exercises for this chapter.

4. 0.28899 5. 0.11739

6. $p(\text{caused by either}) = p(\text{caused by patient}) + p(\text{caused by doctor}) - p(\text{caused by both}) = \dfrac{183}{1080}$

7. $p(\text{selected nearer machine} \mid \text{wins}) = \dfrac{5}{7}$

8. a. $\dfrac{181}{871}$ b. $\dfrac{59}{181}$ c. $\dfrac{59}{334}$

9. $p(\text{orders either}) = \dfrac{647}{693}$ 10. $p(\text{either}) = 0.56$

11. $p(\text{both same color}) = p(\text{both white}) + p(\text{both red}) + p(\text{both black}) = \dfrac{19}{66}$

12. $p(\text{neither speaks foreign language}) = (0.17)(0.34) = 0.0578$

13. a. $p(B \mid A) = \dfrac{0.030}{0.114} = \dfrac{5}{19}$ b. $p(C \mid A) = \dfrac{0.048}{0.114} = \dfrac{8}{19}$

    c. $p(D \mid A) = \dfrac{0.036}{0.114} = \dfrac{6}{19}$

14. Choice (c)   15. Choice (b)   16. Choice (c)   17. Choice (d)
18. Choice (d)

CHAPTER 6

Section 6.2 (pages 295–298)

1. Any positive number greater than 0.
3. No. The sum of all the probabilities is not 1.

5.

Number of Girls, x	$p(x)$
0	$\dfrac{1}{8}$
1	$\dfrac{3}{8}$
2	$\dfrac{3}{8}$
3	$\dfrac{1}{8}$
	$\dfrac{8}{8} = 1$

7.

Number of 1's, x	$p(x)$
0	$\dfrac{25}{36}$
1	$\dfrac{10}{36}$
2	$\dfrac{1}{36}$
	$\dfrac{36}{36} = 1$

9.

Number of Heads, x	$p(x)$
0	$\dfrac{1}{8}$
1	$\dfrac{3}{8}$
2	$\dfrac{3}{8}$
3	$\dfrac{1}{8}$
	$\dfrac{8}{8} = 1$

11.

Number of Yellow Balls, x	$p(x)$
0	$\left(\dfrac{14}{19}\right)\left(\dfrac{14}{19}\right)\left(\dfrac{14}{19}\right) = \dfrac{2744}{6859}$
1	$3\left(\dfrac{5}{19}\right)\left(\dfrac{14}{19}\right)\left(\dfrac{14}{19}\right) = \dfrac{2940}{6859}$
2	$3\left(\dfrac{5}{19}\right)\left(\dfrac{5}{19}\right)\left(\dfrac{14}{19}\right) = \dfrac{1050}{6859}$
3	$\left(\dfrac{5}{19}\right)\left(\dfrac{5}{19}\right)\left(\dfrac{5}{19}\right) = \dfrac{125}{6859}$
	$\dfrac{6859}{6859} = 1$

13.

Number of Defective Spark Plugs, x	$p(x)$
0	$\left(\frac{6}{8}\right)\left(\frac{6}{8}\right) = \frac{36}{64}$
1	$2\left(\frac{2}{8}\right)\left(\frac{6}{8}\right) = \frac{24}{64}$
2	$\left(\frac{2}{8}\right)\left(\frac{2}{8}\right) = \frac{4}{64}$
	$\frac{64}{64} = 1$

15.

x	$p(x)$
0	$\frac{5}{9}$
1	$\frac{4}{9}$
2	$\frac{3}{9}$
3	$\frac{2}{9}$
4	$\frac{1}{9}$
	$\frac{15}{9}$

No, as the sum of the probabilities is not 1.

17. **a.** 0.106   **b.** 0.360   **c.** 1   **d.** 0.699

Section 6.4 (pages 306–310)

1. $\mu = 3.8$

3. Mean $= \mu = 3.90$
   Variance $= \sigma^2 = 4.29$
   Standard deviation $= \sigma = \sqrt{4.29} \approx 2.0712$

5. Mean $= \mu = 3.76$
   Variance $= \sigma^2 = 3.5224$
   Standard deviation $= \sigma = \sqrt{3.5224} \approx 1.8768$

7. Mean $= \mu = 3.09$
   Variance $= \sigma^2 = 12.71 - (3.09)^2 = 3.1619$
   Standard deviation $= \sqrt{3.1619} \approx 1.7782$

9. a.

x	$p(x)$
2	$\frac{1}{36}$
3	$\frac{2}{36}$
4	$\frac{3}{36}$
5	$\frac{4}{36}$
6	$\frac{5}{36}$
7	$\frac{6}{36}$
8	$\frac{5}{36}$
9	$\frac{4}{36}$
10	$\frac{3}{36}$
11	$\frac{2}{36}$
12	$\frac{1}{36}$

b. Mean $= \mu = 7$ **c.** Variance $= \sigma^2 \approx 5.8333$

11. a. $\mu = 3.25$
   b. $\sigma^2 = 1.7275$
   $\sigma = \sqrt{1.7275} \approx 1.3143$

13. a. $\mu = 4.29$
   b. $\sigma^2 = 4.1059$
   $\sigma = \sqrt{4.1059} \approx 2.0263$

Section 6.5 (pages 324–327)

1. $p(\text{exactly 4 work}) = \dfrac{7!}{4!\ 3!}\ (0.70)^4(0.30)^3 = 0.2269$

3. $p(\text{at most 2 women}) = p(0\ \text{women}) + p(1\ \text{woman}) + p(2\ \text{women})$

$$= \frac{10!}{0!\ 10!}\ (0.125)^0(0.875)^{10}$$

$$+ \frac{10!}{1!\ 9!}\ (0.125)^1(0.875)^9 + \frac{10!}{2!\ 8!}\ (0.125)^2(0.875)^8$$

$$= 0.8805$$

5. a. p(at most half will be audited)

$$= p(0) + p(1) + p(2) + p(3)$$

$$= \frac{6!}{0!\,6!}\,(0.08)^0(0.92)^6 + \frac{6!}{1!\,5!}\,(0.08)^1(0.92)^5$$

$$+ \frac{6!}{2!\,4!}\,(0.08)^2(0.92)^4 + \frac{6!}{3!\,3!}\,(0.08)^3(0.92)^3$$

$$= 0.9996$$

b. $p(0) = \dfrac{6!}{0!\,6!}\,(0.03)^0(0.97)^6 = 0.8330$

7. $\dfrac{9!}{3!\,6!}\,(0.28)^3(0.72)^6 = 0.2569$

9. p(at least 4) $= 1 - p(0) - p(1) - p(2) - p(3) = 0.013$

11. $\dfrac{6!}{4!\,2!}\left(\dfrac{3}{5}\right)^4\left(\dfrac{2}{5}\right)^2 = 0.3110$

13. a. $\dfrac{9!}{9!\,0!}\,(0.93)^9(0.07)^0 = 0.5204$ b. $\dfrac{9!}{5!\,4!}\,(0.93)^5(0.07)^4 = 0.0021$

15. a. $\dfrac{8!}{2!6!}(0.31)^2(0.69)^6 = 0.2904$

b. p(at most 2) $= p(0) + p(1) + p(2)$

$$= \frac{8!}{0!\,8!}\,(0.31)^0(0.69)^8 + \frac{8!}{1!\,7!}\,(0.31)^1(0.69)^7 + \frac{8!}{2!\,6!}\,(0.31)^2(0.69)^6$$

$$= 0.5265$$

c. p(at least 2) $= 1 - p(0) - p(1) = 0.7639$

Section 6.6 (pages 330–331)

1. $\mu = 228;\ \sigma \approx 11.889$ 3. $\mu = 173{,}776.5;\ \sigma \approx 377.487$
5. $\mu = 330;\ \sigma \approx 10.488$ 7. $\mu = 1700;\ \sigma \approx 15.9687$
9. $\mu = 768.5;\ \sigma \approx 19.005$ 11. $\mu = 1700;\ \sigma \approx 15.9687$
13. $\mu = 240(0.88) = 211.2$. Since the airline only has 210 available seats, if the average or more people show up it will not have a seat for everyone.

Section 6.7 (pages 334–336)

1. a. p(at most 2 customers) $= p(0) + p(1) + p(2)$

$$= \frac{e^{-5}5^0}{0!} + \frac{e^{-5}5^1}{1!} + \frac{e^{-5}5^2}{2!}$$

$$= 0.12465$$

b. p(at least 2 customers) $= 1 - p(0) - p(1)$
$$= 0.95957$$

c. p(exactly 2 customers) $= \dfrac{e^{-5}5^2}{2!} = 0.08422$

3. p(less than or equal to 4) $= p(0) + p(1) + p(2) + p(3) + p(4)$
$$= \frac{e^{-7}7^0}{0!} + \frac{e^{-7}7^1}{1!} + \frac{e^{-7}7^2}{2!} + \frac{e^{-7}7^3}{3!} + \frac{e^{-7}7^4}{4!}$$
$$= 0.1729$$

5. **a.** p(at least 3) $= 1 - p(0) - p(1) - p(2)$
$$= 1 - \frac{e^{-8}8^0}{0!} - \frac{e^{-8}8^1}{1!} - \frac{e^{-8}8^2}{2!}$$
$$= 0.98625$$

b. p(at most 3) $= p(0) + p(1) + p(2) + p(3)$
$$= 0.04238$$

c. p(exactly 3) $= \dfrac{e^{-8}8^3}{3!} = 0.028626$

d. p(between two and four) $= 0.028626$

7. $\dfrac{e^{-4}4^4}{4!} = 0.1954$

9. **a.** p(at least two) $= 1 - p(0) - p(1)$
$$= 1 - \frac{e^{-4}4^0}{0!} - \frac{e^{-4}4^1}{1!} = 0.90842$$

b. p(no errors) $= \dfrac{e^{-4}4^0}{0!} = 0.018316$

11. p(at most 3) $= p(0) + p(1) + p(2) + p(3)$
$$= \frac{e^{-5}5^0}{0!} + \frac{e^{-5}5^1}{1!} + \frac{e^{-5}5^2}{2!} + \frac{e^{-5}5^3}{3!}$$
$$= 0.26502$$

Section 6.8 (pages 339–340)

1. **a.** $\dfrac{\dbinom{10}{2}\dbinom{90}{8}}{\dbinom{100}{10}} = 0.2015$

b. $\dfrac{\dbinom{10}{0}\dbinom{90}{10} + \dbinom{10}{1}\dbinom{90}{9} + \dbinom{10}{2}\dbinom{90}{8}}{\dbinom{100}{10}} = 0.7429$

3. $\dfrac{\dbinom{48}{0}\dbinom{21}{3} + \dbinom{48}{1}\dbinom{21}{2} + \dbinom{48}{2}\dbinom{21}{1} + \dbinom{48}{3}\dbinom{21}{0}}{\dbinom{69}{6}} = 0.00044$

5. $\dfrac{\dbinom{5}{0}\dbinom{7}{3}}{\dbinom{12}{3}} = \dfrac{35}{220} = \dfrac{7}{44}$

7. a. $\dfrac{\dbinom{9}{0}\dbinom{6}{4}}{\dbinom{15}{4}} = \dfrac{15}{1365}$ b. $\dfrac{\dbinom{9}{1}\dbinom{6}{3}}{\dbinom{15}{4}} = \dfrac{180}{1365}$

c. $\dfrac{\dbinom{9}{2}\dbinom{6}{2}}{\dbinom{15}{4}} = \dfrac{540}{1365}$ d. $\dfrac{\dbinom{9}{3}\dbinom{6}{1}}{\dbinom{15}{4}} = \dfrac{504}{1365}$

e. $\dfrac{\dbinom{9}{4}\dbinom{6}{0}}{\dbinom{15}{4}} = \dfrac{126}{1365}$

9. $\dfrac{\dbinom{18}{0}\dbinom{9}{3}}{\dbinom{27}{3}} = \dfrac{84}{2925}$

Testing Your Understanding of This Chapter's Concepts (pages 345–346)

1. a.

x	p(x)	x · p(x)	x²	x² · p(x)
4	$\dfrac{4}{36}$	$\dfrac{16}{36}$	16	$\dfrac{64}{36}$
5	$\dfrac{4}{36}$	$\dfrac{20}{36}$	25	$\dfrac{100}{36}$
6	$\dfrac{5}{36}$	$\dfrac{30}{36}$	36	$\dfrac{180}{36}$
7	$\dfrac{6}{36}$	$\dfrac{42}{36}$	49	$\dfrac{294}{36}$
8	$\dfrac{7}{36}$	$\dfrac{56}{36}$	64	$\dfrac{448}{36}$
9	$\dfrac{4}{36}$	$\dfrac{36}{36}$	81	$\dfrac{324}{36}$
10	$\dfrac{3}{36}$	$\dfrac{30}{36}$	100	$\dfrac{300}{36}$
11	$\dfrac{2}{36}$	$\dfrac{22}{36}$	121	$\dfrac{242}{36}$
12	$\dfrac{1}{36}$	$\dfrac{12}{36}$	144	$\dfrac{144}{36}$
		$\dfrac{264}{36}$		$\dfrac{2096}{36}$

b. Mean = $\mu = \dfrac{22}{3}$

Variance = $\sigma^2 \approx 4.4444$

Standard deviation = $\sigma = \sqrt{4.4444} \approx 2.108$

2. No.

3. $\dfrac{6!}{0!\,6!}(0.70)^0(0.30)^6 = 0.000729$

4. a. $p(x) = \dfrac{\dbinom{6}{x}\dbinom{3}{5-x}}{\dbinom{9}{5}}$

x	$p(x)$	$x \cdot p(x)$	x^2	$x^2 \cdot p(x)$
2	$\dfrac{15}{126}$	$\dfrac{30}{126}$	4	$\dfrac{60}{126}$
3	$\dfrac{60}{126}$	$\dfrac{180}{126}$	9	$\dfrac{540}{126}$
4	$\dfrac{45}{126}$	$\dfrac{180}{126}$	16	$\dfrac{720}{126}$
5	$\dfrac{6}{126}$	$\dfrac{30}{126}$	25	$\dfrac{150}{126}$
		$\dfrac{420}{126}$		$\dfrac{1470}{126}$

b. Mean $\mu = 3.3333$
Variance $= \sigma^2 \approx 0.5556$
Standard deviation $= \sigma = \sqrt{0.5556} \approx 0.7454$

c. Using Chebyshev's Theorem, the probability that x will fall within $k = 2$ standard deviations of the mean is at least $1 - \dfrac{1}{2^2} = 0.75$.

5.

	Outcome 1	Outcome 2	Possible Outcomes (number of branches)
Start	Success	Success	Success Success
		Failure	Success Failure
	Failure	Success	Failure Success
		Failure	Failure Failure

Thinking Critically (pages 346–347)

1.

x	$p(x)$	$x \cdot p(x)$	x^2	$x^2 \cdot p(x)$
1	$\dfrac{7}{28}$	$\dfrac{7}{28}$	1	$\dfrac{7}{28}$
2	$\dfrac{6}{28}$	$\dfrac{12}{28}$	4	$\dfrac{24}{28}$
3	$\dfrac{5}{28}$	$\dfrac{15}{28}$	9	$\dfrac{45}{28}$
4	$\dfrac{4}{28}$	$\dfrac{16}{28}$	16	$\dfrac{64}{28}$
5	$\dfrac{3}{28}$	$\dfrac{15}{28}$	25	$\dfrac{75}{28}$
6	$\dfrac{2}{28}$	$\dfrac{12}{28}$	36	$\dfrac{72}{28}$
7	$\dfrac{1}{28}$	$\dfrac{7}{28}$	49	$\dfrac{49}{28}$
		$\dfrac{84}{28} = 3$		$\dfrac{336}{28} = 12$

Mean $= \mu = 3$
Variance $= \sigma^2 = 3$
Standard deviation $= \sigma = \sqrt{3} \approx 1.7321$

2. a.

x	$p(x)$	$x \cdot p(x)$	x^2	$x^2 \cdot p(x)$
1	0.05	0.05	1	0.05
2	0.43	0.86	4	1.72
3	0.17	0.51	9	1.53
4	0.25	1.00	16	4.00
5	0.06	0.30	25	1.50
6	0.03	0.18	36	1.08
7	0.01	0.07	49	0.49
		2.97		10.37

Mean $= \mu = 2.97$
Variance $= \sigma^2 = 1.5491$
Standard deviation $= \sigma = \sqrt{1.5491} \approx 1.2446$

b. Using Chebyshev's Theorem, the probability that x will fall within $k = 2$ standard deviations of the mean is at least $1 - \dfrac{1}{2^2} = 0.75$.

3. $\sigma^2 = \Sigma(x - \mu)^2 \cdot p(x)$
$\quad = \Sigma(x^2 - 2\mu x + \mu^2) \cdot p(x)$
$\quad = \Sigma x^2 \cdot p(x) - 2\mu\Sigma x \cdot p(x) + \mu^2\Sigma p(x)$ Note: $\mu = \Sigma x \cdot p(x)$
$\quad = \Sigma x^2 \cdot p(x) - 2\mu\mu + \mu^2$ and $\Sigma p(x) = 1$
$\quad = \Sigma x^2 \cdot p(x) - \mu^2$
$\quad = \Sigma x^2 \cdot p(x) - [\Sigma x \cdot p(x)]^2$

4. $(p + q)^3 = p^3 + 3p^2q + 3pq^2 + q^3$

The first term p^3 is the probability of successes on all three trials when a binomial experiment is performed three times. The second term $3p^2q$ is the probability of two successes in the three trials. The third term $3pq^2$ is the probability of one success in the three trials. The last term q^3 is the probability of no successes in the three trials.

5. a.

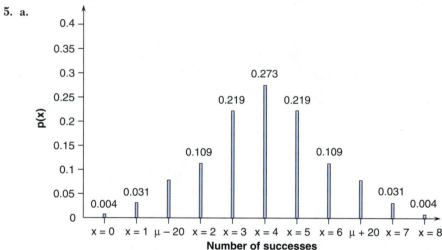

b. $\mu = np = 8\left(\dfrac{1}{2}\right) = 4; \sigma = \sqrt{npq} = \sqrt{8\left(\dfrac{1}{2}\right)\left(\dfrac{1}{2}\right)} = \sqrt{2} \approx 1.4141$

Review Exercises for Chapter 6 (pages 347–348)

1. 0.03 since the sum of all the probabilities must be 1.

2. a. p(at least 5) $= p(5) + p(6) + p(7) + p(8) + p(9)$
Using Table III in the Appendix, we get
$\quad = 0.066 + 0.176 + 0.302 + 0.302 + 0.134$
$\quad = 0.98$

b. p(at most 3) $= p(0) + p(1) + p(2) + p(3) = 0.003$

3. $\mu = 92; \sigma = \approx 8.4167$

4. $\displaystyle\sum_{x=0}^{8} \frac{e^{-5} 5^x}{x!} \geq 0.90$ By trial and error, the owner should have eight or more on hand.

5. Mean $= 4.1$

Standard deviation $= \approx 0.9434$

6. $p(\text{at most } 1) = p(0) + p(1)$

$$= \frac{8!}{0!8!}(0.95)^8(0.05)^0 + \frac{8!}{1!7!}(0.95)^7(0.05)^1$$

$$= 0.9427$$

7. $p(\text{at most } 3) = p(0) + p(1) + p(2) + p(3)$

$$= \frac{e^{-4}4^0}{0!} + \frac{e^{-4}4^1}{1!} + \frac{e^{-4}4^2}{2!} + \frac{e^{-4}4^3}{3!}$$

$$= 0.4335$$

8. $\mu = 281.6$; $\sigma = \approx 5.8131$

9. $310 = n(0.88)$

$352.27 = n$

Accept 353 reservations

10. $\dfrac{4!}{0!4!}\left(\dfrac{15}{25}\right)^0\left(\dfrac{10}{25}\right)^4 = 0.0256$

11. $p(\text{at most } 3) = p(0) + p(1) + p(2) + p(3)$

$$= \frac{12!}{0!12!}(0.20)^0(0.80)^{12} + \frac{12!}{1!11!}(0.20)^1(0.80)^{11}$$

$$+ \frac{12!}{2!10!}(0.20)^2(0.80)^{10} + \frac{12!}{3!9!}(0.20)^3(0.80)^9$$

$$= 0.7946$$

Chapter Test (pages 348–350)

1.

x	$p(x)$
0	$\left(\dfrac{3}{8}\right)\left(\dfrac{3}{8}\right) = \dfrac{9}{64}$
1	$2\left(\dfrac{5}{8}\right)\left(\dfrac{3}{8}\right) = \dfrac{30}{64}$
2	$\left(\dfrac{5}{8}\right)\left(\dfrac{5}{8}\right) = \dfrac{25}{64}$

2. a. $p(\text{at most } 4) = p(0) + p(1) + p(2) + p(3) + p(4)$

$$= \frac{e^{-6}6^0}{0!} + \frac{e^{-6}6^1}{1!} + \frac{e^{-6}6^2}{2!} + \frac{e^{-6}6^3}{3!} + \frac{e^{-6}6^4}{4!}$$

$$= 0.2851$$

b. $p(\text{at least } 2) = 1 - p(0) - p(1) = 0.9826$

3. $p(\text{at least } 1) = 1 - p(0)$

$$= 1 - \frac{10!}{0!10!}(0.004)^0(0.996)^{10} = 0.0393$$

4. a.

x	$p(x)$
0	$\left(\dfrac{40}{52}\right)\left(\dfrac{40}{52}\right)\left(\dfrac{40}{52}\right) = \dfrac{64{,}000}{140{,}608}$
1	$3\left(\dfrac{40}{52}\right)\left(\dfrac{40}{52}\right)\left(\dfrac{12}{52}\right) = \dfrac{57{,}600}{140{,}608}$
2	$3\left(\dfrac{40}{52}\right)\left(\dfrac{12}{52}\right)\left(\dfrac{12}{52}\right) = \dfrac{17{,}280}{140{,}608}$
3	$\left(\dfrac{12}{52}\right)\left(\dfrac{12}{52}\right)\left(\dfrac{12}{52}\right) = \dfrac{1728}{140{,}608}$

b.

x	$p(x)$
0	$\left(\dfrac{40}{52}\right)\left(\dfrac{39}{51}\right)\left(\dfrac{38}{50}\right) = \dfrac{59{,}280}{132{,}600}$
1	$3\left(\dfrac{40}{52}\right)\left(\dfrac{39}{51}\right)\left(\dfrac{12}{50}\right) = \dfrac{56{,}160}{132{,}600}$
2	$3\left(\dfrac{40}{52}\right)\left(\dfrac{12}{51}\right)\left(\dfrac{12}{50}\right) = \dfrac{15{,}840}{132{,}600}$
3	$\left(\dfrac{12}{52}\right)\left(\dfrac{11}{51}\right)\left(\dfrac{10}{50}\right) = \dfrac{1320}{132{,}600}$

5. $\mu = 340$; $\sigma \approx 7.1414$

6. $p(\text{more than } 5) = 1 - p(0) - p(1) - p(2) - p(3) - p(4) - p(5) = 0.0014$

7. a. $\mu = 4.501$ b. $\sigma = \approx 1.0178$

c. Using Chebyshev's Theorem, the probability that x will fall within $k = 2$ standard deviations of the mean is at least $1 - \dfrac{1}{2^2} = 0.75$

8. a. $\dfrac{\dbinom{8}{5}\dbinom{7}{0}}{\dbinom{15}{5}} = \dfrac{56}{3003}$ b. $\dfrac{\dbinom{8}{0}\dbinom{7}{5}}{\dbinom{15}{5}} = \dfrac{21}{3003}$

c. $\dfrac{\dbinom{8}{3}\dbinom{7}{2} + \dbinom{8}{4}\dbinom{7}{1} + \dbinom{8}{5}\dbinom{7}{0}}{\dbinom{15}{5}} = \dfrac{1722}{3003}$

9. $p(\text{at most } 2) = p(0) + p(1) + p(2) = 0.1672$

10. $\dfrac{\binom{11}{0}\binom{34}{8} + \binom{11}{1}\binom{34}{7} + \binom{11}{2}\binom{34}{6} + \binom{11}{3}\binom{34}{5}}{\binom{45}{8}} = 0.9149$

11. 0.238   since the sum of all the probabilities must be 1.
12. Choice (c)   **13.** Choice (d)   **14.** Choice (c)   **15.** Choice (b)
16. Choice (d)

CHAPTER 7

Section 7.3 (pages 370–372)

1. a. 0.4292   **b.** 0.2123   **c.** 0.0268   **d.** 0.2177   **e.** 0.0104
   f. 0.8951   **g.** 0.9604   **h.** 0.1303
3. a. $z = -1.04$ **b.** $z = -0.84$ **c.** $z = -0.67$ **d.** $z = -0.52$
5. a. $z = 2.63$ **b.** $z = 2.17$ **c.** $z = -2.15$
   d. $z = 3.05-$ or $z = 3.06$ or $z = 3.07$ **e.** $z = 2.57$
   f. $z = 2.58$
7. a. 0.99 percentile   **b.** 95.25 percentile   **c.** 0.38 percentile
9. We can use either $z = -1.64$ or $z = -1.65$. Using $z = -1.64$, we have
   $62 = 70 + (-1.64)\sigma$ and $\sigma = 4.878$. Using $z = -1.65$, we have
   $62 = 70 + (-1.65)\sigma$ and $\sigma = 4.848$.
11. $\left.\begin{array}{l} 9.63 = \mu - 1.82\sigma \\ 12.19 = \mu + 1.02\sigma \end{array}\right\}$ Solving simultaneously gives $\mu = 11.2706$ and $\sigma = 0.9014$.
13. a. 0.0702   **b.** 0.0138   **c.** 0.0474   **d.** 0.3386   **e.** 0.0906
   f. 0.1648

Section 7.4 (pages 376–378)

1. 0.5493   **3.** 0.0968   **5.** 0.2420   **7.** 0.7823
9. a. 0.2033   **b.** 51.92   **c.** 78.92
11. 0.6369   **13.** 0.9066   **15.** $\sigma = \dfrac{1}{1.04} = 0.9615$
17. a. $\mu = 7.4935$ **b.** $\mu = 7.0065$

Section 7.5 (pages 385–388)

1. 0.6950   **3.** 0.7123   **5.** 0.0808   **7.** 0.1610   **9.** 0.0170   **11.** 0.1788
13. a. 0.4325   **b.** 0.1350
15. 0.5517   **17.** 0.0668   **19.** 0.0179

Testing Your Understanding of This Chapter's Concepts (page 392)

1. One with a standard deviation of 5.

2. a.

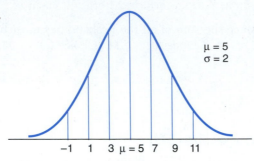

$\mu = 5$
$\sigma = 2$

$-1 \quad 1 \quad 3 \; \mu = 5 \; 7 \quad 9 \quad 11$

b.

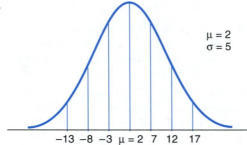

$\mu = 2$
$\sigma = 5$

$-13 \; -8 \; -3 \; \mu = 2 \; 7 \quad 12 \quad 17$

c.

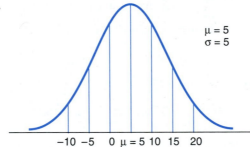

$\mu = 5$
$\sigma = 5$

$-10 \; -5 \quad 0 \; \mu = 5 \; 10 \quad 15 \quad 20$

3. $\mu = 109.68$

4. Not really, as $np = 8\left(\dfrac{1}{2}\right)$ is not greater than 5.

5. 30.2

Thinking Critically (pages 392–393)

1. $\left.\begin{array}{l} 402 = \mu - 1.04\sigma \\ 523 = \mu + 1.14\sigma \end{array}\right\}$ Solving simultaneously gives $\mu = 459.7249$ and $\sigma = 55.5046$.

2. $k \geq 18.2574$ **3.** $0.6687 = \sigma$

4. No. Different values of σ will make the curve wider or narrower. μ locates the center of the distribution.

5. It is $1\dfrac{1}{3}$ standard deviations away from the mean of x.

Review Exercises (pages 393–395)

1. Choice (b)   **2.** 0.9306   **3.** 0.0516   **4.** 0.5557   **5.** 0.8849
6. 0.1820   **7.** 0.1379
8. $\left.\begin{array}{l} 15.948 = \mu + 1.40\sigma \\ 15.2152 = \mu - 0.89\sigma \end{array}\right\}$ Solving simultaneously gives $\sigma = 0.32$ and $\mu = 15.5$.
9. a. 0.1292   **b.** 0.8849
10. 0.1271   **11.** After 1080.15 hours.   **12.** Allow 254.65 minutes.
13. When production exceeds 2246.75 sweaters.   **14.** 0.0329
15. a. 0.3936   **b.** 0.3228   **c.** 0.0951
16. 0.0475

Chapter Test (pages 395–397)

1. $z = -0.77$ **2.** $z = -1.08$ **3.** 65.688 inches   **4.** 85.84   **5.** 0.0478
6. a. 0.0526   **b.** 0.1977   **c.** 0.2451   **d.** 0.0638
7. 0.5670   **8.** 0.3557   **9.** $\mu = 123.09$ **10.** 0.5762   **11.** 0.9573
12. $21.875 = \sigma$ **13.** 0.4052
14. a. $4.54285 = \mu$ **b.** $3.70715 = \mu$
15. $4.0988 = \mu$ **16.** Choice (a)   **17.** Choice (a)   **18.** Choice (c)
19. Choice (b)   **20.** Choice (a)

CHAPTER 8

Section 8.2 (pages 405–406)

1. Pages 104, 223, 241, 421, 375, 289, 94, 103, 71, 510, 23, 10, 521, 70, 486, 541, 326, 293, 24, 296, 7, 53, 5, 259, and 97.
3. Those officers whose numbers are 191, 196, 749, 69, 818, 210, 849, 449, 114, 855, 271, 202, 744, 639, 19, 177, 195, 616, 152, 641, 76, 866, 571, 645, 665, 424, 766, 658, 801, 542, 378, 92, 835, 533, and 212.
5. Those refrigerators whose serial numbers are 36207, 34095, 32081, 57004, 60672, 48840, 60045, 31595, 53900, 65255, 64350, 46104, 41135, 67658, 66134, 64568, 42607, 59920, 69774, 41688, 48413, 49518, 45585, 70002, 69352, 48223, 52666, 30680, 63213, 58678, 70960, 64835, 51132, 30421, 32586, 64760, 43808, 33611, 34952, and 56942.

7. Those whose numbers are 015, 255, 225, 062, 110, 054, 333, 463, 337, 224, 055, 288, 048, 390, 256, 298, 279, and 188.

9. Those containers whose numbers are 4934, 4968, 4401, 2533, 0815, 5218, 3001, 0151, 4944, 0118, 2349, 5185, 3580, 4655, 5000, 4576, 4080, 0599, 1805, and 2816.

Section 8.5 (pages 418–419)

1. $\mu_{\bar{x}} = 7.5$
   $\sigma_{\bar{x}} = \sqrt{1.166667} \approx 1.0801$

3. $\mu_{\bar{x}} = 44.17$
   $\sigma_{\bar{x}} = \sqrt{5.161075} \approx 2.2718$

5. a. $\mu = 15$
   $\sigma = \sqrt{24.4} \approx 4.9396$

   b. Size 2

Number of Arrests, X	Sample Mean, $\bar{X}$	$\bar{X} - \mu_{\bar{x}}$	$(\bar{X} - \mu_{\bar{x}})^2$
9 and 18	13.5	-1.5	2.25
9 and 13	11	-4	16.00
9 and 12	10.5	-4.5	20.25
9 and 23	16	1	1.00
18 and 13	15.5	0.5	0.25
18 and 12	15	0	0.00
18 and 23	20.5	5.5	30.25
13 and 12	12.5	-2.5	6.25
13 and 23	18	3	9.00
12 and 23	17.5	2.5	6.25
	150		91.50

   Size 3

Number of Arrests, X	Sample Mean, $\bar{X}$	$\bar{X} - \mu_{\bar{x}}$	$(\bar{X} - \mu_{\bar{x}})^2$
9, 18, and 13	13.333	-1.6667	2.7779
9, 18, and 12	13	-2	4.0000
9, 18, and 23	16.6667	1.6667	2.7779
9, 13, and 12	11.3333	-3.6667	13.4447
9, 13, and 23	15	0	0.0000
9, 12, and 23	14.6667	-0.3333	0.1111
18, 13, and 12	14.3333	-0.6667	0.4445
18, 13, and 23	18	3	9.0000
18, 12, and 23	17.6667	2.6667	7.1113
13, 12, and 23	16	1	1.0000
	150		40.6674

c.

Size 2	Size 3

$$\mu_{\bar{x}} = \frac{150}{10} = 15 \qquad\qquad \mu_{\bar{x}} = \frac{150}{10} = 15$$

$$\sigma_{\bar{x}} = \sqrt{\frac{91.50}{10}} = \sqrt{9.15} \approx 3.0249 \quad \sigma_{\bar{x}} = \sqrt{\frac{40.6674}{10}} = \sqrt{4.06674} \approx 2.0166$$

d. We use $\sigma_{\bar{x}} = \dfrac{\sigma}{\sqrt{n}} \sqrt{\dfrac{N-n}{N-1}}$. In both cases, $N = 5$.

$$\sigma_{\bar{x}} = \frac{4.9396}{\sqrt{2}} \sqrt{\frac{5-2}{5-1}} \qquad \sigma_{\bar{x}} = \frac{4.9396}{\sqrt{3}} \sqrt{\frac{5-3}{5-1}}$$

$$\approx 3.0249 \qquad\qquad\qquad \approx 2.0166$$

7. a. $\mu_{\bar{x}} = 6428$

$$\sigma_{\bar{x}} = \frac{461}{\sqrt{144}} \approx 38.4167$$

b. $\mu_{\bar{x}} = 6428$

$$\sigma_{\bar{x}} = \frac{461}{\sqrt{64}} \approx 57.625$$

Section 8.7 (pages 424–426)

1. 0.9921   **3.** 0.0110   **5.** 0.0351
7. Between 4.9371 and 5.0629 ounces.
9. 0.9249   **11.** 0.9162

Testing Your Understanding of This Chapter's Concepts (page 430)

1. The relationship is expressed in the formula $\sigma_{\bar{x}} = \dfrac{\sigma}{\sqrt{n}}$ when the sample size is less than 5% of the population size.

2. Yes

3. a. $\mu = \dfrac{70}{5} = 14$

$$\sigma = \sqrt{\frac{122}{5}} = \sqrt{24.4} \approx 4.9396$$

b. Size 2

Number of Stores, x	Sample Mean, $\bar{x}$	$\bar{x} - \mu_{\bar{x}}$	$(\bar{x} - \mu_{\bar{x}})^2$
17 and 12	14.5	0.5	0.25
17 and 22	19.5	5.5	30.25
17 and 8	12.5	-1.5	2.25
17 and 11	14	0	0.00
12 and 22	17	3	9.00
12 and 8	10	-4	16.00
12 and 11	11.5	-2.5	6.25
22 and 8	15	1	1.00
22 and 11	16.5	2.5	6.25
8 and 11	9.5	-4.5	20.25
	140		91.50

Size 3

Number of Stores, x	Sample Mean, $\bar{x}$	$\bar{x} - \mu_{\bar{x}}$	$(\bar{x} - \mu_{\bar{x}})^2$
17, 12, and 22	17	3	9.0000
17, 12, and 8	12.3333	-1.6667	2.7779
17, 12, and 11	13.3333	-0.6667	0.4449
12, 22, and 8	14	0	0.0000
12, 22, and 11	15	1	1.0000
12, 8, and 11	10.3333	-3.6667	13.4447
17, 22, and 8	15.6667	1.6667	2.7779
17, 22, and 11	16.6667	2.6667	7.1113
22, 8, and 11	13.6667	-0.3333	0.1111
17, 8, and 11	12	2	4.0000
	140		40.6678

c. Size 2 $\quad \mu_{\bar{x}} = \dfrac{140}{10} = 14$

$$\sigma_{\bar{x}} = \sqrt{\dfrac{91.5}{10}} = \sqrt{9.15} \approx 3.0249$$

Size 3 $\quad \mu_{\bar{x}} = \dfrac{140}{10} = 14$

$$\sigma_{\bar{x}} = \sqrt{\dfrac{40.6678}{10}} = \sqrt{4.06678} \approx 2.0166$$

d.

Size 2

$$\sigma_x = \dfrac{4.9396}{\sqrt{2}} \sqrt{\dfrac{5-2}{5-1}}$$

$$\approx 3.0249$$

Size 3

$$\sigma_x = \dfrac{4.9396}{\sqrt{3}} \sqrt{\dfrac{5-3}{5-1}}$$

$$\approx 2.0166$$

e.

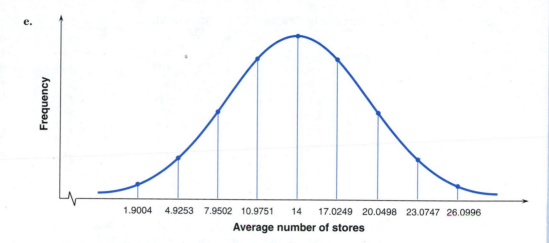

Average number of stores

Thinking Critically (page 430)

1. a. 0.2236   b. Approximately 0.
2. The larger the sample size, the better the estimate of the population mean, since it is based on more data, which encompass more of the population.
3. σ_x represents the standard deviation of the x's, whereas $\sigma_{\bar{x}}$ represents the standard deviation of the sampling distribution of the mean. It is the standard error of the mean. Generally, $\sigma_{\bar{x}}$ will be smaller than σ.

Review Exercises (pages 431–432)

1. The students selected are the first 20 2-digit numbers in column 7 that are between 1 and 58 with no repeats. Select those students whose numbers are 27, 15, 39, 18, 57, 38, 56, 36, 47, 55, 45, 32, 37, 26, 28, 29, 6, 20, 9, and 11.
2. 0.0113   3. 0.9998   4. Between 27.6068 and 28.3932 ounces.
5. 0.9959   6. 0.9719   7. Above 523.
8. Those doctors whose numbers are 06907, 12765, 04213, 04711, 07523, 00358, 10493, 01536, 06243, 11008, 05463, 05597, 04839, 06990, 02011, 07972, 10281, 03427, 08178, 09998, 07351, 12908, 07391, 07856, and 06121.
9. $\mu_{\bar{x}} = 36,000$

$$\sigma_{\bar{x}} = \sqrt{51,200,000} \approx 7155.4175$$

10. 0.0853

Chapter Test (pages 432–434)

1. Between $404.67 and $441.33
2. 0.0002   3. 0.3483   4. 0.9959   5. 0.2265

6. a.

Number of Children, X	Sample Mean, $\bar{x}$	$\bar{x} - \mu_{\bar{x}}$	$(\bar{x} - \mu_{\bar{x}})^2$
4 and 3	3.5	-0.5	0.25
4 and 6	5	1	1.00
4 and 5	4.5	0.5	0.25
4 and 2	3	-1	1.00
3 and 6	4.5	0.5	0.25
3 and 5	4	0	0
3 and 2	2.5	-1.5	2.25
6 and 5	5.5	1.5	2.25
6 and 2	4	0	0.00
5 and 2	3.5	-0.5	0.25
Totals	40		7.50

c. $\mu_{\bar{x}} = 4$ **d.** $\sigma_{\bar{x}} = \sqrt{0.75} \approx 0.866$

7. Those cars whose serial numbers are 64809, 53498, 31016, 59533, 69445, 33488, 52267, 30168, 38005, 40027, 44048, 35126, 48708, 59931, 51038, 47358, 53416, 38857, 34072, 69179, 39440, 60468, 57740, 38867, 56865, 36320, 67869, 47564, 60756, 55322, 45834, 60952, 66556, 32832, and 37937.

8. 0.7642   **9.** Between 34,078.8 and 35,921.2 miles.

10. $\mu_{\bar{x}} = 168$

$$\sigma_{\bar{x}} = \frac{53}{\sqrt{64}} = 6.625$$

11. a. $\mu = 8$

$$\sigma = \sqrt{10.8} \approx 3.2863$$

b. Size 2

Number of Years, X	Sample Mean, $\bar{x}$	$\bar{x} - \mu_{\bar{x}}$	$(\bar{x} - \mu_{\bar{x}})^2$
5 and 10	7.5	-0.5	0.25
5 and 4	4.5	-3.5	12.25
5 and 8	6.5	-1.5	2.25
5 and 13	9	1	1.00
10 and 4	7	-1	1.00
10 and 8	9	1	1.00
10 and 13	11.5	3.5	12.25
4 and 8	6	-2	4.00
4 and 13	8.5	0.5	0.25
8 and 13	10.5	2.5	6.25
	80		40.5

Size 3

Number of Years, X	Sample Mean, $\bar{X}$	$\bar{X} - \mu_{\bar{x}}$	$(\bar{X} - \mu_{\bar{x}})^2$
5, 10, and 4	6.3333	−1.6667	2.7779
5, 10, and 8	7.6667	−0.3333	0.1111
5, 10, and 13	9.3333	1.3333	1.7777
5, 4, and 8	5.6667	−2.3333	5.4443
5, 4, and 13	7.3333	−0.6667	0.4445
5, 8, and 13	8.6667	0.6667	0.4445
10, 4, and 8	7.3333	−0.6667	0.4445
10, 4, and 13	9	1	1.0000
10, 8, and 13	10.3333	2.3333	5.4443
4, 8, and 13	8.3333	0.3333	0.1111
	80		17.9999

c. $\mu_{\bar{x}} = \dfrac{80}{10} = 8$ $\mu_{\bar{x}} = \dfrac{80}{10} = 8$

$\sigma_{\bar{x}} = \sqrt{\dfrac{40.5}{10}} = \sqrt{4.05} \approx 2.0125$

$\sigma_{\bar{x}} = \sqrt{\dfrac{17.9999}{10}} = \sqrt{1.79999} \approx 1.3416$

d. We use $\sigma_{\bar{x}} = \dfrac{\sigma}{\sqrt{n}}\sqrt{\dfrac{N-n}{N-1}}$. In both cases, $N = 5$.

$\sigma_{\bar{x}} = \dfrac{3.2863}{\sqrt{2}}\sqrt{\dfrac{5-2}{5-1}}$ $\sigma_{\bar{x}} = \dfrac{3.2863}{\sqrt{3}}\sqrt{\dfrac{5-3}{5-1}}$

    ≈ 2.0125 ≈ 1.3416

12. Choice (e)   13. Choice (c)   14. Choice (a)   15. Choice (b)
16. Choice (a)   17. Choice (d)

CHAPTER 9

Section 9.3 (pages 445–447)

1. 90% confidence interval: $2.69 to $2.81
3. 95% confidence interval: 9.89 to 10.11 days
5. 95% confidence interval: $14.03 to $15.97
7. 95% confidence interval: between $514.58 and $531.12
9. 85% confidence interval: between $1332.28 and $1417.72
11. a. 95% confidence interval: between 20.34 and 22.16
    b. 99% confidence interval: between 20.05 and 22.45

Section 9.4 (pages 451–454)

1. 99% confidence interval: between $6291.91 and $8108.09
3. 95% confidence interval: between $2217.03 and $2524.97
5. 95% confidence interval: between 25.802 and 33.998 years
7. 90% confidence interval: between 5.226 and 5.974
9. 95% confidence interval: between 25.33 and 28.27
11. 95% confidence interval: between 15.0404 and 20.9596

Section 9.6 (pages 458–459)

1. 95% confidence interval for σ: between 2.38 and 3.06
3. 99% confidence interval for σ: between $6.80 and $10.43
5. Sample size = 12   7. Sample size = 85   9. Sample size = 11

Section 9.7 (pages 464–465)

1. a. 95% confidence interval: between 0.272 and 0.328
   b. 95% confidence interval: between 0.0725 and 0.0875
3. 95% confidence interval: between 0.4165 and 0.4905
5. 0.1660   7. 0.1271 ($\hat{p}$ = 0.4714)
9. 90% confidence interval: between 0.75 and 0.84
11. 0.9881   13. 0.9738 ($\hat{p}$ = 0.194285714)

Testing Your Understanding of This Chapter's Concepts (pages 468–469)

1. True. To double accuracy we cut e in half and formula indicates that we need to quadruple the sample size.
2. Choice (c)
3. 90% confidence interval: between $17.16 and $20.84
4. 0.0150
5. 90% confidence interval: between 4.798 minutes and 5.762 minutes. We are assuming that the population sampled is normally distributed.

Thinking Critically (pages 469–470)

1. a. 90% confidence interval: between 41.2% and 45.4% ($\hat{p}$ = 0.4333 and $\sigma_{\hat{p}}$ = 0.0128)
   b. Sample size = 943, 327.
2. 99.7% confidence interval: between 79.86% and 80.14%
3. 95% confidence interval: between 10.683 and 14.651
4. Large sample size.

5. $df = n - 1 = 9 - 1 = 8$

$$y^2 = \left[1.645\left(\frac{8 \cdot 8 + 3}{8 \cdot 8 + 1}\right) \right]^2 \qquad t = \sqrt{8 \cdot \left(2.718^{\frac{2.8751}{8}} - 1 \right)}$$

$$= 2.8751 \qquad\qquad\qquad = 1.860$$

The $t_{0.05}$ value with 8 degrees of freedom is also 1.860.

Review Exercises (pages 470–471)

1. 95% confidence interval: between 81.4% and 91.2%
2. 0.7657
3. 99% confidence interval: between $13.37 and $23.35
4. 0.0427   5. Sample size = 27
6. 95% confidence interval: between 4.895 and 9.105
7. 95% confidence interval: between 0.650 and 0.730
8. 95% confidence interval: between 28.157 and 29.843
9. 95% confidence interval: between 62.6% and 71.6%
10. 95% confidence interval: between 18.508 and 27.492

Chapter Test (pages 471–474)

1. 0.0007
2. 95% confidence interval: between 0.679 and 0.751
3. 95% confidence interval: between $1.10 and $1.38
4. 90% confidence interval: between 52.46 and 56.14
5. 0.2079
6. 95% confidence interval: between 24.47 and 27.53 years
7. Sample size = 10.   8. 0.0059   9. 0.9941
10. 90% confidence interval: between 33.27 and 36.73 patients
11. 0.0010   12. Sample size = 249
13. 90% confidence interval: between 68.5% and 76.7%
14. Choice (b)   15. Choice (a)   16. Choice (c)   17. Choice (e)
18. Choice (b)   19. Choice (b)

CHAPTER 10

Section 10.4 (pages 493–494)

1. $z = -12.87$. Reject manufacturer's claim.
3. $z = -9.30$. Reject claim. Ajax pays lower than the average hourly rate.
5. $z = 3.90$. Reject union's claim.
7. $z = -1.31$. Do not reject manager's claim.

9. $z = 7.05$. Reject official's claim.

11. $z = -4.76$. Reject manufacturer's claim.

Section 10.5 (pages 497–498)

1. $t = -3.53$. Reject officials' claim.
3. $t = -5.65$. Reject claim. Service has improved.
5. $t = 9.5$. Reject claim. Average loan is more than $14,000.
7. $t = 1.02$. Do not reject claim. Machine is not significantly overfilling cups.
9. $t = -11.299$. Reject null hypothesis.

Section 10.6 (pages 504–507)

1. $z = 2.63$. Reject null hypothesis. There is a significant difference.
3. $z = 2.62$. Reject null hypothesis. There is a significant difference.
5. $z = 5.05$. Reject null hypothesis. There is a significant difference.
7. $z = -1.35$. Do not reject null hypothesis. There is no significant difference.
9. $z = -263.52$. Reject null hypothesis. There is a significant difference.
11. $z = 5.23$. Reject null hypothesis. There is a significant difference.

Section 10.7 (pages 510–512)

1. $s_p \approx 9015.51$ $t = -1.04$
Do not reject null hypothesis. There is not enough evidence to indicate a significant difference.
3. $s_p \approx 1.983$ $t = -1.15$
Do not reject null hypothesis. There is not enough evidence to indicate a significant difference.
5. $s_p \approx 2.675$ $t = 4.06$
Reject null hypothesis. There is evidence to indicate a significant difference.
7. $s_p \approx 1.041$ $t = 4.88$
Reject null hypothesis. There is evidence to indicate a significant difference.
9. $s_p \approx 0.191$ $t = -0.199$
Do not reject null hypothesis. There is not enough evidence to indicate a significant difference.

Section 10.8 (pages 516–518)

1. $z = -2.73$. Reject null hypothesis. Do not accept manufacturer's claim.
3. $z = 1.72$. Do not reject null hypothesis and environmentalist's claim.
5. $z = 0.24$. Do not reject null hypothesis. We cannot reject university's claim.
7. $z = -0.58$. Do not reject null hypothesis. We cannot reject dermatologist's claim.

9. $z = 1.19$. Do not reject null hypothesis. We cannot reject the housing manager's claim.

11. $z = -0.31$. Do not reject null hypothesis. Do not reject the travel agency's claim.

Testing Your Understanding of This Chapter's Concepts (pages 522–524)

1. No. When the null hypothesis is rejected, either a Type I error is made or a correct decision is made. When the null hypothesis is not rejected, either a Type II error or a correct decision is made. Since it is not possible to simultaneously reject and accept the null hypothesis, it is not possible to make both errors at the same time.

2. a. Not for any reasonable level of significance ($\alpha < 0.50$). The rejection region for such a level involves positive values of the test statistic; the given sample mean and hypothesized value for the mean yield a negative value for the test statistic. Therefore the null hypothesis would not be rejected.
   b. Possibly a Type-II error if H_0 is false.

3. If both sample sizes n_1 and n_2 are equal, call them both n. Then

$$s_p = \sqrt{\frac{(n-1)s_1^2 + (n-1)s_2^2}{n+n-2}}$$

$$= \sqrt{\frac{(n-1)(s_1^2 + s_2^2)}{2n-2}}$$

$$= \sqrt{\frac{(n-1)(s_1^2 + s_2^2)}{2(n-1)}}$$

$$= \sqrt{\frac{s_1^2 + s_2^2}{2}}$$

Thus, s_p^2 is just the mean of s_1^2 and s_2^2.

4. p(Type-II error) could be 1.

Thinking Critically (pages 524–525)

1. $s_p = 2.03588$. 90% confidence interval: 2.63 and 5.37 cents.

2. a. **Jonathan Williams Houses**   **Martin Luther King Houses**

   Mean $= \bar{x} = 15691$ Mean $= \bar{x} = 15025$

   Standard deviation $= s = 1631$ Standard deviation $= s = 1910$

   $n = 10$ $n = 12$

   b. $s_p \approx 1789.84$. 95% confidence interval: between -932.62 and $\$2264.63$.
   c. $t = 0.869$. Do not reject null hypothesis. The mean family incomes are not significantly different.

Review Exercises (pages 525–527)

1. $z = 1.74$. Do not reject null hypothesis. There is not enough evidence to suggest that the difference is significant.
2. $z = -14.14$. Reject null hypothesis. Do not accept company's claim.
3. $z = 2.38$. Reject null hypothesis. Do not accept company's claim.
4. $z = 2.80$. Reject null hypothesis. Do not accept the 25% estimate.
5. $z = 3.88$. Reject null hypothesis. The difference is significant.
6. $z = 0.782$. Do not reject null hypothesis. Do not reject fire department claim.
7. $z = 1.61$. Do not reject hypothesis. Do not reject farmer's claim.
8. $z = -4.75$. Reject null hypothesis. The difference is significant.
9. $z = 0.45$. Do not reject null hypothesis. Do not reject inspector's claim.
10. $t = -6.26$. Reject null hypothesis. Service has improved.
11. $z = 1.95$. Do not reject null hypothesis. Do not reject teacher's claim.
12. $z = -2.47$. Reject null hypothesis. Bank is paying below average hourly rate.
13. $z = 4.33$. Reject null hypothesis. The sociologist's claim seems to be accurate.
14. $z = 0.45$. Do not reject null hypothesis. Do not reject Red Cross claim.
15. $z = 6.56$. Reject null hypothesis. Reject court official's claim.
16. $z = 2.76$. Reject null hypothesis. Reject Chamber of Commerce claim.
17. $z = 3.52$. Reject null hypothesis. The difference is significant.

Chapter Test (pages 528–531)

1. $z = -0.89$. Do not reject null hypothesis. We cannot reject sociologist's claim.
2. $z = 1.72$. Reject null hypothesis. Decision is a correct one.
3. $z = -1.54$. Do not reject null hypothesis. We cannot say that there is a significant difference.
4. $z = -3.57$. Reject null hypothesis. Difference between weight loss is significant.
5. $t = -2.12$. Do not reject null hypothesis. Coin does appear to be fair at the 1% level of significance.
6. Choice (b)
7. $t = -8.06$. Reject null hypothesis. Reject manufacturer's claim.
8. $z = -4.12$. Reject null hypothesis. Difference is significant.
9. $z = -1.29$. Do not reject null hypothesis. Do not reject representative's claim.
10. $z = -4.54$. Reject null hypothesis. Reject claim.
11. $z = -9.8$. Reject null hypothesis. Viewing the film reduces connection time significantly.
12. $t = 5.04$. Reject null hypothesis. Reject banking official's claim.
13. $t = 1.07$. Do not reject null hypothesis. There is not enough evidence to indicate a significant difference.
14. $t = 1.707$. Do not reject null hypothesis. There is not enough evidence to indicate a significant difference.
15. Choice (a)   16. Choice (a)   17. Choice (b)   18. Choice (b)
19. Choice (a)

CHAPTER 11

Section 11.3 (pages 545–549)

1. **a.** Positive correlation.
   b. Probably zero correlation. Some people may disagree.
   c. Positive correlation.
   d. Negative correlation.
   e. Negative correlation.
   f. Zero correlation.
   g. Probably zero correlation.
   h. Positive correlation.
   i. Positive correlation.

3. **a.**

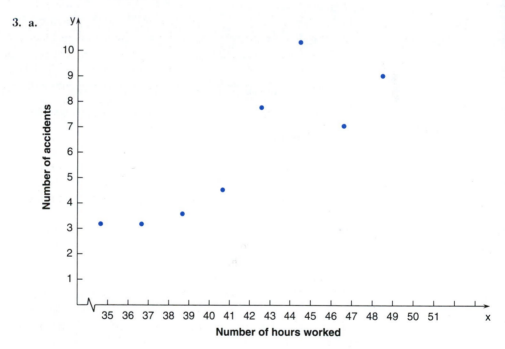

$r = 0.8591$

b. Yes

5. **a.**

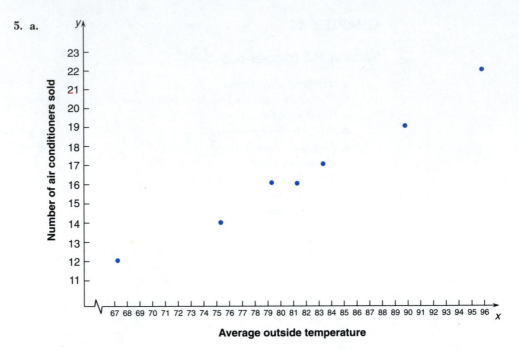

$r = 0.9914$

b. Yes

7. a.

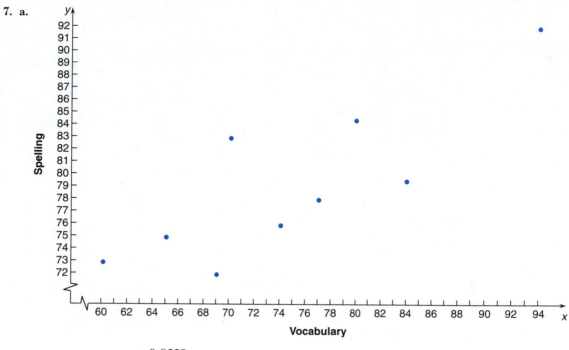

$r = 0.8229$

b. Most likely

9. a.

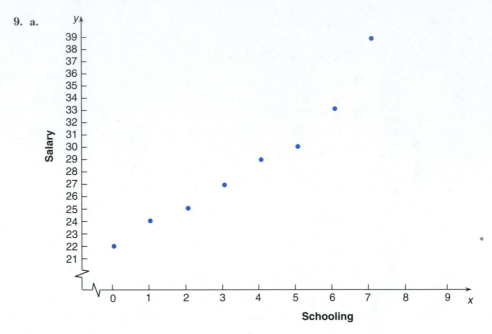

$r = 0.9739$
  b. Yes
11. a. $r = 0.9625$
  b. Yes

Section 11.4 (page 551)

1. Significant   **3.** Not significant   **5.** Significant   **7.** Significant
9. Not significant   **11.** Significant

Section 11.6 (pages 561–566)

1. a. Least-squares prediction equation: $\hat{y} = 7.3045 + 0.9219x$
  b. $\hat{y} = 7.3045 + 0.9219(76) = 77.3689$ inches
3. a. Least-squares prediction equation: $\hat{y} = 460.2278 + 0.7961x$
  b. $\hat{y} = 460.2278 + 0.7961(70) = 515.9548$
5. a. Least-squares prediction equation: $\hat{y} = 10.4503 + 0.8832x$
  b. $\hat{y} = 10.4503 + 0.8832(70) = 72.2743$
7. a. Least-squares prediction equation: $\hat{y} = 68.9722 + 6.3033x$
  b. $\hat{y} = 68.9722 + 6.3033(8) = 119.3986$ heartbeats
9. a. Least-squares prediction equation: $\hat{y} = 27.349 + 0.01285x$
  b. $\hat{y} = 27.349 + 0.01285(33,000) = 451.399$
11. a. Least-squares prediction equation: $\hat{y} = 555.5 + 1.6823x$
  b. $y = 555.5 + 1.6823(600) = \$1564.88$ thousands or $\$1,564,880$

Section 11.7 (page 572)

1. $s_e = \sqrt{\dfrac{7.7661}{8}} \approx 0.9853$ 3. $s_e = \sqrt{9.9146} \approx 3.1487$

5. $s_e = \sqrt{2.54273} \approx 1.5946$
7. $s_e = \sqrt{6.88408} \approx 2.6238$ 9. $s_e = \sqrt{5256.75} \approx 72.5034$
11. $s_e = \sqrt{12836.6} \approx 113.2987$

Section 11.8 (page 578)

1. $t = 12.8835$. Do not accept H_0: $\beta_1 = 0$.
   95% prediction interval: 74.7349 to 80.0029
3. $t = 9.9541$. Do not accept H_0: $\beta_1 = 0$.
   95% prediction interval: 507.5685 to 524.3411
5. $t = 5.7293$. Do not accept H_0: $\beta_1 = 0$.
   95% prediction interval: 68.1218 to 76.4268
7. $t = 16.96597$. Do not accept H_0: $\beta_1 = 0$.
   95% prediction interval: 111.9291 to 126.8681
9. $t = 4.5408$. Do not accept H_0: $\beta_1 = 0$.
   95% prediction interval: 233.7265 to 669.2735
11. $t = 18.116$. Do not accept H_0: $\beta_1 = 0$.
   95% prediction interval: 1252.2257 to 1877.4646

Section 11.10 (pages 581–582)

1. a. Least-squares prediction equation: $\hat{y} = 84.705 + 1.0652x_1 + 2.409x_2$
   b. $\hat{y} = 84.705 + 1.0652(1.8) + 2.409(2.8) = 93.36756$
3. a. Least-squares prediction equation: $\hat{y} = 10053.6 - 432.8x_1 - 0.054929x_2$
   b. $\hat{y} = 10053.6 - 432.8(2.5) - 0.054929(45,000) = \$6499.80$

Testing Your Understanding of This Chapter's Concepts (pages 587–588)

1. Choice (d)   2. Choice (b)   3. Yes
4. a. 0   b. A very large number

Thinking Critically (page 588)

1. Start with the formula for b_1 in Formula 11.2. Divide numerator and denominator by $n - 1$. Using the fact that $s_x^2 = \dfrac{n\Sigma x^2 - (\Sigma x)^2}{n(n - 1)}$, we get the equation for b_1 given in Formula 11.3. Also, the formulas for b_0 are equivalent since $\dfrac{1}{n}\Sigma x = \bar{x}$ and $\dfrac{1}{n}\Sigma y = \bar{y}$.

2. $s_x = \sqrt{\dfrac{n\Sigma x^2 - (\Sigma x)^2}{n(n-1)}}$ and $s_y = \sqrt{\dfrac{n\Sigma y^2 - (\Sigma y)^2}{n(n-1)}}$.   Also,

$s_{xy} = \dfrac{\Sigma(x - \bar{x})(y - \bar{y})}{n-1}$

Thus,

$$\frac{\Sigma(x - \bar{x})(y - \bar{y})(n-1)}{\sqrt{\dfrac{n\Sigma x^2 - (\Sigma x)^2}{n(n-1)}}\sqrt{\dfrac{n\Sigma y^2 - (\Sigma y)^2}{n(n-1)}}} = \frac{s_{xy}}{s_x s_y} = \frac{n(\Sigma xy - (\Sigma x)(\Sigma y))}{\sqrt{n\Sigma x^2 - (\Sigma x)^2}\sqrt{n\Sigma y^2 - (\Sigma y)^2}}$$

This is the formula for r. Thus, $r = \dfrac{s_{xy}}{s_x s_y}$.

3. Correlation analysis determines a correlation coefficient for measuring the strength of a relationship between the variables. Regression analysis involves finding an equation connecting the variables. We determine a regression equation once we know that there is a significant correlation between the variables.

Review Exercises (pages 588–592)

1. a.

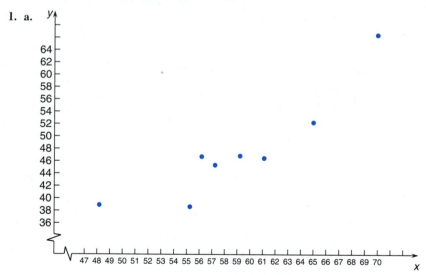

   b. $r = 0.8943$
2. a. Least-squares prediction equation: $\hat{y} = -183.2534 + 25.4641x$
   b. $\hat{y} = -183.2534 + 25.4641(14) = 173.244$
3. 95% prediction interval: between 94.1216 and 252.3664
4. Least-squares prediction equation: $\hat{y} = 44.986 - 0.2838x$
5. Least-squares prediction equation: $\hat{y} = 28.3216 - 0.7643x$
6. Least-squares prediction equation: $\hat{y} = 26.953 - 0.0806x$

7. a. Least-squares prediction equation: $\hat{y} = 29.213 + 0.36172x_1 + 0.5674x_2$
   b. $\hat{y} = 29.213 + 0.36172(128) + 0.5674(18) = 85.726.$
8. Choice (a)
9. a. $r = -0.9832$

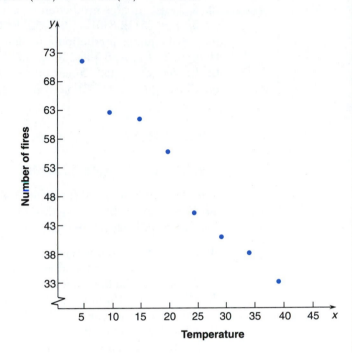

   b. Yes
10. a. Least-squares prediction equation: $\hat{y} = 11185.5492 - 1117.1484x$
   b. $\hat{y} = 11185.5492 - 1117.1484(3.5) = \$7275.53$

Chapter Test (pages 592–596)

1. a. Least-squares prediction equation: $\hat{y} = -9.991 + 1.7429x$
   b. $\hat{y} = -9.991 = 1.7429(15) = 16.1525$
2. $r = 0.9908$
3. a. Least-squares prediction equation: $\hat{y} = 4.4365 - 0.1394x$
   b. $\hat{y} = 4.4365 - 0.1394(21.5) = 1.4394$
4. a. $r = 0.9630$
   b. Least-squares prediction equation: $\hat{y} = 11.8299 + 6.9758x$
   c. $s_e \approx \sqrt{39.2909} \approx 6.2682$
   d. $\hat{y} = 11.8299 + 6.9758(10.5) = 85.0758$ pounds
5. Least-squares prediction equation: $\hat{y} = 8 + 2.375x$
6. a. Least-squares prediction equation: $\hat{y} = 9.0124 + 3.9795x$
   b. $\hat{y} = 9.0124 + 3.9795(10) = 48.8074$
7. 95% prediction interval: between 44.2468 and 53.368

8. a. Least-squares prediction equation: $\hat{y} = 101.3219 - 2.0786x$
   b. $\hat{y} = 101.3219 - 2.0786(12) = 76.3787$
9. a. Least-squares prediction equation: $y = 25112.8401 - 945.5253x$
   b. $y = 25112.8401 - 945.5253(13) = 12821.0112$
10. Least-squares prediction equation: $y = 34.192 - 3.468x_1 + 0.394x_2$
11. $y = 34.192 - 3.468(2.3) + 0.394(4) = 27.7916$
12. Choice (c)   13. Choice (a)   14. Choice (b)   15. Choice (c)
16. Choice (d)   17. Choice (a)   18. Choice (a)

CHAPTER 12

Section 12.2 (pages 605–610)

Note: Although expected frequencies will usually be given to two decimal places and the Chi-square subtotals to three decimal places, the calculations will usually be done using full calculator accuracy.

1. $\chi^2 = 4.425$. Do not reject null hypothesis that there is no significant difference between the corresponding proportion of men in the various age groups who drink alcoholic beverages.
3. $\chi^2 = 0.558$. Do not reject null hypothesis that there is a significant difference between the proportion of male and female passengers who rent headphones.
5. $\chi^2 = 2.171$. Do not reject null hypothesis that there is a significant difference between the families that own a VCR in all three states.
7. $\chi^2 = 10.801$. Do not reject null hypothesis that there is a significant difference between the percentage of homes containing excessive levels of radon gas on all the streets.
9. $\chi^2 = 35.009$. Reject null hypothesis that the percentage of former graduates who contributed money for scholarships is the same in the different graduating classes.
11. $\chi^2 = 0.586$. Do not reject null hypothesis that the percentage of books or periodicals added to the library's holding is the same for all types of math-related books or periodicals. Actually there is one cell whose expected count is less than 5.0.
13. $\chi^2 = 1.950$. Do not reject null hypothesis that the proportion of people in the different geographic areas who believe that the defense budget should be cut is the same.

Section 12.3 (pages 613–616)

1. $\chi^2 = 17.9$. Reject null hypothesis. Die is not fair.
3. $\chi^2 = 11.8$. Reject null hypothesis. Absences do not occur with equal frequency.

5. $\chi^2 = 3.5179$. Do not reject null hypothesis that the frequencies of the types of characteristics obtained by gene-splicing are in the specified ratio.

7. $\chi^2 = 8.1386$. Do not reject null hypothesis. The manager's claim seems to be correct.

9. $\chi^2 = 13.832$. Reject null hypothesis. All TV stations *do not* have equal shares of the viewing audience.

Section 12.4 (pages 620–626)

1. $\chi^2 = 0.512$. Do not reject null hypothesis that the number of defective calculators produced is independent of the production work shift.

3. $\chi^2 = 26.090$. Reject null hypothesis that smoking and drinking are independent.

5. $\chi^2 = 400.264$. Reject null hypothesis that the pregnancy test kit results are independent of the clinical test results.

7. $\chi^2 = 49.637$. Reject null hypothesis that the region in which a person lives is independent of the person's opinion about the proposal.

9. $\chi^2 = 1.445$. Do not reject null hypothesis that the type of traveler is independent of the degree of satisfaction.

11. $\chi^2 = 760.212$. Reject null hypothesis that age is independent of the environmental problem that most concerns an individual.

Testing Your Understanding of This Chapter's Concepts (pages 628–629)

1. Since the null hypothesis is rejected only when the value of the test statistic is too large, the rejection region is always on the right, that is, it is always right-tailed.

2. No

3. **b.** Since this is a contingency table and not a simple chi-square problem, the value of p will be different for each cell.

Thinking Critically (pages 629–630)

1. $\chi^2 = 10.804$. Reject null hypothesis that the type of antitheft device used is independent of the size of the car.

2. $\chi^2 = 1.658$. When the cells are combined, we would not reject the null hypothesis. By combining cells, we arrive at a different conclusion than when we did not combine any cells. Actually, some statisticians believe that the requirement that all expected frequencies should be at least 5 is too restrictive.

3. $\chi^2 = 32.550$. Reject null hypothesis that the proportion of people in each age group who fear heights is the same.

4.

	1	2	Total
1	a	b	$a + b$
2	c	d	$c + d$

Total $a + c$ $b + d$ $a + b + c + d$

Column one, row one expected value: $\dfrac{(a + b)(a + c)}{a + b + c + d}$

Column two, row one expected value: $\dfrac{(a + b)(b + d)}{a + b + c + d}$

Column one, row two expected value: $\dfrac{(a + c)(c + d)}{a + b + c + d}$

Column two, row two expected value: $\dfrac{(c + d)(b + d)}{a + b + c + d}$

$$\chi^2 = \frac{\left[a - \dfrac{(a + b)(a + c)}{a + b + c + d}\right]^2}{\dfrac{(a + b)(a + c)}{a + b + c + d}} + \frac{\left[b - \dfrac{(a + b)(b + d)}{a + b + c + d}\right]^2}{\dfrac{(a + b)(b + d)}{a + b + c + d}}$$

$$+ \frac{\left[c - \dfrac{(a + c)(c + d)}{a + b + c + d}\right]^2}{\dfrac{(a + c)(c + d)}{a + b + c + d}} + \frac{\left[d - \dfrac{(c + d)(b + d)}{a + b + c + d}\right]^2}{\dfrac{(c + d)(b + d)}{a + b + c + d}}$$

This simplifies to $\chi^2 = \dfrac{(a + b + c + d)(ad - bc)^2}{(a + b)(c + d)(b + d)(a + c)}$

Review Exercises (pages 631–633)

1. The expected frequency for all the cells in row one is 5.
The expected frequency for all the cells in row two is 6.67.
The expected frequency for all the cells in row three is 8.33.
The expected frequency for all the cells in row four is 5.

2. $p = \dfrac{62}{125} = 0.496$

3. $\chi^2 = 6.517$

4. $\chi^2 = 0.473$. Do not reject hypothesis that the rating of the meal is independent of the sex of the flier.

5. Expected number of drug users who believe that drug use is harmful $=$ $\dfrac{(650)(910)}{1470} = 402.38$ or 402.4. Choice (a)

6. $\chi^2 = 8.252$. Reject null hypothesis that the proportion of police officers who are not working because of job-related disabilities is the same for all categories.

7. $\chi^2 = 38.535$. Reject null hypothesis that the plea entered is independent of the jail sentence.

8. $\chi^2 = 45.609$. Reject null hypothesis that the type of crime committed is independent of the precinct where it was committed.

9. $\chi^2 = 5.155$. Do not reject null hypothesis that the sex of the juror is independent of his or her tendency to be liberal.

10. $\chi^2 = 6.1879$. Do not reject null hypothesis that the percentages are correct.

Chapter Test (pages 633–637)

1. $\chi^2 = 10.999$. Reject null hypothesis that the amount of any purchase at the store is independent of the time that it is purchased.

2. $\chi^2 = 2.587$. Do not reject null hypothesis that the sex of the viewer is independent of the viewer's opinion.

3. $\chi^2 = 9.094$. Do not reject null hypothesis that the proportion of smokers in the different occupations mentioned is the same.

4. $\chi^2 = 10.666$. Do not reject null hypothesis that the number of arrests for drunken driving is the same for the different days of the week.

5. $\chi^2 = 4.277$. Do not reject null hypothesis that the level of pollution is independent of the river sampled.

6. $\chi^2 = 5.0655$. Do not reject null hypothesis that the sizes of the dresses sold are in the specified ratio.

7. $\chi^2 = 2.126$. Do not reject null hypothesis that family income is independent of the method of heating used.

8. $\chi^2 = 9.707$. Reject null hypothesis that the proportion of Americans who think that our government is doing enough for the homeless is the same for the cities mentioned.

9. $\chi^2 = 9.467$. Do not reject null hypothesis that the time of day when the accident occurs is independent of the day of the week when the accident occurs.

10. $\chi^2 = 1.7221$. Do not reject null hypothesis. The data seem to be consistent with the theory.

11. Choice (a)   12. Choice (b)   13. Choice (c)   14. Choice (d)

15. Choice (a)

CHAPTER 13

Section 13.4 (pages 658–664)

1. Do not reject null hypothesis that there is no significant difference in the average price charged by these gas stations for a gallon of unleaded gas. ($F = 0.17$)
3. Do not reject null hypothesis that the average blood pressure loss is not significantly different for the various blood pressure pills. ($F = 2.06$)
5. Do not accept claim that the differences among the sample means are significant ($F = 1.879$).
7. We do not reject null hypothesis that the average price of gas is the same for all brands ($F = 0.6667$). However, we do reject null hypothesis that average price of gas is the same for all neighborhoods ($F = 4.8889$).
9. Do not reject null hypothesis. The data do not indicate that there is a significant difference in the average time required to complete the transaction at these banks ($F = 1.115$).
11. Do not reject null hypothesis. The data do not indicate that the type of diet used significantly affects average blood serum cholesterol level ($F = 1.33$).
13. The data do not indicate that the difference in the average oil obtained from the various regions is significant ($F = 0.958$). However, the difference in the average oil obtained by the different conversion techniques is significant ($F = 7.195$).

Testing Your Understanding of This Chapter's Concepts (pages 667–668)

1. Assumptions for one-way ANOVA
   a. *Independent samples:* The samples taken from the various populations are independent of one another.
   b. *Normal populations:* The populations from which the samples are obtained are (approximately) normally distributed.
   c. *Equal standard deviations:* The populations from which the samples are obtained all have the same (often unknown) variance, σ^2.
2. $F_{20}^{10}(0.01) = 3.3$ so that the probability that $F \leq 3.37$ is approximately 0.99.
3. ANOVA techniques tell us whether the variation among the sample means is too large to be attributed to chance. This, of course, implies that the differences among the sample means is significant.

Thinking Critically (page 668)

1. A, B, C, D, and E.
2. Reject null hypothesis that the sample means are equal ($F = 8.27$).
3. The 95% confidence interval for the difference between average tar content of Brand X and Brand Y cigarettes is between -5.438 and 3.772.

4. **a.** The 95% confidence interval for the difference between the average tar content of Brand X and Brand Z cigarettes is between -11.438 and -2.228.
   b. The 95% confidence interval for the difference between the average tar content of Brand Y and Brand Z cigarettes is between -10.605 and -1.395.

Review Exercises (pages 668–671)

1. Cell (1) entry: 0.1667 or 0.2   **2.** Cell (2) entry: 142.5   **3.** Cell (3) entry: 142.7
4. Cell (4) entry: 2   **5.** Cell (5) entry: 9   **6.** Cell (6) entry: 11
7. Cell (7) entry: 0.08335 or 0.1   **8.** Cell (8) entry: 15.833   **9.** Cell (9) entry: 0.00526 or 0.01
10. Choice (a) (Technically speaking, Choice (b) is also correct if we are testing differences between variances)
11. Choice (a)   **12.** Choice (c)
13. Do not accept null hypothesis. There is a significant difference in the average number of false alarms reported in the various sections of the city ($F = 14.129$).
14. Do not reject null hypothesis. The data do not indicate that the average number of accidental breakages is significantly different for all the stores ($F = 0.232$)

Chapter Test (pages 671–675)

1. Do not reject null hypothesis. The data do not indicate that there is a significant difference between the average lives of these batteries ($F = 0.628$)
2. Do not reject null hypothesis. The data do not indicate that there is a significant difference in the average production of these employees ($F = 0.023$)
3. Do not reject null hypothesis. The data do not indicate that there is a significant difference in the average number of cases handled by these judges ($F = 0.380$)
4. Do not reject null hypothesis. The data do not indicate that there is a significant difference in the average time needed by each student to learn to overcome the particular speech defect by the different methods ($F = 1.022$)
5. Do not reject null hypothesis. The data do not indicate that there is a significant difference in the average weight gain of an animal over a fixed length of time as a result of the different chemicals ($F = 0.369$)
6. Do not reject null hypothesis. The data do not indicate that there is a significant difference in the average number of errors in the books typeset by each of the companies ($F = 0.855$)
7. Do not reject null hypothesis. The data do not indicate that there is a significant difference in the preparation of the students in each of the teacher's classes ($F = 0.0116$)
8. Do not reject null hypothesis. The data do not indicate that the average reaction time to the different stimuli is significantly different ($F = 0.8189$)
9. Do not reject null hypothesis. The data do not indicate that the average age of

people hospitalized is significantly different for all the age groups. Also, the data do not indicate that the average age of people hospitalized is significantly different for all reasons for hospitalization ($F = 2.266$ and $F = 0.413$)

10. Do not reject null hypothesis. The data do not indicate that the average amount of money in a settled claim is significantly different for all the drive-in centers or that it is significantly different for all the claims examiners ($F = 0.503$ and $F = 0.6796$)

11. Choice (a)   12. Choice (b)   13. Choice (c)   14. Choice (c)
15. Choice (c)   16. Choice (b)

CHAPTER 14

Section 14.3 (pages 687–693)

1. $n = 11$ and the number of minus signs is 5. We do not reject null hypothesis that the median salary is $27,000.
3. $n = 12$ and the number of minus signs is 5. We do not reject null hypothesis that the median age is 19 years.
5. There is not enough evidence to show that the new fares have decreased ridership.
7. Reject null hypothesis. The number of complaints has decreased ($z = 2.33$)
9. Do not reject null hypothesis. The data do not indicate that the new equipment has significantly reduced the number of damaged pieces of mail ($z = 1.51$)
11. Reject null hypothesis. The number of felony arrests has decreased as a result of the number of police cars in the patrol car.

Section 14.4 (pages 698–700)

1. We do not reject null hypothesis. The new policy does not seem to have an effect on drug use at the schools.
3. We do not reject null hypothesis. The data do not indicate that the new policy has a significant effect on reducing burglaries.

Section 14.5 (pages 706–708)

1. Reject null hypothesis. The mean score for both groups is not the same ($z = 3.76$)
3. Do not reject null hypothesis. The data do not indicate that the average number of patients treated at both hospitals is the same ($z = -0.19$)
5. Do not reject null hypothesis. The data do not indicate that there is a significant difference in the average number of home equity loans approved by both banks ($z = 0.351$)

7. Do not reject null hypothesis. The data do not indicate that the average number of abandoned cars towed away by the two disposal agencies is significantly different ($z = 0.80$)

Section 14.7 (pages 715–719)

1. a. $R = 0.1$ is not significant. We cannot conclude that there is a significant correlation between the ratings of scouts 1 and 2.
    b. $R = -0.9$ is not significant. We cannot conclude that there is a significant correlation between the ratings of scouts 1 and 3.
    c. $R = -0.5$ is not significant. We cannot conclude that there is a significant correlation between the ratings of scouts 2 and 4.
3. $R = -0.1429$ is not significant. We cannot conclude that there is a significant correlation between the student's rankings of the teachers.
5. There are 15 runs where there are 15 C's and 10 I's. Do not reject null hypothesis of randomness.
7. There are 8 runs where there are 15 A's and 9 N's. Do not reject null hypothesis of randomness.
9. The median of the numbers is 17.5. There are 6 runs where we have 8 a's and 8 b's. Do not reject null hypothesis of randomness.
11. There are 18 runs where we have 12 D's and 16 R's. Do not reject null hypothesis of randomness.
13. The median of the numbers is 19. There are 8 runs where we have 7 a's and 6 b's. Do not reject null hypothesis of randomness.
15. Between woman 1 and woman 2. $R = 0.18$
    Between woman 1 and woman 3. $R = -0.39$
    Between woman 1 and woman 4. $R = -0.32$
    Between woman 1 and woman 5. $R = -0.50$
    Between woman 2 and woman 3. $R = -0.43$
    Between woman 2 and woman 4. $R = 0.43$
    Between woman 2 and woman 5. $R = -0.57$
    Between woman 3 and woman 4. $R = 0$
    Between woman 3 and woman 5. $R = 0.78571$
    Between woman 4 and woman 5. $R = 0.14$
    There is no significant correlation between the ratings of the women.

Section 14.8 (pages 723–724)

1. We do not reject null hypothesis. The data do not indicate that there is a significant difference in the average time required to assemble the computer. (Test statistic $= 6.266$)
3. Do not reject null hypothesis. The data do not indicate that there is a significant

difference between the average decrease in the blood serum cholesterol levels of people taking each of these types of pills ($x^2 = 4.477$)

Testing Your Understanding of This Chapter's Concepts (page 729)

1. Yes. In the one-tailed test, the alternate hypothesis is that the probability distribution for population A is shifted to the right of that for B. In the two-tailed test, the alternate hypothesis is that the probability distribution for population A is shifted to the left or to the right of that for B.
2. The distributions are assumed to be continuous so that the probability of tied measurements is 0, and each measurement can be assigned a unique rank.
3. Yes. Because large values of the test statistic support the alternative hypothesis that the populations have different probability distributions, the rejection region is located in the upper tail of the x^2 distribution.
4. The lowest value H of the test statistic is 0 and the largest value is 12.5 and occurs when the ranks for sample 1 are 1, 2, 3, 4, and 5. The ranks for sample 2 are 6, 7, 8, 9, and 10, and the ranks for sample 3 are 11, 12, 13, 14, and 15.

Thinking Critically (page 729)

1. $R_a = 12$, $n_a = 13$, and $n_b = 12$.
   $z = -0.61$. Do not reject null hypothesis.
3. Yes

Review Exercises (pages 730–734)

1. Do not reject null hypothesis. The data do not indicate that there is a significant difference in the average time needed to run the marathon by both groups ($z = -0.489$)
2. There are 14 runs where there are 14 R's and 14 H's. Do not reject null hypothesis of randomness.
3. Do not reject null hypothesis. The data do not indicate that the men and women differ significantly on the issue ($z = 1.935$)
4. We cannot conclude that there is a significant correlation between the day of the weeks on which transactions took place and the number of transactions conducted.
5. Do not reject null hypothesis. The data do not indicate that the average number of miles per gallon is significantly different for both cars ($z = -1.32$)

6. There are 12 runs where we have 9 a's and 10 b's. Do not reject null hypothesis of randomness.

7. $R = 0.8857$. The R value is very close to the chart value of 0.886. Thus, we may conclude that there is no significant correlation between the rankings. Actually, more data are needed before we can arrive at a definitive conclusion.

8. Do not reject null hypothesis. The data do not indicate that the special teaching technique is effective.

9. Do not reject null hypothesis. The data do not indicate that drugs have an effect on performance on the anxiety test.

10. There are 9 runs where we have 11 a's and 12 b's. Do not reject null hypothesis of randomness.

11. Do not reject null hypothesis. The data do not indicate that the average tolerance level differs significantly for females and males ($z = -0.89$)

12. Do not reject null hypothesis. The data do not indicate that there is a significant difference in the average results at both ages ($z = 0.462$)

13. There are 13 runs where there are 14 M's and 11 W's. Do not reject null hypotheses of randomness.

14. Reject null hypothesis. Stopping drinking has an effect on a person's weight.

15. There are 11 runs where we have 10 a's and 10 b's. Do not reject null hypothesis of randomness.

Chapter Test (pages 734–739)

1. There is no significant correlation between the rankings of the male and female reviewers.

2. There are 11 runs where we have 17 F's and 10 M's. Do not reject null hypothesis of randomness.

3. Do not reject null hypothesis. The data do not indicate that there is a significant difference in the scores obtained by students using any of these three programs (Test statistic $= 2.547$)

4. Reject null hypothesis. The average price for the computer printer is different in both cities ($z = -2.01$)

5. Do not reject null hypothesis. The data do not indicate that the medication has a significant effect on the number of tasks completed.

6. There are 8 runs where we have 14 b's and 13 a's. Reject null hypothesis of randomness.

7. Do not reject null hypothesis. The data do not indicate that there is a significant difference in the contestant's estimate of the value of both vacations.

8. Reject null hypothesis. The data do indicate that there is a significant difference in the average time needed to complete the questionnaire depending on which pollster gives the instructions (Test statistic $= 7.956$).

9. There is no significant correlation between the rankings.

10. Do not reject null hypothesis. There is no significant difference in the performing evaluations (Test statistic = 0.542).

11. Do not reject null hypothesis. The data do not indicate that there is a significant difference in the average number of push-ups performed by students in both classes ($z = 0.11$)

12. Choice (a)   13. Choice (a)   14. Choice (b)   15. Choice (a)

16. Choice (a)

Index

Frequently Used Formulas

Relative frequency $\dfrac{f_i}{n}$

Mean $\dfrac{\Sigma x}{n} = \dfrac{x_1 + x_2 + \cdots + x_n}{n}$

Weighted mean $\bar{x}_w = \dfrac{\Sigma xw}{\Sigma w}$

Variance $\sigma^2 = \dfrac{\Sigma (x - \mu)^2}{n}$ or $\dfrac{\Sigma x^2}{n} - \dfrac{(\Sigma x)^2}{n^2}$

Standard deviation $\sigma = \sqrt{\dfrac{\Sigma (x - \mu)^2}{n}}$

Average deviation $\dfrac{\Sigma |x - \mu|}{n}$

Sample standard deviation $\sqrt{\dfrac{\Sigma (x - \bar{x})^2}{n - 1}}$

Percentile rank of x $\dfrac{B + \frac{1}{2}E}{n} \cdot 100$

z-score $z = \dfrac{x - \mu}{\sigma}$ Original score $x = \mu + z\sigma$

Probability, p $p = \dfrac{f}{n}$

$_nP_r = \dfrac{n!}{(n - r)!}$ $_nP_n = n!$ $_nC_r = \dfrac{n!}{r!(n - r)!}$

Number of permutations with repetitions $\dfrac{n!}{p!q!r! \cdots}$

Mathematical expectation $m_1p_1 + m_2p_2 + m_3p_3 + \cdots$

Addition rule (for mutually exclusive events) $p(A \text{ or } B) = p(A) + p(B)$

Addition rule (general case) $p(A \text{ or } B) = p(A) + p(B) - p(A \text{ and } B)$

Complement of event A $p(A') = 1 - p(A)$

Conditional probability formula $p(A|B) = \dfrac{p(A \text{ and } B)}{p(B)}$

Multiplication rule $p(A \text{ and } B) = p(A|B) \cdot p(B)$

Multiplication rule (for independent events) $p(A \text{ and } B) = p(A) \cdot p(B)$

Bayes' rule $\dfrac{p(B|A_n)p(A_n)}{p(B|A_1)p(A_1) + p(B|A_2)p(A_2) + \cdots + p(B|A_n)p(A_n)}$